Härdle · Simar

Applied Multivariate Statistical Analysis

AF646097

Wolfgang Härdle · Léopold Simar

Applied Multivariate Statistical Analysis

Second Edition

Springer

Wolfgang Härdle
Humboldt-Universität zu Berlin
CASE - Center for Applied Statistics and Economics
Institut für Statistik und Ökonometrie
Spandauer Straße 1
10178 Berlin, Germany
e-mail: haerdle@wiwi.hu-berlin.de

Léopold Simar
Université Catholique Louvain
Inst. Statistique
voie du Roman Payas 20
1348 Louvain-la-Neuve, Belgium
e-mail: simar@stat.ucl.ac.be

Cover art: Detail of "Portrait of the Franciscanmonk Luca Pacioli, humanist and mathematician (1445-1517)"by Jacopo de Barbari,Museo di Capodimonte, Naples, Italy

Library of Congress Control Number: 2007929787

ISBN 978-3-540-72243-4 Springer Berlin Heidelberg New York
ISBN 978-3-540-03079-9 1st Edition Springer Berlin Heidelberg New York

Mathematics Subject Classification (2000): 62H10, 62H12, 62H15, 62H17, 62H20, 62H25, 62H30, 62F25

This work is subject to copyright. All rights are reserved, whether the whole or part of the material is concerned, specifically the rights of translation, reprinting, reuse of illustrations, recitation, broadcasting, reproduction onmicrofilm or in any other way, and storage in data banks. Duplication of this publication or parts thereof is permitted only under the provisions of the German Copyright Law of September 9, 1965, in its current version, and permission for use must always be obtained from Springer. Violations are liable to prosecution under the German Copyright Law.
Springer is a part of Springer Science+BusinessMedia
springer.com
The use of general descriptive names, registered names, trademarks, etc. in this publication does not imply, even in the absence of a specific statement, that such names are exempt from the relevant protective laws and regulations and therefore free for general use.
Cover-design: WMX Design GmbH, Heidelberg

Härdle & Simar: Applied Multivariate Statistical Analysis

© 2003, 2007 Springer-Verlag Berlin Heidelberg

All rights reserved. No part of this publication may be reproduced, stored in any electronic or mechanical form, including photocopy, recording or otherwise, without the prior written permission of the publisher.

First Indian Reprint 2011

ISBN 978-81-8489-862-0

"This edition is licensed for sale only in India, Pakistan, Bangladesh, Sri Lanka and Nepal. Circulation of this edition outside of these territories is UNAUTHORIZED AND STRICTLY PROHIBITED."

This edition is published by Springer (India) Private Limited, (a part of Springer Science+Business Media), Registered Office: 3rd Floor, Gandharva Mahavidyalaya, 212, Deen Dayal Upadhyaya Marg, New Delhi – 110 002, India

Printed in India by Rakmo Press (P) Ltd., New Delhi.

Preface to the 2nd Edition

The second edition of this book widens the scope of the methods and applications of Applied Multivariate Statistical Analysis. We have introduced more up to date data sets in our examples. These give the text a higher degree of timeliness and add an even more applied flavour. Since multivariate statistical methods are heavily used in quantitative finance and risk management we have put more weight on the presentation of distributions and their densities.

We discuss in detail different families of heavy tailed distributions (Laplace, Generalized Hyperbolic). We also devoted a section on copulae, a new concept of dependency used in the financial risk management and credit scoring. In the chapter on computer intensive methods we have added support vector machines, a new classification technique from statistical learning theory. We apply this method to bankruptcy and rating analysis of firms. The very important CART (Classification and Regression Tree) technique is also now inserted into this chapter. We give an application to rating of companies.

The probably most important step towards readability and user friendliness of this book is that we have translated all Quantlets into the R and Matlab language. The algorithms can be downloaded from the authors' web sites. In the preparation of this 2nd edition, we received helpful output from Anton Andriyashin, Ying Chen, Song Song and Uwe Ziegenhagen. We would like to thank them.

Wolfgang Karl Härdle and Leopold Simar

Berlin and Louvain la Neuve, June 2007

Preface to the 1st Edition

Most of the observable phenomena in the empirical sciences are of a multivariate nature. In financial studies, assets in stock markets are observed simultaneously and their joint development is analyzed to better understand general tendencies and to track indices. In medicine recorded observations of subjects in different locations are the basis of reliable diagnoses and medication. In quantitative marketing consumer preferences are collected in order to construct models of consumer behavior. The underlying theoretical structure of these and many other quantitative studies of applied sciences is multivariate. This book on Applied Multivariate Statistical Analysis presents the tools and concepts of multivariate data analysis with a strong focus on applications.

The aim of the book is to present multivariate data analysis in a way that is understandable for non-mathematicians and practitioners who are confronted by statistical data analysis. This is achieved by focusing on the practical relevance and through the e-book character of this text. All practical examples may be recalculated and modified by the reader using a standard web browser and without reference or application of any specific software.

The book is divided into three main parts. The first part is devoted to graphical techniques describing the distributions of the variables involved. The second part deals with multivariate random variables and presents from a theoretical point of view distributions, estimators and tests for various practical situations. The last part is on multivariate techniques and introduces the reader to the wide selection of tools available for multivariate data analysis. All data sets are given in the appendix and are downloadable from www.md-stat.com. The text contains a wide variety of exercises the solutions of which are given in a separate textbook. In addition a full set of transparencies on www.md-stat.com is provided making it easier for an instructor to present the materials in this book. All transparencies contain hyper links to the statistical web service so that students and instructors alike may recompute all examples via a standard web browser.

The first section on descriptive techniques is on the construction of the boxplot. Here the standard data sets on genuine and counterfeit bank notes and on the Boston housing data are introduced. Flury faces are shown in Section 1.5, followed by the presentation of Andrews curves and parallel coordinate plots. Histograms, kernel densities and scatterplots complete the first part of the book. The reader is introduced to the concept of skewness and correlation from a graphical point of view.

At the beginning of the second part of the book the reader goes on a short excursion into matrix algebra. Covariances, correlation and the linear model are introduced. This section is followed by the presentation of the ANOVA technique and its application to the multiple linear model. In Chapter 4 the multivariate distributions are introduced and thereafter specialized to the multinormal. The theory of estimation and testing ends the discussion on multivariate random variables.

The third and last part of this book starts with a geometric decomposition of data matrices. It is influenced by the French school of analyse de données. This geometric point of view is linked to principal components analysis in Chapter 9. An important discussion on factor analysis follows with a variety of examples from psychology and economics. The section on cluster analysis deals with the various cluster techniques and leads naturally to the problem of discrimination analysis. The next chapter deals with the detection of correspondence between factors. The joint structure of data sets is presented in the chapter on canonical correlation analysis and a practical study on prices and safety features of automobiles is given. Next the important topic of multidimensional scaling is introduced, followed by the tool of conjoint measurement analysis. The conjoint measurement analysis is often used in psychology and marketing in order to measure preference orderings for certain goods. The applications in finance (Chapter 17) are numerous. We present here the CAPM model and discuss efficient portfolio allocations. The book closes with a presentation on highly interactive, computationally intensive techniques.

This book is designed for the advanced bachelor and first year graduate student as well as for the inexperienced data analyst who would like a tour of the various statistical tools in a multivariate data analysis workshop. The experienced reader with a bright knowledge of algebra will certainly skip some sections of the multivariate random variables part but will hopefully enjoy the various mathematical roots of the multivariate techniques. A graduate student might think that the first part on description techniques is well known to him from his training in introductory statistics. The mathematical and the applied parts of the book (II, III) will certainly introduce him into the rich realm of multivariate statistical data analysis modules.

The inexperienced computer user of this e-book is slowly introduced to an interdisciplinary way of statistical thinking and will certainly enjoy the various practical examples. This e-book is designed as an interactive document with various links to other features. The complete e-book may be downloaded from www.xplore-stat.de using the license key given on the last page of this book. Our e-book design offers a complete PDF and HTML file with links to MD*Tech computing servers.

The reader of this book may therefore use all the presented methods and data via the local XploRe Quantlet Server (XQS) without downloading or buying additional software. Such XQ Servers may also be installed in a department or addressed freely on the web (see www.i-xplore.de for more information).

A book of this kind would not have been possible without the help of many friends, colleagues and students. For the technical production of the e-book we would like to thank Jörg Feuerhake, Zdeněk Hlávka, Torsten Kleinow, Sigbert Klinke, Heiko Lehmann, Marlene Müller. The book has been carefully read by Christian Hafner, Mia Huber, Stefan Sperlich, Axel Werwatz. We would also like to thank Pavel Čížek, Isabelle De Macq, Holger Gerhardt, Alena Myšičková and Manh Cuong Vu for the solutions to various statistical problems and exercises. We thank Clemens Heine from Springer Verlag for continuous support and valuable suggestions on the style of writing and on the contents covered.

W. Härdle and L. Simar

Berlin and Louvain-la-Neuve, August 2003

Contents

Part I

Descriptive Techniques

1 Comparison of Batches

Multivariate statistical analysis is concerned with analyzing and understanding data in high dimensions. We suppose that we are given a set $\{x_i\}_{i=1}^n$ of n observations of a variable vector X in $\mathbb{R}^p$. That is, we suppose that each observation x_i has p dimensions:

$$x_i = (x_{i1}, x_{i2}, ..., x_{ip}),$$

and that it is an observed value of a variable vector $X \in \mathbb{R}^p$. Therefore, X is composed of p random variables:

$$X = (X_1, X_2, ..., X_p)$$

where X_j, for $j = 1, \ldots, p$, is a one-dimensional random variable. How do we begin to analyze this kind of data? Before we investigate questions on what inferences we can reach from the data, we should think about how to look at the data. This involves descriptive techniques. Questions that we could answer by descriptive techniques are:

- Are there components of X that are more spread out than others?
- Are there some elements of X that indicate subgroups of the data?
- Are there outliers in the components of X?
- How "normal" is the distribution of the data?
- Are there "low-dimensional" linear combinations of X that show "non-normal" behavior?

One difficulty of descriptive methods for high dimensional data is the human perceptional system. Point clouds in two dimensions are easy to understand and to interpret. With modern interactive computing techniques we have the possibility to see real time 3D rotations and thus to perceive also three-dimensional data. A "sliding technique" as described in Härdle and Scott (1992) may give insight into four-dimensional structures by presenting dynamic 3D density contours as the fourth variable is changed over its range.

A qualitative jump in presentation difficulties occurs for dimensions greater than or equal to 5, unless the high-dimensional structure can be mapped into lower-dimensional components Klinke and Polzehl (1995). Features like clustered subgroups or outliers, however, can be detected using a purely graphical analysis.

In this chapter, we investigate the basic descriptive and graphical techniques allowing simple exploratory data analysis. We begin the exploration of a data set using boxplots. A boxplot is a simple univariate device that detects outliers component by component and that can compare distributions of the data among different groups. Next several multivariate techniques are introduced (Flury faces, Andrews' curves and parallel coordinate plots) which provide graphical displays addressing the questions formulated above. The advantages and the disadvantages of each of these techniques are stressed.

Two basic techniques for estimating densities are also presented: histograms and kernel densities. A density estimate gives a quick insight into the shape of the distribution of the data. We show that kernel density estimates overcome some of the drawbacks of the histograms.

Figure 1.1. An old Swiss 1000-franc bank note.

Finally, scatterplots are shown to be very useful for plotting bivariate or trivariate variables against each other: they help to understand the nature of the relationship among variables in a data set and allow to detect groups or clusters of points. Draftman plots or matrix plots are the visualization of several bivariate scatterplots on the same display. They help detect structures in conditional dependences by *brushing* across the plots. Outliers and observations that need special attention may be discovered with Andrews curves and Parallel Coordinate Plots. This chapter ends with an explanatory analysis of the Boston Housing data.

1.1 Boxplots

EXAMPLE 1.1 *The Swiss bank data (see Appendix, Table B.2) consists of 200 measurements on Swiss bank notes. The first half of these measurements are from genuine bank notes, the other half are from counterfeit bank notes.*

The authorities have measured, as indicated in Figure 1.1,

$$\begin{aligned} X_1 &= \textit{length of the bill} \\ X_2 &= \textit{height of the bill (left)} \\ X_3 &= \textit{height of the bill (right)} \\ X_4 &= \textit{distance of the inner frame to the lower border} \\ X_5 &= \textit{distance of the inner frame to the upper border} \\ X_6 &= \textit{length of the diagonal of the central picture.} \end{aligned}$$

These data are taken from Flury and Riedwyl (1988). The aim is to study how these measurements may be used in determining whether a bill is genuine or counterfeit.

The *boxplot* is a graphical technique that displays the distribution of variables. It helps us see the location, skewness, spread, tail length and outlying points.

City	Country	Pop. (10000)	Order Statistics
Tokyo	Japan	3420	$x_{(15)}$
Mexico City	Mexico	2280	$x_{(14)}$
Seoul	South Korea	2230	$x_{(13)}$
New York	USA	2190	$x_{(12)}$
Sao Paulo	Brazil	2020	$x_{(11)}$
Bombay	India	1985	$x_{(10)}$
Delhi	India	1970	$x_{(9)}$
Shanghai	China	1815	$x_{(8)}$
Los Angeles	USA	1800	$x_{(7)}$
Osaka	Japan	1680	$x_{(6)}$
Jakarta	Indonesia	1655	$x_{(5)}$
Calcutta	India	1565	$x_{(4)}$
Cairo	Egypt	1560	$x_{(3)}$
Manila	Philippines	1495	$x_{(2)}$
Karachi	Pakistan	1430	$x_{(1)}$

Table 1.1. The 15 largest world cities in 2006.

It is particularly useful in comparing different batches. The boxplot is a graphical representation of the *Five Number Summary*. To introduce the Five Number Summary, let us consider for a moment a smaller, one-dimensional data set: the population of the 15 largest world cities in 2006 (Table 1.1).

In the Five Number Summary, we calculate the upper quartile F_U, the lower quartile F_L, the median and the extremes. Recall that order statistics $\{x_{(1)}, x_{(2)}, \ldots, x_{(n)}\}$ are a set of ordered values $x_1, x_2, \ldots, x_n$ where $x_{(1)}$ denotes the minimum and $x_{(n)}$ the maximum. The *median* M typically cuts the set of observations in two equal parts, and is defined as

$$M = \begin{cases} x_{(\frac{n+1}{2})} & n \text{ odd} \\ \frac{1}{2}\left\{x_{(\frac{n}{2})} + x_{(\frac{n}{2}+1)}\right\} & n \text{ even} \end{cases}. \tag{1.1}$$

The quartiles cut the set into four equal parts, which are often called *fourths* (that is why we use the letter F). Using a definition that goes back to Hoaglin, Mosteller and Tukey (1983) the definition of a median can be generalized to fourths, eights, etc. Considering the order statistics we can define the depth of a data value $x_{(i)}$ as $\min\{i, n-i+1\}$. If n is odd, the depth of the median is $\frac{n+1}{2}$. If n is even, $\frac{n+1}{2}$ is a fraction. Thus, the median is determined to be the average between the two data values belonging to the next larger and smaller order statistics, i.e., $M = \frac{1}{2}\left\{x_{(\frac{n}{2})} + x_{(\frac{n}{2}+1)}\right\}$. In our example, we have $n = 15$ hence the median $M = x_{(8)} = 1815$.

We proceed in the same way to get the fourths. Take the depth of the median and calculate

$$\text{depth of fourth} = \frac{[\text{depth of median}] + 1}{2}$$

with $[z]$ denoting the largest integer smaller than or equal to z. In our example this gives 4.5 and thus leads to the two fourths

$$\begin{aligned} F_L &= \frac{1}{2}\left\{x_{(4)} + x_{(5)}\right\} \\ F_U &= \frac{1}{2}\left\{x_{(11)} + x_{(12)}\right\} \end{aligned}$$

#	15	World Cities	
M	8	1815	
F	4.5	1610	2105
	1	1430	3420

Table 1.2. Five number summary.

(recalling that a depth which is a fraction corresponds to the average of the two nearest data values).

The *F-spread*, d_F, is defined as $d_F = F_U - F_L$. The *outside bars*

$$F_U + 1.5d_F \tag{1.2}$$
$$F_L - 1.5d_F \tag{1.3}$$

are the borders beyond which a point is regarded as an outlier. For the number of points outside these bars see Exercise 1.3. For the $n = 15$ data points the fourths are $1,610 = \frac{1}{2}\left\{x_{(4)} + x_{(5)}\right\}$ and $2,105 = \frac{1}{2}\left\{x_{(11)} + x_{(12)}\right\}$. Therefore the *F-spread* and the upper and lower *outside bars* in the above example are calculated as follows:

$$d_F = F_U - F_L = 183.5 - 74 = 495 \tag{1.4}$$
$$F_L - 1.5d_F = 74 - 1.5 \cdot 109.5 = 867.5 \tag{1.5}$$
$$F_U + 1.5d_F = 183.5 + 1.5 \cdot 109.5 = 2847.5. \tag{1.6}$$

Since Tokyo is beyond the outside bars it is considered to be an outlier. The minimum and the maximum are called the *extremes*. The *mean* is defined as

$$\overline{x} = n^{-1} \sum_{i=1}^{n} x_i,$$

which is 1939.7 in our example. The mean is a measure of location. The median (1815), the fourths (1610;2105) and the extremes (1430;3420) constitute basic information about the data. The combination of these five numbers leads to the Five Number Summary as displayed in Table 1.4. The depths of each of the five numbers have been added as an additional column.

Construction of the Boxplot

1. Draw a box with borders (edges) at F_L and F_U (i.e., 50% of the data are in this box).
2. Draw the median as a solid line (|) and the mean as a dotted line (⁞).
3. Draw "whiskers" from each end of the box to the most remote point that is <u>NOT</u> an outlier.
4. Show outliers as either "$\star$" or "$\bullet$" depending on whether they are outside of $F_{UL} \pm 1.5d_F$ or $F_{UL} \pm 3d_F$ respectively. Label them if possible.

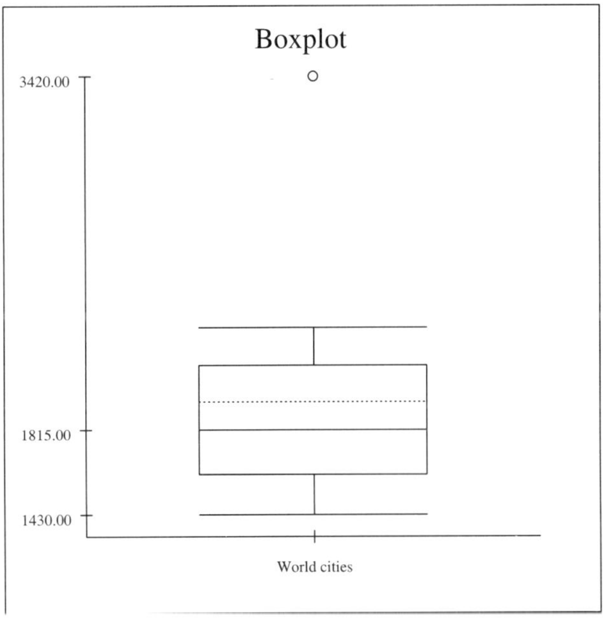

Figure 1.2. Boxplot for world cities. MVAboxcity

In the world cities example the cutoff points (outside bars) are at 867.5 and 2847.5, hence we draw whiskers to Karachi and Mexico City. We can see from Figure 1.2 that the data are very skew: The upper half of the data (above the median) is more spread out than the lower half (below the median). The data contains one outlier marked as a circle. The mean (as a non-robust measure of location) is pulled away from the median.

Boxplots are very useful tools in comparing batches. The relative location of the distribution of different batches tells us a lot about the batches themselves. Before we come back to the Swiss bank data let us compare the fuel economy of vehicles from different countries, see Figure 1.3 and Table B.3.

EXAMPLE 1.2 *The data are from the second column of Table B.3 and show the mileage (miles per gallon) of U.S. American, Japanese and European cars. The five-number summaries for these data sets are* $\{12, 16.8, 18.8, 22, 30\}$, $\{18, 22, 25, 30.5, 35\}$, *and* $\{14, 19, 23, 25, 28\}$ *for American, Japanese, and European cars, respectively. This reflects the information shown in Figure 1.3. The following conclusions can be made:*

- *Japanese cars achieve higher fuel efficiency than U.S. and European cars.*
- *There is one outlier, a very fuel-efficient car (VW-Rabbit Diesel).*
- *The main body of the U.S. car data (the box) lies below the Japanese car data.*
- *The worst Japanese car is more fuel-efficient than almost 50 percent of the U.S. cars.*
- *The spread of the Japanese and the U.S. cars are almost equal.*
- *The median of the Japanese data is above that of the European data and the U.S. data.*

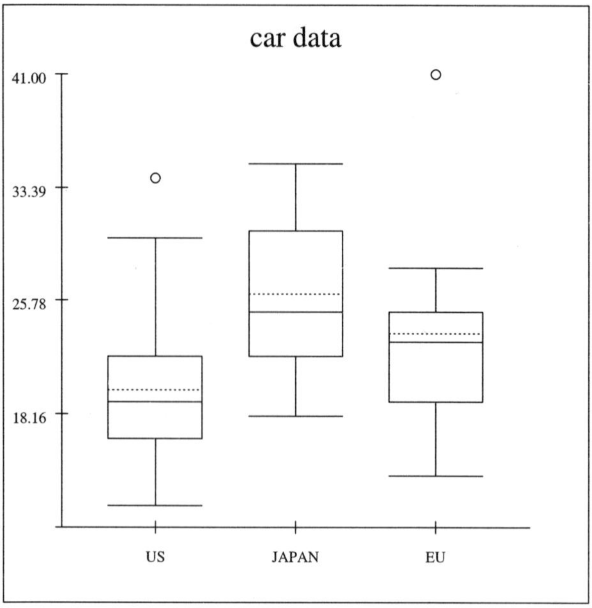

Figure 1.3. Boxplot for the mileage of American, Japanese and European cars (from left to right). `MVAboxcar`

#	100	Genuine Bank Notes		
M	50.5		141.5	
F	25.75	141.25		141.8
	1	140.65		142.4

Table 1.3. Five number summary.

Now let us apply the boxplot technique to the bank data set. In Figure 1.4 we show the parallel boxplot of the diagonal variable X_6. On the left is the value of the genuine bank notes and on the right the value of the counterfeit bank notes.

One sees that the diagonals of the genuine bank notes tend to be larger. It is harder to see a clear distinction when comparing the length of the bank notes X_1, see Figure 1.5. There are a few outliers in both plots. Almost all the observations of the diagonal of the genuine notes are above the ones from the counterfeit. There is one observation in Figure 1.4 of the genuine notes that is almost equal to the median of the counterfeit notes. Can the parallel boxplot technique help us distinguish between the two types of bank notes?

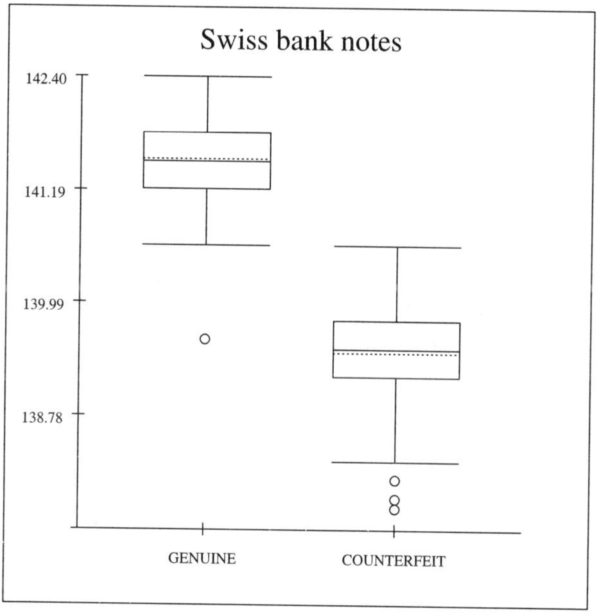

Figure 1.4. The X_6 variable of Swiss bank data (diagonal of bank notes). MVAboxbank6

#	100	Counterfeit Bank Notes		
M	50.5		139.5	
F	25.75	139.2		139.8
	1	138.3		140.65

Table 1.4. Five number summary.

Summary

↪ The median and mean bars are measures of locations.

↪ The relative location of the median (and the mean) in the box is a measure of skewness.

↪ The length of the box and whiskers are a measure of spread.

Summary (continued)
↪ The length of the whiskers indicate the tail length of the distribution.
↪ The outlying points are indicated with a "⋆" or "•" depending on if they are outside of $F_{UL} \pm 1.5 d_F$ or $F_{UL} \pm 3 d_F$ respectively.
↪ The boxplots do not indicate multi modality or clusters.
↪ If we compare the relative size and location of the boxes, we are comparing distributions.

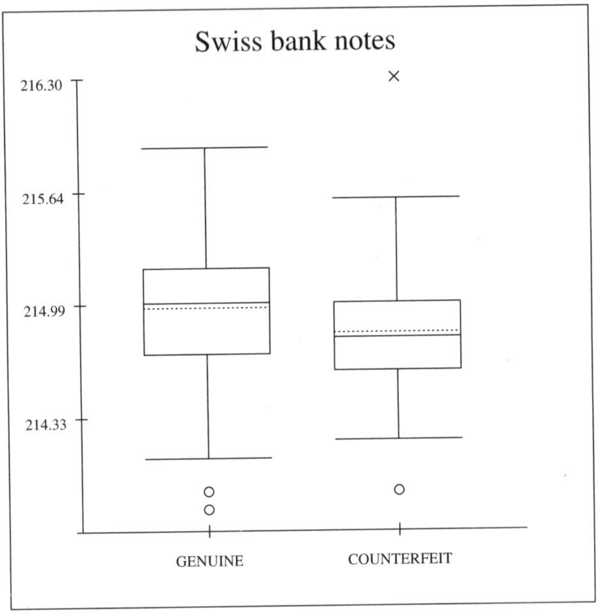

Figure 1.5. The X_1 variable of Swiss bank data (length of bank notes).
MVAboxbank1

1.2 Histograms

Histograms are density estimates. A density estimate gives a good impression of the distribution of the data. In contrast to boxplots, density estimates show possible multimodality of the data. The idea is to locally represent the data density by counting the number of observations in a sequence of consecutive intervals (bins) with origin x_0. Let $B_j(x_0, h)$ denote the *bin* of length h which is the element of a bin grid starting at x_0:

$$B_j(x_0, h) = [x_0 + (j-1)h, x_0 + jh), \quad j \in \mathbb{Z},$$

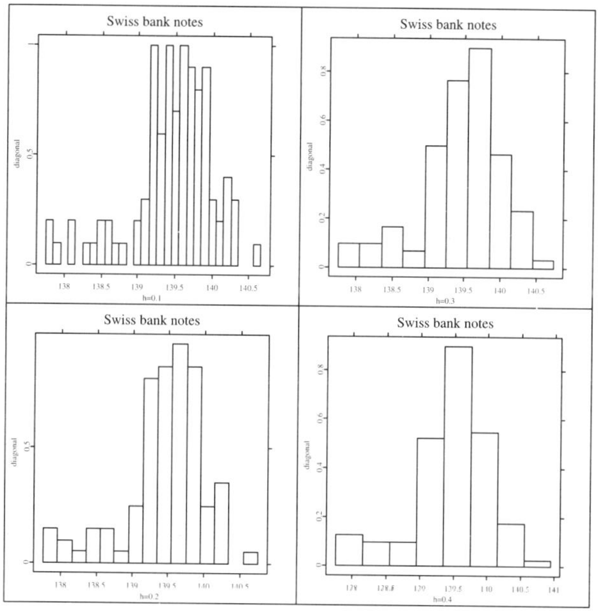

Figure 1.6. Diagonal of counterfeit bank notes. Histograms with $x_0 =$ 137.8 and $h = 0.1$ (upper left), $h = 0.2$ (lower left), $h = 0.3$ (upper right), $h = 0.4$ (lower right). MVAhisbank1

where $[.,.)$ denotes a left closed and right open interval. If $\{x_i\}_{i=1}^n$ is an i.i.d. sample with density f, the histogram is defined as follows:

$$\widehat{f}_h(x) = n^{-1}h^{-1}\sum_{j\in\mathbb{Z}}\sum_{i=1}^{n} \boldsymbol{I}\{x_i \in B_j(x_0,h)\}\,\boldsymbol{I}\{x \in B_j(x_0,h)\}. \tag{1.7}$$

In sum (1.7) the first indicator function $\boldsymbol{I}\{x_i \in B_j(x_0,h)\}$ (see Symbols & Notation in Appendix A) counts the number of observations falling into bin $B_j(x_0,h)$. The second indicator function is responsible for "localizing" the counts around x. The parameter h is a smoothing or localizing parameter and controls the width of the histogram bins. An h that is too large leads to very big blocks and thus to a very unstructured histogram. On the other hand, an h that is too small gives a very variable estimate with many unimportant peaks.

The effect of h is given in detail in Figure 1.6. It contains the histogram (upper left) for the diagonal of the counterfeit bank notes for $x_0 = 137.8$ (the minimum of these observations) and $h = 0.1$. Increasing h to $h = 0.2$ and using the same origin, $x_0 = 137.8$, results in the histogram shown in the lower left of the figure. This density histogram is somewhat smoother due to the larger h. The binwidth is next set to $h = 0.3$ (upper right). From this histogram, one has the impression that the distribution of the diagonal is bimodal with peaks at about 138.5 and 139.9. The detection of modes requires a fine tuning of the binwidth. Using methods from smoothing methodology (Härdle, Müller, Sperlich and Werwatz, 2003) one can find an "optimal" binwidth h for n observations:

$$h_{opt} = \left(\frac{24\sqrt{\pi}}{n}\right)^{1/3}.$$

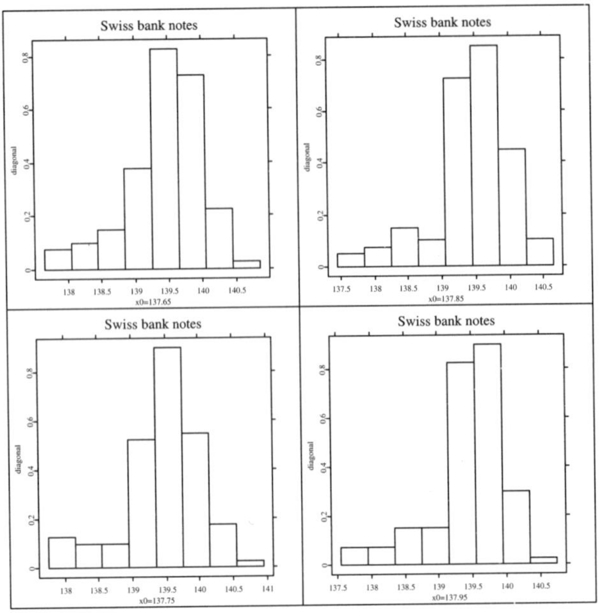

Figure 1.7. Diagonal of counterfeit bank notes. Histogram with $h = 0.4$ and origins $x_0 = 137.65$ (upper left), $x_0 = 137.75$ (lower left), $x_0 = 137.85$ (upper right), $x_0 = 137.95$ (lower right). `MVAhisbank2`

Unfortunately, the binwidth h is not the only parameter determining the shapes of $\widehat{f}$.

In Figure 1.7, we show histograms with $x_0 = 137.65$ (upper left), $x_0 = 137.75$ (lower left), with $x_0 = 137.85$ (upper right), and $x_0 = 137.95$ (lower right). All the graphs have been scaled equally on the y-axis to allow comparison. One sees that—despite the fixed binwidth h—the interpretation is not facilitated. The shift of the origin x_0 (to 4 different locations) created 4 different histograms. This property of histograms strongly contradicts the goal of presenting data features. Obviously, the same data are represented quite differently by the 4 histograms. A remedy has been proposed by Scott (1985): "Average the shifted histograms!". The result is presented in Figure 1.8. Here all bank note observations (genuine and counterfeit) have been used. The (so ccalled) averaged shifted histogram is no longer dependent on the origin and shows a clear bimodality of the diagonals of the Swiss bank notes.

Summary

↪ Modes of the density are detected with a histogram.

↪ Modes correspond to strong peaks in the histogram.

Summary (continued)
↪ Histograms with the same h need not be identical. They also depend on the origin x_0 of the grid.
↪ The influence of the origin x_0 is drastic. Changing x_0 creates different looking histograms.
↪ The consequence of an h that is too large is an unstructured histogram that is too flat.
↪ A binwidth h that is too small results in an unstable histogram.
↪ There is an "optimal" $h = (24\sqrt{\pi}/n)^{1/3}$.
↪ It is recommended to use averaged histograms. They are kernel densities.

1.3 Kernel Densities

The major difficulties of histogram estimation may be summarized in four critiques:

- determination of the binwidth h, which controls the shape of the histogram,
- choice of the bin origin x_0, which also influences to some extent the shape,
- loss of information since observations are replaced by the central point of the interval in which they fall,
- the underlying density function is often assumed to be smooth, but the histogram is not smooth.

Rosenblatt (1956), Whittle (1958), and Parzen (1962) developed an approach which avoids the last three difficulties. First, a smooth kernel function rather than a box is used as the basic building block. Second, the smooth function is centered directly over each observation. Let us study this refinement by supposing that x is the center value of a bin. The histogram can in fact be rewritten as

$$\widehat{f}_h(x) = n^{-1}h^{-1}\sum_{i=1}^{n} \boldsymbol{I}(|x - x_i| \le \frac{h}{2}). \tag{1.8}$$

If we define $K(u) = \boldsymbol{I}(|u| \le \frac{1}{2})$, then (1.8) changes to

$$\widehat{f}_h(x) = n^{-1}h^{-1}\sum_{i=1}^{n} K\left(\frac{x - x_i}{h}\right). \tag{1.9}$$

This is the general form of the kernel estimator. Allowing smoother kernel functions like the quartic kernel,

$$K(u) = \frac{15}{16}(1 - u^2)^2\ \boldsymbol{I}(|u| \le 1),$$

and computing x not only at bin centers gives us the kernel density estimator. Kernel estimators can also be derived via weighted averaging of rounded points (WARPing) or by averaging histograms with different origins, see Scott (1985). Table 1.7 introduces some commonly used kernels.

Different kernels generate different shapes of the estimated density. The most important parameter is the so-called bandwidth h, and can be optimized, for example, by cross-validation;

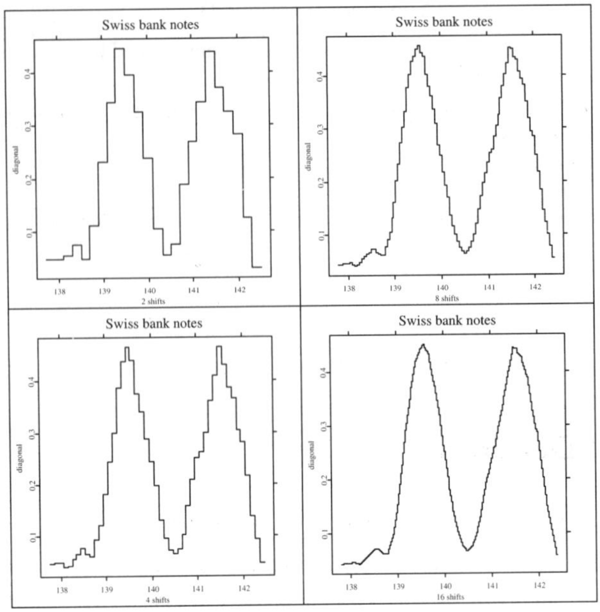

Figure 1.8. Averaged shifted histograms based on all (counterfeit and genuine) Swiss bank notes: there are 2 shifts (upper left), 4 shifts (lower left), 8 shifts (upper right), and 16 shifts (lower right). MVAashbank

$K(\bullet)$	Kernel
$K(u) = \frac{1}{2}\,\boldsymbol{I}(\lvert u\rvert \le 1)$	Uniform
$K(u) = (1-\lvert u\rvert)\,\boldsymbol{I}(\lvert u\rvert \le 1)$	Triangle
$K(u) = \frac{3}{4}(1-u^2)\,\boldsymbol{I}(\lvert u\rvert \le 1)$	Epanechnikov
$K(u) = \frac{15}{16}(1-u^2)^2\,\boldsymbol{I}(\lvert u\rvert \le 1)$	Quartic (Biweight)
$K(u) = \frac{1}{\sqrt{2\pi}}\exp(-\frac{u^2}{2}) = \varphi(u)$	Gaussian

Table 1.7. Kernel functions.

see Härdle (1991) for details. The cross-validation method minimizes the integrated squared error. This measure of discrepancy is based on the squared differences $\left\{\hat{f}_h(x) - f(x)\right\}^2$. Averaging these squared deviations over a grid of points $\{x_l\}_{l=1}^{L}$ leads to

$$L^{-1}\sum_{l=1}^{L}\left\{\hat{f}_h(x_l) - f(x_l)\right\}^2.$$

Asymptotically, if this grid size tends to zero, we obtain the integrated squared error:

$$\int\left\{\hat{f}_h(x) - f(x)\right\}^2 dx.$$

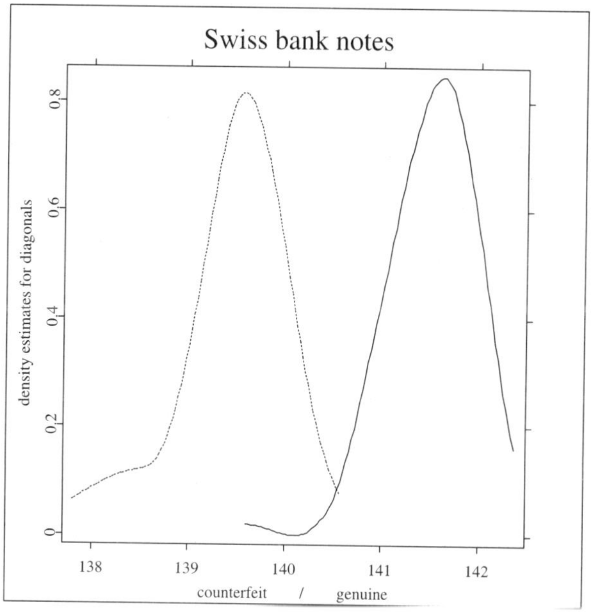

Figure 1.9. Densities of the diagonals of genuine and counterfeit bank notes. Automatic density estimates. MVAdenbank

In practice, it turns out that the method consists of selecting a bandwidth that minimizes the cross-validation function

$$\int \hat{f}_h^2 - 2\sum_{i=1}^{n} \hat{f}_{h,i}(x_i)$$

where $\hat{f}_{h,i}$ is the density estimate obtained by using all datapoints except for the i-th observation. Both terms in the above function involve double sums. Computation may therefore be slow. There are many other density bandwidth selection methods. Probably the fastest way to calculate this is to refer to some reasonable reference distribution. The idea of using the Normal distribution as a reference, for example, goes back to Silverman (1986). The resulting choice of h is called the *rule of thumb*.

For the Gaussian kernel from Table 1.7 and a Normal reference distribution, the rule of thumb is to choose

$$h_G = 1.06\,\widehat{\sigma}\,n^{-1/5} \tag{1.10}$$

where $\widehat{\sigma} = \sqrt{n^{-1}\sum_{i=1}^{n}(x_i - \overline{x})^2}$ denotes the sample standard deviation. This choice of h_G optimizes the integrated squared distance between the estimator and the true density. For the quartic kernel, we need to transform (1.10). The modified rule of thumb is:

$$h_Q = 2.62 \cdot h_G. \tag{1.11}$$

Figure 1.9 shows the automatic density estimates for the diagonals of the counterfeit and genuine bank notes. The density on the left is the density corresponding to the diagonal of the counterfeit data. The separation is clearly visible, but there is also an overlap. The problem of distinguishing between the counterfeit and genuine bank notes is not solved by

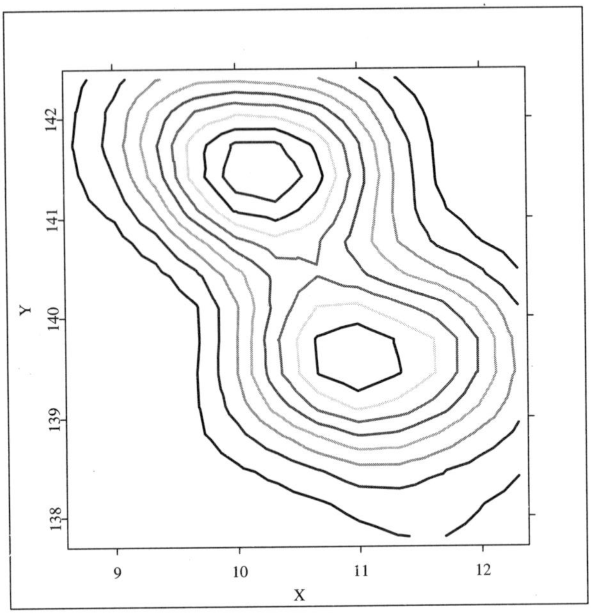

Figure 1.10. Contours of the density of X_5 and X_6 of genuine and counterfeit bank notes. MVAcontbank2

just looking at the diagonals of the notes! The question arises whether a better separation could be achieved using not only the diagonals but one or two more variables of the data set. The estimation of higher dimensional densities is analogous to that of one-dimensional. We show a two dimensional density estimate for X_4 and X_5 in Figure 1.10. The contour lines indicate the height of the density. One sees two separate distributions in this higher dimensional space, but they still overlap to some extent.

We can add one more dimension and give a graphical representation of a three dimensional density estimate, or more precisely an estimate of the joint distribution of X_4, X_5 and X_6. Figure 1.11 shows the contour areas at 3 different levels of the density: 0.2 (light grey), 0.4 (grey), and 0.6 (black) of this three dimensional density estimate. One can clearly recognize two "ellipsoids" (at each level), but as before, they overlap. In Chapter 12 we will learn how to separate the two ellipsoids and how to develop a discrimination rule to distinguish between these data points.

Summary

↪ Kernel densities estimate distribution densities by the kernel method.

↪ The bandwidth h determines the degree of smoothness of the estimate $\widehat{f}$.

Summary (continued)
↪ Kernel densities are smooth functions and they can graphically represent distributions (up to 3 dimensions).
↪ A simple (but not necessarily correct) way to find a good bandwidth is to compute the rule of thumb bandwidth $h_G = 1.06\widehat{\sigma}n^{-1/5}$. This bandwidth is to be used only in combination with a Gaussian kernel φ.
↪ Kernel density estimates are a good descriptive tool for seeing modes, location, skewness, tails, asymmetry, etc.

1.4 Scatterplots

Scatterplots are bivariate or trivariate plots of variables against each other. They help us understand relationships among the variables of a data set. A downward-sloping scatter indicates that as we increase the variable on the horizontal axis, the variable on the vertical axis decreases. An analogous statement can be made for upward-sloping scatters.

Figure 1.12 plots the 5th column (upper inner frame) of the bank data against the 6th column (diagonal). The scatter is downward-sloping. As we already know from the previous section on marginal comparison (e.g., Figure 1.9) a good separation between genuine and counterfeit bank notes is visible for the diagonal variable. The sub-cloud in the upper half (circles) of Figure 1.12 corresponds to the true bank notes. As noted before, this separation is not distinct, since the two groups overlap somewhat.

This can be verified in an interactive computing environment by showing the index and coordinates of certain points in this scatterplot. In Figure 1.12, the 70th observation in the merged data set is given as a thick circle, and it is from a genuine bank note. This observation lies well embedded in the cloud of counterfeit bank notes. One straightforward approach that could be used to tell the counterfeit from the genuine bank notes is to draw a straight line and define notes above this value as genuine. We would of course misclassify the 70th observation, but can we do better?

If we extend the two-dimensional scatterplot by adding a third variable, e.g., X_4 (lower distance to inner frame), we obtain the scatterplot in three-dimensions as shown in Figure 1.13. It becomes apparent from the location of the point clouds that a better separation is obtained. We have rotated the three dimensional data until this satisfactory 3D view was obtained. Later, we will see that rotation is the same as bundling a high-dimensional observation into one or more linear combinations of the elements of the observation vector. In other words, the "separation line" parallel to the horizontal coordinate axis in Figure 1.12 is in Figure 1.13 a plane and no longer parallel to one of the axes. The formula for such a separation plane is a linear combination of the elements of the observation vector:

$$a_1x_1 + a_2x_2 + \ldots + a_6x_6 = \text{const.} \tag{1.12}$$

The algorithm that automatically finds the weights $(a_1, \ldots, a_6)$ will be investigated later on in Chapter 12.

Let us study yet another technique: the scatterplot matrix. If we want to draw all possible two-dimensional scatterplots for the variables, we can create a so-called *draftman's plot* (named after a draftman who prepares drafts for parliamentary discussions). Similar to a draftman's plot the scatterplot matrix helps in creating new ideas and in building knowledge about dependencies and structure.

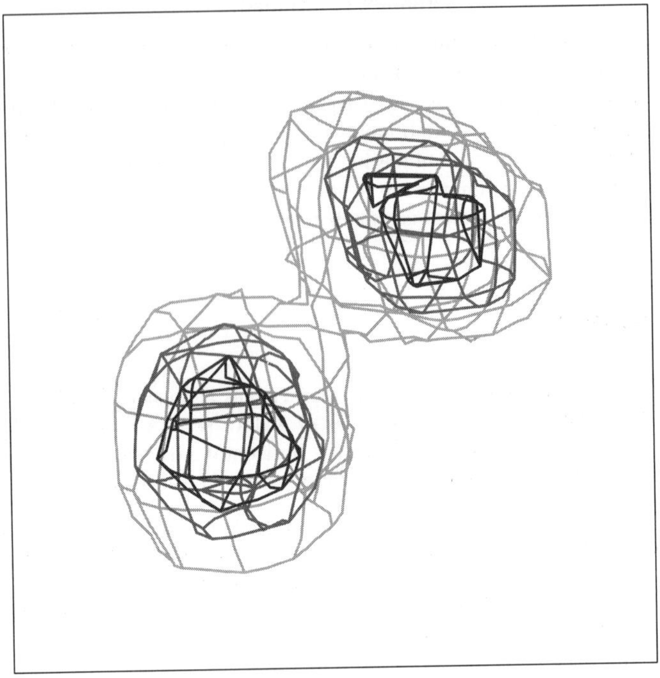

Figure 1.11. Contours of the density of X_4, X_5, X_6 of genuine and counterfeit bank notes. MVAcontbank3

Figure 1.14 shows a draftman plot applied to the last four columns of the full bank data set. For ease of interpretation we have distinguished between the group of counterfeit and genuine bank notes by a different color. As discussed several times before, the separability of the two types of notes is different for different scatterplots. Not only is it difficult to perform this separation on, say, scatterplot X_3 vs. X_4, in addition the "separation line" is no longer parallel to one of the axes. The most obvious separation happens in the scatterplot in the lower right where we show, as in Figure 1.12, X_5 vs. X_6. The separation line here would be upward-sloping with an intercept at about $X_6 = 139$. The upper right half of the draftman plot shows the density contours that we have introduced in Section 1.3.

The power of the draftman plot lies in its ability to show the the internal connections of the scatter diagrams. Define a *brush* as a re-scalable rectangle that we can move via keyboard or mouse over the screen. Inside the brush we can highlight or color observations. Suppose the technique is installed in such a way that as we move the brush in one scatter, the corresponding observations in the other scatters are also highlighted. By moving the brush, we can study conditional dependence.

If we brush (i.e., highlight or color the observation with the brush) the X_5 vs. X_6 plot and move through the upper point cloud, we see that in other plots (e.g., X_3 vs. X_4), the corresponding observations are more embedded in the other sub-cloud.

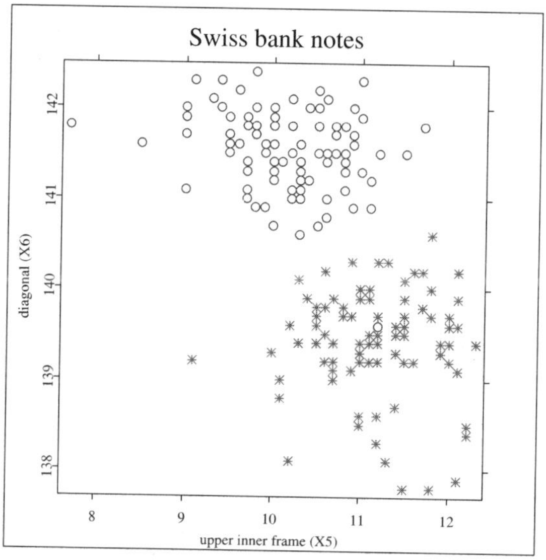

Figure 1.12. 2D scatterplot for X_5 vs. X_6 of the bank notes. Genuine notes are circles, counterfeit notes are stars. MVAscabank56

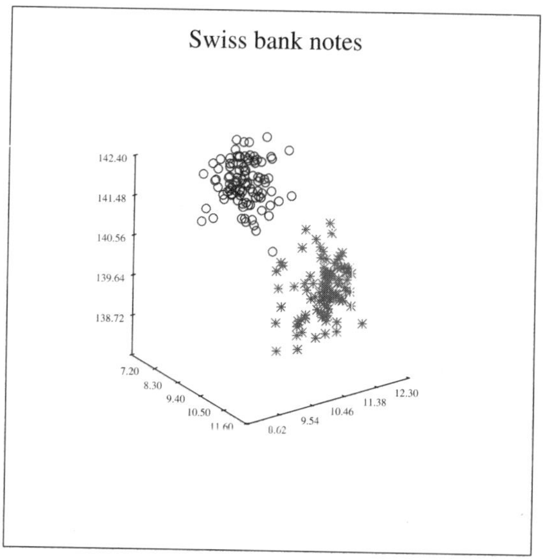

Figure 1.13. 3D Scatterplot of the bank notes for (X_4, X_5, X_6). Genuine notes are circles, counterfeit are stars. MVAscabank456

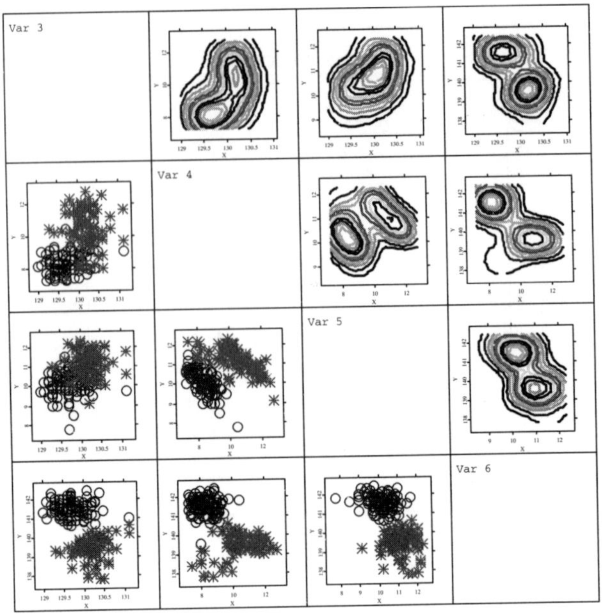

Figure 1.14. Draftman plot of the bank notes. The pictures in the left column show (X_3, X_4), (X_3, X_5) and (X_3, X_6), in the middle we have (X_4, X_5) and (X_4, X_6), and in the lower right is (X_5, X_6). The upper right half contains the corresponding density contour plots. MVAdrafbank4

Summary

↪ Scatterplots in two and three dimensions helps in identifying separated points, outliers or sub-clusters.

↪ Scatterplots help us in judging positive or negative dependencies.

↪ Draftman scatterplot matrices help detect structures conditioned on values of other variables.

↪ As the brush of a scatterplot matrix moves through a point cloud, we can study conditional dependence.

1.5 Chernoff-Flury Faces

If we are given data in numerical form, we tend to display it also numerically. This was done in the preceding sections: an observation $x_1 = (1, 2)$ was plotted as the point $(1, 2)$ in a two-dimensional coordinate system. In multivariate analysis we want to understand data in low dimensions (e.g., on a 2D computer screen) although the structures are hidden in high dimensions. The numerical display of data structures using coordinates therefore ends at dimensions greater than three.

If we are interested in condensing a structure into 2D elements, we have to consider alternative graphical techniques. The Chernoff-Flury faces, for example, provide such a condensation of high-dimensional information into a simple "face". In fact faces are a simple way to graphically display high-dimensional data. The size of the face elements like pupils, eyes, upper and lower hair line, etc., are assigned to certain variables. The idea of using faces goes back to Chernoff (1973) and has been further developed by Bernhard Flury. We follow the design described in Flury and Riedwyl (1988) which uses the following characteristics.

1	right eye size
2	right pupil size
3	position of right pupil
4	right eye slant
5	horizontal position of right eye
6	vertical position of right eye
7	curvature of right eyebrow
8	density of right eyebrow
9	horizontal position of right eyebrow
10	vertical position of right eyebrow
11	right upper hair line
12	right lower hair line
13	right face line
14	darkness of right hair
15	right hair slant
16	right nose line
17	right size of mouth
18	right curvature of mouth
19–36	like 1–18, only for the left side.

First, every variable that is to be coded into a characteristic face element is transformed into a $(0,1)$ scale, i.e., the minimum of the variable corresponds to 0 and the maximum to 1. The extreme positions of the face elements therefore correspond to a certain "grin" or "happy" face element. Dark hair might be coded as 1, and blond hair as 0 and so on.

As an example, consider the observations 91 to 110 of the bank data. Recall that the bank data set consists of 200 observations of dimension 6 where, for example, X_6 is the diagonal of the note. If we assign the six variables to the following face elements

$$
\begin{aligned}
X_1 &= 1, 19 \text{ (eye sizes)}\\
X_2 &= 2, 20 \text{ (pupil sizes)}\\
X_3 &= 4, 22 \text{ (eye slants)}\\
X_4 &= 11, 29 \text{ (upper hair lines)}\\
X_5 &= 12, 30 \text{ (lower hair lines)}\\
X_6 &= 13, 14, 31, 32 \text{ (face lines and darkness of hair)},
\end{aligned}
$$

we obtain Figure 1.15. Also recall that observations 1–100 correspond to the genuine notes, and that observations 101–200 correspond to the counterfeit notes. The counterfeit bank

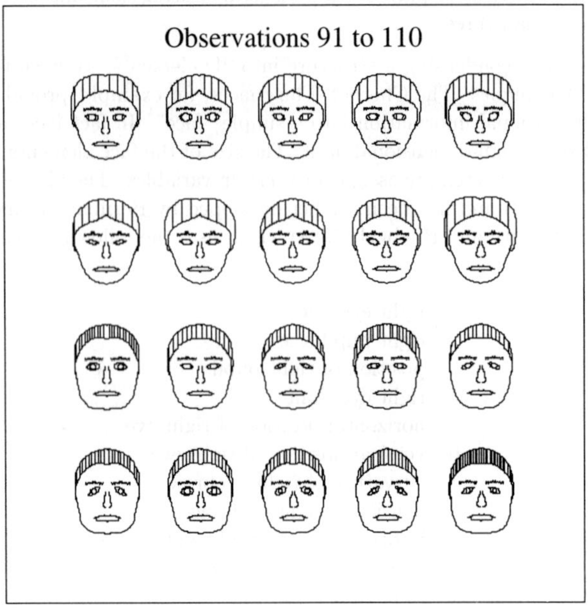

Figure 1.15. Chernoff-Flury faces for observations 91 to 110 of the bank notes. MVAfacebank10

notes then correspond to the upper half of Figure 1.15. In fact the faces for these observations look more grim and less happy. The variable X_6 (diagonal) already worked well in the boxplot on Figure 1.4 in distinguishing between the counterfeit and genuine notes. Here, this variable is assigned to the face line and the darkness of the hair. That is why we clearly see a good separation within these 20 observations.

What happens if we include all 100 genuine and all 100 counterfeit bank notes in the Chernoff-Flury face technique? Figures 1.16 and 1.17 show the faces of the genuine bank notes with the same assignments as used before and Figures 1.18 and 1.19 show the faces of the counterfeit bank notes. Comparing Figure 1.16 and Figure 1.18 one clearly sees that the diagonal (face line) is longer for genuine bank notes. Equivalently coded is the hair darkness (diagonal) which is lighter (shorter) for the counterfeit bank notes. One sees that the faces of the genuine bank notes have a much darker appearance and have broader face lines. The faces in Figures 1.16–1.17 are obviously different from the ones in Figures 1.18–1.19.

Summary

↪ Faces can be used to detect subgroups in multivariate data.

↪ Subgroups are characterized by similar looking faces.

Summary (continued)
↪ Outliers are identified by extreme faces, e.g., dark hair, smile or a happy face.
↪ If one element of X is unusual, the corresponding face element significantly changes in shape.

1.6 Andrews' Curves

The basic problem of graphical displays of multivariate data is the dimensionality. Scatterplots work well up to three dimensions (if we use interactive displays). More than three dimensions have to be coded into displayable 2D or 3D structures (e.g., faces). The idea of coding and representing multivariate data by curves was suggested by Andrews (1972). Each multivariate observation $X_i = (X_{i,1}, .., X_{i,p})$ is transformed into a curve as follows:

$$f_i(t) = \begin{cases} \frac{X_{i,1}}{\sqrt{2}} + X_{i,2}\sin(t) + X_{i,3}\cos(t) + ... + X_{i,p-1}\sin(\frac{p-1}{2}t) + X_{i,p}\cos(\frac{p-1}{2}t) & \text{for } p \text{ odd} \\ \frac{X_{i,1}}{\sqrt{2}} + X_{i,2}\sin(t) + X_{i,3}\cos(t) + ... + X_{i,p}\sin(\frac{p}{2}t) & \text{for } p \text{ even} \end{cases} \tag{1.13}$$

such that the observation represents the coefficients of a so-called Fourier series ($t \in [-\pi, \pi]$).

Suppose that we have three-dimensional observations: $X_1 = (0, 0, 1)$, $X_2 = (1, 0, 0)$ and $X_3 = (0, 1, 0)$. Here $p = 3$ and the following representations correspond to the Andrews' curves:

$$\begin{aligned} f_1(t) &= \cos(t) \\ f_2(t) &= \frac{1}{\sqrt{2}} \quad \text{and} \\ f_3(t) &= \sin(t). \end{aligned}$$

These curves are indeed quite distinct, since the observations X_1, X_2, and X_3 are the 3D unit vectors: each observation has mass only in one of the three dimensions. The order of the variables plays an important role.

EXAMPLE 1.3 *Let us take the 96th observation of the Swiss bank note data set,*

$$X_{96} = (215.6, 129.9, 129.9, 9.0, 9.5, 141.7).$$

The Andrews' curve is by (1.13):

$$f_{96}(t) = \frac{215.6}{\sqrt{2}} + 129.9\sin(t) + 129.9\cos(t) + 9.0\sin(2t) + 9.5\cos(2t) + 141.7\sin(3t).$$

Figure 1.20 shows the Andrews' curves for observations 96–105 of the Swiss bank note data set. We already know that the observations 96–100 represent genuine bank notes, and that the observations 101–105 represent counterfeit bank notes. We see that at least four curves differ from the others, but it is hard to tell which curve belongs to which group.

We know from Figure 1.4 that the sixth variable is an important one. Therefore, the Andrews' curves are calculated again using a reversed order of the variables.

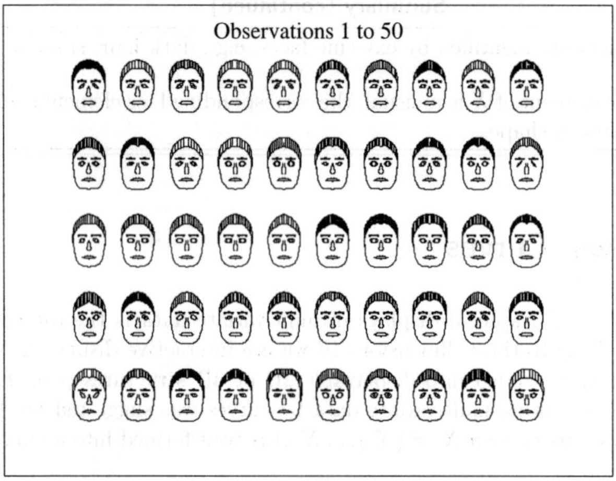

Figure 1.16. Chernoff-Flury faces for observations 1 to 50 of the bank notes. MVAfacebank50

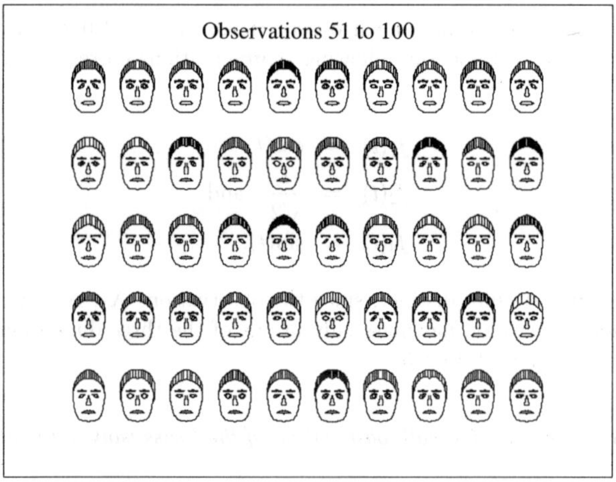

Figure 1.17. Chernoff-Flury faces for observations 51 to 100 of the bank notes. MVAfacebank50

EXAMPLE 1.4 *Let us consider again the 96th observation of the Swiss bank note data set,*

$$X_{96} = (215.6, 129.9, 129.9, 9.0, 9.5, 141.7).$$

The Andrews' curve is computed using the reversed order of variables:

$$f_{96}(t) = \frac{141.7}{\sqrt{2}} + 9.5\sin(t) + 9.0\cos(t) + 129.9\sin(2t) + 129.9\cos(2t) + 215.6\sin(3t).$$

In Figure 1.21 the curves f_{96}–f_{105} for observations 96–105 are plotted. Instead of a difference

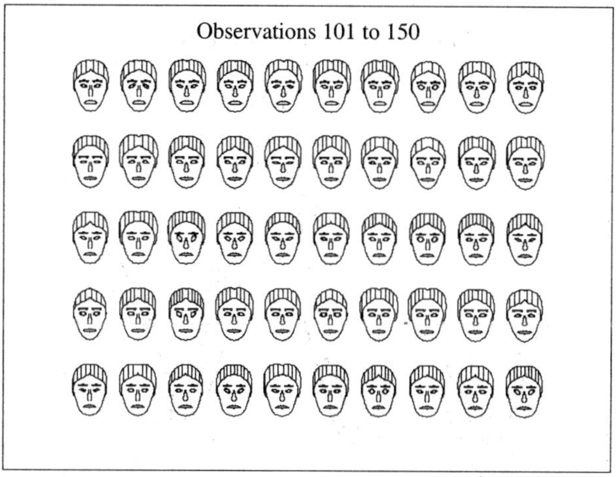

Figure 1.18. Chernoff-Flury faces for observations 101 to 150 of the bank notes. `MVAfacebank50`

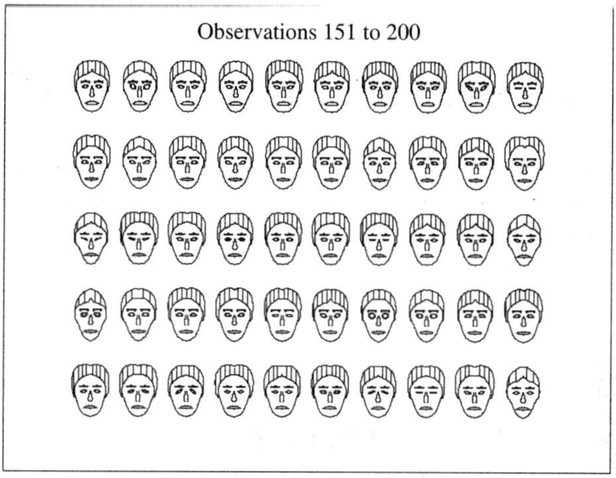

Figure 1.19. Chernoff-Flury faces for observations 151 to 200 of the bank notes. `MVAfacebank50`

in high frequency, now we have a difference in the intercept, which makes it more difficult for us to see the differences in observations.

This shows that the order of the variables plays an important role for the interpretation. If X is high-dimensional, then the last variables will have only a small visible contribution to the curve. They fall into the high frequency part of the curve. To overcome this problem Andrews suggested using an order which is suggested by Principal Component Analysis. This technique will be treated in detail in Chapter 9. In fact, the sixth variable will appear

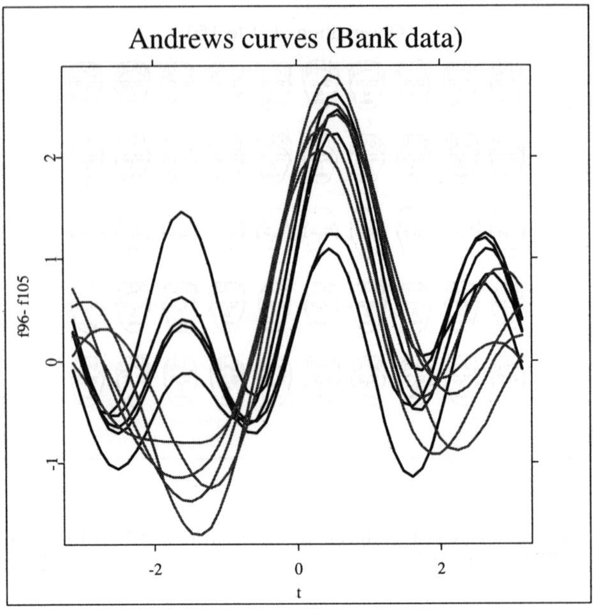

Figure 1.20. Andrews' curves of the observations 96–105 from the Swiss bank note data. The order of the variables is 1,2,3,4,5,6. MVAandcur

there as the most important variable for discriminating between the two groups. If the number of observations is more than 20, there may be too many curves in one graph. This will result in an over plotting of curves or a bad "signal-to-ink-ratio", see Tufte (1983). It is therefore advisable to present multivariate observations via Andrews' curves only for a limited number of observations.

Summary

- ↪ Outliers appear as single Andrews' curves that look different from the rest.
- ↪ A subgroup of data is characterized by a set of simular curves.
- ↪ The order of the variables plays an important role for interpretation.
- ↪ The order of variables may be optimized by Principal Component Analysis.
- ↪ For more than 20 observations we may obtain a bad "signal-to-ink-ratio", i.e., too many curves are overlaid in one picture.

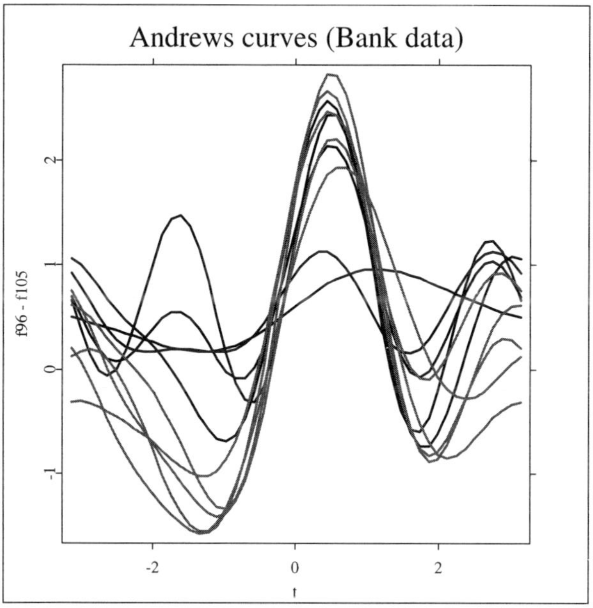

Figure 1.21. Andrews' curves of the observations 96–105 from the Swiss bank note data. The order of the variables is 6,5,4,3,2,1. MVAandcur2

1.7 Parallel Coordinate Plots

Parallel coordinates plots (PCP) constitute a technique that is based on a non-Cartesian coordinate system and therefore allows one to "see" more than four dimensions. The idea is simple: Instead of plotting observations in an orthogonal coordinate system, one draws their coordinates in a system of parallel axes. One first scales all variables to $max = 1$ and $min = 0$. Index j of the coordinate is mapped onto the horizontal axis, and the value $x_j$is mapped onto the vertical axis. This way of representation is very useful for high-dimensional data. It is however also sensitive to the order of the variables, since certain trends in the data can be shown more clearly in one ordering than in another.

EXAMPLE 1.5 *Take once again the observations 96–105 of the Swiss bank notes. These observations are six dimensional, so we can't show them in a six dimensional Cartesian coordinate system. Using the parallel coordinates plot technique, however, they can be plotted on parallel axes. This is shown in Figure 1.22.*

We have already noted in Example 1.3 that the diagonal X_6 plays an important role. This important role is clearly visible from Figure 1.22 The last coordinate X_6 shows two different subgroups. The full bank note data set is displayed in Figure 1.23. One sees an overlap of the coordinate values for indices 1–3 and an increased separability for the indices 4–6.

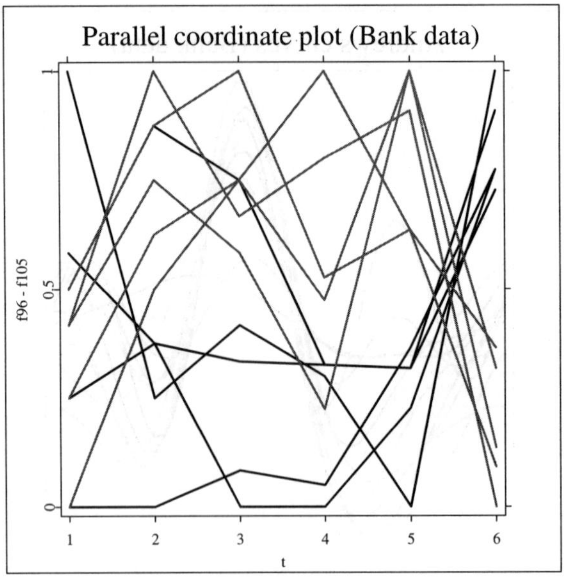

Figure 1.22. Parallel coordinates plot of observations 96–105. MVAparcoo1

Summary

- ↪ Parallel coordinates plots overcome the visualization problem of the Cartesian coordinate system for dimensions greater than 4.
- ↪ Outliers are visible as outlying polygon curves.
- ↪ The order of variables is still important, for example, for detection of subgroups.
- ↪ Subgroups may be screened by selective coloring in an interactive manner.

1.8 Boston Housing

Aim of the Analysis

The Boston Housing data set was analyzed by Harrison and Rubinfeld (1978) who wanted to find out whether "clean air" had an influence on house prices. We will use this data set in this chapter and in most of the following chapters to illustrate the presented methodology. The data are described in Appendix B.1.

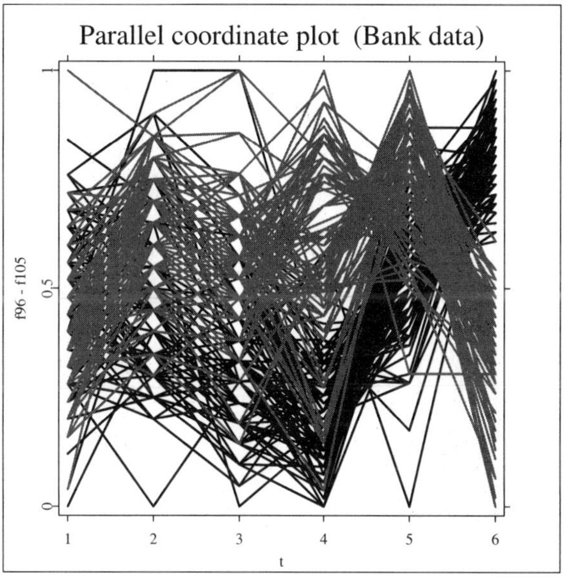

Figure 1.23. The entire bank data set. Genuine bank notes are displayed as black lines. The counterfeit bank notes are shown as red (grey) lines. MVAparcoo2

What Can Be Seen from the PCPs

In order to highlight the relations of X_{14} to the remaining 13 variables we color all of the observations with X_{14} >median(X_{14}) as red lines in Figure 1.24. Some of the variables seem to be strongly related. The most obvious relation is the negative dependence between X_{13} and X_{14}. It can also be argued that there exists a strong dependence between X_{12} and X_{14} since no red lines are drawn in the lower part of X_{12}. The opposite can be said about X_{11}: there are only red lines plotted in the lower part of this variable. Low values of X_{11} induce high values of X_{14}.

For the PCP, the variables have been rescaled over the interval $[0, 1]$ for better graphical representations. The PCP shows that the variables are not distributed in a symmetric manner. It can be clearly seen that the values of X_1 and X_9 are much more concentrated around 0. Therefore it makes sense to consider transformations of the original data.

The Scatterplot Matrix

One characteristic of the PCPs is that many lines are drawn on top of each other. This problem is reduced by depicting the variables in pairs of scatterplots. Including all 14 variables in one large scatterplot matrix is possible, but makes it hard to see anything from the plots. Therefore, for illustratory purposes we will analyze only one such matrix from a subset of the variables in Figure 1.25. On the basis of the PCP and the scatterplot matrix we would like to interpret each of the thirteen variables and their eventual relation to the 14th variable. Included in the figure are images for X_1–X_5 and X_{14}, although each variable

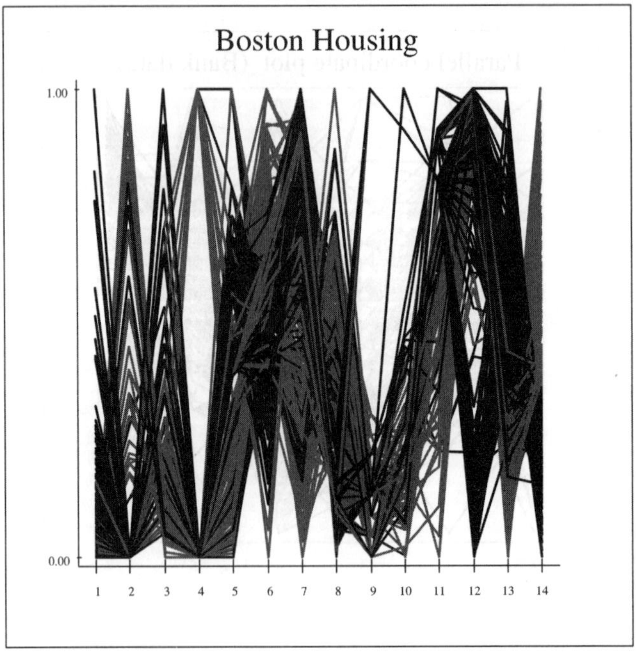

Figure 1.24. Parallel coordinates plot for Boston Housing data. MVApcphousing

is discussed in detail below. All references made to scatterplots in the following refer to Figure 1.25.

Per-capita crime rate X_1

Taking the logarithm makes the variable's distribution more symmetric. This can be seen in the boxplot of $\widetilde{X}_1$ in Figure 1.27 which shows that the median and the mean have moved closer to each other than they were for the original X_1. Plotting the kernel density estimate (KDE) of $\widetilde{X}_1 = \log(X_1)$ would reveal that two subgroups might exist with different mean values. However, taking a look at the scatterplots in Figure 1.26 of the logarithms which include X_1 does not clearly reveal such groups. Given that the scatterplot of $\log(X_1)$ vs. $\log(X_{14})$ shows a relatively strong negative relation, it might be the case that the two subgroups of X_1 correspond to houses with two different price levels. This is confirmed by the two boxplots shown to the right of the X_1 vs. X_2 scatterplot (in Figure 1.25): the right boxplot's shape differs a lot from the black one's, having a much higher median and mean.

Proportion of residential area zoned for large lots X_2

It strikes the eye in Figure 1.25 that there is a large cluster of observations for which X_2 is equal to 0. It also strikes the eye that—as the scatterplot of X_1 vs. X_2 shows—there is a strong, though non-linear, negative relation between X_1 and X_2: Almost all observations for which X_2 is high have an X_1-value close to zero, and vice versa, many observations for which X_2 is zero have quite a high per-capita crime rate X_1. This could be due to the location of

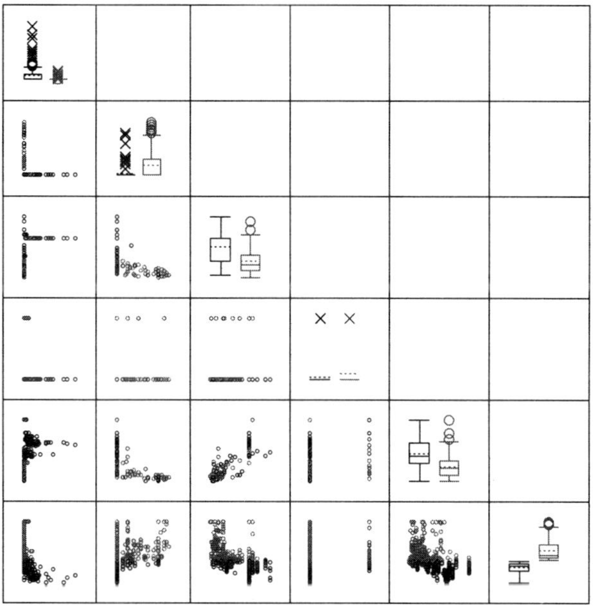

Figure 1.25. Scatterplot matrix for variables $X_1, \ldots, X_5$ and X_{14} of the Boston Housing data. MVAdrafthousing

the areas, e.g., downtown districts might have a higher crime rate and at the same time it is unlikely that any residential land would be zoned in a generous manner.

As far as the house prices are concerned it can be said that there seems to be no clear (linear) relation between X_2 and X_{14}, but it is obvious that the more expensive houses are situated in areas where X_2 is large (this can be seen from the two boxplots on the second position of the diagonal, where the red one has a clearly higher mean/median than the black one).

Proportion of non-retail business acres X_3

The PCP (in Figure 1.24) as well as the scatterplot of X_3 vs. X_{14} shows an obvious negative relation between X_3 and X_{14}. The relationship between the logarithms of both variables seems to be almost linear. This negative relation might be explained by the fact that non-retail business sometimes causes annoying sounds and other pollution. Therefore, it seems reasonable to use X_3 as an explanatory variable for the prediction of X_{14} in a linear-regression analysis.

As far as the distribution of X_3 is concerned it can be said that the kernel density estimate of X_3 clearly has two peaks, which indicates that there are two subgroups. According to the negative relation between X_3 and X_{14} it could be the case that one subgroup corresponds to the more expensive houses and the other one to the cheaper houses.

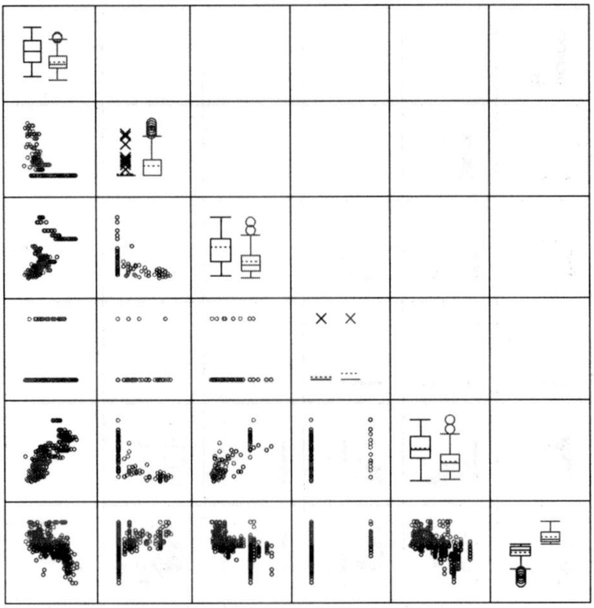

Figure 1.26. Scatterplot matrix for variables $\widetilde{X}_1, \ldots, \widetilde{X}_5$ and $\widetilde{X}_{14}$ of the Boston Housing data. MVAdrafthousingt

Charles River dummy variable X_4

The observation made from the PCP that there are more expensive houses than cheap houses situated on the banks of the Charles River is confirmed by inspecting the scatterplot matrix. Still, we might have some doubt that the proximity to the river influences the house prices. Looking at the original data set, it becomes clear that the observations for which X_4 equals one are districts that are close to each other. Apparently, the Charles River does not flow through too many different districts. Thus, it may be pure coincidence that the more expensive districts are close to the Charles River—their high values might be caused by many other factors such as the pupil/teacher ratio or the proportion of non-retail business acres.

Nitric oxides concentration X_5

The scatterplot of X_5 vs. X_{14} and the separate boxplots of X_5 for more and less expensive houses reveal a clear negative relation between the two variables. As it was the main aim of the authors of the original study to determine whether pollution had an influence on housing prices, it should be considered very carefully whether X_5 can serve as an explanatory variable for the price X_{14}. A possible reason against it being an explanatory variable is that people might not like to live in areas where the emissions of nitric oxides are high. Nitric oxides are emitted mainly by automobiles, by factories and from heating private homes. However, as one can imagine there are many good reasons besides nitric oxides not to live downtown or in industrial areas! Noise pollution, for example, might be a much better explanatory variable for the price of housing units. As the emission of nitric oxides is usually accompanied by noise pollution, using X_5 as an explanatory variable for X_{14} might lead to the false conclusion that people run away from nitric oxides, whereas in reality it is noise pollution that they are trying to escape.

Average number of rooms per dwelling X_6

The number of rooms per dwelling is a possible measure for the size of the houses. Thus we expect X_6 to be strongly correlated with X_{14} (the houses' median price). Indeed—apart from some outliers—the scatterplot of X_6 vs. X_{14} shows a point cloud which is clearly upward-sloping and which seems to be a realisation of a linear dependence of X_{14} on X_6. The two boxplots of X_6 confirm this notion by showing that the quartiles, the mean and the median are all much higher for the red than for the black boxplot.

Proportion of owner-occupied units built prior to 1940 X_7

There is no clear connection visible between X_7 and X_{14}. There could be a weak negative correlation between the two variables, since the (red) boxplot of X_7 for the districts whose price is above the median price indicates a lower mean and median than the (black) boxplot for the district whose price is below the median price. The fact that the correlation is not so clear could be explained by two opposing effects. On the one hand house prices should decrease if the older houses are not in a good shape. On the other hand prices could increase, because people often like older houses better than newer houses, preferring their atmosphere of space and tradition. Nevertheless, it seems reasonable that the houses' age has an influence on their price X_{14}.

Raising X_7 to the power of 2.5 reveals again that the data set might consist of two subgroups. But in this case it is not obvious that the subgroups correspond to more expensive or cheaper houses. One can furthermore observe a negative relation between X_7 and X_8. This could reflect the way the Boston metropolitan area developed over time: the districts with the newer buildings are farther away from employment centres with industrial facilities.

Weighted distance to five Boston employment centres X_8

Since most people like to live close to their place of work, we expect a negative relation between the distances to the employment centres and the houses' price. The scatterplot hardly reveals any dependence, but the boxplots of X_8 indicate that there might be a slightly positive relation as the red boxplot's median and mean are higher than the black one's. Again, there might be two effects in opposite directions at work. The first is that living too close to an employment centre might not provide enough shelter from the pollution created there. The second, as mentioned above, is that people do not travel very far to their workplace.

Index of accessibility to radial highways X_9

The first obvious thing one can observe in the scatterplots, as well in the histograms and the kernel density estimates, is that there are two subgroups of districts containing X_9 values which are close to the respective group's mean. The scatterplots deliver no hint as to what might explain the occurrence of these two subgroups. The boxplots indicate that for the cheaper and for the more expensive houses the average of X_9 is almost the same.

Full-value property tax X_{10}

X_{10} shows a behavior similar to that of X_9: two subgroups exist. A downward-sloping curve seems to underlie the relation of X_{10} and X_{14}. This is confirmed by the two boxplots drawn for X_{10}: the red one has a lower mean and median than the black one.

Pupil/teacher ratio X_{11}

The red and black boxplots of X_{11} indicate a negative relation between X_{11} and X_{14}. This is confirmed by inspection of the scatterplot of X_{11} vs. X_{14}: The point cloud is downward sloping, i.e., the less teachers there are per pupil, the less people pay on median for their dwellings.

Proportion of African American B, $X_{12} = 1000(B - 0.63)^2 \mathbf{I}(B < 0.63)$

Interestingly, X_{12} is negatively—though not linearly—correlated with X_3, X_7 and X_{11}, whereas it is positively related with X_{14}. Having a look at the data set reveals that for almost all districts X_{12} takes on a value around 390. Since B cannot be larger than 0.63, such values can only be caused by B close to zero. Therefore, the higher X_{12} is, the lower the actual proportion of African American is! Among observations 405 through 470 there are quite a few that have a X_{12} that is much lower than 390. This means that in these districts the proportion of African American is above zero. We can observe two clusters of points in the scatterplots of $\log(X_{12})$: one cluster for which X_{12} is close to 390 and a second one for which X_{12} is between 3 and 100. When X_{12} is positively related with another variable, the actual proportion of African American is negatively correlated with this variable and vice versa. This means that African American live in areas where there is a high proportion of non-retail business acres, where there are older houses and where there is a high (i.e., bad) pupil/teacher ratio. It can be observed that districts with housing prices above the median can only be found where the proportion of African American is virtually zero!

Proportion of lower status of the population X_{13}

Of all the variables X_{13} exhibits the clearest negative relation with X_{14}—hardly any outliers show up. Taking the square root of X_{13} and the logarithm of X_{14} transforms the relation into a linear one.

Transformations

Since most of the variables exhibit an asymmetry with a higher density on the left side, the following transformations are proposed:

$$\begin{aligned}
\widetilde{X_1} &= \log(X_1)\\
\widetilde{X_2} &= X_2/10\\
\widetilde{X_3} &= \log(X_3)\\
\widetilde{X_4} & \quad \text{none, since } X_4 \text{ is binary}\\
\widetilde{X_5} &= \log(X_5)\\
\widetilde{X_6} &= \log(X_6)\\
\widetilde{X_7} &= {X_7}^{2.5}/10000\\
\widetilde{X_8} &= \log(X_8)\\
\widetilde{X_9} &= \log(X_9)\\
\widetilde{X_{10}} &= \log(X_{10})\\
\widetilde{X_{11}} &= \exp(0.4 \times X_{11})/1000\\
\widetilde{X_{12}} &= X_{12}/100\\
\widetilde{X_{13}} &= \sqrt{X_{13}}\\
\widetilde{X_{14}} &= \log(X_{14})
\end{aligned}$$

Taking the logarithm or raising the variables to the power of something smaller than one helps to reduce the asymmetry. This is due to the fact that lower values move further away from each other, whereas the distance between greater values is reduced by these transformations.

Figure 1.27 displays boxplots for the original mean variance scaled variables as well as for the proposed transformed variables. The transformed variables' boxplots are more symmetric and have less outliers than the original variables' boxplots.

1.9 Exercises

EXERCISE 1.1 *Is the upper extreme always an outlier?*

EXERCISE 1.2 *Is it possible for the mean or the median to lie outside of the fourths or even outside of the outside bars?*

EXERCISE 1.3 *Assume that the data are normally distributed $N(0,1)$. What percentage of the data do you expect to lie outside the outside bars?*

EXERCISE 1.4 *What percentage of the data do you expect to lie outside the outside bars if we assume that the data are normally distributed $N(0,\sigma^2)$ with unknown variance σ^2?*

EXERCISE 1.5 *How would the five-number summary of the 15 largest U.S. cities differ from that of the 50 largest U.S. cities? How would the five-number summary of 15 observations of $N(0,1)$-distributed data differ from that of 50 observations from the same distribution?*

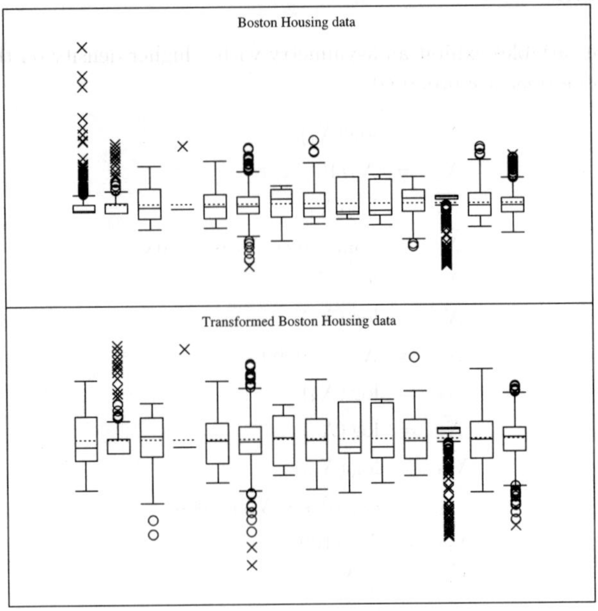

Figure 1.27. Boxplots for all of the variables from the Boston Housing data before and after the proposed transformations. MVAboxbhd

EXERCISE 1.6 *Is it possible that all five numbers of the five-number summary could be equal? If so, under what conditions?*

EXERCISE 1.7 *Suppose we have 50 observations of $X \sim N(0,1)$ and another 50 observations of $Y \sim N(2,1)$. What would the 100 Flury faces look like if you had defined as face elements the face line and the darkness of hair? Do you expect any similar faces? How many faces do you think should look like observations of Y even though they are X observations?*

EXERCISE 1.8 *Draw a histogram for the mileage variable of the car data (Table B.3). Do the same for the three groups (U.S., Japan, Europe). Do you obtain a similar conclusion as in the parallel boxplot on Figure 1.3 for these data?*

EXERCISE 1.9 *Use some bandwidth selection criterion to calculate the optimally chosen bandwidth h for the diagonal variable of the bank notes. Would it be better to have one bandwidth for the two groups?*

EXERCISE 1.10 *In Figure 1.9 the densities overlap in the region of diagonal ≈ 140.4. We partially observed this in the boxplot of Figure 1.4. Our aim is to separate the two groups. Will we be able to do this effectively on the basis of this diagonal variable alone?*

EXERCISE 1.11 *Draw a parallel coordinates plot for the car data.*

EXERCISE 1.12 *How would you identify discrete variables (variables with only a limited number of possible outcomes) on a parallel coordinates plot?*

EXERCISE 1.13 *True or false: the height of the bars of a histogram are equal to the relative frequency with which observations fall into the respective bins.*

EXERCISE 1.14 *True or false: kernel density estimates must always take on a value between 0 and 1. (Hint: Which quantity connected with the density function has to be equal to 1? Does this property imply that the density function has to always be less than 1?)*

EXERCISE 1.15 *Let the following data set represent the heights of 13 students taking the Applied Multivariate Statistical Analysis course:*

$$1.72, 1.83, 1.74, 1.79, 1.94, 1.81, 1.66, 1.60, 1.78, 1.77, 1.85, 1.70, 1.76.$$

1. *Find the corresponding five-number summary.*
2. *Construct the boxplot.*
3. *Draw a histogram for this data set.*

EXERCISE 1.16 *Describe the unemployment data (see Table B.19) that contain unemployment rates of all German Federal States using various descriptive techniques.*

EXERCISE 1.17 *Using yearly population data (see B.20), generate*

1. *a boxplot (choose one of variables)*
2. *an Andrew's Curve (choose ten data points)*
3. *a scatterplot*
4. *a histogram (choose one of the variables)*

What do these graphs tell you about the data and their structure?

EXERCISE 1.18 *Make a draftman plot for the car data with the variables*

$$\begin{aligned} X_1 &= \textit{price}, \\ X_2 &= \textit{mileage}, \\ X_8 &= \textit{weight}, \\ X_9 &= \textit{length}. \end{aligned}$$

Move the brush into the region of heavy cars. What can you say about price, mileage and length? Move the brush onto high fuel economy. Mark the Japanese, European and U.S. American cars. You should find the same condition as in boxplot Figure 1.3.

EXERCISE 1.19 *What is the form of a scatterplot of two independent random variables X_1 and X_2 with standard Normal distribution?*

EXERCISE 1.20 *Rotate a three-dimensional standard normal point cloud in 3D space. Does it "almost look the same from all sides"? Can you explain why or why not?*

Part II

Multivariate Random Variables

2 A Short Excursion into Matrix Algebra

This chapter is a reminder of basic concepts of matrix algebra, which are particularly useful in multivariate analysis. It also introduces the notations used in this book for vectors and matrices. Eigenvalues and eigenvectors play an important role in multivariate techniques. In Sections 2.2 and 2.3, we present the spectral decomposition of matrices and consider the maximization (minimization) of quadratic forms given some constraints.

In analyzing the multivariate normal distribution, partitioned matrices appear naturally. Some of the basic algebraic properties are given in Section 2.5. These properties will be heavily used in Chapters 4 and 5.

The geometry of the multinormal and the geometric interpretation of the multivariate techniques (Part III) intensively uses the notion of angles between two vectors, the projection of a point on a vector and the distances between two points. These ideas are introduced in Section 2.6.

2.1 Elementary Operations

A matrix $\mathcal{A}$ is a system of numbers with n rows and p columns:

$$\mathcal{A} = \begin{pmatrix} a_{11} & a_{12} & \dots & \dots & \dots & a_{1p} \\ \vdots & a_{22} & & & & \vdots \\ \vdots & \vdots & \ddots & & & \vdots \\ \vdots & \vdots & & \ddots & & \vdots \\ \vdots & \vdots & & & \ddots & \vdots \\ a_{n1} & a_{n2} & \dots & \dots & \dots & a_{np} \end{pmatrix}.$$

We also write (a_{ij}) for $\mathcal{A}$ and $\mathcal{A}(n \times p)$ to indicate the numbers of rows and columns. Vectors are matrices with one column and are denoted as x or $x(p \times 1)$. Special matrices and vectors are defined in Table 2.1. Note that we use small letters for scalars as well as for vectors.

Matrix Operations

Elementary operations are summarized below:

$$\begin{aligned} \mathcal{A}^\top &= (a_{ji}) \\ \mathcal{A} + \mathcal{B} &= (a_{ij} + b_{ij}) \\ \mathcal{A} - \mathcal{B} &= (a_{ij} - b_{ij}) \\ c \cdot \mathcal{A} &= (c \cdot a_{ij}) \\ \mathcal{A} \cdot \mathcal{B} &= \mathcal{A}(n \times p)\ \mathcal{B}(p \times m) = \mathcal{C}(n \times m) = \left(\sum_{j=1}^{p} a_{ij} b_{jk} \right). \end{aligned}$$

Properties of Matrix Operations

$$\begin{aligned} \mathcal{A}+\mathcal{B} &= \mathcal{B}+\mathcal{A} \\ \mathcal{A}(\mathcal{B}+\mathcal{C}) &= \mathcal{A}\mathcal{B}+\mathcal{A}\mathcal{C} \\ \mathcal{A}(\mathcal{B}\mathcal{C}) &= (\mathcal{A}\mathcal{B})\mathcal{C} \\ (\mathcal{A}^\top)^\top &= \mathcal{A} \\ (\mathcal{A}\mathcal{B})^\top &= \mathcal{B}^\top\mathcal{A}^\top \end{aligned}$$

Name	Definition	Notation	Example
scalar	$p = n = 1$	a	3
column vector	$p = 1$	a	$\begin{pmatrix}1\\3\end{pmatrix}$
row vector	$n = 1$	$a^\top$	$\begin{pmatrix}1 & 3\end{pmatrix}$
vector of ones	$(\underbrace{1,\ldots,1}_{n})^\top$	1_n	$\begin{pmatrix}1\\1\end{pmatrix}$
vector of zeros	$(\underbrace{0,\ldots,0}_{n})^\top$	0_n	$\begin{pmatrix}0\\0\end{pmatrix}$
square matrix	$n = p$	$\mathcal{A}(p \times p)$	$\begin{pmatrix}2 & 0\\0 & 2\end{pmatrix}$
diagonal matrix	$a_{ij} = 0,\ i \neq j,\ n = p$	$\operatorname{diag}(a_{ii})$	$\begin{pmatrix}1 & 0\\0 & 2\end{pmatrix}$
identity matrix	$\operatorname{diag}(\underbrace{1,\ldots,1}_{p})$	$\mathcal{I}_p$	$\begin{pmatrix}1 & 0\\0 & 1\end{pmatrix}$
unit matrix	$a_{ij} \equiv 1,\ n = p$	$1_n 1_n^\top$	$\begin{pmatrix}1 & 1\\1 & 1\end{pmatrix}$
symmetric matrix	$a_{ij} = a_{ji}$		$\begin{pmatrix}1 & 2\\2 & 3\end{pmatrix}$
null matrix	$a_{ij} = 0$	0	$\begin{pmatrix}0 & 0\\0 & 0\end{pmatrix}$
upper triangular matrix	$a_{ij} = 0, i < j$		$\begin{pmatrix}1 & 2 & 4\\0 & 1 & 3\\0 & 0 & 1\end{pmatrix}$
idempotent matrix	$\mathcal{A}\mathcal{A} = \mathcal{A}$		$\begin{pmatrix}1 & 0 & 0\\0 & \frac{1}{2} & \frac{1}{2}\\0 & \frac{1}{2} & \frac{1}{2}\end{pmatrix}$
orthogonal matrix	$\mathcal{A}^\top\mathcal{A} = \mathcal{I} = \mathcal{A}\mathcal{A}^\top$		$\begin{pmatrix}\frac{1}{\sqrt{2}} & \frac{1}{\sqrt{2}}\\\frac{1}{\sqrt{2}} & -\frac{1}{\sqrt{2}}\end{pmatrix}$

Table 2.1. Special matrices and vectors.

Matrix Characteristics

Rank

The *rank*, $\operatorname{rank}(\mathcal{A})$, of a matrix $\mathcal{A}(n \times p)$ is defined as the maximum number of linearly independent rows (columns). A set of k rows a_j of $\mathcal{A}(n\times p)$ are said to be linearly independent if $\sum_{j=1}^{k} c_j a_j = 0_p$ implies $c_j = 0, \forall j$, where $c_1, \ldots, c_k$ are scalars. In other words no rows in this set can be expressed as a linear combination of the $(k-1)$ remaining rows.

Trace

The *trace* of a matrix is the sum of its diagonal elements

$$\operatorname{tr}(\mathcal{A}) = \sum_{i=1}^{p} a_{ii}.$$

Determinant

The *determinant* is an important concept of matrix algebra. For a square matrix $\mathcal{A}$, it is defined as:

$$\det(\mathcal{A}) = |\mathcal{A}| = \sum (-1)^{|\tau|}\, a_{1\tau(1)} \dots a_{p\tau(p)},$$

the summation is over all permutations τ of $\{1, 2, \dots, p\}$, and $|\tau| = 0$ if the permutation can be written as a product of an even number of transpositions and $|\tau| = 1$ otherwise.

EXAMPLE 2.1 *In the case of* $p = 2$, $\mathcal{A} = \begin{pmatrix} a_{11} & a_{12} \\ a_{21} & a_{22} \end{pmatrix}$ *and we can permute the digits "1" and "2" once or not at all. So,*

$$|\mathcal{A}| = a_{11}\, a_{22} - a_{12}\, a_{21}.$$

Transpose

For $\mathcal{A}(n \times p)$ and $\mathcal{B}(p \times n)$

$$(\mathcal{A}^\top)^\top = \mathcal{A}, \text{ and } (\mathcal{A}\mathcal{B})^\top = \mathcal{B}^\top \mathcal{A}^\top.$$

Inverse

If $|\mathcal{A}| \neq 0$ and $\mathcal{A}(p \times p)$, then the inverse $\mathcal{A}^{-1}$ exists:

$$\mathcal{A}\, \mathcal{A}^{-1} = \mathcal{A}^{-1}\, \mathcal{A} = \mathcal{I}_p.$$

For small matrices, the inverse of $\mathcal{A} = (a_{ij})$ can be calculated as

$$\mathcal{A}^{-1} = \frac{\mathcal{C}}{|\mathcal{A}|},$$

where $\mathcal{C} = (c_{ij})$ is the adjoint matrix of $\mathcal{A}$. The elements c_{ji} of $\mathcal{C}^\top$ are the co-factors of $\mathcal{A}$:

$$c_{ji} = (-1)^{i+j} \begin{vmatrix} a_{11} & \dots & a_{1(j-1)} & a_{1(j+1)} & \dots & a_{1p} \\ \vdots & & & & & \\ a_{(i-1)1} & \dots & a_{(i-1)(j-1)} & a_{(i-1)(j+1)} & \dots & a_{(i-1)p} \\ a_{(i+1)1} & \dots & a_{(i+1)(j-1)} & a_{(i+1)(j+1)} & \dots & a_{(i+1)p} \\ \vdots & & & & & \\ a_{p1} & \dots & a_{p(j-1)} & a_{p(j+1)} & \dots & a_{pp} \end{vmatrix}.$$

G-inverse

A more general concept is the *G-inverse* (Generalized Inverse) $\mathcal{A}^-$ which satisfies the following:

$$\mathcal{A}\,\mathcal{A}^-\mathcal{A} = \mathcal{A}.$$

Later we will see that there may be more than one G-inverse.

EXAMPLE 2.2 *The generalized inverse can also be calculated for singular matrices. We have:*

$$\begin{pmatrix} 1 & 0 \\ 0 & 0 \end{pmatrix}\begin{pmatrix} 1 & 0 \\ 0 & 0 \end{pmatrix}\begin{pmatrix} 1 & 0 \\ 0 & 0 \end{pmatrix} = \begin{pmatrix} 1 & 0 \\ 0 & 0 \end{pmatrix},$$

which means that the generalized inverse of $\mathcal{A} = \begin{pmatrix} 1 & 0 \\ 0 & 0 \end{pmatrix}$ *is* $\mathcal{A}^- = \begin{pmatrix} 1 & 0 \\ 0 & 0 \end{pmatrix}$ *even though the inverse matrix of* $\mathcal{A}$ *does not exist in this case.*

Eigenvalues, Eigenvectors

Consider a $(p \times p)$ matrix $\mathcal{A}$. If there exists a scalar λ and a vector γ such that

$$\mathcal{A}\gamma = \lambda\gamma, \tag{2.1}$$

then we call

λ	an eigenvalue
γ	an eigenvector.

It can be proven that an eigenvalue λ is a root of the p-th order polynomial $|\mathcal{A} - \lambda\mathcal{I}_p| = 0$. Therefore, there are up to p eigenvalues $\lambda_1, \lambda_2, \ldots, \lambda_p$ of $\mathcal{A}$. For each eigenvalue λ_j, there exists a corresponding eigenvector γ_j given by equation (2.1) . Suppose the matrix $\mathcal{A}$ has the eigenvalues $\lambda_1, \ldots, \lambda_p$. Let $\Lambda = \operatorname{diag}(\lambda_1, \ldots, \lambda_p)$.

The determinant $|\mathcal{A}|$ and the trace $\operatorname{tr}(\mathcal{A})$ can be rewritten in terms of the eigenvalues:

$$|\mathcal{A}| = |\Lambda| = \prod_{j=1}^{p} \lambda_j \tag{2.2}$$

$$\operatorname{tr}(\mathcal{A}) = \operatorname{tr}(\Lambda) = \sum_{j=1}^{p} \lambda_j. \tag{2.3}$$

An idempotent matrix $\mathcal{A}$ (see the definition in Table 2.1) can only have eigenvalues in $\{0, 1\}$ therefore $\operatorname{tr}(\mathcal{A}) = \operatorname{rank}(\mathcal{A}) =$ number of eigenvalues $\neq 0$.

EXAMPLE 2.3 *Let us consider the matrix* $\mathcal{A} = \begin{pmatrix} 1 & 0 & 0 \\ 0 & \frac{1}{2} & \frac{1}{2} \\ 0 & \frac{1}{2} & \frac{1}{2} \end{pmatrix}$. *It is easy to verify that* $\mathcal{A}\mathcal{A} = \mathcal{A}$ *which implies that the matrix* $\mathcal{A}$ *is idempotent.*

We know that the eigenvalues of an idempotent matrix are equal to 0 or 1. In this case, the eigenvalues of $\mathcal{A}$ *are* $\lambda_1 = 1$, $\lambda_2 = 1$, *and* $\lambda_3 = 0$ *since* $\begin{pmatrix} 1 & 0 & 0 \\ 0 & \frac{1}{2} & \frac{1}{2} \\ 0 & \frac{1}{2} & \frac{1}{2} \end{pmatrix}\begin{pmatrix} 1 \\ 0 \\ 0 \end{pmatrix} = 1\begin{pmatrix} 1 \\ 0 \\ 0 \end{pmatrix}$,

$$\begin{pmatrix} 1 & 0 & 0 \\ 0 & \frac{1}{2} & \frac{1}{2} \\ 0 & \frac{1}{2} & \frac{1}{2} \end{pmatrix}\begin{pmatrix} 0 \\ \frac{\sqrt{2}}{2} \\ \frac{\sqrt{2}}{2} \end{pmatrix} = 1\begin{pmatrix} 0 \\ \frac{\sqrt{2}}{2} \\ \frac{\sqrt{2}}{2} \end{pmatrix}, \text{ and } \begin{pmatrix} 1 & 0 & 0 \\ 0 & \frac{1}{2} & \frac{1}{2} \\ 0 & \frac{1}{2} & \frac{1}{2} \end{pmatrix}\begin{pmatrix} 0 \\ \frac{\sqrt{2}}{2} \\ -\frac{\sqrt{2}}{2} \end{pmatrix} = 0\begin{pmatrix} 0 \\ \frac{\sqrt{2}}{2} \\ -\frac{\sqrt{2}}{2} \end{pmatrix}.$$

Using formulas (2.2) and (2.3), we can calculate the trace and the determinant of $\mathcal{A}$ *from the eigenvalues:* $\operatorname{tr}(\mathcal{A}) = \lambda_1 + \lambda_2 + \lambda_3 = 2$, $|\mathcal{A}| = \lambda_1\lambda_2\lambda_3 = 0$, *and* $\operatorname{rank}(\mathcal{A}) = 2$.

Properties of Matrix Characteristics

$\mathcal{A}(n \times n),\ \mathcal{B}(n \times n),\ c \in \mathbb{R}$

$$\begin{aligned} \operatorname{tr}(\mathcal{A}+\mathcal{B}) &= \operatorname{tr}\mathcal{A} + \operatorname{tr}\mathcal{B} && (2.4)\\ \operatorname{tr}(c\mathcal{A}) &= c\operatorname{tr}\mathcal{A} && (2.5)\\ |c\mathcal{A}| &= c^n|\mathcal{A}| && (2.6)\\ |\mathcal{A}\mathcal{B}| &= |\mathcal{B}\mathcal{A}| = |\mathcal{A}||\mathcal{B}| && (2.7) \end{aligned}$$

$\mathcal{A}(n \times p),\ \mathcal{B}(p \times n)$

$$\begin{aligned} \operatorname{tr}(\mathcal{A}\cdot\mathcal{B}) &= \operatorname{tr}(\mathcal{B}\cdot\mathcal{A}) && (2.8)\\ \operatorname{rank}(\mathcal{A}) &\le \min(n,p) && \\ \operatorname{rank}(\mathcal{A}) &\ge 0 && (2.9)\\ \operatorname{rank}(\mathcal{A}) &= \operatorname{rank}(\mathcal{A}^\top) && (2.10)\\ \operatorname{rank}(\mathcal{A}^\top\mathcal{A}) &= \operatorname{rank}(\mathcal{A}) && (2.11)\\ \operatorname{rank}(\mathcal{A} \mid \mathcal{B}) &\le \operatorname{rank}(\mathcal{A}) + \operatorname{rank}(\mathcal{B}) && (2.12)\\ \operatorname{rank}(\mathcal{A}\mathcal{B}) &\le \min\{\operatorname{rank}(\mathcal{A}), \operatorname{rank}(\mathcal{B})\} && (2.13) \end{aligned}$$

$\mathcal{A}(n \times p),\ \mathcal{B}(p \times q),\ \mathcal{C}(q \times n)$

$$\begin{aligned} \operatorname{tr}(\mathcal{A}\mathcal{B}\mathcal{C}) &= \operatorname{tr}(\mathcal{B}\mathcal{C}\mathcal{A}) && \\ &= \operatorname{tr}(\mathcal{C}\mathcal{A}\mathcal{B}) && (2.14)\\ \operatorname{rank}(\mathcal{A}\mathcal{B}\mathcal{C}) &= \operatorname{rank}(\mathcal{B}) \quad \text{for nonsingular } \mathcal{A}, \mathcal{C} && (2.15) \end{aligned}$$

$\mathcal{A}(p \times p)$

$$\begin{aligned} |\mathcal{A}^{-1}| &= |\mathcal{A}|^{-1} && (2.16)\\ \operatorname{rank}(\mathcal{A}) &= p \quad \text{if and only if } \mathcal{A} \text{ is nonsingular.} && (2.17) \end{aligned}$$

Summary

- ↪ The determinant $|\mathcal{A}|$ is the product of the eigenvalues of $\mathcal{A}$.
- ↪ The inverse of a matrix $\mathcal{A}$ exists if $|\mathcal{A}| \neq 0$.
- ↪ The trace $\operatorname{tr}(\mathcal{A})$ is the sum of the eigenvalues of $\mathcal{A}$.
- ↪ The sum of the traces of two matrices equals the trace of the sum of the two matrices.
- ↪ The trace $\operatorname{tr}(\mathcal{A}\mathcal{B})$ equals $\operatorname{tr}(\mathcal{B}\mathcal{A})$.
- ↪ The $\operatorname{rank}(\mathcal{A})$ is the maximal number of linearly independent rows (columns) of $\mathcal{A}$.

2.2 Spectral Decompositions

The computation of eigenvalues and eigenvectors is an important issue in the analysis of matrices. The spectral decomposition or Jordan decomposition links the structure of a matrix to the eigenvalues and the eigenvectors.

THEOREM 2.1 (Jordan Decomposition) *Each symmetric matrix $\mathcal{A}(p \times p)$ can be written as*

$$\mathcal{A} = \Gamma\ \Lambda\ \Gamma^\top = \sum_{j=1}^{p} \lambda_j \gamma_j \gamma_j^\top \tag{2.18}$$

where

$$\Lambda = \operatorname{diag}(\lambda_1, \ldots, \lambda_p)$$

and where

$$\Gamma = (\gamma_1, \gamma_2, \ldots, \gamma_p)$$

is an orthogonal matrix consisting of the eigenvectors γ_j of $\mathcal{A}$.

EXAMPLE 2.4 *Suppose that $\mathcal{A} = \left(\begin{smallmatrix} 1 & 2 \\ 2 & 3 \end{smallmatrix}\right)$. The eigenvalues are found by solving $|\mathcal{A} - \lambda\mathcal{I}| = 0$. This is equivalent to*

$$\begin{vmatrix} 1-\lambda & 2 \\ 2 & 3-\lambda \end{vmatrix} = (1-\lambda)(3-\lambda) - 4 = 0.$$

Hence, the eigenvalues are $\lambda_1 = 2 + \sqrt{5}$ and $\lambda_2 = 2 - \sqrt{5}$. The eigenvectors are $\gamma_1 = (0.5257, 0.8506)^\top$ and $\gamma_2 = (0.8506, -0.5257)^\top$. They are orthogonal since $\gamma_1^\top \gamma_2 = 0$.

Using spectral decomposition, we can define powers of a matrix $\mathcal{A}(p \times p)$. Suppose $\mathcal{A}$ is a symmetric matrix. Then by Theorem 2.1

$$\mathcal{A} = \Gamma\Lambda\Gamma^\top,$$

and we define for some $\alpha \in \mathbb{R}$

$$\mathcal{A}^\alpha = \Gamma\Lambda^\alpha\Gamma^\top, \tag{2.19}$$

where $\Lambda^\alpha = \operatorname{diag}(\lambda_1^\alpha, \ldots, \lambda_p^\alpha)$. In particular, we can easily calculate the inverse of the matrix $\mathcal{A}$. Suppose that the eigenvalues of $\mathcal{A}$ are positive. Then with $\alpha = -1$, we obtain the inverse of $\mathcal{A}$ from

$$\mathcal{A}^{-1} = \Gamma\Lambda^{-1}\Gamma^\top. \tag{2.20}$$

Another interesting decomposition which is later used is given in the following theorem.

THEOREM 2.2 (Singular Value Decomposition) *Each matrix $\mathcal{A}(n \times p)$ with rank r can be decomposed as*

$$\mathcal{A} = \Gamma\ \Lambda\ \Delta^\top,$$

where $\Gamma(n \times r)$ and $\Delta(p \times r)$. Both Γ and Δ are column orthonormal, i.e., $\Gamma^\top\Gamma = \Delta^\top\Delta = \mathcal{I}_r$ and $\Lambda = \operatorname{diag}\left(\lambda_1^{1/2}, \ldots, \lambda_r^{1/2}\right)$, $\lambda_j > 0$. The values $\lambda_1, \ldots, \lambda_r$ are the non-zero eigenvalues of the matrices $\mathcal{A}\mathcal{A}^\top$ and $\mathcal{A}^\top\mathcal{A}$. Γ and Δ consist of the corresponding r eigenvectors of these matrices.

This is obviously a generalization of Theorem 2.1 (Jordan decomposition). With Theorem 2.2, we can find a G-inverse $\mathcal{A}^-$ of $\mathcal{A}$. Indeed, define $\mathcal{A}^- = \Delta\,\Lambda^{-1}\,\Gamma^\top$. Then $\mathcal{A}\,\mathcal{A}^-\,\mathcal{A} = \Gamma\,\Lambda\,\Delta^\top = \mathcal{A}$. Note that the G-inverse is not unique.

EXAMPLE 2.5 *In Example 2.2, we showed that the generalized inverse of* $\mathcal{A} = \begin{pmatrix} 1 & 0 \\ 0 & 0 \end{pmatrix}$ *is* $\mathcal{A}^- \begin{pmatrix} 1 & 0 \\ 0 & 0 \end{pmatrix}$. *The following also holds*

$$\begin{pmatrix} 1 & 0 \\ 0 & 0 \end{pmatrix}\begin{pmatrix} 1 & 0 \\ 0 & 8 \end{pmatrix}\begin{pmatrix} 1 & 0 \\ 0 & 0 \end{pmatrix} = \begin{pmatrix} 1 & 0 \\ 0 & 0 \end{pmatrix}$$

which means that the matrix $\begin{pmatrix} 1 & 0 \\ 0 & 8 \end{pmatrix}$ *is also a generalized inverse of* $\mathcal{A}$.

Summary

- ↪ The Jordan decomposition gives a representation of a symmetric matrix in terms of eigenvalues and eigenvectors.
- ↪ The eigenvectors belonging to the largest eigenvalues indicate the "main direction" of the data.
- ↪ The Jordan decomposition allows one to easily compute the power of a symmetric matrix $\mathcal{A}$: $\mathcal{A}^\alpha = \Gamma\Lambda^\alpha\Gamma^\top$.
- ↪ The singular value decomposition (SVD) is a generalization of the Jordan decomposition to non-quadratic matrices.

2.3 Quadratic Forms

A quadratic form $Q(x)$ is built from a symmetric matrix $\mathcal{A}(p \times p)$ and a vector $x \in \mathbb{R}^p$:

$$Q(x) = x^\top\,\mathcal{A}\,x = \sum_{i=1}^{p}\sum_{j=1}^{p} a_{ij}x_i x_j. \tag{2.21}$$

Definiteness of Quadratic Forms and Matrices

$Q(x) > 0$ for all $x \neq 0$	*positive definite*
$Q(x) \geq 0$ for all $x \neq 0$	*positive semidefinite*

A matrix $\mathcal{A}$ is called positive definite (semidefinite) if the corresponding quadratic form $Q(.)$ is positive definite (semidefinite). We write $\mathcal{A} > 0$ (≥ 0).

Quadratic forms can always be diagonalized, as the following result shows.

THEOREM 2.3 *If $\mathcal{A}$ is symmetric and $Q(x) = x^\top \mathcal{A} x$ is the corresponding quadratic form, then there exists a transformation $x \mapsto \Gamma^\top x = y$ such that*

$$x^\top \mathcal{A} x = \sum_{i=1}^{p} \lambda_i y_i^2,$$

where λ_i are the eigenvalues of $\mathcal{A}$.

Proof:
$\mathcal{A} = \Gamma \, \Lambda \, \Gamma^\top$. By Theorem 2.1 and $y = \Gamma^\top \alpha$ we have that $x^\top \mathcal{A} x = x^\top \Gamma \Lambda \Gamma^\top x = y^\top \Lambda y = \sum_{i=1}^p \lambda_i y_i^2$. $\square$

Positive definiteness of quadratic forms can be deduced from positive eigenvalues.

THEOREM 2.4 $\mathcal{A} > 0$ *if and only if all* $\lambda_i > 0$, $i = 1, \ldots, p$.

Proof:
$0 < \lambda_1 y_1^2 + \cdots + \lambda_p y_p^2 = x^\top \mathcal{A} x$ for all $x \neq 0$ by Theorem 2.3. $\square$

COROLLARY 2.1 *If* $\mathcal{A} > 0$, *then* $\mathcal{A}^{-1}$ *exists and* $|\mathcal{A}| > 0$.

EXAMPLE 2.6 *The quadratic form $Q(x) = x_1^2 + x_2^2$ corresponds to the matrix $\mathcal{A} = \begin{pmatrix} 1 & 0 \\ 0 & 1 \end{pmatrix}$ with eigenvalues $\lambda_1 = \lambda_2 = 1$ and is thus positive definite. The quadratic form $Q(x) = (x_1 - x_2)^2$ corresponds to the matrix $\mathcal{A} = \begin{pmatrix} 1 & -1 \\ -1 & 1 \end{pmatrix}$ with eigenvalues $\lambda_1 = 2, \lambda_2 = 0$ and is positive semidefinite. The quadratic form $Q(x) = x_1^2 - x_2^2$ with eigenvalues $\lambda_1 = 1, \lambda_2 = -1$ is indefinite.*

In the statistical analysis of multivariate data, we are interested in maximizing quadratic forms given some constraints.

THEOREM 2.5 *If $\mathcal{A}$ and $\mathcal{B}$ are symmetric and $\mathcal{B} > 0$, then the maximum of $\frac{x^\top \mathcal{A} x}{x^\top \mathcal{B} x}$ is given by the largest eigenvalue of $\mathcal{B}^{-1}\mathcal{A}$. More generally,*

$$\max_x \frac{x^\top \mathcal{A} x}{x^\top \mathcal{B} x} = \lambda_1 \geq \lambda_2 \geq \cdots \geq \lambda_p = \min_x \frac{x^\top \mathcal{A} x}{x^\top \mathcal{B} x},$$

where $\lambda_1, \ldots, \lambda_p$ denote the eigenvalues of $\mathcal{B}^{-1}\mathcal{A}$. The vector which maximizes (minimizes) $\frac{x^\top \mathcal{A} x}{x^\top \mathcal{B} x}$ is the eigenvector of $\mathcal{B}^{-1}\mathcal{A}$ which corresponds to the largest (smallest) eigenvalue of $\mathcal{B}^{-1}\mathcal{A}$. If $x^\top \mathcal{B} x = 1$, we get

$$\max_x x^\top \mathcal{A} x = \lambda_1 \geq \lambda_2 \geq \cdots \geq \lambda_p = \min_x x^\top \mathcal{A} x$$

Proof:
By definition, $\mathcal{B}^{1/2} = \Gamma_{\mathcal{B}} \, \Lambda_{\mathcal{B}}^{1/2} \, \Gamma_{\mathcal{B}}^\top$ and symmetric. Then $x^\top \mathcal{B} x = \left\| x^\top \mathcal{B}^{1/2} \right\|^2 = \left\| \mathcal{B}^{1/2} x \right\|^2$. Set $y = \frac{\mathcal{B}^{1/2} x}{\| \mathcal{B}^{1/2} x \|}$, then

$$\max_x \frac{x^\top \mathcal{A} x}{x^\top \mathcal{B} x} = \max_{\{y : y^\top y = 1\}} y^\top \mathcal{B}^{-1/2} \, \mathcal{A} \mathcal{B}^{-1/2} y. \tag{2.22}$$

From Theorem 2.1, let

$$\mathcal{B}^{-1/2}\ \mathcal{A}\ \mathcal{B}^{-1/2} = \Gamma\ \Lambda\ \Gamma^{\top}$$

be the spectral decomposition of $\mathcal{B}^{-1/2}\ \mathcal{A}\ \mathcal{B}^{-1/2}$. Set

$$z = \Gamma^{\top} y\ \Rightarrow\ z^{\top} z = y^{\top} \Gamma\ \Gamma^{\top}\ y = y^{\top} y.$$

Thus (2.22) is equivalent to

$$\max_{\{z:z^{\top}z=1\}} z^{\top}\ \Lambda\ z = \max_{\{z:z^{\top}z=1\}} \sum_{i=1}^{p} \lambda_i z_i^2.$$

But

$$\max_z \sum \lambda_i z_i^2 \le \lambda_1 \underbrace{\max_z \sum z_i^2}_{=1} = \lambda_1.$$

The maximum is thus obtained by $z = (1, 0, \dots, 0)^{\top}$, i.e.,

$$y = \gamma_{1} \Rightarrow x = \mathcal{B}^{-1/2}\gamma_{1}.$$

Since $\mathcal{B}^{-1}\mathcal{A}$ and $\mathcal{B}^{-1/2}\ \mathcal{A}\ \mathcal{B}^{-1/2}$ have the same eigenvalues, the proof is complete.
To maximizes (minimizes) $x^{\top}\mathcal{A}x$ under $x^{\top}\mathcal{B}x = 1$, below is another proof using the Lagrange method.

$$\max_x x^{\top}\mathcal{A}x = \max_x [x^{\top}\mathcal{A}x - \lambda\left(x^{\top}\mathcal{B}x - 1\right)].$$

The first derivative of it respect to x is equal to 0:

$$2\mathcal{A}x - 2\lambda\mathcal{B}x = 0.$$

so

$$\mathcal{B}^{-1}\mathcal{A}x = \lambda x$$

By the definition of eigenvector and eigenvalue, our maximizer x^* is $\mathcal{B}^{-1}\mathcal{A}'s$ eigenvector corresponding to eigenvalue λ. So

$$\max_{\{x:x^{\top}\mathcal{B}x=1\}} x^{\top}\mathcal{A}x = \max_{\{x:x^{\top}\mathcal{B}x=1\}} x^{\top}\mathcal{B}\mathcal{B}^{-1}\mathcal{A}x = \max_{\{x:x^{\top}\mathcal{B}x=1\}} x^{\top}\mathcal{B}\lambda x = \max \lambda$$

which is just the maximum eigenvalue of $\mathcal{B}^{-1}\mathcal{A}$, and we choose the corresponding eigenvector as our maximizer x^*. □

EXAMPLE 2.7 *Consider the following matrices*

$$\mathcal{A} = \begin{pmatrix} 1 & 2 \\ 2 & 3 \end{pmatrix} \quad \textit{and} \quad \mathcal{B} = \begin{pmatrix} 1 & 0 \\ 0 & 1 \end{pmatrix}.$$

We calculate

$$\mathcal{B}^{-1}\mathcal{A} = \begin{pmatrix} 1 & 2 \\ 2 & 3 \end{pmatrix}.$$

The biggest eigenvalue of the matrix $\mathcal{B}^{-1}\mathcal{A}$ *is* $2 + \sqrt{5}$. *This means that the maximum of* $x^{\top}\mathcal{A}x$ *under the constraint* $x^{\top}\mathcal{B}x = 1$ *is* $2 + \sqrt{5}$.

Notice that the constraint $x^{\top}\mathcal{B}x = 1$ *corresponds, with our choice of* $\mathcal{B}$, *to the points which lie on the unit circle* $x_1^2 + x_2^2 = 1$.

Summary

↪ A quadratic form can be described by a symmetric matrix $\mathcal{A}$.

↪ Quadratic forms can always be diagonalized.

↪ Positive definiteness of a quadratic form is equivalent to positiveness of the eigenvalues of the matrix $\mathcal{A}$.

↪ The maximum and minimum of a quadratic form given some constraints can be expressed in terms of eigenvalues.

2.4 Derivatives

For later sections of this book, it will be useful to introduce matrix notation for derivatives of a scalar function of a vector x with respect to x. Consider $f : \mathbb{R}^p \longrightarrow \mathbb{R}$ and a $(p \times 1)$ vector x, then $\frac{\partial f(x)}{\partial x}$ is the column vector of partial derivatives $\left\{\frac{\partial f(x)}{\partial x_j}\right\}, j = 1, \ldots, p$ and $\frac{\partial f(x)}{\partial x^\top}$ is the row vector of the same derivative ($\frac{\partial f(x)}{\partial x}$ is called the *gradient* of f).

We can also introduce second order derivatives: $\frac{\partial^2 f(x)}{\partial x \partial x^\top}$ is the $(p \times p)$ matrix of elements $\frac{\partial^2 f(x)}{\partial x_i \partial x_j}, i = 1, \ldots, p$ and $j = 1, \ldots, p$. ($\frac{\partial^2 f(x)}{\partial x \partial x^\top}$ is called the *Hessian* of f).

Suppose that a is a $(p \times 1)$ vector and that $\mathcal{A} = \mathcal{A}^\top$ is a $(p \times p)$ matrix. Then

$$\frac{\partial a^\top x}{\partial x} = \frac{\partial x^\top a}{\partial x} = a, \tag{2.23}$$

$$\frac{\partial x^\top \mathcal{A} x}{\partial x} = 2\mathcal{A}x. \tag{2.24}$$

The Hessian of the quadratic form $Q(x) = x^\top \mathcal{A} x$ is:

$$\frac{\partial^2 x^\top \mathcal{A} x}{\partial x \partial x^\top} = 2\mathcal{A}. \tag{2.25}$$

EXAMPLE 2.8 *Consider the matrix*

$$\mathcal{A} = \begin{pmatrix} 1 & 2 \\ 2 & 3 \end{pmatrix}.$$

From formulas (2.24) and (2.25) it immediately follows that the gradient of $Q(x) = x^\top \mathcal{A} x$ *is*

$$\frac{\partial x^\top \mathcal{A} x}{\partial x} = 2\mathcal{A}x = 2\begin{pmatrix} 1 & 2 \\ 2 & 3 \end{pmatrix} x = \begin{pmatrix} 2x & 4x \\ 4x & 6x \end{pmatrix}$$

and the Hessian is

$$\frac{\partial^2 x^\top \mathcal{A} x}{\partial x \partial x^\top} = 2\mathcal{A} = 2\begin{pmatrix} 1 & 2 \\ 2 & 3 \end{pmatrix} = \begin{pmatrix} 2 & 4 \\ 4 & 6 \end{pmatrix}.$$

2.5 Partitioned Matrices

Very often we will have to consider certain groups of rows and columns of a matrix $\mathcal{A}(n \times p)$. In the case of two groups, we have

$$\mathcal{A} = \begin{pmatrix} \mathcal{A}_{11} & \mathcal{A}_{12} \\ \mathcal{A}_{21} & \mathcal{A}_{22} \end{pmatrix}$$

where $\mathcal{A}_{ij}(n_i \times p_j),\ i,j = 1,2,\ n_1 + n_2 = n$ and $p_1 + p_2 = p$.

If $\mathcal{B}(n \times p)$ is partitioned accordingly, we have:

$$\begin{aligned}
\mathcal{A} + \mathcal{B} &= \begin{pmatrix} \mathcal{A}_{11} + \mathcal{B}_{11} & \mathcal{A}_{12} + \mathcal{B}_{12} \\ \mathcal{A}_{21} + \mathcal{B}_{21} & \mathcal{A}_{22} + \mathcal{B}_{22} \end{pmatrix} \\
\mathcal{B}^{\top} &= \begin{pmatrix} \mathcal{B}_{11}^{\top} & \mathcal{B}_{21}^{\top} \\ \mathcal{B}_{12}^{\top} & \mathcal{B}_{22}^{\top} \end{pmatrix} \\
\mathcal{A}\mathcal{B}^{\top} &= \begin{pmatrix} \mathcal{A}_{11}\mathcal{B}_{11}^{\top} + \mathcal{A}_{12}\mathcal{B}_{12}^{\top} & \mathcal{A}_{11}\mathcal{B}_{21}^{\top} + \mathcal{A}_{12}\mathcal{B}_{22}^{\top} \\ \mathcal{A}_{21}\mathcal{B}_{11}^{\top} + \mathcal{A}_{22}\mathcal{B}_{12}^{\top} & \mathcal{A}_{21}\mathcal{B}_{21}^{\top} + \mathcal{A}_{22}\mathcal{B}_{22}^{\top} \end{pmatrix}.
\end{aligned}$$

An important particular case is the square matrix $\mathcal{A}(p \times p)$, partitioned such that $\mathcal{A}_{11}$ and $\mathcal{A}_{22}$ are both square matrices (i.e., $n_j = p_j, j = 1,2$). It can be verified that when $\mathcal{A}$ is non-singular ($\mathcal{A}\mathcal{A}^{-1} = \mathcal{I}_p$):

$$\mathcal{A}^{-1} = \begin{pmatrix} \mathcal{A}^{11} & \mathcal{A}^{12} \\ \mathcal{A}^{21} & \mathcal{A}^{22} \end{pmatrix} \tag{2.26}$$

where

$$\left\{ \begin{aligned}
\mathcal{A}^{11} &= (\mathcal{A}_{11} - \mathcal{A}_{12}\mathcal{A}_{22}^{-1}\mathcal{A}_{21})^{-1} \stackrel{\text{def}}{=} (\mathcal{A}_{11\cdot 2})^{-1} \\
\mathcal{A}^{12} &= -(\mathcal{A}_{11\cdot 2})^{-1}\mathcal{A}_{12}\mathcal{A}_{22}^{-1} \\
\mathcal{A}^{21} &= -\mathcal{A}_{22}^{-1}\mathcal{A}_{21}(\mathcal{A}_{11\cdot 2})^{-1} \\
\mathcal{A}^{22} &= \mathcal{A}_{22}^{-1} + \mathcal{A}_{22}^{-1}\mathcal{A}_{21}(\mathcal{A}_{11\cdot 2})^{-1}\mathcal{A}_{12}\mathcal{A}_{22}^{-1}
\end{aligned} \right. .$$

An alternative expression can be obtained by reversing the positions of $\mathcal{A}_{11}$ and $\mathcal{A}_{22}$ in the original matrix.

The following results will be useful if $\mathcal{A}_{11}$ is non-singular:

$$|\mathcal{A}| = |\mathcal{A}_{11}||\mathcal{A}_{22} - \mathcal{A}_{21}\mathcal{A}_{11}^{-1}\mathcal{A}_{12}| = |\mathcal{A}_{11}||\mathcal{A}_{22\cdot 1}|. \tag{2.27}$$

If $\mathcal{A}_{22}$ is non-singular, we have that:

$$|\mathcal{A}| = |\mathcal{A}_{22}||\mathcal{A}_{11} - \mathcal{A}_{12}\mathcal{A}_{22}^{-1}\mathcal{A}_{21}| = |\mathcal{A}_{22}||\mathcal{A}_{11\cdot 2}|. \tag{2.28}$$

A useful formula is derived from the alternative expressions for the inverse and the determinant. For instance let

$$\mathcal{B} = \begin{pmatrix} 1 & b^{\top} \\ a & \mathcal{A} \end{pmatrix}$$

where a and b are $(p \times 1)$ vectors and $\mathcal{A}$ is non-singular. We then have:

$$|\mathcal{B}| = |\mathcal{A} - ab^{\top}| = |\mathcal{A}||1 - b^{\top}\mathcal{A}^{-1}a| \tag{2.29}$$

and equating the two expressions for $\mathcal{B}^{22}$, we obtain the following:

$$(\mathcal{A} - ab^{\top})^{-1} = \mathcal{A}^{-1} + \frac{\mathcal{A}^{-1}ab^{\top}\mathcal{A}^{-1}}{1 - b^{\top}\mathcal{A}^{-1}a}. \tag{2.30}$$

EXAMPLE 2.9 *Let's consider the matrix*

$$\mathcal{A} = \begin{pmatrix} 1 & 2 \\ 2 & 2 \end{pmatrix}.$$

We can use formula (2.26) to calculate the inverse of a partitioned matrix, i.e., $\mathcal{A}^{11} = -1, \mathcal{A}^{12} = \mathcal{A}^{21} = 1, \mathcal{A}^{22} = -1/2$. *The inverse of* $\mathcal{A}$ *is*

$$\mathcal{A}^{-1} = \begin{pmatrix} -1 & 1 \\ 1 & -0.5 \end{pmatrix}.$$

It is also easy to calculate the determinant of $\mathcal{A}$*:*

$$|\mathcal{A}| = |1||2-4| = -2.$$

Let $\mathcal{A}(n \times p)$ and $\mathcal{B}(p \times n)$ be any two matrices and suppose that $n \geq p$. From (2.27) and (2.28) we can conclude that

$$\begin{vmatrix} -\lambda\mathcal{I}_n & -\mathcal{A} \\ \mathcal{B} & \mathcal{I}_p \end{vmatrix} = (-\lambda)^{n-p}|\mathcal{B}\mathcal{A} - \lambda\mathcal{I}_p| = |\mathcal{A}\mathcal{B} - \lambda\mathcal{I}_n|. \tag{2.31}$$

Since both determinants on the right-hand side of (2.31) are polynomials in λ, we find that the n eigenvalues of $\mathcal{AB}$ yield the p eigenvalues of $\mathcal{BA}$ plus the eigenvalue 0, $n-p$ times.

The relationship between the eigenvectors is described in the next theorem.

THEOREM 2.6 *For* $\mathcal{A}(n \times p)$ *and* $\mathcal{B}(p \times n)$*, the non-zero eigenvalues of* $\mathcal{AB}$ *and* $\mathcal{BA}$ *are the same and have the same multiplicity. If* x *is an eigenvector of* $\mathcal{AB}$ *for an eigenvalue* $\lambda \neq 0$*, then* $y = \mathcal{B}x$ *is an eigenvector of* $\mathcal{BA}$*.*

COROLLARY 2.2 *For* $\mathcal{A}(n \times p)$*,* $\mathcal{B}(q \times n)$*,* $a(p \times 1)$*, and* $b(q \times 1)$ *we have*

$$\operatorname{rank}(\mathcal{A}ab^{\top}\mathcal{B}) \leq 1.$$

The non-zero eigenvalue, if it exists, equals $b^{\top}\mathcal{BA}a$ *(with eigenvector* $\mathcal{A}a$*).*

Proof:
Theorem 2.6 asserts that the eigenvalues of $\mathcal{A}ab^{\top}\mathcal{B}$ are the same as those of $b^{\top}\mathcal{BA}a$. Note that the matrix $b^{\top}\mathcal{BA}a$ is a scalar and hence it is its own eigenvalue λ_1.

Applying $\mathcal{A}ab^{\top}\mathcal{B}$ to $\mathcal{A}a$ yields

$$(\mathcal{A}ab^{\top}\mathcal{B})(\mathcal{A}a) = (\mathcal{A}a)(b^{\top}\mathcal{BA}a) = \lambda_1\mathcal{A}a.$$

□

2.6 Geometrical Aspects

Distance

Let $x, y \in \mathbb{R}^p$. A distance d is defined as a function

$$d : \mathbb{R}^{2p} \to \mathbb{R}_+ \quad \text{which fulfills} \quad \begin{cases} d(x,y) > 0 & \forall x \neq y \\ d(x,y) = 0 & \text{if and only if } x = y \\ d(x,y) \leq d(x,z) + d(z,y) & \forall x,y,z \end{cases}.$$

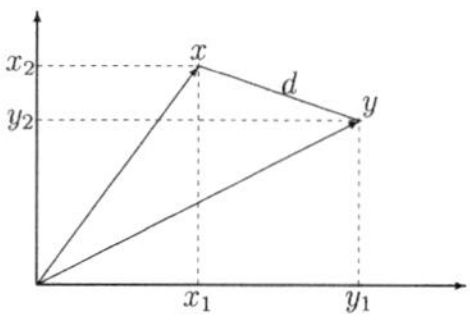

Figure 2.1. Distance d.

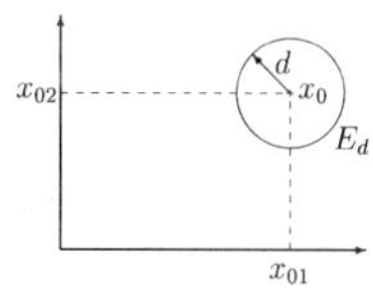

Figure 2.2. Iso–distance sphere.

A *Euclidean distance* d between two points x and y is defined as

$$d^2(x, y) = (x - y)^T \mathcal{A}(x - y) \tag{2.32}$$

where $\mathcal{A}$ is a positive definite matrix ($\mathcal{A} > 0$). $\mathcal{A}$ is called a *metric*.

EXAMPLE 2.10 *A particular case is when* $\mathcal{A} = \mathcal{I}_p$, *i.e.,*

$$d^2(x, y) = \sum_{i=1}^{p} (x_i - y_i)^2. \tag{2.33}$$

Figure 2.1 illustrates this definition for $p = 2$.

Note that the sets $E_d = \{x \in \mathbb{R}^p \mid (x - x_0)^\top (x - x_0) = d^2\}$, i.e., the spheres with radius d and center x_0, are the Euclidean $\mathcal{I}_p$ *iso-distance* curves from the point x_0 (see Figure 2.2).

The more general distance (2.32) with a positive definite matrix $\mathcal{A}$ ($\mathcal{A} > 0$) leads to the iso-distance curves

$$E_d = \{x \in \mathbb{R}^p \mid (x - x_0)^\top \mathcal{A}(x - x_0) = d^2\}, \tag{2.34}$$

i.e., ellipsoids with center x_0, matrix $\mathcal{A}$ and constant d (see Figure 2.3).

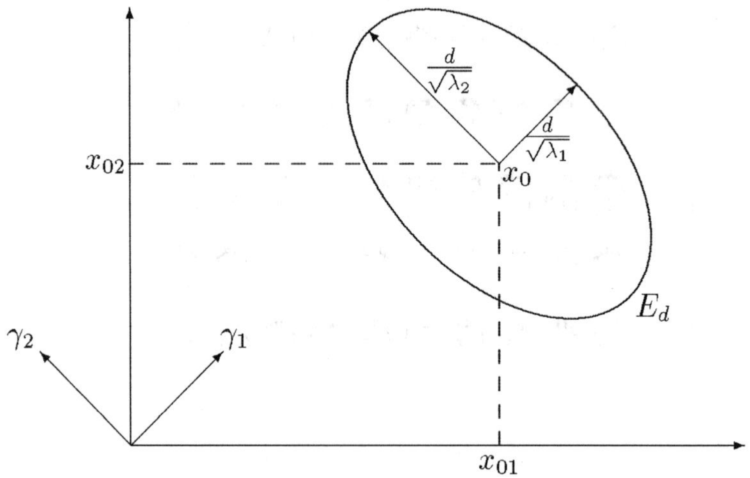

Figure 2.3. Iso–distance ellipsoid.

Let $\gamma_1, \gamma_2, ..., \gamma_p$ be the orthonormal eigenvectors of $\mathcal{A}$ corresponding to the eigenvalues $\lambda_1 \geq \lambda_2 \geq ... \geq \lambda_p$. The resulting observations are given in the next theorem.

THEOREM 2.7 *(i) The principal axes of E_d are in the direction of γ_i; $i = 1, \ldots, p$.*

(ii) The half-lengths of the axes are $\sqrt{\frac{d^2}{\lambda_i}}$; $i = 1, \ldots, p$.

(iii) The rectangle surrounding the ellipsoid E_d is defined by the following inequalities:

$$x_{0i} - \sqrt{d^2 a^{ii}} \leq x_i \leq x_{0i} + \sqrt{d^2 a^{ii}}, \quad i = 1, \ldots, p,$$

where a^{ii} is the (i, i) element of $\mathcal{A}^{-1}$. By the rectangle surrounding the ellipsoid E_d we mean the rectangle whose sides are parallel to the coordinate axis.

It is easy to find the coordinates of the tangency points between the ellipsoid and its surrounding rectangle parallel to the coordinate axes. Let us find the coordinates of the tangency point that are in the direction of the j-th coordinate axis (positive direction).

For ease of notation, we suppose the ellipsoid is centered around the origin ($x_0 = 0$). If not, the rectangle will be shifted by the value of x_0.

The coordinate of the tangency point is given by the solution to the following problem:

$$x = \arg \max_{x^\top \mathcal{A} x = d^2} e_j^\top x \tag{2.35}$$

where $e_j^\top$ is the j-th column of the identity matrix $\mathcal{I}_p$. The coordinate of the tangency point in the negative direction would correspond to the solution of the min problem: by symmetry, it is the opposite value of the former.

The solution is computed via the Lagrangian $L = e_j^\top x - \lambda(x^\top \mathcal{A}x - d^2)$ which by (2.23) leads to the following system of equations:

$$\frac{\partial L}{\partial x} = e_j - 2\lambda \mathcal{A}x = 0 \tag{2.36}$$

$$\frac{\partial L}{\partial \lambda} = x^T \mathcal{A}x - d^2 = 0. \tag{2.37}$$

This gives $x = \frac{1}{2\lambda}\mathcal{A}^{-1}e_j$, or componentwise

$$x_i = \frac{1}{2\lambda}a^{ij}, \ i = 1, \dots, p \tag{2.38}$$

where a^{ij} denotes the (i,j)-th element of $\mathcal{A}^{-1}$.

Premultiplying (2.36) by $x^\top$, we have from (2.37):

$$x_j = 2\lambda d^2.$$

Comparing this to the value obtained by (2.38), for $i = j$ we obtain $2\lambda = \sqrt{\frac{a^{jj}}{d^2}}$. We choose the positive value of the square root because we are maximizing $e_j^\top x$. A minimum would correspond to the negative value. Finally, we have the coordinates of the tangency point between the ellipsoid and its surrounding rectangle in the positive direction of the j-th axis:

$$x_i = \sqrt{\frac{d^2}{a^{jj}}}\, a^{ij}, \ i = 1, \dots, p. \tag{2.39}$$

The particular case where $i = j$ provides statement (iii) in Theorem 2.7.

Remark: usefulness of Theorem 2.7

Theorem 2.7 will prove to be particularly useful in many subsequent chapters. First, it provides a helpful tool for graphing an ellipse in two dimensions. Indeed, knowing the slope of the principal axes of the ellipse, their half-lengths and drawing the rectangle inscribing the ellipse allows one to quickly draw a rough picture of the shape of the ellipse.

In Chapter 7, it is shown that the confidence region for the vector μ of a multivariate normal population is given by a particular ellipsoid whose parameters depend on sample characteristics. The rectangle inscribing the ellipsoid (which is much easier to obtain) will provide the simultaneous confidence intervals for all of the components in μ.

In addition it will be shown that the contour surfaces of the multivariate normal density are provided by ellipsoids whose parameters depend on the mean vector and on the covariance matrix. We will see that the tangency points between the contour ellipsoids and the surrounding rectangle are determined by regressing one component on the $(p-1)$ other components. For instance, in the direction of the j-th axis, the tangency points are given by the intersections of the ellipsoid contours with the regression line of the vector of $(p-1)$ variables (all components except the j-th) on the j-th component.

Norm of a Vector

Consider a vector $x \in \mathbb{R}^p$. The norm or length of x (with respect to the metric $\mathcal{I}_p$) is defined as

$$\|x\| = d(0, x) = \sqrt{x^\top x}.$$

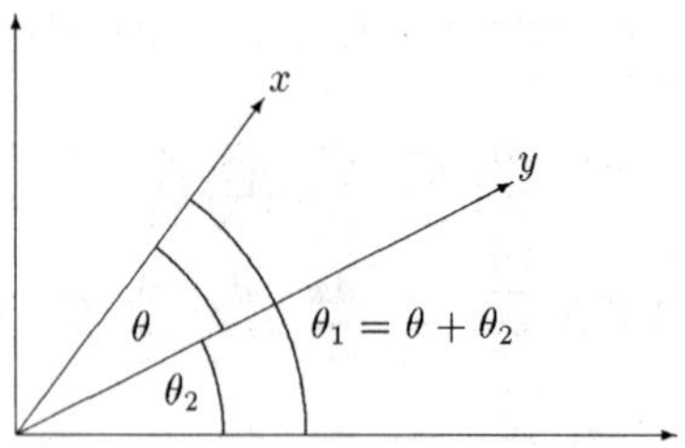

Figure 2.4. Angle between vectors.

If $\|x\| = 1$, x is called a *unit vector.* A more general norm can be defined with respect to the metric $\mathcal{A}$:

$$\|x\|_{\mathcal{A}} = \sqrt{x^\top \mathcal{A} x}.$$

Angle between two Vectors

Consider two vectors x and $y \in \mathbb{R}^p$. The angle θ between x and y is defined by the cosine of θ:

$$\cos\theta = \frac{x^\top y}{\|x\| \ \|y\|}, \tag{2.40}$$

see Figure 2.4. Indeed for $p = 2$, $x = \begin{pmatrix} x_1 \\ x_2 \end{pmatrix}$ and $y = \begin{pmatrix} y_1 \\ y_2 \end{pmatrix}$, we have

$$\begin{array}{rclcrcl} \|x\| \cos\theta_1 & = & x_1 \ ; & & \|y\| \cos\theta_2 & = & y_1 \\ \|x\| \sin\theta_1 & = & x_2 \ ; & & \|y\| \sin\theta_2 & = & y_2, \end{array} \tag{2.41}$$

therefore,

$$\cos\theta = \cos\theta_1 \cos\theta_2 + \sin\theta_1 \sin\theta_2 = \frac{x_1 y_1 + x_2 y_2}{\|x\| \ \|y\|} = \frac{x^\top y}{\|x\| \ \|y\|} \ .$$

REMARK 2.1 *If $x^\top y = 0$, then the angle θ is equal to $\frac{\pi}{2}$. From trigonometry, we know that the cosine of θ equals the length of the base of a triangle ($||p_x||$) divided by the length of the hypotenuse ($||x||$). Hence, we have*

$$||p_x|| = ||x|| |\cos\theta| = \frac{|x^\top y|}{\|y\|}, \tag{2.42}$$

where p_x is the projection of x on y (which is defined below). It is the coordinate of x on the y vector, see Figure 2.5.

The angle can also be defined with respect to a general metric $\mathcal{A}$

$$\cos\theta = \frac{x^\top \mathcal{A} y}{\|x\|_{\mathcal{A}} \ \|y\|_{\mathcal{A}}}. \tag{2.43}$$

If $\cos\theta = 0$ then x is orthogonal to y with respect to the metric $\mathcal{A}$.

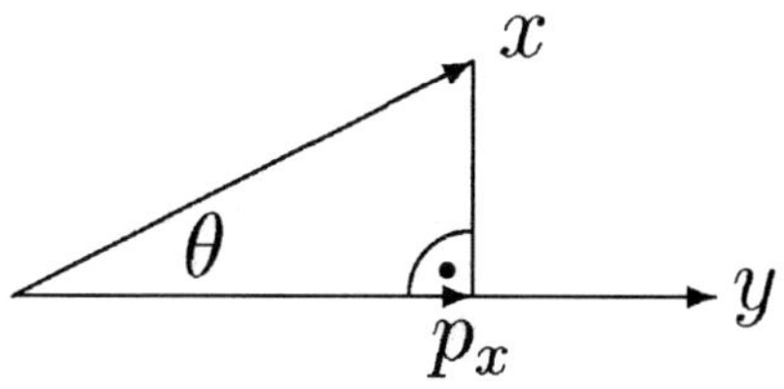

Figure 2.5. Projection.

EXAMPLE 2.11 *Assume that there are two centered (i.e., zero mean) data vectors. The cosine of the angle between them is equal to their correlation (defined in (3.8))! Indeed for x and y with $\overline{x} = \overline{y} = 0$ we have*

$$r_{XY} = \frac{\sum x_i y_i}{\sqrt{\sum x_i^2 \sum y_i^2}} = \cos\theta$$

according to formula (2.40).

Rotations

When we consider a point $x \in \mathbb{R}^p$, we generally use a p-coordinate system to obtain its geometric representation, like in Figure 2.1 for instance. There will be situations in multivariate techniques where we will want to rotate this system of coordinates by the angle θ.

Consider for example the point P with coordinates $x = (x_1, x_2)^\top$ in $\mathbb{R}^2$ with respect to a given set of orthogonal axes. Let Γ be a (2×2) orthogonal matrix where

$$\Gamma = \begin{pmatrix} \cos\theta & \sin\theta \\ -\sin\theta & \cos\theta \end{pmatrix}. \tag{2.44}$$

If the axes are rotated about the origin through an angle θ in a clockwise direction, the new coordinates of P will be given by the vector y

$$y = \Gamma\, x, \tag{2.45}$$

and a rotation through the same angle in a counterclockwise direction gives the new coordinates as

$$y = \Gamma^\top x. \tag{2.46}$$

More generally, premultiplying a vector x by an orthogonal matrix Γ geometrically corresponds to a rotation of the system of axes, so that the first new axis is determined by the first row of Γ. This geometric point of view will be exploited in Chapters 9 and 10.

Column Space and Null Space of a Matrix

Define for $\mathcal{X}(n \times p)$

$$Im(\mathcal{X}) \stackrel{def}{=} C(\mathcal{X}) = \{x \in \mathbb{R}^n \mid \exists a \in \mathbb{R}^p \text{ so that } \mathcal{X}a = x\},$$

the space generated by the columns of $\mathcal{X}$ or the *column space* of $\mathcal{X}$. Note that $C(\mathcal{X}) \subseteq \mathbb{R}^n$ and $\dim\{C(\mathcal{X})\} = \operatorname{rank}(\mathcal{X}) = r \le \min(n, p)$.

$$Ker(\mathcal{X}) \stackrel{def}{=} N(\mathcal{X}) = \{y \in \mathbb{R}^p \mid \mathcal{X}y = 0\}$$

is the *null space* of $\mathcal{X}$. Note that $N(\mathcal{X}) \subseteq \mathbb{R}^p$ and that $\dim\{N(\mathcal{X})\} = p - r$.

REMARK 2.2 *$N(\mathcal{X}^\top)$ is the orthogonal complement of $C(\mathcal{X})$ in $\mathbb{R}^n$, i.e., given a vector $b \in \mathbb{R}^n$ it will hold that $x^\top b = 0$ for all $x \in C(\mathcal{X})$, if and only if $b \in N(\mathcal{X}^\top)$.*

EXAMPLE 2.12 *Let* $\mathcal{X} = \begin{pmatrix} 2 & 3 & 5 \\ 4 & 6 & 7 \\ 6 & 8 & 6 \\ 8 & 2 & 4 \end{pmatrix}$. *It is easy to show (e.g. by calculating the determinant of $\mathcal{X}$) that* $\operatorname{rank}(\mathcal{X}) = 3$. *Hence, the columns space of $\mathcal{X}$ is $C(\mathcal{X}) = \mathbb{R}^3$. The null space of $\mathcal{X}$ contains only the zero vector $(0,0,0)^\top$ and its dimension is equal to* $\operatorname{rank}(\mathcal{X}) - 3 = 0$.

For $\mathcal{X} = \begin{pmatrix} 2 & 3 & 1 \\ 4 & 6 & 2 \\ 6 & 8 & 3 \\ 8 & 2 & 4 \end{pmatrix}$, *the third column is a multiple of the first one and the matrix $\mathcal{X}$ cannot be of full rank. Noticing that the first two columns of $\mathcal{X}$ are independent, we see that* $\operatorname{rank}(\mathcal{X}) = 2$. *In this case, the dimension of the columns space is 2 and the dimension of the null space is 1.*

Projection Matrix

A matrix $\mathcal{P}(n \times n)$ is called an (orthogonal) projection matrix in $\mathbb{R}^n$ if and only if $\mathcal{P} = \mathcal{P}^\top = \mathcal{P}^2$ ($\mathcal{P}$ is idempotent). Let $b \in \mathbb{R}^n$. Then $a = \mathcal{P}b$ is the projection of b on $C(\mathcal{P})$.

Projection on $C(\mathcal{X})$

Consider $\mathcal{X}(n \times p)$ and let

$$\mathcal{P} = \mathcal{X}(\mathcal{X}^\top\mathcal{X})^{-1}\mathcal{X}^\top \tag{2.47}$$

and $\mathcal{Q} = \mathcal{I}_n - \mathcal{P}$. It's easy to check that $\mathcal{P}$ and $\mathcal{Q}$ are idempotent and that

$$\mathcal{P}\mathcal{X} = \mathcal{X} \text{ and } \mathcal{Q}\mathcal{X} = 0. \tag{2.48}$$

Since the columns of $\mathcal{X}$ are projected onto themselves, the projection matrix $\mathcal{P}$ projects any vector $b \in \mathbb{R}^n$ onto $C(\mathcal{X})$. Similarly, the projection matrix $\mathcal{Q}$ projects any vector $b \in \mathbb{R}^n$ onto the orthogonal complement of $C(\mathcal{X})$.

THEOREM 2.8 *Let $\mathcal{P}$ be the projection (2.47) and $\mathcal{Q}$ its orthogonal complement. Then:*

(i) $x = \mathcal{P}b \Rightarrow x \in C(\mathcal{X})$,

(ii) $y = \mathcal{Q}b \Rightarrow y^\top x = 0 \;\forall x \in C(\mathcal{X})$.

Proof:
(i) holds, since $x = \mathcal{X}(\mathcal{X}^\top\mathcal{X})^{-1}\mathcal{X}^\top b = \mathcal{X}a$, where $a = (\mathcal{X}^\top\mathcal{X})^{-1}\mathcal{X}^\top b \in \mathbb{R}^p$.
(ii) follows from $y = b - \mathcal{P}b$ and $x = \mathcal{X}a \Rightarrow y^\top x = b^\top\mathcal{X}a - b^\top\mathcal{X}(\mathcal{X}^\top\mathcal{X})^{-1}\mathcal{X}^\top\mathcal{X}a = 0$. □

REMARK 2.3 *Let $x, y \in \mathbb{R}^n$ and consider $p_x \in \mathbb{R}^n$, the projection of x on y (see Figure 2.5). With $\mathcal{X} = y$ we have from (2.47)*

$$p_x = y(y^\top y)^{-1}y^\top x = \frac{y^\top x}{\|y\|^2}\, y \tag{2.49}$$

and we can easily verify that

$$\|p_x\| = \sqrt{p_x^\top p_x} = \frac{|y^\top x|}{\|y\|}.$$

See again Remark 2.1.

Summary

↪ A distance between two p-dimensional points x and y is a quadratic form $(x-y)^\top\mathcal{A}(x-y)$ in the vectors of differences $(x-y)$. A distance defines the norm of a vector.

↪ Iso-distance curves of a point x_0 are all those points that have the same distance from x_0. Iso-distance curves are ellipsoids whose principal axes are determined by the direction of the eigenvectors of $\mathcal{A}$. The half-length of principal axes is proportional to the inverse of the roots of the eigenvalues of $\mathcal{A}$.

↪ The angle between two vectors x and y is given by $\cos\theta = \frac{x^\top\mathcal{A}y}{\|x\|_\mathcal{A}\,\|y\|_\mathcal{A}}$ w.r.t. the metric $\mathcal{A}$.

↪ For the Euclidean distance with $\mathcal{A} = \mathcal{I}$ the correlation between two centered data vectors x and y is given by the cosine of the angle between them, i.e., $\cos\theta = r_{XY}$.

↪ The projection $\mathcal{P} = \mathcal{X}(\mathcal{X}^\top\mathcal{X})^{-1}\mathcal{X}^\top$ is the projection onto the column space $C(\mathcal{X})$ of $\mathcal{X}$.

↪ The projection of $x \in \mathbb{R}^n$ on $y \in \mathbb{R}^n$ is given by $p_x = \frac{y^\top x}{\|y\|^2}y$.

2.7 Exercises

EXERCISE 2.1 *Compute the determinant for a (3×3) matrix.*

EXERCISE 2.2 *Suppose that $|\mathcal{A}| = 0$. Is it possible that all eigenvalues of $\mathcal{A}$ are positive?*

EXERCISE 2.3 *Suppose that all eigenvalues of some (square) matrix $\mathcal{A}$ are different from zero. Does the inverse $\mathcal{A}^{-1}$ of $\mathcal{A}$ exist?*

EXERCISE 2.4 *Write a program that calculates the Jordan decomposition of the matrix*

$$\mathcal{A} = \begin{pmatrix} 1 & 2 & 3 \\ 2 & 1 & 2 \\ 3 & 2 & 1 \end{pmatrix}.$$

Check Theorem 2.1 numerically.

EXERCISE 2.5 *Prove (2.23), (2.24) and (2.25).*

EXERCISE 2.6 *Show that a projection matrix only has eigenvalues in* $\{0, 1\}$.

EXERCISE 2.7 *Draw some iso-distance ellipsoids for the metric* $\mathcal{A} = \Sigma^{-1}$ *of Example 3.13.*

EXERCISE 2.8 *Find a formula for* $|\mathcal{A} + aa^\top|$ *and for* $(\mathcal{A} + aa^\top)^{-1}$. *(Hint: use the inverse partitioned matrix with* $\mathcal{B} = \begin{pmatrix} 1 & -a^\top \\ a & \mathcal{A} \end{pmatrix}$*.)*

EXERCISE 2.9 *Prove the Binomial inverse theorem for two non-singular matrices* $\mathcal{A}(p \times p)$ *and* $\mathcal{B}(p \times p)$: $(\mathcal{A} + \mathcal{B})^{-1} = \mathcal{A}^{-1} - \mathcal{A}^{-1}(\mathcal{A}^{-1} + \mathcal{B}^{-1})^{-1}\mathcal{A}^{-1}$. *(Hint: use (2.26) with* $\mathcal{C} = \begin{pmatrix} \mathcal{A} & I_p \\ -I_p & \mathcal{B}^{-1} \end{pmatrix}$*.)*

3 Moving to Higher Dimensions

We have seen in the previous chapters how very simple graphical devices can help in understanding the structure and dependency of data. The graphical tools were based on either univariate (bivariate) data representations or on "slick" transformations of multivariate information perceivable by the human eye. Most of the tools are extremely useful in a modelling step, but unfortunately, do not give the full picture of the data set. One reason for this is that the graphical tools presented capture only certain dimensions of the data and do not necessarily concentrate on those dimensions or subparts of the data under analysis that carry the maximum structural information. In Part III of this book, powerful tools for reducing the dimension of a data set will be presented. In this chapter, as a starting point, simple and basic tools are used to describe dependency. They are constructed from elementary facts of probability theory and introductory statistics (for example, the covariance and correlation between two variables).

Sections 3.1 and 3.2 show how to handle these concepts in a multivariate setup and how a simple test on correlation between two variables can be derived. Since linear relationships are involved in these measures, Section 3.4 presents the simple linear model for two variables and recalls the basic t-test for the slope. In Section 3.5, a simple example of one-factorial analysis of variance introduces the notations for the well known F-test.

Due to the power of matrix notation, all of this can easily be extended to a more general multivariate setup. Section 3.3 shows how matrix operations can be used to define summary statistics of a data set and for obtaining the empirical moments of linear transformations of the data. These results will prove to be very useful in most of the chapters in Part III.

Finally, matrix notation allows us to introduce the flexible multiple linear model, where more general relationships among variables can be analyzed. In Section 3.6, the least squares adjustment of the model and the usual test statistics are presented with their geometric interpretation. Using these notations, the ANOVA model is just a particular case of the multiple linear model.

3.1 Covariance

Covariance is a measure of dependency between random variables. Given two (random) variables X and Y the (theoretical) covariance is defined by:

$$\sigma_{XY} = Cov(X, Y) = E(XY) - (EX)(EY). \tag{3.1}$$

The precise definition of expected values is given in Chapter 4. If X and Y are independent of each other, the covariance $Cov(X, Y)$ is necessarily equal to zero, see Theorem 3.1. The converse is not true. The covariance of X with itself is the variance:

$$\sigma_{XX} = Var(X) = Cov(X, X).$$

If the variable X is p-dimensional multivariate, e.g., $X = \begin{pmatrix} X_1 \\ \vdots \\ X_p \end{pmatrix}$, then the theoretical covariances among all the elements are put into matrix form, i.e., the covariance matrix:

$$\Sigma = \begin{pmatrix} \sigma_{X_1X_1} & \dots & \sigma_{X_1X_p} \\ \vdots & \ddots & \vdots \\ \sigma_{X_pX_1} & \dots & \sigma_{X_pX_p} \end{pmatrix}.$$

Properties of covariance matrices will be detailed in Chapter 4. Empirical versions of these quantities are:

$$s_{XY} = \frac{1}{n}\sum_{i=1}^{n}(x_i - \overline{x})(y_i - \overline{y}) \tag{3.2}$$

$$s_{XX} = \frac{1}{n}\sum_{i=1}^{n}(x_i - \overline{x})^2. \tag{3.3}$$

For small n, say $n \leq 20$, we should replace the factor $\frac{1}{n}$ in (3.2) and (3.3) by $\frac{1}{n-1}$ in order to correct for a small bias. For a p-dimensional random variable, one obtains the empirical covariance matrix (see Section 3.3 for properties and details)

$$\mathcal{S} = \begin{pmatrix} s_{X_1X_1} & \dots & s_{X_1X_p} \\ \vdots & \ddots & \vdots \\ s_{X_pX_1} & \dots & s_{X_pX_p} \end{pmatrix}.$$

For a scatterplot of two variables the covariances measure "how close the scatter is to a line". Mathematical details follow but it should already be understood here that in this sense covariance measures only "linear dependence".

EXAMPLE 3.1 *If $\mathcal{X}$ is the entire bank data set, one obtains the covariance matrix $\mathcal{S}$ as indicated below:*

$$\mathcal{S} = \begin{pmatrix} 0.14 & 0.03 & 0.02 & -0.10 & -0.01 & 0.08 \\ 0.03 & 0.12 & 0.10 & 0.21 & 0.10 & -0.21 \\ 0.02 & 0.10 & 0.16 & 0.28 & 0.12 & -0.24 \\ -0.10 & 0.21 & 0.28 & 2.07 & 0.16 & -1.03 \\ -0.01 & 0.10 & 0.12 & 0.16 & 0.64 & -0.54 \\ 0.08 & -0.21 & -0.24 & -1.03 & -0.54 & 1.32 \end{pmatrix}. \tag{3.4}$$

The empirical covariance between X_4 and X_5, i.e., $s_{X_4X_5}$, is found in row 4 and column 5. The value is $s_{X_4X_5} = 0.16$. Is it obvious that this value is positive? In Exercise 3.1 we will discuss this question further.

If $\mathcal{X}_f$ denotes the counterfeit bank notes, we obtain:

$$\mathcal{S}_f = \begin{pmatrix} 0.123 & 0.031 & 0.023 & -0.099 & 0.019 & 0.011 \\ 0.031 & 0.064 & 0.046 & -0.024 & -0.012 & -0.005 \\ 0.024 & 0.046 & 0.088 & -0.018 & 0.000 & 0.034 \\ -0.099 & -0.024 & -0.018 & 1.268 & -0.485 & 0.236 \\ 0.019 & -0.012 & 0.000 & -0.485 & 0.400 & -0.022 \\ 0.011 & -0.005 & 0.034 & 0.236 & -0.022 & 0.308 \end{pmatrix}. \tag{3.5}$$

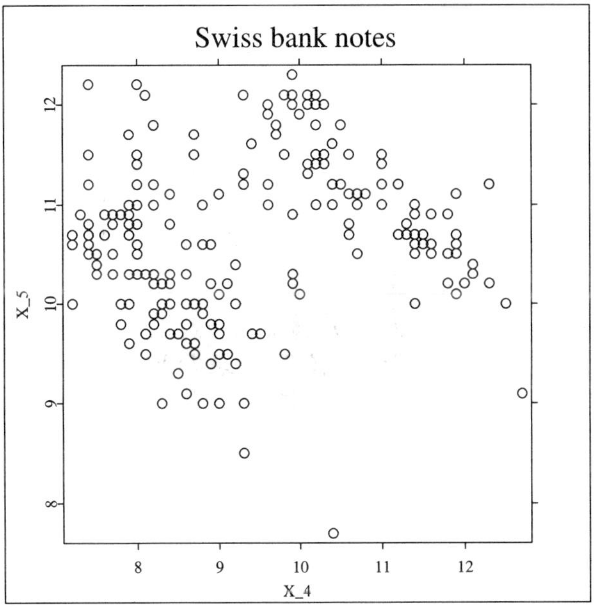

Figure 3.1. Scatterplot of variables X_4 vs. X_5 of the entire bank data set.
MVAscabank45

For the genuine, $\mathcal{X}_g$, we have:

$$\mathcal{S}_g = \begin{pmatrix} 0.149 & 0.057 & 0.057 & 0.056 & 0.014 & 0.005 \\ 0.057 & 0.131 & 0.085 & 0.056 & 0.048 & -0.043 \\ 0.057 & 0.085 & 0.125 & 0.058 & 0.030 & -0.024 \\ 0.056 & 0.056 & 0.058 & 0.409 & -0.261 & -0.000 \\ 0.014 & 0.049 & 0.030 & -0.261 & 0.417 & -0.074 \\ 0.005 & -0.043 & -0.024 & -0.000 & -0.074 & 0.198 \end{pmatrix}. \tag{3.6}$$

Note that the covariance between X_4 (distance of the frame to the lower border) and X_5 (distance of the frame to the upper border) is negative in both (3.5) and (3.6)! Why would this happen? In Exercise 3.2 we will discuss this question in more detail.

At first sight, the matrices $\mathcal{S}_f$ and $\mathcal{S}_g$ look different, but they create almost the same scatterplots (see the discussion in Section 1.4). Similarly, the common principal component analysis in Chapter 9 suggests a joint analysis of the covariance structure as in Flury and Riedwyl (1988).

Scatterplots with point clouds that are "upward-sloping", like the one in the upper left of Figure 1.14, show variables with positive covariance. Scatterplots with "downward-sloping" structure have negative covariance. In Figure 3.1 we show the scatterplot of X_4 vs. X_5 of the entire bank data set. The point cloud is upward-sloping. However, the two sub-clouds of counterfeit and genuine bank notes are downward-sloping.

EXAMPLE 3.2 *A textile shop manager is studying the sales of "classic blue" pullovers over 10 different periods. He observes the number of pullovers sold (X_1), variation in price (X_2,*

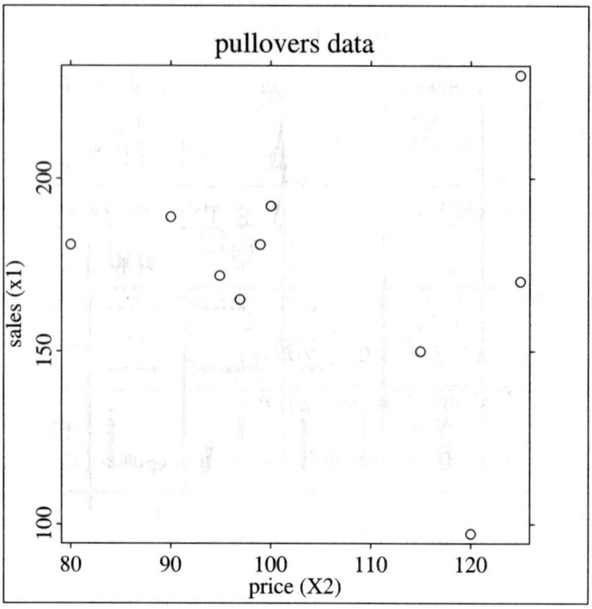

Figure 3.2. Scatterplot of variables X_2 vs. X_1 of the pullovers data set.
MVAscapull1

in EUR), the advertisement costs in local newspapers (X_3, in EUR) and the presence of a sales assistant (X_4, in hours per period). Over the periods, he observes the following data matrix:

$$\mathcal{X} = \begin{pmatrix} 230 & 125 & 200 & 109 \\ 181 & 99 & 55 & 107 \\ 165 & 97 & 105 & 98 \\ 150 & 115 & 85 & 71 \\ 97 & 120 & 0 & 82 \\ 192 & 100 & 150 & 103 \\ 181 & 80 & 85 & 111 \\ 189 & 90 & 120 & 93 \\ 172 & 95 & 110 & 86 \\ 170 & 125 & 130 & 78 \end{pmatrix}.$$

He is convinced that the price must have a large influence on the number of pullovers sold. So he makes a scatterplot of X_2 vs. X_1, see Figure 3.2. A rough impression is that the cloud is somewhat downward-sloping. A computation of the empirical covariance yields

$$s_{X_1X_2} = \frac{1}{9}\sum_{i=1}^{10}\left(X_{1i} - \bar{X}_1\right)\left(X_{2i} - \bar{X}_2\right) = -80.02,$$

a negative value as expected.

Note: The covariance function is scale dependent. Thus, if the prices in this example were in Japanese Yen (JPY), we would obtain a different answer (see Exercise 3.16). A measure of (linear) dependence independent of the scale is the correlation, which we introduce in the next section.

Summary

- ↪ The covariance is a measure of dependence.
- ↪ Covariance measures only linear dependence.
- ↪ Covariance is scale dependent.
- ↪ There are nonlinear dependencies that have zero covariance.
- ↪ Zero covariance does not imply independence.
- ↪ Independence implies zero covariance.
- ↪ Negative covariance corresponds to downward-sloping scatterplots.
- ↪ Positive covariance corresponds to upward-sloping scatterplots.
- ↪ The covariance of a variable with itself is its variance $Cov(X,X) = \sigma_{XX} = \sigma_X^2$.
- ↪ For small n, we should replace the factor $\frac{1}{n}$ in the computation of the covariance by $\frac{1}{n-1}$.

3.2 Correlation

The correlation between two variables X and Y is defined from the covariance as the following:

$$\rho_{XY} = \frac{Cov(X,Y)}{\sqrt{Var(X)\,Var(Y)}}. \tag{3.7}$$

The advantage of the correlation is that it is independent of the scale, i.e., changing the variables' scale of measurement does not change the value of the correlation. Therefore, the correlation is more useful as a measure of association between two random variables than the covariance. The empirical version of ρ_{XY} is as follows:

$$r_{XY} = \frac{s_{XY}}{\sqrt{s_{XX}s_{YY}}}. \tag{3.8}$$

The correlation is in absolute value always less than 1. It is zero if the covariance is zero and vice-versa. For p-dimensional vectors $(X_1, \ldots, X_p)^\top$ we have the theoretical correlation matrix

$$\mathcal{P} = \begin{pmatrix} \rho_{X_1X_1} & \cdots & \rho_{X_1X_p} \\ \vdots & \ddots & \vdots \\ \rho_{X_pX_1} & \cdots & \rho_{X_pX_p} \end{pmatrix},$$

and its empirical version, the empirical correlation matrix which can be calculated from the observations,

$$\mathcal{R} = \begin{pmatrix} r_{X_1X_1} & \cdots & r_{X_1X_p} \\ \vdots & \ddots & \vdots \\ r_{X_pX_1} & \cdots & r_{X_pX_p} \end{pmatrix}.$$

EXAMPLE 3.3 *We obtain the following correlation matrix for the genuine bank notes:*

$$\mathcal{R}_g = \begin{pmatrix} 1.00 & 0.41 & 0.41 & 0.22 & 0.05 & 0.03 \\ 0.41 & 1.00 & 0.66 & 0.24 & 0.20 & -0.25 \\ 0.41 & 0.66 & 1.00 & 0.25 & 0.13 & -0.14 \\ 0.22 & 0.24 & 0.25 & 1.00 & -0.63 & -0.00 \\ 0.05 & 0.20 & 0.13 & -0.63 & 1.00 & -0.25 \\ 0.03 & -0.25 & -0.14 & -0.00 & -0.25 & 1.00 \end{pmatrix}, \tag{3.9}$$

and for the counterfeit bank notes:

$$\mathcal{R}_f = \begin{pmatrix} 1.00 & 0.35 & 0.24 & -0.25 & 0.08 & 0.06 \\ 0.35 & 1.00 & 0.61 & -0.08 & -0.07 & -0.03 \\ 0.24 & 0.61 & 1.00 & -0.05 & 0.00 & 0.20 \\ -0.25 & -0.08 & -0.05 & 1.00 & -0.68 & 0.37 \\ 0.08 & -0.07 & 0.00 & -0.68 & 1.00 & -0.06 \\ 0.06 & -0.03 & 0.20 & 0.37 & -0.06 & 1.00 \end{pmatrix}. \tag{3.10}$$

As noted before for $Cov(X_4, X_5)$, the correlation between X_4 (distance of the frame to the lower border) and X_5 (distance of the frame to the upper border) is negative. This is natural, since the covariance and correlation always have the same sign (see also Exercise 3.17).

Why is the correlation an interesting statistic to study? It is related to independence of random variables, which we shall define more formally later on. For the moment we may think of independence as the fact that one variable has no influence on another.

THEOREM 3.1 *If X and Y are independent, then $\rho(X,Y) = Cov(X,Y) = 0$.*

⚠ In general, the converse is not true, as the following example shows.

EXAMPLE 3.4 *Consider a standard normally-distributed random variable X and a random variable $Y = X^2$, which is surely not independent of X. Here we have*

$$Cov(X,Y) = E(XY) - E(X)E(Y) = E(X^3) = 0$$

(because $E(X) = 0$ and $E(X^2) = 1$). Therefore $\rho(X,Y) = 0$, as well. This example also shows that correlations and covariances measure only linear dependence. The quadratic dependence of $Y = X^2$ on X is not reflected by these measures of dependence.

REMARK 3.1 *For two normal random variables, the converse of Theorem 3.1 is true: zero covariance for two normally-distributed random variables implies independence. This will be shown later in Corollary 5.2.*

Theorem 3.1 enables us to check for independence between the components of a bivariate normal random variable. That is, we can use the correlation and test whether it is zero. The distribution of r_{XY} for an arbitrary (X, Y) is unfortunately complicated. The distribution of r_{XY} will be more accessible if (X, Y) are jointly normal (see Chapter 5). If we transform the correlation by Fisher's Z-transformation,

$$W = \frac{1}{2} \log \left(\frac{1 + r_{XY}}{1 - r_{XY}} \right), \tag{3.11}$$

we obtain a variable that has a more accessible distribution. Under the hypothesis that $\rho = 0$, W has an asymptotic normal distribution. Approximations of the expectation and variance of W are given by the following:

$$\begin{array}{rcl} E(W) & \approx & \frac{1}{2} \log \left(\frac{1+\rho_{XY}}{1-\rho_{XY}} \right) \\ Var(W) & \approx & \frac{1}{(n-3)}. \end{array} \tag{3.12}$$

The distribution is given in Theorem 3.2.

THEOREM 3.2

$$Z = \frac{W - E(W)}{\sqrt{Var(W)}} \stackrel{\mathcal{L}}{\longrightarrow} N(0, 1). \tag{3.13}$$

The symbol "$\stackrel{\mathcal{L}}{\longrightarrow}$" denotes convergence in distribution, which will be explained in more detail in Chapter 4.

Theorem 3.2 allows us to test different hypotheses on correlation. We can fix the level of significance α (the probability of rejecting a true hypothesis) and reject the hypothesis if the difference between the hypothetical value and the calculated value of Z is greater than the corresponding critical value of the normal distribution. The following example illustrates the procedure.

EXAMPLE 3.5 *Let's study the correlation between mileage (X_2) and weight (X_8) for the car data set (B.3) where $n = 74$. We have $r_{X_2X_8} = -0.823$. Our conclusions from the boxplot in Figure 1.3 ("Japanese cars generally have better mileage than the others") needs to be revised. From Figure 3.3 and $r_{X_2X_8}$, we can see that mileage is highly correlated with weight, and that the Japanese cars in the sample are in fact all lighter than the others!*

If we want to know whether $\rho_{X_2X_8}$ is significantly different from $\rho_0 = 0$, we apply Fisher's Z-transform (3.11). This gives us

$$w = \frac{1}{2} \log \left(\frac{1 + r_{X_2X_8}}{1 - r_{X_2X_8}} \right) = -1.166 \quad \textit{and} \quad z = \frac{-1.166 - 0}{\sqrt{\frac{1}{71}}} = -9.825,$$

i.e., a highly significant value to reject the hypothesis that $\rho = 0$ (the 2.5% and 97.5% quantiles of the normal distribution are -1.96 and 1.96, respectively). If we want to test the hypothesis that, say, $\rho_0 = -0.75$, we obtain:

$$z = \frac{-1.166 - (-0.973)}{\sqrt{\frac{1}{71}}} = -1.627.$$

This is a nonsignificant value at the $\alpha = 0.05$ level for z since it is between the critical values at the 5% significance level (i.e., $-1.96 < z < 1.96$).

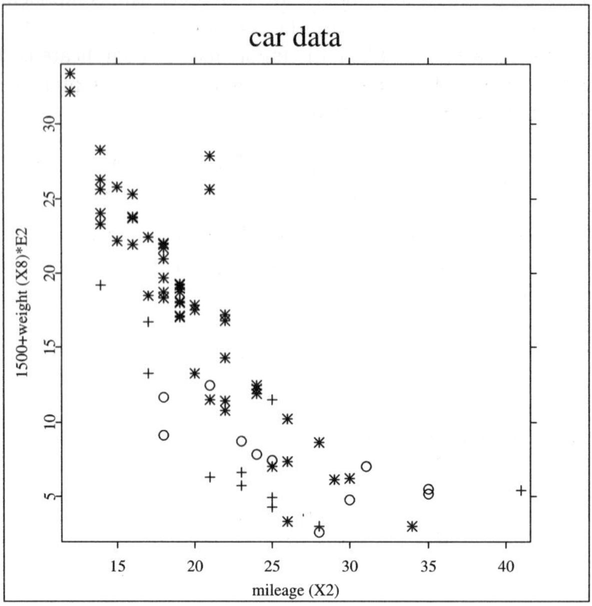

Figure 3.3. Mileage (X_2) vs. weight (X_8) of U.S. (star), European (plus signs) and Japanese (circle) cars. MVAscacar

EXAMPLE 3.6 *Let us consider again the pullovers data set from example 3.2. Consider the correlation between the presence of the sales assistants (X_4) vs. the number of sold pullovers (X_1) (see Figure 3.4). Here we compute the correlation as*

$$r_{X_1X_4} = 0.633.$$

The Z-transform of this value is

$$w = \frac{1}{2}\log_e\left(\frac{1+r_{X_1X_4}}{1-r_{X_1X_4}}\right) = 0.746. \tag{3.14}$$

The sample size is $n = 10$, so for the hypothesis $\rho_{X_1X_4} = 0$, the statistic to consider is:

$$z = \sqrt{7}(0.746 - 0) = 1.974 \tag{3.15}$$

which is just statistically significant at the 5% level (i.e., 1.974 is just a little larger than 1.96).

REMARK 3.2 *The normalizing and variance stabilizing properties of W are asymptotic. In addition the use of W in small samples (for $n \leq 25$) is improved by Hotelling's transform (Hotelling, 1953):*

$$W^* = W - \frac{3W + \tanh(W)}{4(n-1)} \quad with \quad Var(W^*) = \frac{1}{n-1}.$$

The transformed variable W^* is asymptotically distributed as a normal distribution.

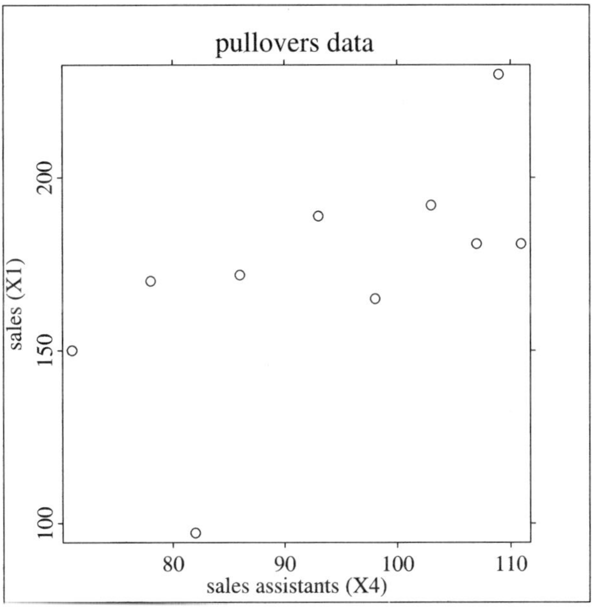

Figure 3.4. Hours of sales assistants (X_4) vs. sales (X_1) of pullovers. `MVAscapull2`

EXAMPLE 3.7 *From the preceding remark, we obtain* $w^* = 0.6663$ *and* $\sqrt{10-1}w^* = 1.9989$ *for the preceding Example 3.6. This value is significant at the* 5% *level.*

REMARK 3.3 *Note that the Fisher's Z-transform is the inverse of the hyperbolic tangent function:* $W = \tanh^{-1}(r_{XY})$*; equivalently* $r_{XY} = \tanh(W) = \frac{e^{2W}-1}{e^{2W}+1}$.

REMARK 3.4 *Under the assumptions of normality of* X *and* Y*, we may test their independence* ($\rho_{XY} = 0$) *using the exact* t*-distribution of the statistic*

$$T = r_{XY}\sqrt{\frac{n-2}{1-r_{XY}^2}} \overset{\rho_{XY}=0}{\sim} t_{n-2}.$$

Setting the probability of the first error type to α*, we reject the null hypothesis* $\rho_{XY} = 0$ *if* $|T| \geq t_{1-\alpha/2;n-2}$.

Summary

↪ The correlation is a standardized measure of dependence.

Summary (continued)
↪ The absolute value of the correlation is always less than one.
↪ Correlation measures only linear dependence.
↪ There are nonlinear dependencies that have zero correlation.
↪ Zero correlation does not imply independence.
↪ Independence implies zero correlation.
↪ Negative correlation corresponds to downward-sloping scatterplots.
↪ Positive correlation corresponds to upward-sloping scatterplots.
↪ Fisher's Z-transform helps us in testing hypotheses on correlation.
↪ For small samples, Fisher's Z-transform can be improved by the transformation $W^* = W - \frac{3W+\tanh(W)}{4(n-1)}$.

3.3 Summary Statistics

This section focuses on the representation of basic summary statistics (means, covariances and correlations) in matrix notation, since we often apply linear transformations to data. The matrix notation allows us to derive instantaneously the corresponding characteristics of the transformed variables. The Mahalanobis transformation is a prominent example of such linear transformations.

Assume that we have observed n realizations of a p-dimensional random variable; we have a data matrix $\mathcal{X}(n \times p)$:

$$\mathcal{X} = \begin{pmatrix} x_{11} & \cdots & x_{1p} \\ \vdots & & \vdots \\ \vdots & & \vdots \\ x_{n1} & \cdots & x_{np} \end{pmatrix}. \tag{3.16}$$

The rows $x_i = (x_{i1}, \ldots, x_{ip}) \in \mathbb{R}^p$ denote the i-th observation of a p-dimensional random variable $X \in \mathbb{R}^p$.

The statistics that were briefly introduced in Section 3.1 and 3.2 can be rewritten in matrix form as follows. The "center of gravity" of the n observations in $\mathbb{R}^p$ is given by the vector $\overline{x}$ of the means $\overline{x}_j$ of the p variables:

$$\overline{x} = \begin{pmatrix} \overline{x}_1 \\ \vdots \\ \overline{x}_p \end{pmatrix} = n^{-1}\mathcal{X}^\top 1_n. \tag{3.17}$$

The dispersion of the n observations can be characterized by the covariance matrix of the p variables. The empirical covariances defined in (3.2) and (3.3) are the elements of the following matrix:

$$\mathcal{S} = n^{-1}\mathcal{X}^\top\mathcal{X} - \overline{x}\,\overline{x}^\top = n^{-1}(\mathcal{X}^\top\mathcal{X} - n^{-1}\mathcal{X}^\top 1_n 1_n^\top \mathcal{X}). \tag{3.18}$$

Note that this matrix is equivalently defined by

$$\mathcal{S} = \frac{1}{n}\sum_{i=1}^{n}(x_i - \overline{x})(x_i - \overline{x})^\top.$$

The covariance formula (3.18) can be rewritten as $\mathcal{S} = n^{-1}\mathcal{X}^\top\mathcal{H}\mathcal{X}$ with the *centering matrix*

$$\mathcal{H} = \mathcal{I}_n - n^{-1}1_n1_n^\top. \tag{3.19}$$

Note that the centering matrix is symmetric and idempotent. Indeed,

$$\begin{aligned}\mathcal{H}^2 &= (\mathcal{I}_n - n^{-1}1_n1_n^\top)(\mathcal{I}_n - n^{-1}1_n1_n^\top)\\ &= \mathcal{I}_n - n^{-1}1_n1_n^\top - n^{-1}1_n1_n^\top + (n^{-1}1_n1_n^\top)(n^{-1}1_n1_n^\top)\\ &= \mathcal{I}_n - n^{-1}1_n1_n^\top = \mathcal{H}.\end{aligned}$$

As a consequence $\mathcal{S}$ is positive semidefinite, i.e.

$$\mathcal{S} \geq 0. \tag{3.20}$$

Indeed for all $a \in \mathbb{R}^p$,

$$\begin{aligned}a^\top\mathcal{S}a &= n^{-1}a^\top\mathcal{X}^\top\mathcal{H}\mathcal{X}a\\ &= n^{-1}(a^\top\mathcal{X}^\top\mathcal{H}^\top)(\mathcal{H}\mathcal{X}a) \quad \text{since } \mathcal{H}^\top\mathcal{H} = \mathcal{H},\\ &= n^{-1}y^\top y = n^{-1}\sum_{j=1}^{p} y_j^2 \geq 0\end{aligned}$$

for $y = \mathcal{H}\mathcal{X}a$. It is well known from the one-dimensional case that $n^{-1}\sum_{i=1}^{n}(x_i - \overline{x})^2$ as an estimate of the variance exhibits a bias of the order n^{-1} (Breiman, 1973). In the multidimensional case, $\mathcal{S}_u = \frac{n}{n-1}\mathcal{S}$ is an unbiased estimate of the true covariance. (This will be shown in Example 4.15.)

The sample correlation coefficient between the i-th and j-th variables is $r_{X_iX_j}$, see (3.8). If $\mathcal{D} = \text{diag}(s_{X_iX_i})$, then the correlation matrix is

$$\mathcal{R} = \mathcal{D}^{-1/2}\mathcal{S}\mathcal{D}^{-1/2}, \tag{3.21}$$

where $\mathcal{D}^{-1/2}$ is a diagonal matrix with elements $(s_{X_iX_i})^{-1/2}$ on its main diagonal.

EXAMPLE 3.8 *The empirical covariances are calculated for the pullover data set.*

The vector of the means of the four variables in the dataset is $\overline{x} = (172.7, 104.6, 104.0, 93.8)^\top$.

The sample covariance matrix is $\mathcal{S} = \begin{pmatrix} 1037.2 & -80.2 & 1430.7 & 271.4\\ -80.2 & 219.8 & 92.1 & -91.6\\ 1430.7 & 92.1 & 2624 & 210.3\\ 271.4 & -91.6 & 210.3 & 177.4 \end{pmatrix}$.

The unbiased estimate of the variance (n =10) is equal to

$$\mathcal{S}_u = \frac{10}{9}\mathcal{S} = \begin{pmatrix} 1152.5 & -88.9 & 1589.7 & 301.6\\ -88.9 & 244.3 & 102.3 & -101.8\\ 1589.7 & 102.3 & 2915.6 & 233.7\\ 301.6 & -101.8 & 233.7 & 197.1 \end{pmatrix}.$$

The sample correlation matrix is $\mathcal{R} = \begin{pmatrix} 1 & -0.17 & 0.87 & 0.63\\ -0.17 & 1 & 0.12 & -0.46\\ 0.87 & 0.12 & 1 & 0.31\\ 0.63 & -0.46 & 0.31 & 1 \end{pmatrix}$.

Linear Transformation

In many practical applications we need to study linear transformations of the original data. This motivates the question of how to calculate summary statistics after such linear transformations.

Let $\mathcal{A}$ be a $(q \times p)$ matrix and consider the transformed data matrix

$$\mathcal{Y} = \mathcal{X}\mathcal{A}^\top = (y_1, \dots, y_n)^\top. \tag{3.22}$$

The row $y_i = (y_{i1}, \dots, y_{iq}) \in \mathbb{R}^q$ can be viewed as the i-th observation of a q-dimensional random variable $Y = \mathcal{A}X$. In fact we have $y_i = x_i\mathcal{A}^\top$. We immediately obtain the mean and the empirical covariance of the variables (columns) forming the data matrix $\mathcal{Y}$:

$$\overline{y} = \frac{1}{n}\mathcal{Y}^\top 1_n = \frac{1}{n}\mathcal{A}\mathcal{X}^\top 1_n = \mathcal{A}\overline{x} \tag{3.23}$$

$$\mathcal{S}_\mathcal{Y} = \frac{1}{n}\mathcal{Y}^\top\mathcal{H}\mathcal{Y} = \frac{1}{n}\mathcal{A}\mathcal{X}^\top\mathcal{H}\mathcal{X}\mathcal{A}^\top = \mathcal{A}\mathcal{S}_\mathcal{X}\mathcal{A}^\top. \tag{3.24}$$

Note that if the linear transformation is nonhomogeneous, i.e.,

$$y_i = \mathcal{A}x_i + b \qquad \text{where} \quad b(q \times 1),$$

only (3.23) changes: $\overline{y} = \mathcal{A}\overline{x} + b$. The formula (3.23) and (3.24) are useful in the particular case of $q = 1$, i.e., $y = \mathcal{X}a \Leftrightarrow y_i = a^\top x_i; i = 1, \dots, n$:

$$\begin{aligned} \overline{y} &= a^\top\overline{x} \\ \mathcal{S}_y &= a^\top\mathcal{S}_\mathcal{X}a. \end{aligned}$$

EXAMPLE 3.9 *Suppose that $\mathcal{X}$ is the pullover data set. The manager wants to compute his mean expenses for advertisement (X_3) and sales assistant (X_4).*

Suppose that the sales assistant charges an hourly wage of 10 EUR. Then the shop manager calculates the expenses Y as $Y = X_3 + 10X_4$. Formula (3.22) says that this is equivalent to defining the matrix $\mathcal{A}(4 \times 1)$ as:

$$\mathcal{A} = (0, 0, 1, 10).$$

Using formulas (3.23) and (3.24), it is now computationally very easy to obtain the sample mean $\overline{y}$ and the sample variance $\mathcal{S}_y$ of the overall expenses:

$$\overline{y} = \mathcal{A}\overline{x} = (0, 0, 1, 10)\begin{pmatrix} 172.7 \\ 104.6 \\ 104.0 \\ 93.8 \end{pmatrix} = 1042.0$$

$$\mathcal{S}_\mathcal{Y} = \mathcal{A}\mathcal{S}_\mathcal{X}\mathcal{A}^\top = (0, 0, 1, 10)\begin{pmatrix} 1152.5 & -88.9 & 1589.7 & 301.6 \\ -88.9 & 244.3 & 102.3 & -101.8 \\ 1589.7 & 102.3 & 2915.6 & 233.7 \\ 301.6 & -101.8 & 233.7 & 197.1 \end{pmatrix}\begin{pmatrix} 0 \\ 0 \\ 1 \\ 10 \end{pmatrix}$$

$$= 2915.6 + 4674 + 19710 = 27299.6.$$

Mahalanobis Transformation

A special case of this linear transformation is

$$z_i = \mathcal{S}^{-1/2}(x_i - \overline{x}), \quad i = 1, \ldots, n. \tag{3.25}$$

Note that for the transformed data matrix $\mathcal{Z} = (z_1, \ldots, z_n)^\top$,

$$\mathcal{S}_{\mathcal{Z}} = n^{-1}\mathcal{Z}^\top \mathcal{H}\mathcal{Z} = \mathcal{I}_p. \tag{3.26}$$

So the Mahalanobis transformation eliminates the correlation between the variables and standardizes the variance of each variable. If we apply (3.24) using $\mathcal{A} = \mathcal{S}^{-1/2}$, we obtain the identity covariance matrix as indicated in (3.26).

Summary

- ↪ The center of gravity of a data matrix is given by its mean vector $\overline{x} = n^{-1}\mathcal{X}^\top 1_n$.
- ↪ The dispersion of the observations in a data matrix is given by the empirical covariance matrix $\mathcal{S} = n^{-1}\mathcal{X}^\top\mathcal{H}\mathcal{X}$.
- ↪ The empirical correlation matrix is given by $\mathcal{R} = \mathcal{D}^{-1/2}\mathcal{S}\mathcal{D}^{-1/2}$.
- ↪ A linear transformation $\mathcal{Y} = \mathcal{X}\mathcal{A}^\top$ of a data matrix $\mathcal{X}$ has mean $\mathcal{A}\overline{x}$ and empirical covariance $\mathcal{A}\mathcal{S}_{\mathcal{X}}\mathcal{A}^\top$.
- ↪ The Mahalanobis transformation is a linear transformation $z_i = \mathcal{S}^{-1/2}(x_i - \overline{x})$ which gives a standardized, uncorrelated data matrix $\mathcal{Z}$.

3.4 Linear Model for Two Variables

We have looked many times now at downward- and upward-sloping scatterplots. What does the eye define here as slope? Suppose that we can construct a line corresponding to the general direction of the cloud. The sign of the slope of this line would correspond to the upward and downward directions. Call the variable on the vertical axis Y and the one on the horizontal axis X. A slope line is a linear relationship between X and Y:

$$y_i = \alpha + \beta x_i + \varepsilon_i, \; i = 1, \ldots, n. \tag{3.27}$$

Here, α is the intercept and β is the slope of the line. The errors (or deviations from the line) are denoted as ε_i and are assumed to have zero mean and finite variance σ^2. The task of finding (α, β) in (3.27) is referred to as a linear adjustment.

In Section 3.6 we shall derive estimators for α and β more formally, as well as accurately describe what a "good" estimator is. For now, one may try to find a "good" estimator $(\widehat{\alpha}, \widehat{\beta})$ via graphical techniques. A very common numerical and statistical technique is to use those $\widehat{\alpha}$ and $\widehat{\beta}$ that minimize:

$$(\widehat{\alpha}, \widehat{\beta}) = \arg\min_{(\alpha,\beta)} \sum_{i=1}^{n} (y_i - \alpha - \beta x_i)^2. \tag{3.28}$$

The solutions to this task are the estimators:

$$\widehat{\beta} = \frac{s_{XY}}{s_{XX}} \tag{3.29}$$

$$\widehat{\alpha} = \overline{y} - \widehat{\beta}\overline{x}. \tag{3.30}$$

The variance of $\widehat{\beta}$ is:

$$Var(\widehat{\beta}) = \frac{\sigma^2}{n \cdot s_{XX}}. \tag{3.31}$$

The standard error (SE) of the estimator is the square root of (3.31),

$$SE(\widehat{\beta}) = \{Var(\widehat{\beta})\}^{1/2} = \frac{\sigma}{(n \cdot s_{XX})^{1/2}}. \tag{3.32}$$

We can use this formula to test the hypothesis that β=0. In an application the variance σ^2 has to be estimated by an estimator $\widehat{\sigma}^2$ that will be given below. Under a normality assumption of the errors, the t-test for the hypothesis $\beta = 0$ works as follows.

One computes the statistic

$$t = \frac{\widehat{\beta}}{SE(\widehat{\beta})} \tag{3.33}$$

and rejects the hypothesis at a 5% significance level if $\mid t \mid \geq t_{0.975;n-2}$, where the 97.5% quantile of the Student's t_{n-2} distribution is clearly the 95% critical value for the two-sided test. For $n \geq 30$, this can be replaced by 1.96, the 97.5% quantile of the normal distribution. An estimator $\widehat{\sigma}^2$ of σ^2 will be given in the following.

EXAMPLE 3.10 *Let us apply the linear regression model (3.27) to the "classic blue" pullovers. The sales manager believes that there is a strong dependence on the number of sales as a function of price. He computes the regression line as shown in Figure 3.5.*

How good is this fit? This can be judged via goodness-of-fit measures. Define

$$\widehat{y}_i = \widehat{\alpha} + \widehat{\beta}x_i, \tag{3.34}$$

as the predicted value of y as a function of x. With $\widehat{y}$ the textile shop manager in the above example can predict sales as a function of prices x. The variation in the response variable is:

$$ns_{YY} = \sum_{i=1}^{n}(y_i - \overline{y})^2. \tag{3.35}$$

The variation explained by the linear regression (3.27) with the predicted values (3.34) is:

$$\sum_{i=1}^{n}(\widehat{y}_i - \overline{y})^2. \tag{3.36}$$

The residual sum of squares, the minimum in (3.28), is given by:

$$RSS = \sum_{i=1}^{n}(y_i - \widehat{y}_i)^2. \tag{3.37}$$

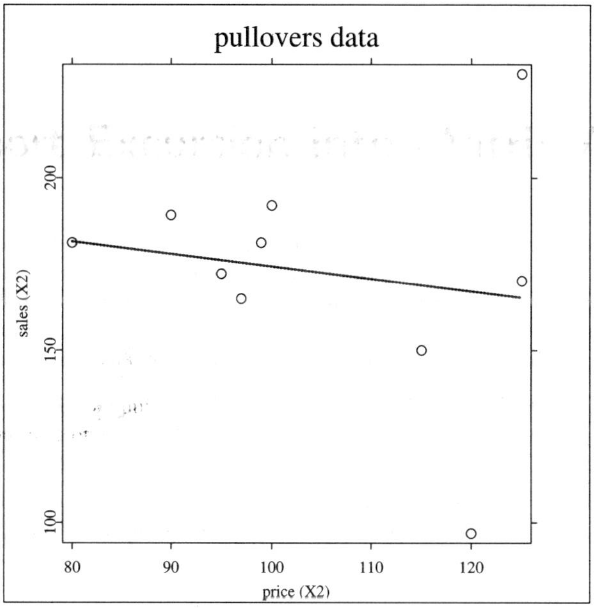

Figure 3.5. Regression of sales (X_1) on price (X_2) of pullovers. MVAregpull

An unbiased estimator $\widehat{\sigma}^2$ of σ^2 is given by $RSS/(n-2)$.
The following relation holds between (3.35)–(3.37):

$$\begin{aligned} \sum_{i=1}^{n}(y_i-\overline{y})^2 &= \sum_{i=1}^{n}(\widehat{y}_i-\overline{y})^2+\sum_{i=1}^{n}(y_i-\widehat{y}_i)^2, \\ \textit{total variation} &= \textit{explained variation} + \textit{unexplained variation}. \end{aligned} \tag{3.38}$$

The *coefficient of determination* is r^2:

$$r^2 = \frac{\sum_{i=1}^{n}(\widehat{y}_i-\overline{y})^2}{\sum_{i=1}^{n}(y_i-\overline{y})^2} = \frac{\textit{explained variation}}{\textit{total variation}}. \tag{3.39}$$

The coefficient of determination increases with the proportion of explained variation by the linear relation (3.27). In the extreme cases where $r^2 = 1$, all of the variation is explained by the linear regression (3.27). The other extreme, $r^2 = 0$, is where the empirical covariance is $s_{XY} = 0$. The coefficient of determination can be rewritten as

$$r^2 = 1 - \frac{\sum_{i=1}^{n}(y_i-\widehat{y}_i)^2}{\sum_{i=1}^{n}(y_i-\overline{y})^2}. \tag{3.40}$$

From (3.39), it can be seen that in the linear regression (3.27), $r^2 = r_{XY}^2$ is the square of the correlation between X and Y.

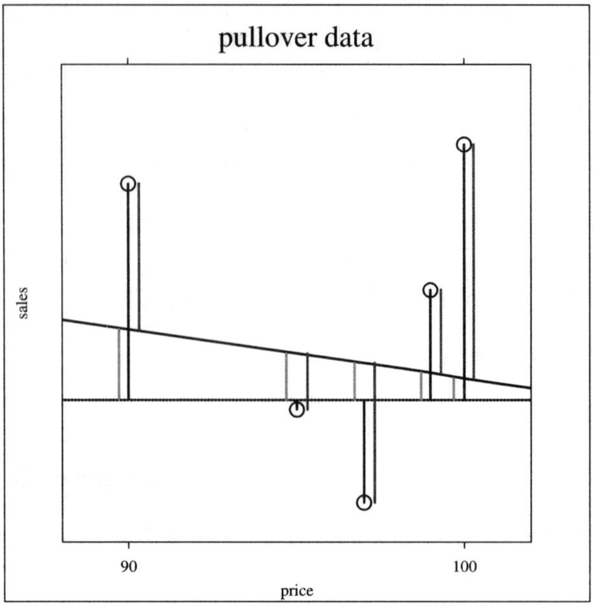

Figure 3.6. Regression of sales (X_1) on price (X_2) of pullovers. The overall mean is given by the dashed line. MVAregzoom

EXAMPLE 3.11 *For the above pullover example, we estimate*

$$\widehat{\alpha} = 210.774 \quad \textit{and} \quad \widehat{\beta} = -0.364.$$

The coefficient of determination is

$$r^2 = 0.028.$$

The textile shop manager concludes that sales are not influenced very much by the price (in a linear way).

The geometrical representation of formula (3.38) can be graphically evaluated using Figure 3.6. This plot shows a section of the linear regression of the "sales" on "price" for the pullovers data. The distance between any point and the overall mean is given by the distance between the point and the regression line and the distance between the regression line and the mean. The sums of these two distances represent the total variance (solid blue lines from the observations to the overall mean), i.e., the explained variance (distance from the regression curve to the mean) and the unexplained variance (distance from the observation to the regression line), respectively.

In general the regression of Y on X is different from that of X on Y. We will demonstrate this using once again the Swiss bank notes data.

EXAMPLE 3.12 *The least squares fit of the variables X_4 (X) and X_5 (Y) from the genuine bank notes are calculated. Figure 3.7 shows the fitted line if X_5 is approximated by a linear function of X_4. In this case the parameters are*

$$\widehat{\alpha} = 15.464 \quad \textit{and} \quad \widehat{\beta} = -0.638.$$

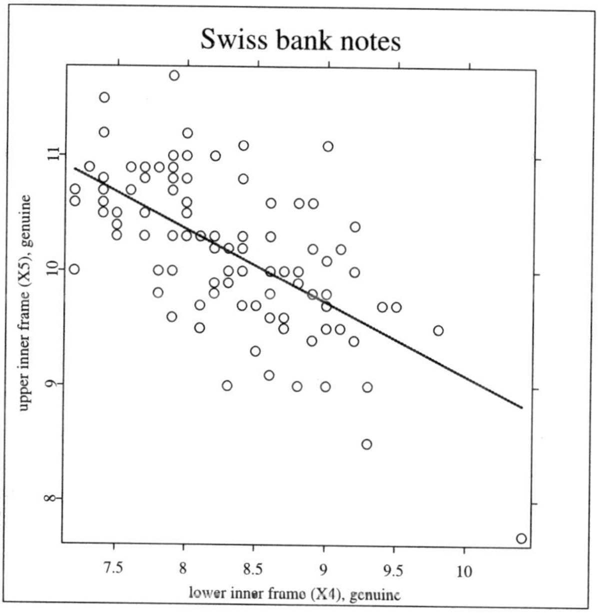

Figure 3.7. Regression of X_5 (upper inner frame) on X_4 (lower inner frame) for genuine bank notes. MVAregbank

If we predict X_4 by a function of X_5 instead, we would arrive at a different intercept and slope

$$\widehat{\alpha} = 14.666 \quad \textit{and} \quad \widehat{\beta} = -0.626.$$

The linear regression of Y on X is given by minimizing (3.28), i.e., the vertical errors ε_i. The linear regression of X on Y does the same but here the errors to be minimized in the least squares sense are measured horizontally. As seen in Example 3.12, the two least squares lines are different although both measure (in a certain sense) the slope of the cloud of points.

As shown in the next example, there is still one other way to measure the main direction of a cloud of points: it is related to the spectral decomposition of covariance matrices.

EXAMPLE 3.13 *Suppose that we have the following covariance matrix:*

$$\Sigma = \begin{pmatrix} 1 & \rho \\ \rho & 1 \end{pmatrix}.$$

Figure 3.8 shows a scatterplot of a sample of two normal random variables with such a covariance matrix (with $\rho = 0.8$).

The eigenvalues of Σ are, as was shown in Example 2.4, solutions to:

$$\begin{vmatrix} 1-\lambda & \rho \\ \rho & 1-\lambda \end{vmatrix} = 0.$$

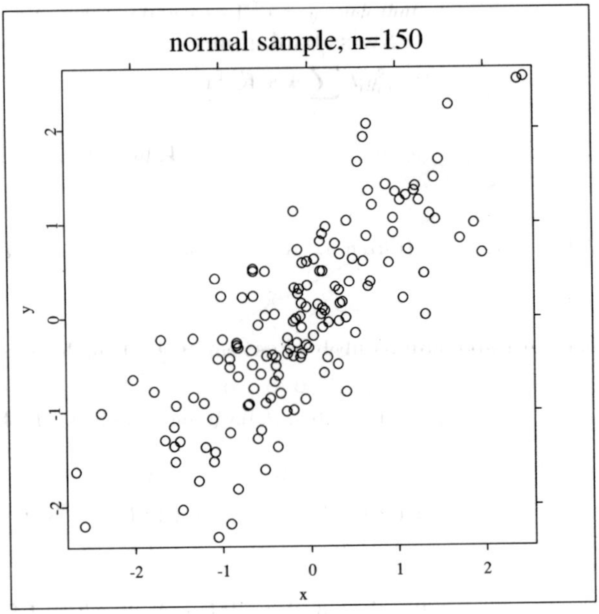

Figure 3.8. Scatterplot for a sample of two correlated normal random variables (sample size $n = 150$, $\rho = 0.8$). MVAcorrnorm

Hence, $\lambda_1 = 1 + \rho$ *and* $\lambda_2 = 1 - \rho$. *Therefore* $\Lambda = \operatorname{diag}(1 + \rho, 1 - \rho)$. *The eigenvector corresponding to* $\lambda_1 = 1 + \rho$ *can be computed from the system of linear equations:*

$$\begin{pmatrix} 1 & \rho \\ \rho & 1 \end{pmatrix} \begin{pmatrix} x_1 \\ x_2 \end{pmatrix} = (1 + \rho) \begin{pmatrix} x_1 \\ x_2 \end{pmatrix}$$

or

$$\begin{aligned} x_1 + \rho x_2 &= x_1 + \rho x_1 \\ \rho x_1 + x_2 &= x_2 + \rho x_2 \end{aligned}$$

and thus

$$x_1 = x_2.$$

The first (standardized) eigenvector is

$$\gamma_1 = \begin{pmatrix} 1/\sqrt{2} \\ 1/\sqrt{2} \end{pmatrix}.$$

The direction of this eigenvector is the diagonal in Figure 3.8 and captures the main variation in this direction. We shall come back to this interpretation in Chapter 9. The second eigenvector (orthogonal to γ_1*) is*

$$\gamma_2 = \begin{pmatrix} 1/\sqrt{2} \\ -1/\sqrt{2} \end{pmatrix}.$$

So finally

$$\Gamma = (\gamma_1, \gamma_2) = \begin{pmatrix} 1/\sqrt{2} & 1/\sqrt{2} \\ 1/\sqrt{2} & -1/\sqrt{2} \end{pmatrix}$$

and we can check our calculation by

$$\Sigma = \Gamma \, \Lambda \, \Gamma^{\top} \, .$$

The first eigenvector captures the main direction of a point cloud. The linear regression of Y on X and X on Y accomplished, in a sense, the same thing. In general the direction of the eigenvector and the least squares slope are different. The reason is that the least squares estimator minimizes either vertical or horizontal errors (in 3.28), whereas the first eigenvector corresponds to a minimization that is orthogonal to the eigenvector (see Chapter 9).

Summary

- $\hookrightarrow$ The linear regression $y = \alpha + \beta x + \varepsilon$ models a linear relation between two one-dimensional variables.
- $\hookrightarrow$ The sign of the slope $\widehat{\beta}$ is the same as that of the covariance and the correlation of x and y.
- $\hookrightarrow$ A linear regression predicts values of Y given a possible observation x of X.
- $\hookrightarrow$ The coefficient of determination r^2 measures the amount of variation in Y which is explained by a linear regression on X.
- $\hookrightarrow$ If the coefficient of determination is $r^2 = 1$, then all points lie on one line.
- $\hookrightarrow$ The regression line of X on Y and the regression line of Y on X are in general different.
- $\hookrightarrow$ The t-test for the hypothesis $\beta = 0$ is $t = \frac{\widehat{\beta}}{SE(\widehat{\beta})}$, where $SE(\widehat{\beta}) = \frac{\hat{\sigma}}{(n \cdot s_{XX})^{1/2}}$.
- $\hookrightarrow$ The t-test rejects the null hypothesis $\beta = 0$ at the level of significance α if $\mid t \mid \geq t_{1-\alpha/2;n-2}$ where $t_{1-\alpha;n-2}$ is the $1 - \alpha/2$ quantile of the Student's t-distribution with $(n-2)$ degrees of freedom.
- $\hookrightarrow$ The standard error $SE(\widehat{\beta})$ increases/decreases with less/more spread in the X variables.
- $\hookrightarrow$ The direction of the first eigenvector of the covariance matrix of a two-dimensional point cloud is different from the least squares regression line.

3.5 Simple Analysis of Variance

In a simple (i.e., one–factorial) analysis of variance (ANOVA), it is assumed that the average values of the response variable y are induced by one simple factor. Suppose that this factor takes on p values and that for each factor level, we have $m = n/p$ observations. The sample is of the form given in Table 3.1, where all of the observations are independent.

The goal of a simple ANOVA is to analyze the observation structure

$$y_{kl} = \mu_l + \varepsilon_{kl} \text{ for } k = 1, \ldots, m, \text{ and } l = 1, \ldots, p. \tag{3.41}$$

Each factor has a mean value μ_l. Each observation y_{kl} is assumed to be a sum of the corresponding factor mean value μ_l and a zero mean random error ε_{kl}. The linear regression model falls into this scheme with $m = 1$, $p = n$ and $\mu_i = \alpha + \beta x_i$, where x_i is the i-th level value of the factor.

EXAMPLE 3.14 *The "classic blue" pullover company analyzes the effect of three marketing strategies*

1 advertisement in local newspaper,
2 presence of sales assistant,
3 luxury presentation in shop windows.

All of these strategies are tried in 10 different shops. The resulting sale observations are given in Table 3.2.

There are $p = 3$ factors and $n = mp = 30$ observations in the data. The "classic blue" pullover company wants to know whether all three marketing strategies have the same mean effect or whether there are differences. Having the same effect means that all μ_l in (3.41) equal one value, μ. The hypothesis to be tested is therefore

$$H_0 : \mu_l = \mu \text{ for } l = 1, \dots, p.$$

The alternative hypothesis, that the marketing strategies have different effects, can be formulated as

$$H_1 : \mu_l \neq \mu_{l'} \text{ for some } l \text{ and } l'.$$

This means that one marketing strategy is better than the others.

sample element	factor levels l				
1	y_{11}	$\cdots$	y_{1l}	$\cdots$	y_{1p}
2	$\vdots$		$\vdots$		$\vdots$
$\vdots$	$\vdots$		$\vdots$		$\vdots$
k	y_{k1}	$\cdots$	y_{kl}	$\cdots$	y_{kp}
$\vdots$	$\vdots$		$\vdots$		$\vdots$
$m = n/p$	y_{m1}	$\cdots$	y_{ml}	$\cdots$	y_{mp}

Table 3.1. Observation structure of a simple ANOVA.

shop k	marketing strategy *factor* l		
	1	2	3
1	9	10	18
2	11	15	14
3	10	11	17
4	12	15	9
5	7	15	14
6	11	13	17
7	12	7	16
8	10	15	14
9	11	13	17
10	13	10	15

Table 3.2. Pullover sales as function of marketing strategy.

The method used to test this problem is to compute as in (3.38) the total variation and to decompose it into the sources of variation. This gives:

$$\sum_{l=1}^{p}\sum_{k=1}^{m}(y_{kl}-\bar{y})^2 = m\sum_{l=1}^{p}(\bar{y}_l-\bar{y})^2 + \sum_{l=1}^{p}\sum_{k=1}^{m}(y_{kl}-\bar{y}_l)^2 \tag{3.42}$$

The total variation (sum of squares=SS) is:

$$SS(\text{reduced}) = \sum_{l=1}^{p}\sum_{k=1}^{m}(y_{kl}-\bar{y})^2 \tag{3.43}$$

where $\bar{y} = n^{-1}\sum_{l=1}^{p}\sum_{k=1}^{m} y_{kl}$ is the overall mean. Here the total variation is denoted as $SS(\text{reduced})$, since in comparison with the model under the alternative H_1, we have a reduced set of parameters. In fact there is 1 parameter $\mu = \mu_l$ under H_0. Under H_1, the "full" model, we have three parameters, namely the three different means μ_l.

The variation under H_1 is therefore:

$$SS(\text{full}) = \sum_{l=1}^{p}\sum_{k=1}^{m}(y_{kl}-\bar{y}_l)^2 \tag{3.44}$$

where $\bar{y}_l = m^{-1}\sum_{k=1}^{m} y_{kl}$ is the mean of each factor l. The hypothetical model H_0 is called reduced, since it has (relative to H_1) fewer parameters.

The F-test of the linear hypothesis is used to compare the difference in the variations under the reduced model H_0 (3.43) and the full model H_1 (3.44) to the variation under the full model H_1:

$$F = \frac{\{SS(\text{reduced}) - SS(\text{full})\}/\{df(r) - df(f)\}}{SS(\text{full})/df(f)}. \tag{3.45}$$

Here $df(f)$ and $df(r)$ denote the degrees of freedom under the full model and the reduced model respectively. The degrees of freedom are essential in specifying the shape of the F-distribution. They have a simple interpretation: $df(\cdot)$ is equal to the number of observations minus the number of parameters in the model.

From Example 3.14, $p = 3$ parameters are estimated under the full model, i.e., $df(f) = n - p = 30 - 3 = 27$. Under the reduced model, there is one parameter to estimate, namely the overall mean, i.e., $df(r) = n - 1 = 29$. We can compute

$$SS(\text{reduced}) = 260.3$$

and

$$SS(\text{full}) = 157.7.$$

The F-statistic (3.45) is therefore

$$F = \frac{(260.3 - 157.7)/2}{157.7/27} = 8.78.$$

This value needs to be compared to the quantiles of the $F_{2,27}$ distribution. Looking up the critical values in a F-distribution shows that the test statistic above is highly significant. We conclude that the marketing strategies have different effects.

The F-test in a linear regression model

The t-test of a linear regression model can be put into this framework. For a linear regression model (3.27), the reduced model is the one with $\beta = 0$:

$$y_i = \alpha + 0 \cdot x_i + \varepsilon_i.$$

The reduced model has $n-1$ degrees of freedom and one parameter, the intercept α.

The full model is given by $\beta \neq 0$,

$$y_i = \alpha + \beta \cdot x_i + \varepsilon_i,$$

and has $n-2$ degrees of freedom, since there are two parameters (α, β).

The SS(reduced) equals

$$SS(\text{reduced}) = \sum_{i=1}^{n} (y_i - \bar{y})^2 = \textit{total variation}.$$

The SS(full) equals

$$SS(\text{full}) = \sum_{i=1}^{n} (y_i - \hat{y}_i)^2 = \text{RSS} = \textit{unexplained variation}.$$

The F-test is therefore, from (3.45),

$$F = \frac{(\textit{total variation - unexplained variation})\,/1}{(\textit{unexplained variation})/(n-2)} \tag{3.46}$$

$$= \frac{\textit{explained variation}}{(\textit{unexplained variation})/(n-2)}. \tag{3.47}$$

Using the estimators $\hat{\alpha}$ and $\widehat{\beta}$ the explained variation is:

$$\begin{aligned}
\sum_{i=1}^{n} (\hat{y}_i - \bar{y})^2 &= \sum_{i=1}^{n} \left(\hat{\alpha} + \widehat{\beta} x_i - \bar{y}\right)^2 \\
&= \sum_{i=1}^{n} \left\{(\bar{y} - \widehat{\beta}\bar{x}) + \widehat{\beta} x_i - \bar{y}\right\}^2 \\
&= \sum_{i=1}^{n} \widehat{\beta}^2 (x_i - \bar{x})^2 \\
&= \widehat{\beta}^2 n s_{XX}.
\end{aligned}$$

From (3.32) the F-ratio (3.46) is therefore:

$$F = \frac{\widehat{\beta}^2 n s_{XX}}{RSS/(n-2)} \tag{3.48}$$

$$= \left(\frac{\widehat{\beta}}{SE(\widehat{\beta})}\right)^2. \tag{3.49}$$

The t-test statistic (3.33) is just the square root of the F- statistic (3.49).

Note, using (3.39) the F-statistic can be rewritten as

$$F = \frac{r^2/1}{(1-r^2)/(n-2)}.$$

In the pullover Example 3.11, we obtain $F = \frac{0.028}{0.972}\frac{8}{1} = 0.2305$, so that the null hypothesis $\beta = 0$ cannot be rejected. We conclude therefore that there is only a minor influence of prices on sales.

Summary

- ↪ Simple ANOVA models an output Y as a function of one factor.
- ↪ The reduced model is the hypothesis of equal means.
- ↪ The full model is the alternative hypothesis of different means.
- ↪ The F-test is based on a comparison of the sum of squares under the full and the reduced models.
- ↪ The degrees of freedom are calculated as the number of observations minus the number of parameters.
- ↪ The F-statistic is

$$F = \frac{\{SS(\text{reduced}) - SS(\text{full})\}/\{df(r) - df(f)\}}{SS(\text{full})/df(f)}.$$

- ↪ The F-test rejects the null hypothesis if the F-statistic is larger than the 95% quantile of the $F_{df(r)-df(f),df(f)}$ distribution.
- ↪ The F-test statistic for the slope of the linear regression model $y_i = \alpha + \beta x_i + \varepsilon_i$ is the square of the t-test statistic.

3.6 Multiple Linear Model

The simple linear model and the analysis of variance model can be viewed as a particular case of a more general linear model where the variations of one variable y are explained by p explanatory variables x respectively. Let $y\,(n \times 1)$ and $\mathcal{X}\,(n \times p)$ be a vector of observations on the response variable and a data matrix on the p explanatory variables. An important application of the developed theory is the least squares fitting. The idea is to approximate y by a linear combination $\widehat{y}$ of columns of $\mathcal{X}$, i.e., $\widehat{y} \in C(\mathcal{X})$. The problem is to find $\widehat{\beta} \in \mathbb{R}^p$ such that $\widehat{y} = \mathcal{X}\widehat{\beta}$ is the best fit of y in the least-squares sense. The linear model can be written as

$$y = \mathcal{X}\beta + \varepsilon, \tag{3.50}$$

where ε are the errors. The least squares solution is given by $\widehat{\beta}$:

$$\widehat{\beta} = \arg\min_{\beta}\, (y - \mathcal{X}\beta)^{\top}(y - \mathcal{X}\beta) = \arg\min_{\beta}\, \varepsilon^{\top}\varepsilon. \tag{3.51}$$

Suppose that $(\mathcal{X}^\top\mathcal{X})$ is of full rank and thus invertible. Minimizing the expression (3.51) with respect to β yields:

$$\widehat{\beta} = (\mathcal{X}^\top\mathcal{X})^{-1}\mathcal{X}^\top y. \tag{3.52}$$

The fitted value $\widehat{y} = \mathcal{X}\widehat{\beta} = \mathcal{X}(\mathcal{X}^\top\mathcal{X})^{-1}\mathcal{X}^\top y = \mathcal{P}y$ is the projection of y onto $C(\mathcal{X})$ as computed in (2.47).

The least squares residuals are

$$e = y - \widehat{y} = y - \mathcal{X}\widehat{\beta} = \mathcal{Q}y = (\mathcal{I}_n - \mathcal{P})y.$$

The vector e is the projection of y onto the orthogonal complement of $C(\mathcal{X})$.

REMARK 3.5 *A linear model with an intercept α can also be written in this framework. The approximating equation is:*

$$y_i = \alpha + \beta_1 x_{i1} + \ldots + \beta_p x_{ip} + \varepsilon_i\ ;\ i = 1, \ldots, n.$$

This can be written as:

$$y = \mathcal{X}^*\beta^* + \varepsilon$$

where $\mathcal{X}^* = (1_n\ \mathcal{X})$ (we add a column of ones to the data). We have by (3.52):

$$\widehat{\beta}^* = \begin{pmatrix} \widehat{\alpha} \\ \widehat{\beta} \end{pmatrix} = (\mathcal{X}^{*\top}\mathcal{X}^*)^{-1}\mathcal{X}^{*\top}y.$$

EXAMPLE 3.15 *Let us come back to the "classic blue" pullovers example. In Example 3.11, we considered the regression fit of the sales X_1 on the price X_2 and concluded that there was only a small influence of sales by changing the prices. A linear model incorporating all three variables allows us to approximate sales as a linear function of price (X_2), advertisement (X_3) and presence of sales assistants (X_4) simultaneously. Adding a column of ones to the data (in order to estimate the intercept α) leads to*

$$\widehat{\alpha} = 65.670 \text{ and } \widehat{\beta}_1 = -0.216,\ \widehat{\beta}_2 = 0.485,\ \widehat{\beta}_3 = 0.844.$$

The coefficient of determination is computed as before in (3.40) and is:

$$r^2 = 1 - \frac{e^\top e}{\sum (y_i - \overline{y})^2} = 0.907.$$

We conclude that the variation of X_1 is well approximated by the linear relation.

REMARK 3.6 *The coefficient of determination is influenced by the number of regressors. For a given sample size n, the r^2 value will increase by adding more regressors into the linear model. The value of r^2 may therefore be high even if possibly irrelevant regressors are included. A corrected coefficient of determination for p regressors and a constant intercept ($p+1$ parameters) is*

$$r^2_{adj} = r^2 - \frac{p(1-r^2)}{n-(p+1)}. \tag{3.53}$$

EXAMPLE 3.16 *The corrected coefficient of determination for Example 3.15 is*

$$\begin{aligned} r^2_{adj} &= 0.907 - \frac{3(1-0.907^2)}{10-3-1} \\ &= 0.818. \end{aligned}$$

This means that 81.8% of the variation of the response variable is explained by the explanatory variables.

Note that the linear model (3.50) is very flexible and can model nonlinear relationships between the response y and the explanatory variables x. For example, a quadratic relation in one variable x could be included. Then $y_i = \alpha + \beta_1 x_i + \beta_2 x_i^2 + \varepsilon_i$ could be written in matrix notation as in (3.50), $y = \mathcal{X}\beta + \varepsilon$ where

$$\mathcal{X} = \begin{pmatrix} 1 & x_1 & x_1^2 \\ 1 & x_2 & x_2^2 \\ \vdots & \vdots & \vdots \\ 1 & x_n & x_n^2 \end{pmatrix}.$$

Properties of $\widehat{\beta}$

When y_i is the i-th observation of a random variable Y, the errors are also random. Under standard assumptions (independence, zero mean and constant variance σ^2), inference can be conducted on β. Using the properties of Chapter 4, it is easy to prove:

$$E(\widehat{\beta}) = \beta$$
$$\text{Var}(\widehat{\beta}) = \sigma^2(\mathcal{X}^\top\mathcal{X})^{-1}.$$

The analogue of the t-test for the multivariate linear regression situation is

$$t = \frac{\widehat{\beta}_j}{SE(\widehat{\beta}_j)}.$$

The standard error of each coefficient $\widehat{\beta}_j$ is given by the square root of the diagonal elements of the matrix $Var(\widehat{\beta})$. In standard situations, the variance σ^2 of the error ε is not known. One may estimate it by

$$\hat{\sigma}^2 = \frac{1}{n-(p+1)}(y-\hat{y})^\top(y-\hat{y}),$$

where $(p+1)$ is the dimension of β. In testing $\beta_j = 0$ we reject the hypothesis at the significance level α if $|t| \geq t_{1-\alpha/2;n-(p+1)}$. More general issues on testing linear models are addressed in Chapter 7.

The ANOVA Model in Matrix Notation

The simple ANOVA problem (Section 3.5) may also be rewritten in matrix terms. Recall the definition of a vector of ones from (2.1) and define a vector of zeros as 0_n. Then construct the following $(n \times p)$ matrix, (here $p = 3$),

$$\mathcal{X} = \begin{pmatrix} 1_m & 0_m & 0_m \\ 0_m & 1_m & 0_m \\ 0_m & 0_m & 1_m \end{pmatrix}, \tag{3.54}$$

where $m = 10$. Equation (3.41) then reads as follows.

The parameter vector is $\beta = (\mu_1, \mu_2, \mu_3)^\top$. The data set from Example 3.14 can therefore be written as a linear model $y = \mathcal{X}\beta + \varepsilon$ where $y \in \mathbb{R}^n$ with $n = m \cdot p$ is the stacked vector of the columns of Table 3.1. The projection into the column space $C(\mathcal{X})$ of (3.54) yields the least-squares estimator $\hat{\beta} = (\mathcal{X}^\top\mathcal{X})^{-1}\mathcal{X}^\top y$. Note that $(\mathcal{X}^\top\mathcal{X})^{-1} = (1/10)\mathcal{I}_3$ and that $\mathcal{X}^\top y = (106, 124, 151)^\top$ is the sum $\sum_{k=1}^{m} y_{kj}$ for each factor, i.e., the 3 column sums of Table 3.1.

The least squares estimator is therefore the vector $\hat{\beta}_{H_1} = (\hat{\mu}_1, \hat{\mu}_2, \hat{\mu}_3) = (10.6, 12.4, 15.1)^\top$ of sample means for each factor level $j = 1, 2, 3$. Under the null hypothesis of equal mean values $\mu_1 = \mu_2 = \mu_3 = \mu$, we estimate the parameters under the same constraints. This can be put into the form of a linear constraint:

$$\begin{aligned} -\mu_1 + \mu_2 &= 0 \\ -\mu_1 + \mu_3 &= 0. \end{aligned}$$

This can be written as $\mathcal{A}\beta = a$, where

$$a = \begin{pmatrix} 0 \\ 0 \end{pmatrix}$$

and

$$\mathcal{A} = \begin{pmatrix} -1 & 1 & 0 \\ -1 & 0 & 1 \end{pmatrix}.$$

The constrained least-squares solution can be shown (Exercise 3.24) to be given by:

$$\hat{\beta}_{H_0} = \hat{\beta}_{H_1} - (\mathcal{X}^\top\mathcal{X})^{-1}\mathcal{A}^\top\{\mathcal{A}(\mathcal{X}^\top\mathcal{X})^{-1}\mathcal{A}^\top\}^{-1}(\mathcal{A}\hat{\beta}_{H_1} - a). \tag{3.55}$$

It turns out that (3.55) amounts to simply calculating the overall mean $\bar{y} = 12.7$ of the response variable y: $\hat{\beta}_{H_0} = (12.7, 12.7, 12.7)^\top$.

The F-test that has already been applied in Example 3.14 can be written as

$$F = \frac{\{||y - \mathcal{X}\hat{\beta}_{H_0}||^2 - ||y - \mathcal{X}\hat{\beta}_{H_1}||^2\}/2}{||y - \mathcal{X}\hat{\beta}_{H_1}||^2/27} \tag{3.56}$$

which gives the same significant value 8.78. Note that again we compare the RSS_{H_0} of the reduced model to the RSS_{H_1} of the full model. It corresponds to comparing the lengths of projections into different column spaces. This general approach in testing linear models is described in detail in Chapter 7.

Summary

- ↪ The relation $y = \mathcal{X}\beta + e$ models a linear relation between a one-dimensional variable Y and a p-dimensional variable X. $\mathcal{P}y$ gives the best linear regression fit of the vector y onto $C(\mathcal{X})$. The least squares parameter estimator is $\widehat{\beta} = (\mathcal{X}^\top\mathcal{X})^{-1}\mathcal{X}^\top y$.
- ↪ The simple ANOVA model can be written as a linear model.
- ↪ The ANOVA model can be tested by comparing the length of the projection vectors.
- ↪ The test statistic of the F-Test can be written as

$$\frac{\{||y - \mathcal{X}\hat{\beta}_{H_0}||^2 - ||y - \mathcal{X}\hat{\beta}_{H_1}||^2\}/\{df(r) - df(f)\}}{||y - \mathcal{X}\hat{\beta}_{H_1}||^2/df(f)}.$$

Summary (continued)

↪ The adjusted coefficient of determination is

$$r^2_{\text{adj}} = r^2 - \frac{p(1-r^2)}{n-(p+1)}.$$

3.7 Boston Housing

X	$\overline{x}$	median(X)	$Var(X)$	std(X)
X_1	3.61	0.26	73.99	8.60
X_2	11.36	0.00	543.94	23.32
X_3	11.14	9.69	47.06	6.86
X_4	0.07	0.00	0.06	0.25
X_5	0.55	0.54	0.01	0.12
X_6	6.28	6.21	0.49	0.70
X_7	68.57	77.50	792.36	28.15
X_8	3.79	3.21	4.43	2.11
X_9	9.55	5.00	75.82	8.71
X_{10}	408.24	330.00	28405.00	168.54
X_{11}	18.46	19.05	4.69	2.16
X_{12}	356.67	391.44	8334.80	91.29
X_{13}	12.65	11.36	50.99	7.14
X_{14}	22.53	21.20	84.59	9.20

Table 3.3. Descriptive statistics for the Boston Housing data set. MVAdescbh

The main statistics presented so far can be computed for the data matrix $\mathcal{X}(506 \times 14)$ from our Boston Housing data set. The sample means and the sample medians of each variable are displayed in Table 3.3. The table also provides the unbiased estimates of the variance of each variable and the corresponding standard deviations. The comparison of the means and the medians confirms the assymmetry of the components of $\mathcal{X}$ that was pointed out in Section 1.8.

The (unbiased) sample covariance matrix is given by the following (14×14) matrix $\mathcal{S}_n$:

$$\begin{pmatrix}
73.99 & -40.22 & 23.99 & -0.12 & 0.42 & -1.33 & 85.41 & -6.88 & 46.85 & 844.82 & 5.40 & -302.38 & 27.99 & -30.72 \\
-40.22 & 543.94 & -85.41 & -0.25 & -1.40 & 5.11 & -373.90 & 32.63 & -63.35 & -1236.45 & -19.78 & 373.72 & -68.78 & 77.32 \\
23.99 & -85.41 & 47.06 & 0.11 & 0.61 & -1.89 & 124.51 & -10.23 & 35.55 & 833.36 & 5.69 & -223.58 & 29.58 & -30.52 \\
-0.12 & -0.25 & 0.11 & 0.06 & 0.00 & 0.02 & 0.62 & -0.05 & -0.02 & -1.52 & -0.07 & 1.13 & -0.10 & 0.41 \\
0.42 & -1.40 & 0.61 & 0.00 & 0.01 & -0.02 & 2.39 & -0.19 & 0.62 & 13.05 & 0.05 & -4.02 & 0.49 & -0.46 \\
-1.33 & 5.11 & -1.89 & 0.02 & -0.02 & 0.49 & -4.75 & 0.30 & -1.28 & -34.58 & -0.54 & 8.22 & -3.08 & 4.49 \\
85.41 & -373.90 & 124.51 & 0.62 & 2.39 & -4.75 & 792.36 & -44.33 & 111.77 & 2402.69 & 15.94 & -702.94 & 121.08 & -97.59 \\
-6.88 & 32.63 & -10.23 & -0.05 & -0.19 & 0.30 & -44.33 & 4.43 & -9.07 & -189.66 & -1.06 & 56.04 & -7.47 & 4.84 \\
46.85 & -63.35 & 35.55 & -0.02 & 0.62 & -1.28 & 111.77 & -9.07 & 75.82 & 1335.76 & 8.76 & -353.28 & 30.39 & -30.56 \\
844.82 & -1236.45 & 833.36 & -1.52 & 13.05 & -34.58 & 2402.69 & -189.66 & 1335.76 & 28404.76 & 168.15 & -6797.91 & 654.71 & -726.26 \\
5.40 & -19.78 & 5.69 & -0.07 & 0.05 & -0.54 & 15.94 & -1.06 & 8.76 & 168.15 & 4.69 & -35.06 & 5.78 & -10.11 \\
-302.38 & 373.72 & -223.58 & 1.13 & -4.02 & 8.22 & -702.94 & 56.04 & -353.28 & -6797.91 & -35.06 & 8334.75 & -238.67 & 279.99 \\
27.99 & -68.78 & 29.58 & -0.10 & 0.49 & -3.08 & 121.08 & -7.47 & 30.39 & 654.71 & 5.78 & -238.67 & 50.99 & -48.45 \\
-30.72 & 77.32 & -30.52 & 0.41 & -0.46 & 4.49 & -97.59 & 4.84 & -30.56 & -726.26 & -10.11 & 279.99 & -48.45 & 84.59
\end{pmatrix},$$

and the corresponding correlation matrix $\mathcal{R}(14 \times 14)$ is:

$$\begin{pmatrix}
1.00 & -0.20 & 0.41 & -0.06 & 0.42 & -0.22 & 0.35 & -0.38 & 0.63 & 0.58 & 0.29 & -0.39 & 0.46 & -0.39 \\
-0.20 & 1.00 & -0.53 & -0.04 & -0.52 & 0.31 & -0.57 & 0.66 & -0.31 & -0.31 & -0.39 & 0.18 & -0.41 & 0.36 \\
0.41 & -0.53 & 1.00 & 0.06 & 0.76 & -0.39 & 0.64 & -0.71 & 0.60 & 0.72 & 0.38 & -0.36 & 0.60 & -0.48 \\
-0.06 & -0.04 & 0.06 & 1.00 & 0.09 & 0.09 & 0.09 & -0.10 & -0.01 & -0.04 & -0.12 & 0.05 & -0.05 & 0.18 \\
0.42 & -0.52 & 0.76 & 0.09 & 1.00 & -0.30 & 0.73 & -0.77 & 0.61 & 0.67 & 0.19 & -0.38 & 0.59 & -0.43 \\
-0.22 & 0.31 & -0.39 & 0.09 & -0.30 & 1.00 & -0.24 & 0.21 & -0.21 & -0.29 & -0.36 & 0.13 & -0.61 & 0.70 \\
0.35 & -0.57 & 0.64 & 0.09 & 0.73 & -0.24 & 1.00 & -0.75 & 0.46 & 0.51 & 0.26 & -0.27 & 0.60 & -0.38 \\
-0.38 & 0.66 & -0.71 & -0.10 & -0.77 & 0.21 & -0.75 & 1.00 & -0.49 & -0.53 & -0.23 & 0.29 & -0.50 & 0.25 \\
0.63 & -0.31 & 0.60 & -0.01 & 0.61 & -0.21 & 0.46 & -0.49 & 1.00 & 0.91 & 0.46 & -0.44 & 0.49 & -0.38 \\
0.58 & -0.31 & 0.72 & -0.04 & 0.67 & -0.29 & 0.51 & -0.53 & 0.91 & 1.00 & 0.46 & -0.44 & 0.54 & -0.47 \\
0.29 & -0.39 & 0.38 & -0.12 & 0.19 & -0.36 & 0.26 & -0.23 & 0.46 & 0.46 & 1.00 & -0.18 & 0.37 & -0.51 \\
-0.39 & 0.18 & -0.36 & 0.05 & -0.38 & 0.13 & -0.27 & 0.29 & -0.44 & -0.44 & -0.18 & 1.00 & -0.37 & 0.33 \\
0.46 & -0.41 & 0.60 & -0.05 & 0.59 & -0.61 & 0.60 & -0.50 & 0.49 & 0.54 & 0.37 & -0.37 & 1.00 & -0.74 \\
-0.39 & 0.36 & -0.48 & 0.18 & -0.43 & 0.70 & -0.38 & 0.25 & -0.38 & -0.47 & -0.51 & 0.33 & -0.74 & 1.00
\end{pmatrix}.$$

Analyzing $\mathcal{R}$ confirms most of the comments made from examining the scatterplot matrix in Chapter 1. In particular, the correlation between X_{14} (the value of the house) and all the other variables is given by the last row (or column) of $\mathcal{R}$. The highest correlations (in absolute values) are in decreasing order $X_{13}, X_6, X_{11}, X_{10}$, etc.

Using the Fisher's Z-transform on each of the correlations between X_{14} and the other variables would confirm that all are significantly different from zero, except the correlation between X_{14} and X_4 (the indicator variable for the Charles River). We know, however, that the correlation and Fisher's Z-transform are not appropriate for binary variable.

The same descriptive statistics can be calculated for the transformed variables (transformations were motivated in Section 1.8). The results are given in Table 3.4 and as can be seen

$\widetilde{X}$	$\widetilde{\overline{x}}$	median$(\widetilde{X})$	$Var(\widetilde{X})$	std$(\widetilde{X})$
$\widetilde{X}_1$	−0.78	−1.36	4.67	2.16
$\widetilde{X}_2$	1.14	0.00	5.44	2.33
$\widetilde{X}_3$	2.16	2.27	0.60	0.78
$\widetilde{X}_4$	0.07	0.00	0.06	0.25
$\widetilde{X}_5$	−0.61	−0.62	0.04	0.20
$\widetilde{X}_6$	1.83	1.83	0.01	0.11
$\widetilde{X}_7$	5.06	5.29	12.72	3.57
$\widetilde{X}_8$	1.19	1.17	0.29	0.54
$\widetilde{X}_9$	1.87	1.61	0.77	0.87
$\widetilde{X}_{10}$	5.93	5.80	0.16	0.40
$\widetilde{X}_{11}$	2.15	2.04	1.86	1.36
$\widetilde{X}_{12}$	3.57	3.91	0.83	0.91
$\widetilde{X}_{13}$	3.42	3.37	0.97	0.99
$\widetilde{X}_{14}$	3.03	3.05	0.17	0.41

Table 3.4. Descriptive statistics for the Boston Housing data set after the transformation. `MVAdescbh`

most of the variables are now more symmetric. Note that the covariances and the correla-

Variable	$\hat{\beta}_j$	$SE(\hat{\beta}_j)$	t	p-value
constant	4.1769	0.3790	11.020	0.0000
$\widetilde{X}_1$	−0.0146	0.0117	−1.254	0.2105
$\widetilde{X}_2$	0.0014	0.0056	0.247	0.8051
$\widetilde{X}_3$	−0.0127	0.0223	−0.570	0.5692
$\widetilde{X}_4$	0.1100	0.0366	3.002	0.0028
$\widetilde{X}_5$	−0.2831	0.1053	−2.688	0.0074
$\widetilde{X}_6$	0.4211	0.1102	3.822	0.0001
$\widetilde{X}_7$	0.0064	0.0049	1.317	0.1885
$\widetilde{X}_8$	−0.1832	0.0368	−4.977	0.0000
$\widetilde{X}_9$	0.0684	0.0225	3.042	0.0025
$\widetilde{X}_{10}$	−0.2018	0.0484	−4.167	0.0000
$\widetilde{X}_{11}$	−0.0400	0.0081	−4.946	0.0000
$\widetilde{X}_{12}$	0.0445	0.0115	3.882	0.0001
$\widetilde{X}_{13}$	−0.2626	0.0161	−16.320	0.0000

Table 3.5. Linear regression results for all variables of Boston Housing data set. MVAlinregbh

tions are sensitive to these nonlinear transformations. For example, the correlation matrix is now

$$\begin{pmatrix}
1.00 & -0.52 & 0.74 & 0.03 & 0.81 & -0.32 & 0.70 & -0.74 & 0.84 & 0.81 & 0.45 & -0.48 & 0.62 & -0.57 \\
-0.52 & 1.00 & -0.66 & -0.04 & -0.57 & 0.31 & -0.53 & 0.59 & -0.35 & -0.31 & -0.35 & 0.18 & -0.45 & 0.36 \\
0.74 & -0.66 & 1.00 & 0.08 & 0.75 & -0.43 & 0.66 & -0.73 & 0.58 & 0.66 & 0.46 & -0.33 & 0.62 & -0.55 \\
0.03 & -0.04 & 0.08 & 1.00 & 0.08 & 0.08 & 0.07 & -0.09 & 0.01 & -0.04 & -0.13 & 0.05 & -0.06 & 0.16 \\
0.81 & -0.57 & 0.75 & 0.08 & 1.00 & -0.32 & 0.78 & -0.86 & 0.61 & 0.67 & 0.34 & -0.38 & 0.61 & -0.52 \\
-0.32 & 0.31 & -0.43 & 0.08 & -0.32 & 1.00 & -0.28 & 0.28 & -0.21 & -0.31 & -0.32 & 0.13 & -0.64 & 0.61 \\
0.70 & -0.53 & 0.66 & 0.07 & 0.78 & -0.28 & 1.00 & -0.80 & 0.47 & 0.54 & 0.38 & -0.29 & 0.64 & -0.48 \\
-0.74 & 0.59 & -0.73 & -0.09 & -0.86 & 0.28 & -0.80 & 1.00 & -0.54 & -0.60 & -0.32 & 0.32 & -0.56 & 0.41 \\
0.84 & -0.35 & 0.58 & 0.01 & 0.61 & -0.21 & 0.47 & -0.54 & 1.00 & 0.82 & 0.40 & -0.41 & 0.46 & -0.43 \\
0.81 & -0.31 & 0.66 & -0.04 & 0.67 & -0.31 & 0.54 & -0.60 & 0.82 & 1.00 & 0.48 & -0.43 & 0.53 & -0.56 \\
0.45 & -0.35 & 0.46 & -0.13 & 0.34 & -0.32 & 0.38 & -0.32 & 0.40 & 0.48 & 1.00 & -0.20 & 0.43 & -0.51 \\
-0.48 & 0.18 & -0.33 & 0.05 & -0.38 & 0.13 & -0.29 & 0.32 & -0.41 & -0.43 & -0.20 & 1.00 & -0.36 & 0.40 \\
0.62 & -0.45 & 0.62 & -0.06 & 0.61 & -0.64 & 0.64 & -0.56 & 0.46 & 0.53 & 0.43 & -0.36 & 1.00 & -0.83 \\
-0.57 & 0.36 & -0.55 & 0.16 & -0.52 & 0.61 & -0.48 & 0.41 & -0.43 & -0.56 & -0.51 & 0.40 & -0.83 & 1.00
\end{pmatrix}.$$

Notice that some of the correlations between $\widetilde{X}_{14}$ and the other variables have increased.

If we want to explain the variations of the price $\widetilde{X}_{14}$ by the variation of all the other variables $\widetilde{X}_1, \ldots, \widetilde{X}_{13}$ we could estimate the linear model

$$\widetilde{X}_{14} = \beta_0 + \sum_{j=1}^{13} \beta_j \widetilde{X}_j + \varepsilon. \tag{3.57}$$

The result is given in Table 3.5.

The value of r^2 (0.765) and r^2_{adj} (0.759) show that most of the variance of X_{14} is explained by the linear model (3.57).

Again we see that the variations of $\widetilde{X}_{14}$ are mostly explained by (in decreasing order of the absolute value of the t-statistic) $\widetilde{X}_{13}, \widetilde{X}_8, \widetilde{X}_{11}, \widetilde{X}_{10}, \widetilde{X}_{12}, \widetilde{X}_6, \widetilde{X}_9, \widetilde{X}_4$ and $\widetilde{X}_5$. The other variables $\widetilde{X}_1, \widetilde{X}_2, \widetilde{X}_3$ and $\widetilde{X}_7$ seem to have little influence on the variations of $\widetilde{X}_{14}$. This will be confirmed by the testing procedures that will be developed in Chapter 7.

3.8 Exercises

EXERCISE 3.1 *The covariance $s_{X_4X_5}$ between X_4 and X_5 for the entire bank data set is positive. Given the definitions of X_4 and X_5, we would expect a negative covariance. Using Figure 3.1 can you explain why $s_{X_4X_5}$ is positive?*

EXERCISE 3.2 *Consider the two sub-clouds of counterfeit and genuine bank notes in Figure 3.1 separately. Do you still expect $s_{X_4X_5}$ (now calculated separately for each cloud) to be positive?*

EXERCISE 3.3 *We remarked that for two normal random variables, zero covariance implies independence. Why does this remark not apply to Example 3.4?*

EXERCISE 3.4 *Compute the covariance between the variables*

$$\begin{aligned} X_2 &= \textit{miles per gallon}, \\ X_8 &= \textit{weight} \end{aligned}$$

from the car data set (Table B.3). What sign do you expect the covariance to have?

EXERCISE 3.5 *Compute the correlation matrix of the variables in Example 3.2. Comment on the sign of the correlations and test the hypothesis*

$$\rho_{X_1X_2} = 0.$$

EXERCISE 3.6 *Suppose you have observed a set of observations $\{x_i\}_{i=1}^n$ with $\overline{x} = 0$, $s_{XX} = 1$ and $n^{-1}\sum_{i=1}^n (x_i - \overline{x})^3 = 0$. Define the variable $y_i = x_i^2$. Can you immediately tell whether $r_{XY} \neq 0$?*

EXERCISE 3.7 *Find formulas (3.29) and (3.30) for $\widehat{\alpha}$ and $\widehat{\beta}$ by differentiating the objective function in (3.28) w.r.t. α and β.*

EXERCISE 3.8 *How many sales does the textile manager expect with a "classic blue" pullover price of $x = 105$?*

EXERCISE 3.9 *What does a scatterplot of two random variables look like for $r^2 = 1$ and $r^2 = 0$?*

EXERCISE 3.10 *Prove the variance decomposition (3.38) and show that the coefficient of determination is the square of the simple correlation between X and Y.*

EXERCISE 3.11 *Make a boxplot for the residuals $\varepsilon_i = y_i - \widehat{\alpha} - \widehat{\beta} x_i$ for the "classic blue" pullovers data. If there are outliers, identify them and run the linear regression again without them. Do you obtain a stronger influence of price on sales?*

EXERCISE 3.12 *Under what circumstances would you obtain the same coefficients from the linear regression lines of Y on X and of X on Y?*

EXERCISE 3.13 *Treat the design of Example 3.14 as if there were thirty shops and not ten. Define x_i as the index of the shop, i.e., $x_i = i, i = 1, 2, \ldots, 30$. The null hypothesis is a constant regression line, $EY = \mu$. What does the alternative regression curve look like?*

EXERCISE 3.14 *Perform the test in Exercise 3.13 for the shop example with a 0.99 significance level. Do you still reject the hypothesis of equal marketing strategies?*

EXERCISE 3.15 *Compute an approximate confidence interval for $\rho_{X_2X_8}$ in Example (3.2). Hint: start from a confidence interval for $\tanh^{-1}(\rho_{X_2X_8})$ and then apply the inverse transformation.*

EXERCISE 3.16 *In Example 3.2, using the exchange rate of 1 EUR = 106 JPY, compute the same empirical covariance using prices in Japanese Yen rather than in Euros. Is there a significant difference? Why?*

EXERCISE 3.17 *Why does the correlation have the same sign as the covariance?*

EXERCISE 3.18 *Show that* $\mathrm{rank}(\mathcal{H}) = \mathrm{tr}(\mathcal{H}) = n - 1$.

EXERCISE 3.19 *Show that $\mathcal{X}_* = \mathcal{H}\mathcal{X}\mathcal{D}^{-1/2}$ is the standardized data matrix, i.e., $\overline{x}_* = 0$ and $\mathcal{S}_{\mathcal{X}_*} = \mathcal{R}_{\mathcal{X}}$.*

EXERCISE 3.20 *Compute for the pullovers data the regression of X_1 on X_2, X_3 and of X_1 on X_2, X_4. Which one has the better coefficient of determination?*

EXERCISE 3.21 *Compare for the pullovers data the coefficient of determination for the regression of X_1 on X_2 (Example 3.11), of X_1 on X_2, X_3 (Exercise 3.20) and of X_1 on X_2, X_3, X_4 (Example 3.15). Observe that this coefficient is increasing with the number of predictor variables. Is this always the case?*

EXERCISE 3.22 *Consider the ANOVA problem (Section 3.5) again. Establish the constraint Matrix $\mathcal{A}$ for testing $\mu_1 = \mu_2$. Test this hypothesis via an analog of (3.55) and (3.56).*

EXERCISE 3.23 *Prove (3.52). (Hint, let $f(\beta) = (y - x\beta)^\top (y - x\beta)$ and solve $\frac{\partial f(\beta)}{\partial \beta} = 0$).*

EXERCISE 3.24 *Consider the linear model $Y = \mathcal{X}\beta + \varepsilon$ where $\hat{\beta} = \arg\min_{\beta} \varepsilon^\top \varepsilon$ is subject to the linear constraints $\mathcal{A}\widehat{\beta} = a$ where $\mathcal{A}(q \times p), (q \leq p)$ is of rank q and a is of dimension $(q \times 1)$. Show that $\widehat{\beta} = \widehat{\beta}_{OLS} - (\mathcal{X}^\top\mathcal{X})^{-1}\mathcal{A}^\top \left(\mathcal{A}(\mathcal{X}^\top\mathcal{X})^{-1}\mathcal{A}^\top\right)^{-1} \left(\mathcal{A}\widehat{\beta}_{OLS} - a\right)$ where $\widehat{\beta}_{OLS} = (\mathcal{X}^\top\mathcal{X})^{-1}\mathcal{X}^\top y$. (Hint, let $f(\beta, \lambda) = (y - x\beta)^\top (y - x\beta) - \lambda^\top(\mathcal{A}\beta - a)$ where $\lambda \in \mathbb{R}^q$ and solve $\frac{\partial f(\beta,\lambda)}{\partial \beta} = 0$ and $\frac{\partial f(\beta,\lambda)}{\partial \lambda} = 0$).*

EXERCISE 3.25 *Compute the covariance matrix $\mathcal{S} = Cov(\mathcal{X})$ where $\mathcal{X}$ denotes the matrix of observations on the counterfeit bank notes. Make a Jordan decomposition of $\mathcal{S}$. Why are all of the eigenvalues positive?*

EXERCISE 3.26 *Compute the covariance of the counterfeit notes after they are linearly transformed by the vector $a = (1, 1, 1, 1, 1, 1)^\top$.*

4 Multivariate Distributions

The preceeding chapter showed that by using the two first moments of a multivariate distribution (the mean and the covariance matrix), a lot of information on the relationship between the variables can be made available. Only basic statistical theory was used to derive tests of independence or of linear relationships. In this chapter we give an introduction to the basic probability tools useful in statistical multivariate analysis.

Means and covariances share many interesting and useful properties, but they represent only part of the information on a multivariate distribution. Section 4.1 presents the basic probability tools used to describe a multivariate random variable, including marginal and conditional distributions and the concept of independence. In Section 4.2, basic properties on means and covariances (marginal and conditional ones) are derived.

Since many statistical procedures rely on transformations of a multivariate random variable, Section 4.3 proposes the basic techniques needed to derive the distribution of transformations with a special emphasis on linear transforms. As an important example of a multivariate random variable, Section 4.4 defines the multinormal distribution. It will be analyzed in more detail in Chapter 5 along with most of its "companion" distributions that are useful in making multivariate statistical inferences.

The normal distribution plays a central role in statistics because it can be viewed as an approximation and limit of many other distributions. The basic justification relies on the central limit theorem presented in Section 4.5. We present this central theorem in the framework of sampling theory. A useful extension of this theorem is also given: it is an approximate distribution to transformations of asymptotically normal variables. The increasing power of the computers today makes it possible to consider alternative approximate sampling distributions. These are based on resampling techniques and are suitable for many general situations. Section 4.8 gives an introduction to the ideas behind bootstrap approximations.

4.1 Distribution and Density Function

Let $X = (X_1, X_2, \ldots, X_p)^\top$ be a random vector. The cumulative distribution function (cdf) of X is defined by

$$F(x) = P(X \leq x) = P(X_1 \leq x_1, X_2 \leq x_2, \ldots, X_p \leq x_p).$$

For continuous X, there exists a nonnegative probability density function (pdf) f, such that

$$F(x) = \int_{-\infty}^{x} f(u)du. \tag{4.1}$$

Note that

$$\int_{-\infty}^{\infty} f(u)\,du = 1.$$

Most of the integrals appearing below are multidimensional. For instance, $\int_{-\infty}^{x} f(u)du$ means $\int_{-\infty}^{x_p} \cdots \int_{-\infty}^{x_1} f(u_1, \ldots, u_p)du_1 \cdots du_p$. Note also that the cdf F is differentiable with

$$f(x) = \frac{\partial^p F(x)}{\partial x_1 \cdots \partial x_p}.$$

For discrete X, the values of this random variable are concentrated on a countable or finite set of points $\{c_j\}_{j \in J}$, the probability of events of the form $\{X \in D\}$ can then be computed as

$$P(X \in D) = \sum_{\{j: c_j \in D\}} P(X = c_j).$$

If we partition X as $X = (X_1, X_2)^\top$ with $X_1 \in \mathbb{R}^k$ and $X_2 \in \mathbb{R}^{p-k}$, then the function

$$F_{X_1}(x_1) = P(X_1 \leq x_1) = F(x_{11}, \ldots, x_{1k}, \infty, \ldots, \infty) \tag{4.2}$$

is called the *marginal cdf*. $F = F(x)$ is called the joint cdf. For continuous X the marginal pdf can be computed from the joint density by "integrating out" the variable not of interest.

$$f_{X_1}(x_1) = \int_{-\infty}^{\infty} f(x_1, x_2)dx_2. \tag{4.3}$$

The conditional pdf of X_2 given $X_1 = x_1$ is given as

$$f(x_2 \mid x_1) = \frac{f(x_1, x_2)}{f_{X_1}(x_1)}. \tag{4.4}$$

EXAMPLE 4.1 *Consider the pdf*

$$f(x_1, x_2) = \begin{cases} \frac{1}{2}x_1 + \frac{3}{2}x_2 & 0 \leq x_1, x_2 \leq 1, \\ 0 & \textit{otherwise}. \end{cases}$$

$f(x_1, x_2)$ is a density since

$$\int f(x_1, x_2)dx_1 dx_2 = \frac{1}{2}\left[\frac{x_1^2}{2}\right]_0^1 + \frac{3}{2}\left[\frac{x_2^2}{2}\right]_0^1 = \frac{1}{4} + \frac{3}{4} = 1.$$

The marginal densities are

$$\begin{aligned} f_{X_1}(x_1) &= \int f(x_1, x_2)dx_2 = \int_0^1 \left(\frac{1}{2}x_1 + \frac{3}{2}x_2\right) dx_2 = \frac{1}{2}x_1 + \frac{3}{4}; \\ f_{X_2}(x_2) &= \int f(x_1, x_2)dx_1 = \int_0^1 \left(\frac{1}{2}x_1 + \frac{3}{2}x_2\right) dx_1 = \frac{3}{2}x_2 + \frac{1}{4}. \end{aligned}$$

The conditional densities are therefore

$$f(x_2 \mid x_1) = \frac{\frac{1}{2}x_1 + \frac{3}{2}x_2}{\frac{1}{2}x_1 + \frac{3}{4}} \quad \textit{and} \quad f(x_1 \mid x_2) = \frac{\frac{1}{2}x_1 + \frac{3}{2}x_2}{\frac{3}{2}x_2 + \frac{1}{4}}.$$

Note that these conditional pdf's are nonlinear in x_1 and x_2 although the joint pdf has a simple (linear) structure.

Independence of two random variables is defined as follows.

DEFINITION 4.1 *X_1 and X_2 are independent iff $f(x) = f(x_1, x_2) = f_{X_1}(x_1)f_{X_2}(x_2)$.*

That is, X_1 and X_2 are independent if the conditional pdf's are equal to the marginal densities, i.e., $f(x_1 \mid x_2) = f_{X_1}(x_1)$ and $f(x_2 \mid x_1) = f_{X_2}(x_2)$. Independence can be interpreted as follows: knowing $X_2 = x_2$ does not change the probability assessments on X_1, and conversely.

⚠ Different joint pdf's may have the same marginal pdf's.

EXAMPLE 4.2 *Consider the pdf's*

$$f(x_1, x_2) = 1, \quad 0 < x_1, x_2 < 1,$$

and

$$f(x_1, x_2) = 1 + \alpha(2x_1 - 1)(2x_2 - 1), \quad 0 < x_1,\ x_2 < 1, \quad -1 \le \alpha \le 1.$$

We compute in both cases the marginal pdf's as

$$f_{X_1}(x_1) = 1, \quad f_{X_2}(x_2) = 1.$$

Indeed

$$\int_0^1 1 + \alpha(2x_1 - 1)(2x_2 - 1)dx_2 = 1 + \alpha(2x_1 - 1)[x_2^2 - x_2]_0^1 = 1.$$

Hence we obtain identical marginals from different joint distributions!

Let us study the concept of independence using the bank notes example. Consider the variables X_4 (lower inner frame) and X_5 (upper inner frame). From Chapter 3, we already know that they have significant correlation, so they are almost surely not independent. Kernel estimates of the marginal densities, $\widehat{f}_{X_4}$ and $\widehat{f}_{X_5}$, are given in Figure 4.1. In Figure 4.2 (left) we show the product of these two densities. The kernel density technique was presented in Section 1.3. If X_4 and X_5 are independent, this product $\widehat{f}_{X_4} \cdot \widehat{f}_{X_5}$ should be roughly equal to $\widehat{f}(x_4, x_5)$, the estimate of the joint density of (X_4, X_5). Comparing the two graphs in Figure 4.2 reveals that the two densities are different. The two variables X_4 and X_5 are therefore not independent.

An elegant concept of connecting marginals with joint cdfs is given by *copulas*. Copulas are important in Value-at-Risk calculations and are an essential tool in quantitative finance (Härdle, Kleinow and Stahl, 2002).

For simplicity of presentation we concentrate on the $p = 2$ dimensional case. A 2-dimensional copula is a function $C : [0,1]^2 \rightarrow [0,1]$ with the following properties:

- For every $u \in [0,1]$: $C(0,u) = C(u,0) = 0$.
- For every $u \in [0,1]$: $C(u,1) = u$ and $C(1,u) = u$.
- For every $(u_1, u_2), (v_1, v_2) \in [0,1] \times [0,1]$ with $u_1 \le v_1$ and $u_2 \le v_2$:

$$C(v_1, v_2) - C(v_1, u_2) - C(u_1, v_2) + C(u_1, u_2) \ge 0\,.$$

The usage of the name "copula" for the function C is explained by the following theorem.

THEOREM 4.1 (Sklar's theorem) *Let F be a joint distribution function with marginal distribution functions F_{X_1} and F_{X_2}. Then there exists a copula C with*

$$F(x_1, x_2) = C\{F_{X_1}(x_1), F_{X_2}(x_2)\} \tag{4.5}$$

for every $x_1, x_2 \in \mathbb{R}$. If F_{X_1} and F_{X_2} are continuous, then C is unique. On the other hand, if C is a copula and F_{X_1} and F_{X_2} are distribution functions, then the function F defined by (4.5) is a joint distribution function with marginals F_{X_1} and F_{X_2}.

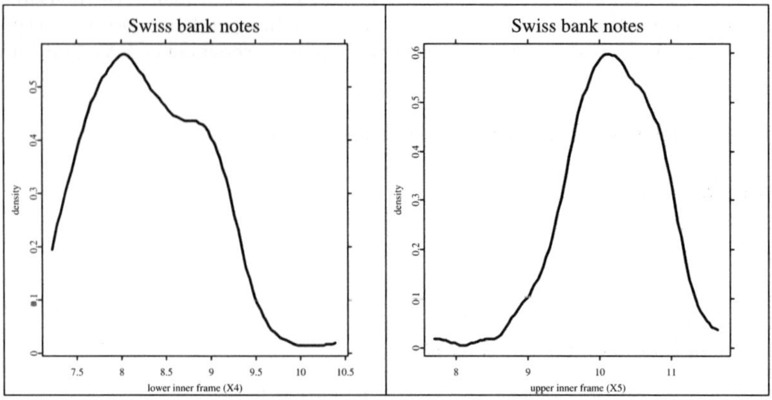

Figure 4.1. Univariate estimates of the density of X_4 (left) and X_5 (right) of the bank notes. MVAdenbank2

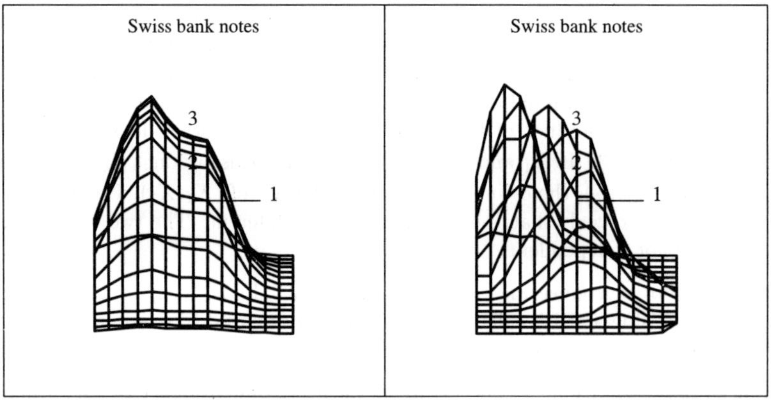

Figure 4.2. The product of univariate density estimates (left) and the joint density estimate (right) for X_4 (left) and X_5 of the bank notes. MVAdenbank3

With Sklar's Theorem, the use of the name "copula" becomes obvious. It was chosen to describe "a function that links a multidimensional distribution to its one-dimensional margins" and appeared in the mathematical literature for the first time in Sklar (1959).

EXAMPLE 4.3 *The structure of independence implies that the product of the distribution functions F_{X_1} and F_{X_2} equals their joint distribution function F,*

$$F(x_1, x_2) = F_{X_1}(x_1) \cdot F_{X_2}(x_2). \tag{4.6}$$

Thus, we obtain the independence copula $C = \Pi$ from

$$\Pi(u_1, \dots, u_n) = \prod_{i=1}^{n} u_i .$$

THEOREM 4.2 *Let X_1 and X_2 be random variables with continuous distribution functions F_{X_1} and F_{X_2} and the joint distribution function F. Then X_1 and X_2 are independent if and only if $C_{X_1,X_2} = \Pi$.*

Proof:
From Sklar's Theorem we know that there exists an unique copula C with

$$P(X_1 \le x_1, X_2 \le x_2) = F(x_1, x_2) = C\{F_{X_1}(x_1), F_{X_2}(x_2)\} . \tag{4.7}$$

Independence can be seen using (4.5) for the joint distribution function F and the definition of Π,

$$F(x_1, x_2) = C\{F_{X_1}(x_1), F_{X_2}(x_2)\} = F_{X_1}(x_1)F_{X_2}(x_2) . \tag{4.8}$$

□

EXAMPLE 4.4 *The Gumbel-Hougaard family of copulae (Nelsen, 1999) is given by the function*

$$C_\theta(u, v) = \exp\left\{- \left[(-\ln u)^\theta + (-\ln v)^\theta\right]^{1/\theta}\right\} . \tag{4.9}$$

The parameter θ may take all values in the interval $[1, \infty)$. The Gumbel-Hougaard copulae are suited to describe bivariate extreme value distributions.

For $\theta = 1$, the expression (4.9) reduces to the product copula, i.e., $C_1(u, v) = \Pi(u, v) = u\,v$. For $\theta \to \infty$ one finds for the Gumbel-Hougaard copula:

$$C_\theta(u, v) \longrightarrow \min(u, v) = M(u, v),$$

where the function M is also a copula such that $C(u, v) \le M(u, v)$ for arbitrary copula C. The copula M is called the Fréchet-Hoeffding upper bound.

Similarly, we obtain the Fréchet-Hoeffding lower bound $W(u, v) = \max(u + v - 1, 0)$ which satisfies $W(u, v) \le C(u, v)$ for any other copula C.

Summary

↪ The cumulative distribution function (cdf) is defined as $F(x) = P(X < x)$.

↪ If a probability density function (pdf) f exists then $F(x) = \int_{-\infty}^{x} f(u)du$.

↪ The pdf integrates to one, i.e., $\int_{-\infty}^{\infty} f(x)dx = 1$.

↪ Let $X = (X_1, X_2)^\top$ be partitioned into sub-vectors X_1 and X_2 with joint cdf F. Then $F_{X_1}(x_1) = P(X_1 \le x_1)$ is the marginal cdf of X_1. The marginal pdf of X_1 is obtained by $f_{X_1}(x_1) = \int_{-\infty}^{\infty} f(x_1, x_2)dx_2$. Different joint pdf's may have the same marginal pdf's.

Summary (continued)
↪ The conditional pdf of X_2 given $X_1 = x_1$ is defined as $f(x_2 \mid x_1) = \frac{f(x_1, x_2)}{f_{X_1}(x_1)}$.
↪ Two random variables X_1 and X_2 are called independent iff $f(x_1, x_2) = f_{X_1}(x_1) f_{X_2}(x_2)$. This is equivalent to $f(x_2 \mid x_1) = f_{X_2}(x_2)$.
↪ Different joint pdf's may have identical marginal pdf's.

4.2 Moments and Characteristic Functions

Moments—Expectation and Covariance Matrix

If X is a random vector with density $f(x)$ then the expectation of X is

$$EX = \begin{pmatrix} EX_1 \\ \vdots \\ EX_p \end{pmatrix} = \int x f(x) dx = \begin{pmatrix} \int x_1 f(x) dx \\ \vdots \\ \int x_p f(x) dx \end{pmatrix} = \mu. \tag{4.10}$$

Accordingly, the expectation of a matrix of random elements has to be understood component by component. The operation of forming expectations is linear:

$$E(\alpha X + \beta Y) = \alpha EX + \beta EY. \tag{4.11}$$

If $\mathcal{A}(q \times p)$ is a matrix of real numbers, we have:

$$E(\mathcal{A}X) = \mathcal{A}EX. \tag{4.12}$$

When X and Y are independent,

$$E(XY^{\top}) = EXEY^{\top}. \tag{4.13}$$

The matrix

$$Var(X) = \Sigma = E(X - \mu)(X - \mu)^{\top} \tag{4.14}$$

is the (theoretical) covariance matrix. We write for a vector X with mean vector μ and covariance matrix Σ,

$$X \sim (\mu, \Sigma). \tag{4.15}$$

The $(p \times q)$ matrix

$$\Sigma_{XY} = Cov(X, Y) = E(X - \mu)(Y - \nu)^{\top} \tag{4.16}$$

is the covariance matrix of $X \sim (\mu, \Sigma_{XX})$ and $Y \sim (\nu, \Sigma_{YY})$. Note that $\Sigma_{XY} = \Sigma_{YX}^{\top}$ and that $Z = \binom{X}{Y}$ has covariance $\Sigma_{ZZ} = \begin{pmatrix} \Sigma_{XX} & \Sigma_{XY} \\ \Sigma_{YX} & \Sigma_{YY} \end{pmatrix}$. From

$$Cov(X, Y) = E(XY^{\top}) - \mu\nu^{\top} = E(XY^{\top}) - EXEY^{\top} \tag{4.17}$$

it follows that $Cov(X, Y) = 0$ in the case where X and Y are independent. We often say that $\mu = E(X)$ is the first order moment of X and that $E(XX^{\top})$ provides the second order moments of X:

$$E(XX^{\top}) = \{E(X_i X_j)\}, \text{ for } i = 1, \ldots, p \text{ and } j = 1, \ldots, p. \tag{4.18}$$

Properties of the Covariance Matrix $\Sigma = Var(X)$

$$\Sigma = (\sigma_{X_iX_j}), \quad \sigma_{X_iX_j} = Cov(X_i, X_j), \quad \sigma_{X_iX_i} = Var(X_i) \tag{4.19}$$
$$\Sigma = E(XX^\top) - \mu\mu^\top \tag{4.20}$$
$$\Sigma \geq 0 \tag{4.21}$$

Properties of Variances and Covariances

$$Var(a^\top X) = a^\top Var(X)a = \sum_{i,j} a_i a_j \sigma_{X_iX_j} \tag{4.22}$$
$$Var(\mathcal{A}X + b) = \mathcal{A}\, Var(X)\mathcal{A}^\top \tag{4.23}$$
$$Cov(X + Y, Z) = Cov(X, Z) + Cov(Y, Z) \tag{4.24}$$
$$Var(X + Y) = Var(X) + Cov(X, Y) + Cov(Y, X) + Var(Y) \tag{4.25}$$
$$Cov(\mathcal{A}X, \mathcal{B}Y) = \mathcal{A}\, Cov(X, Y)\mathcal{B}^\top. \tag{4.26}$$

Let us compute these quantities for a specific joint density.

EXAMPLE 4.5 *Consider the pdf of Example 4.1. The mean vector* $\mu = \binom{\mu_1}{\mu_2}$ *is*

$$\begin{aligned}
\mu_1 &= \int\int x_1 f(x_1, x_2) dx_1 dx_2 = \int_0^1 \int_0^1 x_1 \left(\frac{1}{2}x_1 + \frac{3}{2}x_2\right) dx_1 dx_2 \\
&= \int_0^1 x_1 \left(\frac{1}{2}x_1 + \frac{3}{4}\right) dx_1 = \frac{1}{2}\left[\frac{x_1^3}{3}\right]_0^1 + \frac{3}{4}\left[\frac{x_1^2}{2}\right]_0^1 \\
&= \frac{1}{6} + \frac{3}{8} = \frac{4+9}{24} = \frac{13}{24},
\end{aligned}$$

$$\begin{aligned}
\mu_2 &= \int\int x_2 f(x_1, x_2) dx_1 dx_2 = \int_0^1 \int_0^1 x_2 \left(\frac{1}{2}x_1 + \frac{3}{2}x_2\right) dx_1 dx_2 \\
&= \int_0^1 x_2 \left(\frac{1}{4} + \frac{3}{2}x_2\right) dx_2 = \frac{1}{4}\left[\frac{x_2^2}{2}\right]_0^1 + \frac{3}{2}\left[\frac{x_2^3}{3}\right]_0^1 \\
&= \frac{1}{8} + \frac{1}{2} = \frac{1+4}{8} = \frac{5}{8}.
\end{aligned}$$

The elements of the covariance matrix are

$$\begin{aligned}
\sigma_{X_1X_1} &= EX_1^2 - \mu_1^2 \quad \textit{with} \\
EX_1^2 &= \int_0^1 \int_0^1 x_1^2 \left(\frac{1}{2}x_1 + \frac{3}{2}x_2\right) dx_1 dx_2 = \frac{1}{2}\left[\frac{x_1^4}{4}\right]_0^1 + \frac{3}{4}\left[\frac{x_1^3}{3}\right]_0^1 = \frac{3}{8}
\end{aligned}$$

$$\begin{aligned}
\sigma_{X_2X_2} &= EX_2^2 - \mu_2^2 \quad \textit{with} \\
EX_2^2 &= \int_0^1 \int_0^1 x_2^2 \left(\frac{1}{2}x_1 + \frac{3}{2}x_2\right) dx_1 dx_2 = \frac{1}{4}\left[\frac{x_2^3}{3}\right]_0^1 + \frac{3}{2}\left[\frac{x_2^4}{4}\right]_0^1 = \frac{11}{24}
\end{aligned}$$

$$\begin{aligned}
\sigma_{X_1X_2} &= E(X_1X_2) - \mu_1\mu_2 \quad \textit{with} \\
E(X_1X_2) &= \int_0^1 \int_0^1 x_1 x_2 \left(\frac{1}{2}x_1 + \frac{3}{2}x_2\right) dx_1 dx_2 = \int_0^1 \left(\frac{1}{6}x_2 + \frac{3}{4}x_2^2\right) dx_2 \\
&= \frac{1}{6}\left[\frac{x_2^2}{2}\right]_0^1 + \frac{3}{4}\left[\frac{x_2^3}{3}\right]_0^1 = \frac{1}{3}.
\end{aligned}$$

Hence the covariance matrix is

$$\Sigma = \begin{pmatrix} 0.0815 & 0.0052 \\ 0.0052 & 0.0677 \end{pmatrix}.$$

Conditional Expectations

The conditional expectations are

$$E(X_2 \mid x_1) = \int x_2 f(x_2 \mid x_1)\, dx_2 \quad \text{and} \quad E(X_1 \mid x_2) = \int x_1 f(x_1 \mid x_2)\, dx_1. \tag{4.27}$$

$E(X_2|x_1)$ represents the location parameter of the conditional pdf of X_2 given that $X_1 = x_1$. In the same way, we can define $Var(X_2|X_1 = x_1)$ as a measure of the dispersion of X_2 given that $X_1 = x_1$. We have from (4.20) that

$$Var(X_2|X_1 = x_1) = E(X_2\, X_2^\top|X_1 = x_1) - E(X_2|X_1 = x_1)\, E(X_2^\top|X_1 = x_1).$$

Using the conditional covariance matrix, the conditional correlations may be defined as:

$$\rho_{X_2\, X_3|X_1=x_1} = \frac{Cov(X_2, X_3|X_1 = x_1)}{\sqrt{Var(X_2|X_1 = x_1)\; Var(X_3|X_1 = x_1)}}.$$

These conditional correlations are known as partial correlations between X_2 and X_3, conditioned on X_1 being equal to x_1.

EXAMPLE 4.6 *Consider the following pdf*

$$f(x_1, x_2, x_3) = \frac{2}{3}(x_1 + x_2 + x_3) \text{ where } 0 < x_1, x_2, x_3 < 1.$$

Note that the pdf is symmetric in x_1, x_2 and x_3 which facilitates the computations. For instance,

$$\begin{aligned} f(x_1, x_2) &= \tfrac{2}{3}(x_1 + x_2 + \tfrac{1}{2}) && 0 < x_1, x_2 < 1 \\ f(x_1) &= \tfrac{2}{3}(x_1 + 1) && 0 < x_1 < 1 \end{aligned}$$

and the other marginals are similar. We also have

$$\begin{aligned} f(x_1, x_2|x_3) &= \frac{x_1 + x_2 + x_3}{x_3 + 1}, && 0 < x_1, x_2 < 1 \\ f(x_1|x_3) &= \frac{x_1 + x_3 + \frac{1}{2}}{x_3 + 1}, && 0 < x_1 < 1. \end{aligned}$$

It is easy to compute the following moments:

$E(X_i) = \frac{5}{9}$; $E(X_i^2) = \frac{7}{18}$; $E(X_iX_j) = \frac{11}{36}$ $(i \neq j$ *and* $i, j = 1, 2, 3)$

$E(X_1|X_3 = x_3) = E(X_2|X_3 = x_3) = \frac{1}{12}\left(\frac{6x_3+7}{x_3+1}\right)$;

$E(X_1^2|X_3 = x_3) = E(X_2^2|X_3 = x_3) = \frac{1}{12}\left(\frac{4x_3+5}{x_3+1}\right)$

and

$E(X_1X_2|X_3 = x_3) = \frac{1}{12}\left(\frac{3x_3+4}{x_3+1}\right)$.

Note that the conditional means of X_1 and of X_2, given $X_3 = x_3$, are not linear in x_3. From these moments we obtain:

$$\Sigma = \begin{pmatrix} \frac{13}{162} & -\frac{1}{324} & -\frac{1}{324} \\ -\frac{1}{324} & \frac{13}{162} & -\frac{1}{324} \\ -\frac{1}{324} & -\frac{1}{324} & \frac{13}{162} \end{pmatrix} \text{ in particular } \rho_{X_1X_2} = -\frac{1}{26} \approx -0.0385.$$

The conditional covariance matrix of X_1 and X_2, given $X_3 = x_3$ is

$$Var\left(\begin{pmatrix} X_1 \\ X_2 \end{pmatrix} \mid X_3 = x_3\right) = \begin{pmatrix} \frac{12x_3^2+24x_3+11}{144(x_3+1)^2} & \frac{-1}{144(x_3+1)^2} \\ \frac{-1}{144(x_3+1)^2} & \frac{12x_3^2+24x_3+11}{144(x_3+1)^2} \end{pmatrix}.$$

In particular, the partial correlation between X_1 and X_2, given that X_3 is fixed at x_3, is given by $\rho_{X_1X_2|X_3=x_3} = -\frac{1}{12x_3^2+24x_3+11}$ which ranges from -0.0909 to -0.0213 when x_3 goes from 0 to 1. Therefore, in this example, the partial correlation may be larger or smaller than the simple correlation, depending on the value of the condition $X_3 = x_3$.

EXAMPLE 4.7 *Consider the following joint pdf*

$$f(x_1, x_2, x_3) = 2x_2(x_1 + x_3);\ 0 < x_1, x_2, x_3 < 1.$$

Note the symmetry of x_1 and x_3 in the pdf and that X_2 is independent of (X_1, X_3). It immediately follows that

$$f(x_1, x_3) = (x_1 + x_3) \qquad 0 < x_1, x_3 < 1$$

$$\begin{aligned} f(x_1) &= x_1 + \frac{1}{2}; \\ f(x_2) &= 2x_2; \\ f(x_3) &= x_3 + \frac{1}{2}. \end{aligned}$$

Simple computations lead to

$$E(X) = \begin{pmatrix} \frac{7}{12} \\ \frac{2}{3} \\ \frac{7}{12} \end{pmatrix} \text{ and } \Sigma = \begin{pmatrix} \frac{11}{144} & 0 & -\frac{1}{144} \\ 0 & \frac{1}{18} & 0 \\ -\frac{1}{144} & 0 & \frac{11}{144} \end{pmatrix}.$$

Let us analyze the conditional distribution of (X_1, X_2) given $X_3 = x_3$. We have

$$f(x_1, x_2|x_3) = \frac{4(x_1 + x_3)x_2}{2x_3 + 1} \qquad 0 < x_1, x_2 < 1$$

$$f(x_1|x_3) = 2\left(\frac{x_1 + x_3}{2x_3 + 1}\right) \qquad 0 < x_1 < 1$$

$$f(x_2|x_3) = f(x_2) = 2x_2 \qquad 0 < x_2 < 1$$

so that again X_1 and X_2 are independent conditional on $X_3 = x_3$. In this case

$$\begin{aligned} E\left(\begin{pmatrix} X_1 \\ X_2 \end{pmatrix} |X_3 = x_3\right) &= \begin{pmatrix} \frac{1}{3}\left(\frac{2+3x_3}{1+2x_3}\right) \\ \frac{2}{3} \end{pmatrix} \\ Var\left(\begin{pmatrix} X_1 \\ X_2 \end{pmatrix} |X_3 = x_3\right) &= \begin{pmatrix} \frac{1}{18}\left(\frac{6x_3^2+6x_3+1}{(2x_3+1)^2}\right) & 0 \\ 0 & \frac{1}{18} \end{pmatrix}. \end{aligned}$$

Properties of Conditional Expectations

Since $E(X_2|X_1 = x_1)$ is a function of x_1, say $h(x_1)$, we can define the random variable $h(X_1) = E(X_2|X_1)$. The same can be done when defining the random variable $Var(X_2|X_1)$. These two random variables share some interesting properties:

$$E(X_2) = E\{E(X_2|X_1)\} \tag{4.28}$$

$$Var(X_2) = E\{Var(X_2|X_1)\} + Var\{E(X_2|X_1)\}. \tag{4.29}$$

EXAMPLE 4.8 *Consider the following pdf*

$$f(x_1, x_2) = 2e^{-\frac{x_2}{x_1}}; \; 0 < x_1 < 1, x_2 > 0.$$

It is easy to show that

$$f(x_1) = 2x_1 \;\; for \;\; 0 < x_1 < 1; \quad E(X_1) = \frac{2}{3} \;\; and \;\; Var(X_1) = \frac{1}{18}$$

$$f(x_2|x_1) = \frac{1}{x_1} e^{-\frac{x_2}{x_1}} \;\; for \;\; x_2 > 0; \quad E(X_2|X_1) = X_1 \;\; and \;\; Var(X_2|X_1) = X_1^2.$$

Without explicitly computing $f(x_2)$, we can obtain:

$$E(X_2) = E\left(E(X_2|X_1)\right) = E(X_1) = \frac{2}{3}$$

$$Var(X_2) = E\left(Var(X_2|X_1)\right) + Var\left(E(X_2|X_1)\right) = E(X_1^2) + Var(X_1) = \frac{2}{4} + \frac{1}{18} = \frac{10}{18}.$$

The conditional expectation $E(X_2|X_1)$ viewed as a function $h(X_1)$ of X_1 (known as the regression function of X_2 on X_1), can be interpreted as a conditional approximation of X_2 by a function of X_1. The error term of the approximation is then given by:

$$U = X_2 - E(X_2|X_1).$$

THEOREM 4.3 *Let $X_1 \in \mathbb{R}^k$ and $X_2 \in \mathbb{R}^{p-k}$ and $U = X_2 - E(X_2|X_1)$. Then we have:*

(1) $E(U) = 0$

(2) *$E(X_2|X_1)$ is the best approximation of X_2 by a function $h(X_1)$ of X_1 where $h : \mathbb{R}^k \longrightarrow \mathbb{R}^{p-k}$. "Best" is the minimum mean squared error (MSE), where*

$$MSE(h) = E[\{X_2 - h(X_1)\}^\top \{X_2 - h(X_1)\}].$$

Characteristic Functions

The characteristic function (cf) of a random vector $X \in \mathbb{R}^p$ (respectively its density $f(x)$) is defined as

$$\varphi_X(t) = E(e^{\mathbf{i}t^\top X}) = \int e^{\mathbf{i}t^\top x} f(x)\, dx, \quad t \in \mathbb{R}^p,$$

where $\mathbf{i}$ is the complex unit: $\mathbf{i}^2 = -1$. The cf has the following properties:

$$\varphi_X(0) = 1 \;\text{ and }\; |\varphi_X(t)| \le 1. \tag{4.30}$$

If φ is absolutely integrable, i.e., the integral $\int_{-\infty}^{\infty} |\varphi(x)| dx$ exists and is finite, then

$$f(x) = \frac{1}{(2\pi)^p} \int_{-\infty}^{\infty} e^{-\mathbf{i}t^\top x} \varphi_X(t)\, dt. \tag{4.31}$$

If $X = (X_1, X_2, \ldots, X_p)^\top$, then for $t = (t_1, t_2, \ldots, t_p)^\top$

$$\varphi_{X_1}(t_1) = \varphi_X(t_1, 0, \ldots, 0), \quad \ldots \quad , \varphi_{X_p}(t_p) = \varphi_X(0, \ldots, 0, t_p). \tag{4.32}$$

If $X_1, \ldots, X_p$ are independent random variables, then for $t = (t_1, t_2, \ldots, t_p)^\top$

$$\varphi_X(t) = \varphi_{X_1}(t_1) \cdot \ldots \cdot \varphi_{X_p}(t_p). \tag{4.33}$$

If $X_1, \ldots, X_p$ are independent random variables, then for $t \in \mathbb{R}$

$$\varphi_{X_1+\ldots+X_p}(t) = \varphi_{X_1}(t) \cdot \ldots \cdot \varphi_{X_p}(t). \tag{4.34}$$

The characteristic function can recover all the cross-product moments of any order: $\forall j_k \geq 0, k = 1, \ldots, p$ and for $t = (t_1, \ldots, t_p)^\top$ we have

$$E\left(X_1^{j_1} \cdot \ldots \cdot X_p^{j_p}\right) = \frac{1}{\mathbf{i}^{j_1+\ldots+j_p}} \left[\frac{\partial \varphi_X(t)}{\partial t_1^{j_1} \ldots \partial t_p^{j_p}} \right]_{t=0}. \tag{4.35}$$

EXAMPLE 4.9 *The cf of the density in example 4.5 is given by*

$$\begin{aligned}
\varphi_X(t) &= \int_0^1 \int_0^1 e^{\mathbf{i}t^\top x} f(x) dx \\
&= \int_0^1 \int_0^1 \{\cos(t_1 x_1 + t_2 x_2) + \mathbf{i} \sin(t_1 x_1 + t_2 x_2)\} \left(\frac{1}{2} x_1 + \frac{3}{2} x_2\right) dx_1 dx_2, \\
&= \frac{0.5\, e^{\mathbf{i}\, t_1} \left(3\, \mathbf{i}\, t_1 - 3\, \mathbf{i}\, e^{\mathbf{i}\, t_2}\, t_1 + \mathbf{i}\, t_2 - \mathbf{i}\, e^{\mathbf{i}\, t_2}\, t_2 + t_1\, t_2 - 4\, e^{\mathbf{i}\, t_2}\, t_1\, t_2\right)}{t_1{}^2\, t_2{}^2} \\
&\quad - \frac{0.5\, \left(3\, \mathbf{i}\, t_1 - 3\, \mathbf{i}\, e^{\mathbf{i}\, t_2}\, t_1 + \mathbf{i}\, t_2 - \mathbf{i}\, e^{\mathbf{i}\, t_2}\, t_2 - 3\, e^{\mathbf{i}\, t_2}\, t_1\, t_2\right)}{t_1{}^2\, t_2{}^2}.
\end{aligned}$$

EXAMPLE 4.10 *Suppose $X \in \mathbb{R}^1$ follows the density of the standard normal distribution*

$$f_X(x) = \frac{1}{\sqrt{2\pi}} \exp\left(-\frac{x^2}{2}\right)$$

(see Section 4.4) then the cf can be computed via

$$\begin{aligned}
\varphi_X(t) &= \frac{1}{\sqrt{2\pi}} \int_{-\infty}^{\infty} e^{\mathbf{i}tx} \exp\left(-\frac{x^2}{2}\right) dx \\
&= \frac{1}{\sqrt{2\pi}} \int_{-\infty}^{\infty} \exp\left\{-\frac{1}{2}(x^2 - 2\mathbf{i}tx + \mathbf{i}^2 t^2)\right\} \exp\left\{\frac{1}{2}\mathbf{i}^2 t^2\right\} dx \\
&= \exp\left(-\frac{t^2}{2}\right) \int_{-\infty}^{\infty} \frac{1}{\sqrt{2\pi}} \exp\left\{-\frac{(x - \mathbf{i}t)^2}{2}\right\} dx \\
&= \exp\left(-\frac{t^2}{2}\right),
\end{aligned}$$

since $\mathbf{i}^2 = -1$ and $\int \frac{1}{\sqrt{2\pi}} \exp\left\{-\frac{(x-\mathbf{i}t)^2}{2}\right\} dx = 1$.

	pdf	cf
Uniform	$f(x) = \boldsymbol{I}(x \in [a,b])/(b-a)$	$\varphi_X(t) = (e^{\mathbf{i}bt} - e^{\mathbf{i}at})/(b-a)\mathbf{i}t$
$N_1(\mu, \sigma^2)$	$f(x) = (2\pi\sigma^2)^{-1/2} exp\{-(x-\mu)^2/2\sigma^2\}$	$\varphi_X(t) = e^{\mathbf{i}\mu t - \sigma^2 t^2/2}$
$\chi^2(n)$	$f(x) = \boldsymbol{I}(x > 0) x^{n/2-1} e^{-x/2}/\{\Gamma(n/2) 2^{n/2}\}$	$\varphi_X(t) = (1 - 2\mathbf{i}t)^{-n/2}$
$N_p(\mu, \Sigma)$	$f(x) = \lvert 2\pi\Sigma\rvert^{-1/2} exp\{-(x-\mu)^\top \Sigma(x-\mu)/2\}$	$\varphi_X(t) = e^{\mathbf{i}t^\top \mu - t^\top \Sigma t/2}$

Table 4.1. Characteristic functions for some common distributions.

A variety of distributional characteristics can be computed from $\varphi_X(t)$. The standard normal distribution has a very simple cf, as was seen in Example 4.10. Deviations from normal covariance structures can be measured by the deviations from the cf (or characteristics of it). In Table 4.1 we give an overview of the cf's for a variety of distributions.

THEOREM 4.4 (Cramer-Wold) *The distribution of $X \in \mathbb{R}^p$ is completely determined by the set of all (one-dimensional) distributions of $t^\top X$ where $t \in \mathbb{R}^p$.*

This theorem says that we can determine the distribution of X in $\mathbb{R}^p$ by specifying all of the one-dimensional distributions of the linear combinations

$$\sum_{j=1}^{p} t_j X_j = t^\top X, \quad t = (t_1, t_2, \ldots, t_p)^\top.$$

Cumulant functions

Moments $m_k = \int x^k f(x) dx$ often help in describing distributional characteristics. The normal distribution in $d = 1$ dimension is completely characterized by its standard normal density $f = \varphi$ and the moment parameters are $\mu = m_1$ and $\sigma^2 = m_2 - m_1^2$. Another helpful class of parameters are the cumulants or semi-invariants of a distribution. In order to simplify notation we concentrate here on the one-dimensional ($d = 1$) case.

For a given one dimensional random variable X with density f and finite moments of order k the characteristic function $\varphi_X(t) = E(e^{\mathbf{i}tX})$ has the derivative

$$\frac{1}{\mathbf{i}^j} \left[\frac{\partial^j \log\{\varphi_X(t)\}}{\partial t^j} \right]_{t=0} = \kappa_j, \qquad j = 1, \ldots, k.$$

The values κ_j are called cumulants or semi-invariants since κ_j does not change (for $j > 1$) under a shift transformation $X \mapsto X + a$. The cumulants are natural parameters for dimension reduction methods, in particular the Projection Pursuit method (see Section 18.2).

The relationship between the first k moments $m_1, \ldots, m_k$ and the cumulants is given by

$$\kappa_k = (-1)^{k-1} \begin{vmatrix} m_1 & 1 & \ldots & 0 \\ m_2 & \binom{1}{0} m_1 & \ldots & \\ \vdots & \vdots & \ddots & \vdots \\ m_k & \binom{k-1}{0} m_{k-1} & \ldots & \binom{k-1}{k-2} m_1 \end{vmatrix}. \tag{4.36}$$

EXAMPLE 4.11 *Suppose that $k = 1$, then formula (4.36) above yields*

$$\kappa_1 = m_1.$$

For $k = 2$ we obtain

$$\kappa_2 = -\begin{vmatrix} m_1 & 1 \\ m_2 & \binom{1}{0} m_1 \end{vmatrix} = m_2 - m_1^2.$$

For $k = 3$ we have to calculate

$$\kappa_3 = \begin{vmatrix} m_1 & 1 & 0 \\ m_2 & m_1 & 1 \\ m_3 & m_2 & 2m_1 \end{vmatrix}.$$

Calculating the determinant we have:

$$\begin{aligned} \kappa_3 &= m_1 \begin{vmatrix} m_1 & 1 \\ m_2 & 2m_1 \end{vmatrix} - m_2 \begin{vmatrix} 1 & 0 \\ m_2 & 2m_1 \end{vmatrix} + m_3 \begin{vmatrix} 1 & 0 \\ m_1 & 1 \end{vmatrix} \\ &= m_1(2m_1^2 - m_2) - m_2(2m_1) + m_3 \\ &= m_3 - 3m_1m_2 + 2m_1^3. \end{aligned} \tag{4.37}$$

Similarly one calculates

$$\kappa_4 = m_4 - 4m_3m_1 - 3m_2^2 + 12m_2m_1^2 - 6m_1^4. \tag{4.38}$$

The same type of process is used to find the moments from the cumulants:

$$\begin{aligned} m_1 &= \kappa_1 \\ m_2 &= \kappa_2 + \kappa_1^2 \\ m_3 &= \kappa_3 + 3\kappa_2\kappa_1 + \kappa_1^3 \\ m_4 &= \kappa_4 + 4\kappa_3\kappa_1 + 3\kappa_2^2 + 6\kappa_2\kappa_1^2 + \kappa_1^4. \end{aligned} \tag{4.39}$$

A very simple relationship can be observed between the semi-invariants and the central moments $\mu_k = E(X - \mu)^k$, where $\mu = m_1$ as defined before. In fact, $\kappa_2 = \mu_2$, $\kappa_3 = \mu_3$ and $\kappa_4 = \mu_4 - 3\mu_2^2$.

Skewness γ_3 and kurtosis γ_4 are defined as:

$$\begin{aligned} \gamma_3 &= E(X - \mu)^3/\sigma^3 \\ \gamma_4 &= E(X - \mu)^4/\sigma^4. \end{aligned} \tag{4.40}$$

The skewness and kurtosis determine the shape of one-dimensional distributions. The skewness of a normal distribution is 0 and the kurtosis equals 3. The relation of these parameters to the cumulants is given by:

$$\gamma_3 = \frac{\kappa_3}{\kappa_2^{3/2}} \tag{4.41}$$

From (4.39) and Example 4.11

$$\gamma_4 = \frac{\kappa_4 + 3\kappa_2^2 + \kappa_1^4 - m_1^4}{\sigma^4} = \frac{\kappa_4 + 3\kappa_2^2}{\kappa_2^2} = \frac{\kappa_4}{\kappa_2^2} + 3. \tag{4.42}$$

These relations will be used later in Section 18.2 on Projection Pursuit to determine deviations from normality.

Summary

- ↪ The expectation of a random vector X is $\mu = \int x f(x)\,dx$, the covariance matrix $\Sigma = Var(X) = E(X-\mu)(X-\mu)^{\top}$. We denote $X \sim (\mu, \Sigma)$.
- ↪ Expectations are linear, i.e., $E(\alpha X + \beta Y) = \alpha EX + \beta EY$. If X and Y are independent, then $E(XY^{\top}) = EXEY^{\top}$.
- ↪ The covariance between two random vectors X and Y is $\Sigma_{XY} = Cov(X,Y) = E(X-EX)(Y-EY)^{\top} = E(XY^{\top}) - EXEY^{\top}$. If X and Y are independent, then $Cov(X,Y) = 0$.
- ↪ The characteristic function (cf) of a random vector X is $\varphi_X(t) = E(e^{\mathrm{i}t^{\top}X})$.
- ↪ The distribution of a p-dimensional random variable X is completely determined by all one-dimensional distributions of $t^{\top}X$ where $t \in \mathbb{R}^p$ (Theorem of Cramer-Wold).
- ↪ The conditional expectation $E(X_2|X_1)$ is the MSE best approximation of X_2 by a function of X_1.

4.3 Transformations

Suppose that X has pdf $f_X(x)$. What is the pdf of $Y = 3X$? Or if $X = (X_1, X_2, X_3)^{\top}$, what is the pdf of

$$Y = \begin{pmatrix} 3X_1 \\ X_1 - 4X_2 \\ X_3 \end{pmatrix}?$$

This is a special case of asking for the pdf of Y when

$$X = u(Y) \tag{4.43}$$

for a one-to-one transformation u: $\mathbb{R}^p \to \mathbb{R}^p$. Define the Jacobian of u as

$$\mathcal{J} = \left(\frac{\partial x_i}{\partial y_j}\right) = \left(\frac{\partial u_i(y)}{\partial y_j}\right)$$

and let abs($|\mathcal{J}|$) be the absolute value of the determinant of this Jacobian. The pdf of Y is given by

$$f_Y(y) = \text{abs}(|\mathcal{J}|) \cdot f_X\{u(y)\}. \tag{4.44}$$

Using this we can answer the introductory questions, namely

$$(x_1, \ldots, x_p)^{\top} = u(y_1, \ldots, y_p) = \frac{1}{3}(y_1, \ldots, y_p)^{\top}$$

with

$$\mathcal{J} = \begin{pmatrix} \frac{1}{3} & & 0 \\ & \ddots & \\ 0 & & \frac{1}{3} \end{pmatrix}$$

and hence abs($|\mathcal{J}|$) $= \left(\frac{1}{3}\right)^p$. So the pdf of Y is $\dfrac{1}{3^p} f_X\left(\dfrac{y}{3}\right)$.

This introductory example is a special case of

$$Y = \mathcal{A}X + b, \text{ where } \mathcal{A} \text{ is nonsingular.}$$

The inverse transformation is

$$X = \mathcal{A}^{-1}(Y - b).$$

Therefore

$$\mathcal{J} = \mathcal{A}^{-1},$$

and hence

$$f_Y(y) = \text{abs}(|\mathcal{A}|^{-1}) f_X\{\mathcal{A}^{-1}(y - b)\}. \tag{4.45}$$

EXAMPLE 4.12 *Consider* $X = (X_1, X_2) \in \mathbb{R}^2$ *with density* $f_X(x) = f_X(x_1, x_2)$,

$$\mathcal{A} = \begin{pmatrix} 1 & 1 \\ 1 & -1 \end{pmatrix}, \quad b = \begin{pmatrix} 0 \\ 0 \end{pmatrix}.$$

Then

$$Y = \mathcal{A}X + b = \begin{pmatrix} X_1 + X_2 \\ X_1 - X_2 \end{pmatrix}$$

and

$$|\mathcal{A}| = -2, \quad \text{abs}(|\mathcal{A}|^{-1}) = \frac{1}{2}, \quad \mathcal{A}^{-1} = -\frac{1}{2}\begin{pmatrix} -1 & -1 \\ -1 & 1 \end{pmatrix}.$$

Hence

$$\begin{aligned} f_Y(y) &= \text{abs}(|\mathcal{A}|^{-1}) \cdot f_X(\mathcal{A}^{-1}y) \\ &= \frac{1}{2} f_X\left\{\frac{1}{2}\begin{pmatrix} 1 & 1 \\ 1 & -1 \end{pmatrix}\begin{pmatrix} y_1 \\ y_2 \end{pmatrix}\right\} \\ &= \frac{1}{2} f_X\left\{\frac{1}{2}(y_1 + y_2), \frac{1}{2}(y_1 - y_2)\right\}. \end{aligned} \tag{4.46}$$

EXAMPLE 4.13 *Consider* $X \in \mathbb{R}^1$ *with density* $f_X(x)$ *and* $Y = \exp(X)$. *According to (4.43)* $x = u(y) = \log(y)$ *and hence the Jacobian is*

$$\mathcal{J} = \frac{dx}{dy} = \frac{1}{y}.$$

The pdf of Y *is therefore:*

$$f_Y(y) = \frac{1}{y} f_X\{\log(y)\}.$$

Summary

↪ If X has pdf $f_X(x)$, then a transformed random vector Y, i.e., $X = u(Y)$, has pdf $f_Y(y) = \text{abs}(|\mathcal{J}|) \cdot f_X\{u(y)\}$, where $\mathcal{J}$ denotes the Jacobian $\mathcal{J} = \left(\frac{\partial u(y_i)}{\partial y_j}\right)$.

↪ In the case of a linear relation $Y = \mathcal{A}X + b$ the pdf's of X and Y are related via $f_Y(y) = \text{abs}(|\mathcal{A}|^{-1}) f_X\{\mathcal{A}^{-1}(y - b)\}$.

4.4 The Multinormal Distribution

The multinormal distribution with mean μ and covariance $\Sigma > 0$ has the density

$$f(x) = |2\pi\Sigma|^{-1/2} \exp\left\{-\frac{1}{2}(x-\mu)^{\top}\Sigma^{-1}(x-\mu)\right\}. \tag{4.47}$$

We write $X \sim N_p(\mu, \Sigma)$.

How is this multinormal distribution with mean μ and covariance Σ related to the multivariate standard normal $N_p(0, \mathcal{I}_p)$? Through a linear transformation using the results of Section 4.3, as shown in the next theorem.

THEOREM 4.5 *Let $X \sim N_p(\mu, \Sigma)$ and $Y = \Sigma^{-1/2}(X - \mu)$ (Mahalanobis transformation). Then*

$$Y \sim N_p(0, \mathcal{I}_p),$$

i.e., the elements $Y_j \in \mathbb{R}$ are independent, one-dimensional $N(0,1)$ variables.

Proof:
Note that $(X-\mu)^{\top}\Sigma^{-1}(X-\mu) = Y^{\top}Y$. Application of (4.45) gives $\mathcal{J} = \Sigma^{1/2}$, hence

$$f_Y(y) = (2\pi)^{-p/2} \exp\left(-\frac{1}{2}y^{\top}y\right) \tag{4.48}$$

which is by (4.47) the pdf of a $N_p(0, \mathcal{I}_p)$. □

Note that the above Mahalanobis transformation yields in fact a random variable $Y = (Y_1, \ldots, Y_p)^{\top}$ composed of independent one-dimensional $Y_j \sim N_1(0,1)$ since

$$\begin{aligned} f_Y(y) &= \frac{1}{(2\pi)^{p/2}} \exp\left(-\frac{1}{2}y^{\top}y\right) \\ &= \prod_{j=1}^{p} \frac{1}{\sqrt{2\pi}} \exp\left(-\frac{1}{2}y_j^2\right) \\ &= \prod_{j=1}^{p} f_{Y_j}(y_j). \end{aligned}$$

Here each $f_{Y_j}(y)$ is a standard normal density $\frac{1}{\sqrt{2\pi}} \exp\left(-\frac{y^2}{2}\right)$. From this it is clear that $E(Y) = 0$ and $Var(Y) = \mathcal{I}_p$.

How can we create $N_p(\mu, \Sigma)$ variables on the basis of $N_p(0, \mathcal{I}_p)$ variables? We use the inverse linear transformation

$$X = \Sigma^{1/2}Y + \mu. \tag{4.49}$$

Using (4.11) and (4.23) we can also check that $E(X) = \mu$ and $Var(X) = \Sigma$. The following theorem is useful because it presents the distribution of a variable after it has been linearly transformed. The proof is left as an exercise.

THEOREM 4.6 *Let $X \sim N_p(\mu, \Sigma)$ and $\mathcal{A}(p \times p)$, $c \in \mathbb{R}^p$, where $\mathcal{A}$ is nonsingular.*

Then $Y = \mathcal{A}X + c$ is again a p-variate Normal, i.e.,

$$Y \sim N_p(\mathcal{A}\mu + c, \mathcal{A}\Sigma\mathcal{A}^{\top}). \tag{4.50}$$

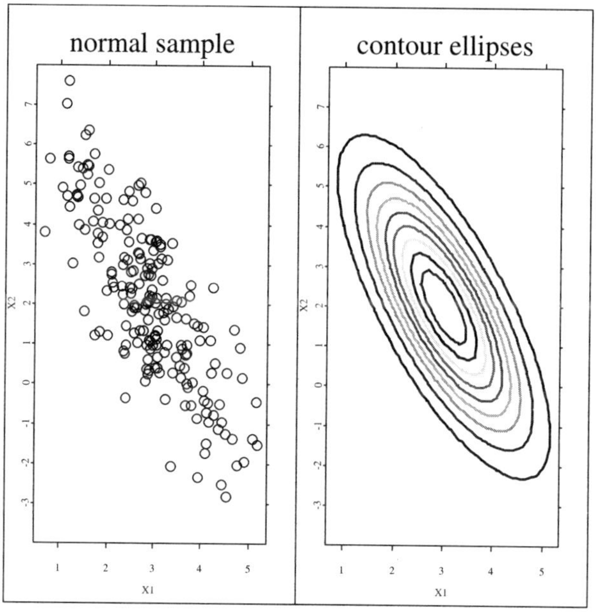

Figure 4.3. Scatterplot of a normal sample and contour ellipses for $\mu = \binom{3}{2}$ and $\Sigma = \begin{pmatrix} 1 & -1.5 \\ -1.5 & 4 \end{pmatrix}$. MVAcontnorm

Geometry of the $N_p(\mu, \Sigma)$ Distribution

From (4.47) we see that the density of the $N_p(\mu, \Sigma)$ distribution is constant on ellipsoids of the form

$$(x - \mu)^\top \Sigma^{-1} (x - \mu) = d^2. \tag{4.51}$$

EXAMPLE 4.14 *Figure 4.3 shows the contour ellipses of a two-dimensional normal distribution. Note that these contour ellipses are the iso-distance curves (2.34) from the mean of this normal distribution corresponding to the metric Σ^{-1}.*

According to Theorem 2.7 in Section 2.6 the half-lengths of the axes in the contour ellipsoid are $\sqrt{d^2\lambda_i}$ where λ_i are the eigenvalues of Σ. If Σ is a diagonal matrix, the rectangle circumscribing the contour ellipse has sides with length $2d\sigma_i$ and is thus naturally proportional to the standard deviations of X_i $(i = 1, 2)$.

The distribution of the quadratic form in (4.51) is given in the next theorem.

THEOREM 4.7 *If $X \sim N_p(\mu, \Sigma)$, then the variable $U = (X - \mu)^\top \Sigma^{-1} (X - \mu)$ has a χ^2_p distribution.*

THEOREM 4.8 *The characteristic function (cf) of a multinormal $N_p(\mu, \Sigma)$ is given by*

$$\varphi_X(t) = \exp(\mathbf{i}\, t^\top \mu - \frac{1}{2} t^\top \Sigma t). \tag{4.52}$$

We can check Theorem 4.8 by transforming the cf back:

$$\begin{aligned} f(x) &= \frac{1}{(2\pi)^p} \int \exp\left(-\mathbf{i}t^\top x + \mathbf{i}t^\top \mu - \frac{1}{2}t^\top \Sigma t\right) dt \\ &= \frac{1}{|2\pi\Sigma^{-1}|^{1/2}|2\pi\Sigma|^{1/2}} \int \exp\left[-\frac{1}{2}\{t^\top \Sigma t + 2\mathbf{i}t^\top(x-\mu) - (x-\mu)^\top\Sigma^{-1}(x-\mu)\}\right] \\ &\qquad \cdot \exp\left[-\frac{1}{2}\{(x-\mu)^\top\Sigma^{-1}(x-\mu)\}\right] dt \\ &= \frac{1}{|2\pi\Sigma|^{1/2}} \exp\left[-\frac{1}{2}\{(x-\mu)^\top\Sigma(x-\mu)\}\right] \end{aligned}$$

since

$$\begin{aligned} &\int \frac{1}{|2\pi\Sigma^{-1}|^{1/2}} \exp\left[-\frac{1}{2}\{t^\top \Sigma t + 2\mathbf{i}t^\top(x-\mu) - (x-\mu)^\top\Sigma^{-1}(x-\mu)\}\right] dt \\ &= \int \frac{1}{|2\pi\Sigma^{-1}|^{1/2}} \exp\left[-\frac{1}{2}\{(t+\mathbf{i}\Sigma^{-1}(x-\mu))^\top\Sigma(t+\mathbf{i}\Sigma^{-1}(x-\mu))\}\right] dt \\ &= 1. \end{aligned}$$

Note that if $Y \sim N_p(0, \mathcal{I}_p)$ (e.g., the Mahalanobis-transform), then

$$\begin{aligned} \varphi_Y(t) &= \exp\left(-\frac{1}{2}t^\top \mathcal{I}_p t\right) = \exp\left(-\frac{1}{2}\sum_{i=1}^p t_i^2\right) \\ &= \varphi_{Y_1}(t_1) \cdot \ldots \cdot \varphi_{Y_p}(t_p) \end{aligned}$$

which is consistent with (4.33).

Singular Normal Distribution

Suppose that we have $\text{rank}(\Sigma) = k < p$, where p is the dimension of X. We define the (singular) density of X with the aid of the G-Inverse Σ^- of Σ,

$$f(x) = \frac{(2\pi)^{-k/2}}{(\lambda_1 \cdots \lambda_k)^{1/2}} \exp\left\{-\frac{1}{2}(x-\mu)^\top\Sigma^-(x-\mu)\right\} \tag{4.53}$$

where

(1) x lies on the hyperplane $\mathcal{N}^\top(x-\mu) = 0$ with $\mathcal{N}(p\times(p-k)) : \mathcal{N}^\top\Sigma = 0$ and $\mathcal{N}^\top\mathcal{N} = \mathcal{I}_k$.

(2) Σ^- is the G-Inverse of Σ, and $\lambda_1, \ldots, \lambda_k$ are the nonzero eigenvalues of Σ.

What is the connection to a multinormal with k-dimensions? If

$$Y \sim N_k(0, \Lambda_1) \quad \text{and} \quad \Lambda_1 = \text{diag}(\lambda_1, \ldots, \lambda_k), \tag{4.54}$$

then there exists an orthogonal matrix $\mathcal{B}(p\times k)$ with $\mathcal{B}^\top\mathcal{B} = \mathcal{I}_k$ so that $X = \mathcal{B}Y + \mu$ where X has a singular pdf of the form (4.53).

Gaussian Copula

In example 4.3 and 4.4 we have introduced copulae. Another important copula is the *Gaussian* or *normal copula*,

$$C_\rho(u,v) = \int_{-\infty}^{\Phi_1^{-1}(u)} \int_{-\infty}^{\Phi_2^{-1}(v)} f_\rho(x_1,x_2)dx_2dx_1\ , \tag{4.55}$$

see Embrechts, McNeil and Straumann (1999). In (4.55), f_ρ denotes the bivariate normal density function with correlation ρ for $n = 2$. The functions Φ_1 and Φ_2 in (4.55) refer to the corresponding one-dimensional standard normal cdfs of the margins.

In the case of vanishing correlation, $\rho = 0$, the Gaussian copula becomes

$$\begin{aligned} C_0(u,v) &= \int_{-\infty}^{\Phi_1^{-1}(u)} f_{X_1}(x_1)dx_1 \int_{-\infty}^{\Phi_2^{-1}(v)} f_{X_2}(x_2)dx_2 \\ &= u\,v \\ &= \Pi(u,v)\,. \end{aligned}$$

Summary

↪ The pdf of a p-dimensional multinormal $X \sim N_p(\mu, \Sigma)$ is

$$f(x) = |2\pi\Sigma|^{-1/2} \exp\left\{-\frac{1}{2}(x-\mu)^\top\Sigma^{-1}(x-\mu)\right\}.$$

The contour curves of a multinormal are ellipsoids with half-lengths proportional to $\sqrt{\lambda_i}$, where λ_i denotes the eigenvalues of Σ ($i = 1, \ldots, p$).

↪ The Mahalanobis transformation transforms $X \sim N_p(\mu, \Sigma)$ to $Y = \Sigma^{-1/2}(X - \mu) \sim N_p(0, \mathcal{I}_p)$. Going the other direction, one can create a $X \sim N_p(\mu, \Sigma)$ from $Y \sim N_p(0, \mathcal{I}_p)$ via $X = \Sigma^{1/2}Y + \mu$.

↪ If the covariance matrix Σ is singular (i.e., $\operatorname{rank}(\Sigma) < p$), then it defines a singular normal distribution.

↪ The gaussian copula is given by

$$C_\rho(u,v) = \int_{-\infty}^{\Phi_1^{-1}(u)} \int_{-\infty}^{\Phi_2^{-1}(v)} f_\rho(x_1, x_2)dx_2dx_1\,.$$

↪ The density of a singular normal distribution is given by

$$\frac{(2\pi)^{-k/2}}{(\lambda_1 \cdots \lambda_k)^{1/2}} \exp\left\{-\frac{1}{2}(x-\mu)^\top\Sigma^{-}(x-\mu)\right\}.$$

4.5 Sampling Distributions and Limit Theorems

In multivariate statistics, we observe the values of a multivariate random variable X and obtain a sample $\{x_i\}_{i=1}^n$, as described in Chapter 3. Under random sampling, these observations are considered to be realizations of a sequence of i.i.d. random variables $X_1, \ldots, X_n$, where each X_i is a p-variate random variable which replicates the *parent* or *population* random variable X. Some notational confusion is hard to avoid: X_i is not the ith component of X, but rather the ith replicate of the p-variate random variable X which provides the ith observation x_i of our sample.

For a given random sample $X_1, \ldots, X_n$, the idea of statistical inference is to analyze the properties of the population variable X. This is typically done by analyzing some characteristic θ of its distribution, like the mean, covariance matrix, etc. Statistical inference in a multivariate setup is considered in more detail in Chapters 6 and 7.

Inference can often be performed using some observable function of the sample $X_1, \ldots, X_n$, i.e., a *statistics*. Examples of such statistics were given in Chapter 3: the sample mean $\bar{x}$, the sample covariance matrix $\mathcal{S}$. To get an idea of the relationship between a statistics and the corresponding population characteristic, one has to derive the sampling distribution of the statistic. The next example gives some insight into the relation of $(\overline{x}, S)$ to (μ, Σ).

EXAMPLE 4.15 *Consider an iid sample of n random vectors $X_i \in \mathbb{R}^p$ where $E(X_i) = \mu$ and $Var(X_i) = \Sigma$. The sample mean $\bar{x}$ and the covariance matrix $\mathcal{S}$ have already been defined in Section 3.3. It is easy to prove the following results*

$$\begin{aligned} E(\bar{x}) &= \tfrac{1}{n}\sum_{i=1}^{n} E(X_i) = \mu \\ Var(\bar{x}) &= \tfrac{1}{n^2}\sum_{i=1}^{n} Var(X_i) = \tfrac{1}{n}\Sigma = E(\bar{x}\bar{x}^\top) - \mu\mu^\top \end{aligned}$$

$$\begin{aligned} E(\mathcal{S}) &= \tfrac{1}{n}E\left\{\sum_{i=1}^{n}(X_i - \bar{x})(X_i - \bar{x})^\top\right\} \\ &= \tfrac{1}{n}E\left\{\sum_{i=1}^{n} X_i X_i^\top - n\bar{x}\bar{x}^\top\right\} \\ &= \tfrac{1}{n}\left\{n\left(\Sigma + \mu\mu^\top\right) - n\left(\tfrac{\Sigma}{n} + \mu\mu^\top\right)\right\} \\ &= \tfrac{n-1}{n}\Sigma. \end{aligned}$$

This shows in particular that $\mathcal{S}$ is a biased estimator of Σ. By contrast, $\mathcal{S}_u = \frac{n}{n-1}\mathcal{S}$ is an unbiased estimator of Σ.

Statistical inference often requires more than just the mean and/or the variance of a statistic. We need the sampling distribution of the statistics to derive confidence intervals or to define rejection regions in hypothesis testing for a given significance level. Theorem 4.9 gives the distribution of the sample mean for a multinormal population.

THEOREM 4.9 *Let $X_1, \ldots, X_n$ be i.i.d. with $X_i \sim N_p(\mu, \Sigma)$. Then $\bar{x} \sim N_p(\mu, \frac{1}{n}\Sigma)$.*

Proof:
$\bar{x} = (1/n)\sum_{i=1}^{n} X_i$ is a linear combination of independent normal variables, so it has a normal distribution (see chapter 5). The mean and the covariance matrix were given in the preceding example. □

With multivariate statistics, the sampling distributions of the statistics are often more difficult to derive than in the preceding Theorem. In addition they might be so complicated that approximations have to be used. These approximations are provided by limit theorems. Since they are based on asymptotic limits, the approximations are only valid when the sample size is large enough. In spite of this restriction, they make complicated situations rather simple. The following central limit theorem shows that even if the parent distribution is not normal, when the sample size n is large, the sample mean $\bar{x}$ has an approximate normal distribution.

THEOREM 4.10 (Central Limit Theorem (CLT)) *Let* $X_1, X_2, \ldots, X_n$ *be i.i.d. with* $X_i \sim (\mu, \Sigma)$. *Then the distribution of* $\sqrt{n}(\overline{x} - \mu)$ *is asymptotically* $N_p(0, \Sigma)$, *i.e.,*

$$\sqrt{n}(\overline{x} - \mu) \xrightarrow{\mathcal{L}} N_p(0, \Sigma) \qquad as \quad n \longrightarrow \infty.$$

The symbol "$\xrightarrow{\mathcal{L}}$" denotes *convergence in distribution* which means that the distribution function of the random vector $\sqrt{n}(\bar{x} - \mu)$ converges to the distribution function of $N_p(0, \Sigma)$.

EXAMPLE 4.16 *Assume that* $X_1, \ldots, X_n$ *are i.i.d. and that they have Bernoulli distributions where* $p = \frac{1}{2}$ *(this means that* $P(X_i = 1) = \frac{1}{2}$, $P(X_i = 0) = \frac{1}{2}$*). Then* $\mu = p = \frac{1}{2}$ *and* $\Sigma = p(1-p) = \frac{1}{4}$. *Hence,*

$$\sqrt{n}\left(\overline{x} - \frac{1}{2}\right) \xrightarrow{\mathcal{L}} N_1\left(0, \frac{1}{4}\right) \qquad as \quad n \longrightarrow \infty.$$

The results are shown in Figure 4.4 for varying sample sizes.

EXAMPLE 4.17 *Now consider a two-dimensional random sample* $X_1, \ldots, X_n$ *that is i.i.d. and created from two independent Bernoulli distributions with* $p = 0.5$. *The joint distribution is given by* $P(X_i = (0,0)^\top) = \frac{1}{4}, P(X_i = (0,1)^\top) = \frac{1}{4}, P(X_i = (1,0)^\top) = \frac{1}{4}, P(X_i = (1,1)^\top) = \frac{1}{4}$. *Here we have*

$$\sqrt{n}\left\{\overline{x} - \begin{pmatrix}\frac{1}{2}\\ \frac{1}{2}\end{pmatrix}\right\} = N_2\left(\begin{pmatrix}0\\0\end{pmatrix}, \begin{pmatrix}\frac{1}{4} & 0\\ 0 & \frac{1}{4}\end{pmatrix}\right) \qquad as \quad n \longrightarrow \infty.$$

Figure 4.5 displays the estimated two-dimensional density for different sample sizes.

The asymptotic normal distribution is often used to construct confidence intervals for the unknown parameters. A confidence interval at the level $1 - \alpha$, $\alpha \in (0, 1)$, is an interval that covers the true parameter with probability $1 - \alpha$:

$$P(\theta \in [\widehat{\theta}_l, \widehat{\theta}_u]) = 1 - \alpha,$$

where θ denotes the (unknown) parameter and $\widehat{\theta}_l$ and $\widehat{\theta}_u$ are the lower and upper confidence bounds respectively.

EXAMPLE 4.18 *Consider the i.i.d. random variables* $X_1, \ldots, X_n$ *with* $X_i \sim (\mu, \sigma^2)$ *and* σ^2 *known. Since we have* $\sqrt{n}(\bar{x} - \mu) \xrightarrow{\mathcal{L}} N(0, \sigma^2)$ *from the CLT, it follows that*

$$P(-u_{1-\alpha/2} \le \sqrt{n}\frac{(\bar{x} - \mu)}{\sigma} \le u_{1-\alpha/2}) \longrightarrow 1 - \alpha, \qquad as \quad n \longrightarrow \infty$$

where $u_{1-\alpha/2}$ *denotes the* $(1 - \alpha/2)$*-quantile of the standard normal distribution. Hence the interval*

$$\left[\bar{x} - \frac{\sigma}{\sqrt{n}} u_{1-\alpha/2}, \ \bar{x} + \frac{\sigma}{\sqrt{n}} u_{1-\alpha/2}\right]$$

is an approximate $(1 - \alpha)$*-confidence interval for* μ.

But what can we do if we do not know the variance σ^2? The following corollary gives the answer.

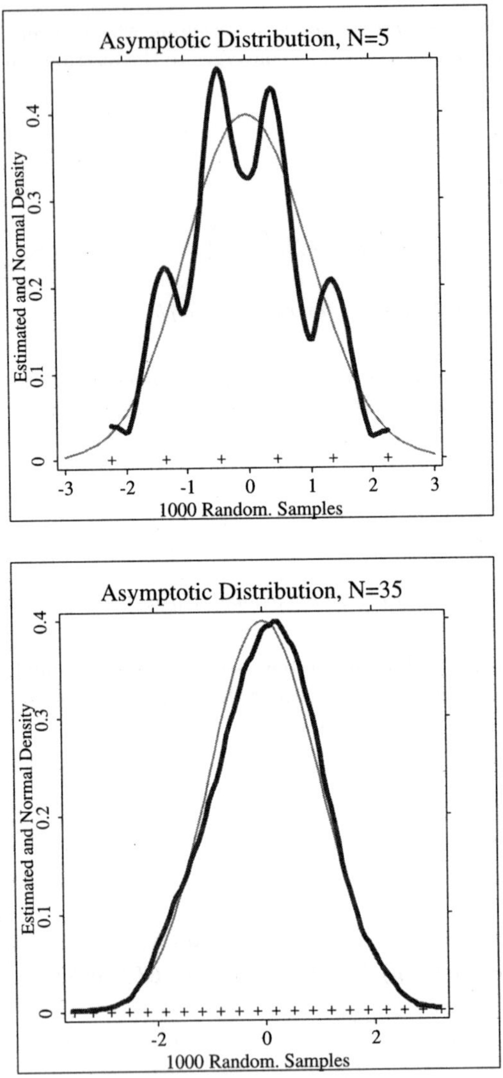

Figure 4.4. The CLT for Bernoulli distributed random variables. Sample size $n = 5$ (left) and $n = 35$ (right). MVAcltbern

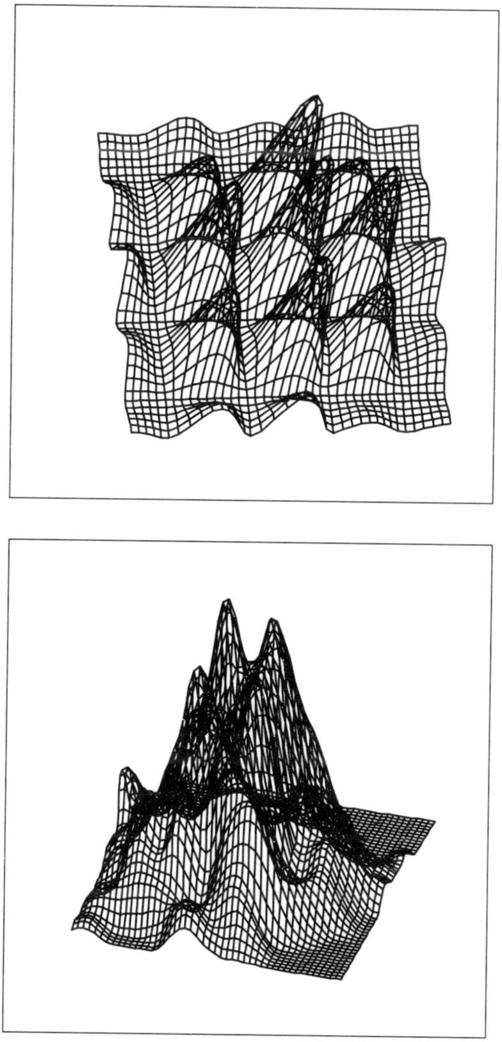

Figure 4.5. The CLT in the two-dimensional case. Sample size $n = 5$ (left) and $n = 85$ (right). MVAcltbern2

COROLLARY 4.1 *If $\widehat{\Sigma}$ is a consistent estimate for Σ, then the CLT still holds, namely*

$$\sqrt{n}\,\widehat{\Sigma}^{-1/2}(\bar{x}-\mu) \xrightarrow{\mathcal{L}} N_p(0,\mathcal{I}) \qquad \text{as} \quad n \longrightarrow \infty.$$

EXAMPLE 4.19 *Consider the i.i.d. random variables $X_1,\ldots,X_n$ with $X_i \sim (\mu,\sigma^2)$, and now with an unknown variance σ^2. From Corollary 4.1 using $\widehat{\sigma}^2 = \frac{1}{n}\sum_{i=1}^n (x_i-\bar{x})^2$ we obtain*

$$\sqrt{n}\left(\frac{\bar{x}-\mu}{\widehat{\sigma}}\right) \xrightarrow{\mathcal{L}} N(0,1) \qquad \text{as} \quad n \longrightarrow \infty.$$

Hence we can construct an approximate $(1-\alpha)$-confidence interval for μ using the variance estimate $\widehat{\sigma}^2$:

$$C_{1-\alpha} = \left[\bar{x} - \frac{\widehat{\sigma}}{\sqrt{n}}u_{1-\alpha/2},\ \bar{x} + \frac{\widehat{\sigma}}{\sqrt{n}}u_{1-\alpha/2}\right].$$

Note that by the CLT

$$P(\mu \in C_{1-\alpha}) \longrightarrow 1-\alpha \qquad \text{as} \quad n \longrightarrow \infty.$$

REMARK 4.1 *One may wonder how large should n be in practice to provide reasonable approximations. There is no definite answer to this question: it mainly depends on the problem at hand (the shape of the distribution of the X_i and the dimension of X_i). If the X_i are normally distributed, the normality of $\bar{x}$ is achieved from $n = 1$. In most situations, however, the approximation is valid in one-dimensional problems for n larger than, say, 50.*

Transformation of Statistics

Often in practical problems, one is interested in a function of parameters for which one has an asymptotically normal statistic. Suppose for instance that we are interested in a cost function depending on the mean μ of the process: $f(\mu) = \mu^\top \mathcal{A}\mu$ where $\mathcal{A} > 0$ is given. To estimate μ we use the asymptotically normal statistic $\bar{x}$. The question is: how does $f(\bar{x})$ behave? More generally, what happens to a statistic t that is asymptotically normal when we transform it by a function $f(t)$? The answer is given by the following theorem.

THEOREM 4.11 *If $\sqrt{n}(t-\mu) \xrightarrow{\mathcal{L}} N_p(0,\Sigma)$ and if $f = (f_1,\ldots,f_q)^\top : \mathbb{R}^p \to \mathbb{R}^q$ are real valued functions which are differentiable at $\mu \in \mathbb{R}^p$, then $f(t)$ is asymptotically normal with mean $f(\mu)$ and covariance $\mathcal{D}^\top\Sigma\mathcal{D}$, i.e.,*

$$\sqrt{n}\{f(t)-f(\mu)\} \xrightarrow{\mathcal{L}} N_q(0,\mathcal{D}^\top\Sigma\mathcal{D}) \qquad \text{for} \quad n \longrightarrow \infty, \tag{4.56}$$

where

$$\mathcal{D} = \left(\frac{\partial f_j}{\partial t_i}\right)(t)\bigg|_{t=\mu}$$

is the $(p \times q)$ matrix of all partial derivatives.

EXAMPLE 4.20 *We are interested in seeing how $f(\bar{x}) = \bar{x}^\top\mathcal{A}\bar{x}$ behaves asymptotically with respect to the quadratic cost function of μ, $f(\mu) = \mu^\top\mathcal{A}\mu$, where $\mathcal{A} > 0$.*

$$D = \frac{\partial f(\bar{x})}{\partial \bar{x}}\bigg|_{\bar{x}=\mu} = 2\mathcal{A}\mu.$$

By Theorem 4.11 we have

$$\sqrt{n}(\bar{x}^\top\mathcal{A}\bar{x} - \mu^\top\mathcal{A}\mu) \xrightarrow{\mathcal{L}} N_1\left(0, 4\mu^\top\mathcal{A}\Sigma\mathcal{A}\mu\right).$$

EXAMPLE 4.21 *Suppose*

$$X_i \sim (\mu, \Sigma); \quad \mu = \begin{pmatrix} 0 \\ 0 \end{pmatrix}, \quad \Sigma = \begin{pmatrix} 1 & 0.5 \\ 0.5 & 1 \end{pmatrix}, \quad p = 2.$$

We have by the CLT (Theorem 4.10) for $n \to \infty$ that

$$\sqrt{n}(\overline{x} - \mu) \xrightarrow{\mathcal{L}} N(0, \Sigma).$$

Suppose that we would like to compute the distribution of $\begin{pmatrix} \overline{x}_1^2 - \overline{x}_2 \\ \overline{x}_1 + 3\overline{x}_2 \end{pmatrix}$. *According to Theorem 4.11 we have to consider $f = (f_1, f_2)^\top$ with*

$$f_1(x_1, x_2) = x_1^2 - x_2, \quad f_2(x_1, x_2) = x_1 + 3x_2, \quad q = 2.$$

Given this $f(\mu) = \binom{0}{0}$ *and*

$$\mathcal{D} = (d_{ij}), \quad d_{ij} = \left(\frac{\partial f_j}{\partial x_i}\right)\bigg|_{x=\mu} = \begin{pmatrix} 2x_1 & 1 \\ -1 & 3 \end{pmatrix}\bigg|_{x=0}.$$

Thus

$$\mathcal{D} = \begin{pmatrix} 0 & 1 \\ -1 & 3 \end{pmatrix}.$$

The covariance is

$$\underbrace{\begin{pmatrix} 0 & -1 \\ 1 & 3 \end{pmatrix}}_{\mathcal{D}^\top} \underbrace{\begin{pmatrix} 1 & \frac{1}{2} \\ \frac{1}{2} & 1 \end{pmatrix}}_{\Sigma} \underbrace{\begin{pmatrix} 0 & 1 \\ -1 & 3 \end{pmatrix}}_{\mathcal{D}} = \underbrace{\begin{pmatrix} 0 & -1 \\ 1 & 3 \end{pmatrix}}_{\mathcal{D}^\top} \underbrace{\begin{pmatrix} -\frac{1}{2} & \frac{5}{2} \\ -1 & \frac{7}{2} \end{pmatrix}}_{\Sigma\mathcal{D}} = \underbrace{\begin{pmatrix} 1 & -\frac{7}{2} \\ -\frac{7}{2} & 13 \end{pmatrix}}_{\mathcal{D}^\top\Sigma\mathcal{D}},$$

which yields

$$\sqrt{n}\begin{pmatrix} \overline{x}_1^2 - \overline{x}_2 \\ \overline{x}_1 + 3\overline{x}_2 \end{pmatrix} \xrightarrow{\mathcal{L}} N_2\left(\begin{pmatrix} 0 \\ 0 \end{pmatrix}, \begin{pmatrix} 1 & -\frac{7}{2} \\ -\frac{7}{2} & 13 \end{pmatrix}\right).$$

EXAMPLE 4.22 *Let us continue the previous example by adding one more component to the function f. Since $q = 3 > p = 2$, we might expect a singular normal distribution. Consider $f = (f_1, f_2, f_3)^\top$ with*

$$f_1(x_1, x_2) = x_1^2 - x_2, \quad f_2(x_1, x_2) = x_1 + 3x_2, \quad f_3 = x_2^3, \quad q = 3.$$

From this we have that

$$\mathcal{D} = \begin{pmatrix} 0 & 1 & 0 \\ -1 & 3 & 0 \end{pmatrix} \quad \text{and thus} \quad \mathcal{D}^\top\Sigma\mathcal{D} = \begin{pmatrix} 1 & -\frac{7}{2} & 0 \\ -\frac{7}{2} & 13 & 0 \\ 0 & 0 & 0 \end{pmatrix}.$$

The limit is in fact a singular normal distribution!

Summary

↪ If $X_1, \ldots, X_n$ are i.i.d. random vectors with $X_i \sim N_p(\mu, \Sigma)$, then $\bar{x} \sim N_p(\mu, \frac{1}{n}\Sigma)$.

↪ If $X_1, \ldots, X_n$ are i.i.d. random vectors with $X_i \sim (\mu, \Sigma)$, then the distribution of $\sqrt{n}(\bar{x} - \mu)$ is asymptotically $N(0, \Sigma)$ (Central Limit Theorem).

↪ If $X_1, \ldots, X_n$ are i.i.d. random variables with $X_i \sim (\mu, \sigma)$, then an asymptotic confidence interval can be constructed by the CLT: $\bar{x} \pm \frac{\widehat{\sigma}}{\sqrt{n}} u_{1-\alpha/2}$.

↪ If t is a statistic that is asymptotically normal, i.e., $\sqrt{n}(t-\mu) \xrightarrow{\mathcal{L}} N_p(0, \Sigma)$, then this holds also for a function $f(t)$, i.e., $\sqrt{n}\{f(t) - f(\mu)\}$ is asymptotically normal.

4.6 Heavy-Tailed Distributions

Heavy-tailed distributions were first introduced by the Italian-born Swiss economist Pareto and extensively studied by Paul Lévy. Although in the beginning these distributions were mainly studied theoretically, nowadays they have found many applications in areas as diverse as finance, medicine, seismology, structural engineering. More concretely, they have been used to model returns of assets in financial markets, stream flow in hydrology, precipitation and hurricane damage in meteorology, earthquake prediction in seismology, pollution, material strength, teletraffic and many others.

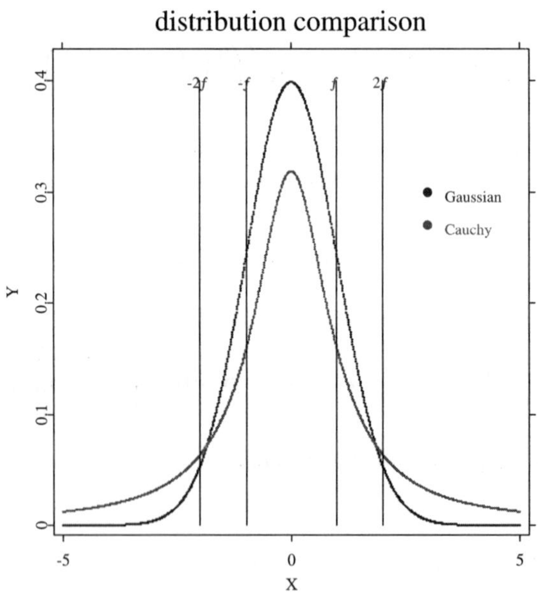

Figure 4.6. Comparison of the pdf of a standard Gaussian (blue) and a Cauchy distribution (red) with location parameter 0 and scale parameter 1. MVAgausscauchy

A distribution is called heavy-tailed if it has higher probability density in its tail area compared with a normal distribution with same mean μ and variance σ^2. Figure 4.6 demonstrates the differences of the pdf curves of a standard Gaussian distribution and a Cauchy distribution with location parameter $\mu = 0$ and scale parameter $\sigma = 1$. The graphic shows that the probability density of the Cauchy distribution is much higher than that of the Gaussian in the tail part, while in the area around the center, the probability density of the Cauchy distribution is much lower.

In terms of kurtosis, a heavy-tailed distribution has kurtosis greater than 3 (See Chapter 4, formula (4.40)), which is called leptokurtic, in contrast to mesokurtic distribution (kurtosis = 3) and platykurtic distribution (kurtosis < 3). Since univariate heavy-tailed distributions serve as basics for their multivariate counterparts and their density properties have been proved useful even in multivariate cases, we will start from introducing some univariate heavy-tailed distributions. Then we will move on to analyze their multivariate counterparts, and their tail behavior.

Generalized Hyperbolic Distribution

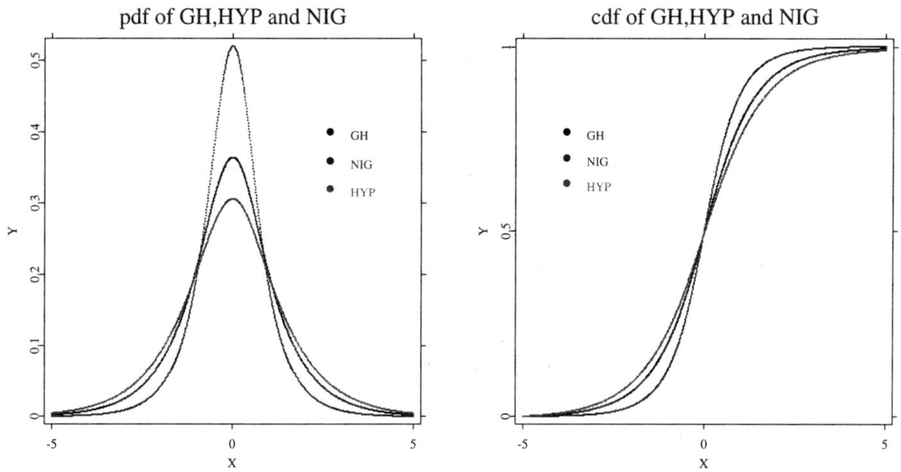

Figure 4.7. pdf (left) and cdf (right) of GH ($\lambda = 0.5$), HYP and NIG with $\alpha = 1, \beta = 0, \delta = 1, \mu = 0$ MVAghdis

The generalized hyperbolic distribution was introduced by Barndorff-Nielsen and at first applied to model grain size distributions of wind blown sands. Today one of its most important uses is in stock price modeling and market risk measurement. The name of the distribution is derived from the fact that its log-density forms a hyperbola, while the log-density of the normal distribution is a parabola.

The density of a one-dimensional generalized hyperbolic (GH) distribution for $x \in \mathbb{R}$ is

$$f_{GH}(x; \lambda, \alpha, \beta, \delta, \mu)$$

$$= \frac{(\sqrt{\alpha^2 - \beta^2}/\delta)^\lambda}{\sqrt{2\pi} K_\lambda(\delta\sqrt{\alpha^2 - \beta^2})} \frac{K_{\lambda-1/2}\{\alpha\sqrt{\delta^2 + (x-\mu)^2}\}}{(\sqrt{\delta^2 + (x-\mu)^2}/\alpha)^{1/2-\lambda}} e^{\beta(x-\mu)} \qquad (4.57)$$

where K_λ is a modified Bessel function of the third kind with index λ

$$K_\lambda(x) = \frac{1}{2}\int_0^\infty y^{\lambda-1} e^{-\frac{x}{2}(y+y^{-1})} dy \tag{4.58}$$

The domain of variation of the parameters is $\mu \in \mathbb{R}$ and

$$\begin{array}{ll} \delta \geq 0, |\beta| < \alpha, & \text{if} \quad \lambda > 0 \\ \delta > 0, |\beta| < \alpha, & \text{if} \quad \lambda = 0 \\ \delta > 0, |\beta| \leq \alpha, & \text{if} \quad \lambda < 0 \end{array}$$

The generalized hyperbolic distribution has the following mean and variance

$$\mathrm{E}[X] = \mu + \frac{\delta\beta}{\sqrt{\alpha^2+\beta^2}} \frac{K_{\lambda+1}(\delta\sqrt{\alpha^2+\beta^2})}{K_\lambda(\delta\sqrt{\alpha^2+\beta^2})} \tag{4.59}$$

$$\begin{aligned} \mathrm{Var}[X] = \delta^2 \Bigg[& \frac{K_{\lambda+1}(\delta\sqrt{\alpha^2+\beta^2})}{\delta\sqrt{\alpha^2+\beta^2} K_\lambda(\delta\sqrt{\alpha^2+\beta^2})} + \frac{\beta^2}{\alpha^2+\beta^2}\Bigg[\frac{K_{\lambda+2}(\delta\sqrt{\alpha^2+\beta^2})}{K_\lambda(\delta\sqrt{\alpha^2+\beta^2})} \\ & - \left\{\frac{K_{\lambda+1}(\delta\sqrt{\alpha^2+\beta^2})}{K_\lambda(\delta\sqrt{\alpha^2+\beta^2})}\right\}^2\Bigg]\Bigg] \end{aligned} \tag{4.60}$$

Where μ and δ play important roles in the density's location and scale respectively. With specific values of λ, we obtain different subclasses of GH such as hyperbolic (HYP) or normal-inverse Gaussian (NIG) distribution.

For λ=1 we obtain the hyperbolic distributions (HYP)

$$f_{HYP}(x;\alpha,\beta,\delta,\mu) = \frac{\sqrt{\alpha^2-\beta^2}}{2\alpha\delta K_1(\delta\sqrt{\alpha^2-\beta^2})} e^{\{-\alpha\sqrt{\delta^2+(x-\mu)^2}+\beta(x-\mu)\}} \tag{4.61}$$

where $x, \mu \in \mathbb{R}, \delta \geq 0$ and $|\beta| < \alpha$

For $\lambda = -1/2$ we obtain the normal-inverse Gaussian distribution (NIG)

$$f_{NIG}(x;\alpha,\beta,\delta,\mu) = \frac{\alpha\delta}{\pi} \frac{K_1\left(\alpha\sqrt{(\delta^2+(x-\mu)^2)}\right)}{\sqrt{\delta^2+(x-\mu)^2}} e^{\{\delta\sqrt{\alpha^2-\beta^2}+\beta(x-\mu)\}} \tag{4.62}$$

Student's t-distribution

The t-distribution was first analyzed by Gosset (1908). He published his results under his pseudonym "Student" by request of his employer. Let X be a normally distributed random variable with mean μ and variance σ^2, and Y be the random variable such that Y^2/σ^2 has a chi-square distribution with n degrees of freedom. Assume that X and Y are independent, then

$$t \stackrel{\text{def}}{=} \frac{X\sqrt{n}}{Y} \tag{4.63}$$

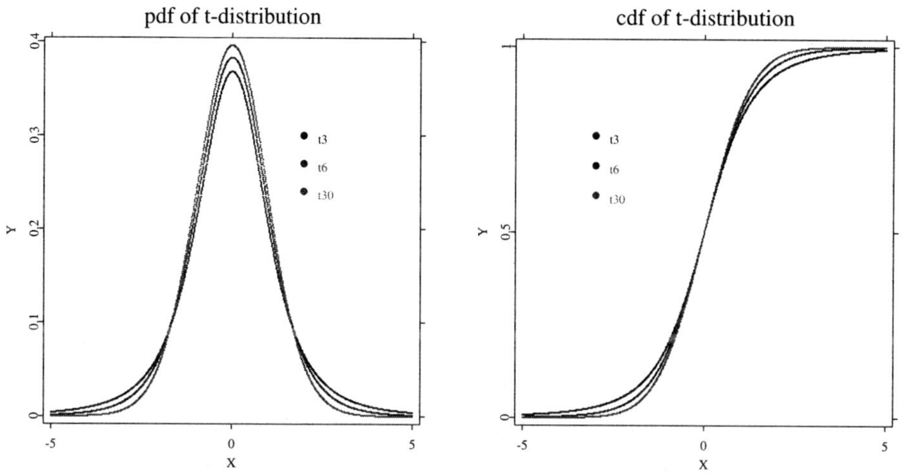

Figure 4.8. pdf (left) and cdf (right) of t-distribution with different degrees of freedom (t3 stands for t-distribution with degree of freedom 3) `MVAtdis`

is distributed as Student's t with n degrees of freedom. The t-distribution has the following density function

$$f_t(x;n) = \frac{\Gamma\left(\frac{n+1}{2}\right)}{\sqrt{n\pi}\Gamma\left(\frac{n}{2}\right)}\left(1+\frac{x^2}{n}\right)^{-\frac{n+1}{2}} \tag{4.64}$$

where n is the number of degrees of freedom, $-\infty < x < \infty$, and Γ is the gamma function, Giri (1996).

$$\Gamma(\alpha) = \int_0^\infty x^{\alpha-1}e^{-x}dx \tag{4.65}$$

The mean, variance, skewness, and kurtosis of Student's t-distribution ($n > 4$) are:

$$\begin{aligned} \mu &= 0 \\ \sigma^2 &= \frac{n}{n-2} \\ \mathit{Skewness} &= 0 \\ \mathit{Kurtosis} &= 3 + \frac{6}{n-4} \end{aligned}$$

The t-distribution is symmetric around 0, which is consistent with the fact that its mean is 0 and skewness is also 0.

Student's t-distribution approaches the normal distribution as n increases, since

$$\lim_{n\to\infty} f_t(x;n) = \frac{1}{\sqrt{2\pi}}e^{-\frac{x^2}{2}} \tag{4.66}$$

In practice the t-distribution is widely used, but its flexibility of modeling is restricted because of the integer-valued tail index.

In the tail area of the t-distribution, x is proportional to $|x|^{-(n+1)}$. In Figure 4.13 we compared the tail-behavior of t-distribution with different degrees of freedom. The left panel contains tails of pdf curves of t-distributions. With higher degree of freedom, the t-distribution decays faster. The approximation of the tail behavior $|x|^{-(n+1)}$ (right panel) shows the same result.

Laplace Distribution

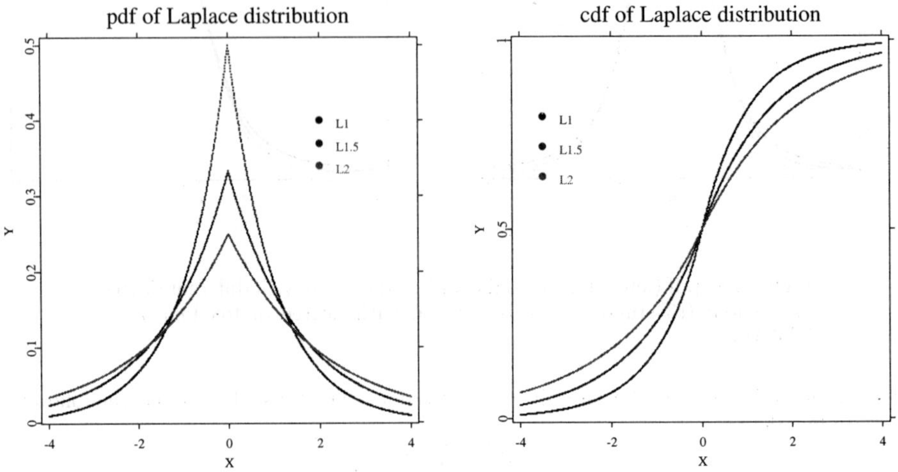

Figure 4.9. pdf (left) and cdf (right) of Laplace distribution with zero mean and different scale parameters (L1 stands for Laplace distribution with $\theta = 1$) MVAlaplacedis

The univariate Laplace distribution with mean zero was introduced by Laplace (1774). The Laplace distribution can be defined as the distribution of differences between two independent variates with identical exponential distributions. Therefore it is also called the double exponential distribution.

The Laplace distribution with mean μ and scale parameter θ has the pdf

$$f_{Laplace}(x;\mu,\theta) = \frac{1}{2\theta} e^{-\frac{|x-\mu|}{\theta}} \tag{4.67}$$

and the cdf

$$F_{Laplace}(x;\mu,\theta) = \frac{1}{2}\{1 + sign(x-\mu)(1 - e^{-\frac{|x-\mu|}{\theta}})\}, \tag{4.68}$$

where *sign* is sign function. The mean, variance, skewness, and kurtosis of the Laplace distribution are

$$\begin{aligned} \mu &= \mu \\ \sigma^2 &= 2\theta^2 \\ Skewness &= 0 \\ Kurtosis &= 6 \end{aligned}$$

With mean 0 and $\theta = 1$, we obtain the standard Laplace distribution

$$f(x) = \frac{e^{-|x|}}{2} \tag{4.69}$$

$$F(x) = \begin{cases} \frac{e^{x}}{2} & \text{for} \quad x < 0 \\ 1 - \frac{e^{-x}}{2} & \text{for} \quad x \geq 0 \end{cases} \tag{4.70}$$

Cauchy Distribution

The Cauchy distribution is motivated by the following example.

EXAMPLE 4.23 *A gangster just robbed a bank. When he finally ran to a point which is s meters away from the wall of the bank, a policeman reached the crime scene. The robber turned back and started to shoot. But he is such a poor shooter that the angle he shot (marked in Figure 4.10 as α) is uniformly distributed. The bullet hits the wall at distance x (from the center). Obviously the distribution of x, the random variable where the bullet hits the wall is of vital knowledge for the policeman.*

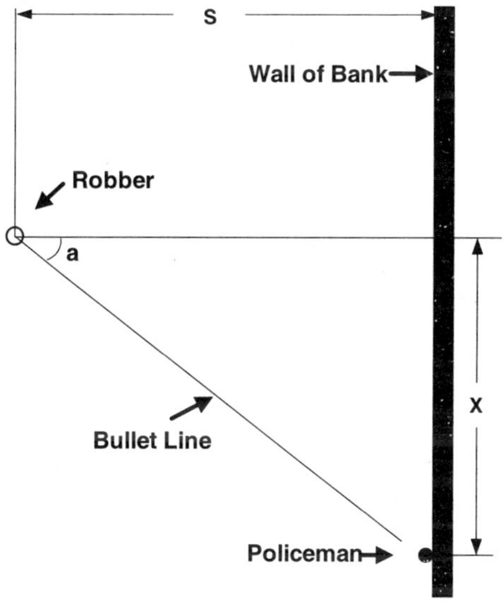

Figure 4.10. Introduction to Cauchy distribution - robber vs. policeman

Since α is uniformly distributed:

$$f(\alpha) = \frac{1}{\pi} \qquad (-\frac{\pi}{2} \leq \alpha \leq \frac{\pi}{2})$$

and

$$\begin{aligned} \tan \alpha &= \frac{x}{s} \\ \alpha &= \arctan\left(\frac{x}{s}\right) \\ d\alpha &= \frac{1}{s}\frac{1}{1+(\frac{x}{s})^2}dx \end{aligned}$$

For a small interval $d\alpha$, the probability is given by

$$\begin{aligned} f(\alpha)d\alpha &= \frac{1}{\pi}d\alpha \\ &= \frac{1}{s\pi}\frac{1}{1+(\frac{x}{s})^2}dx \end{aligned}$$

with

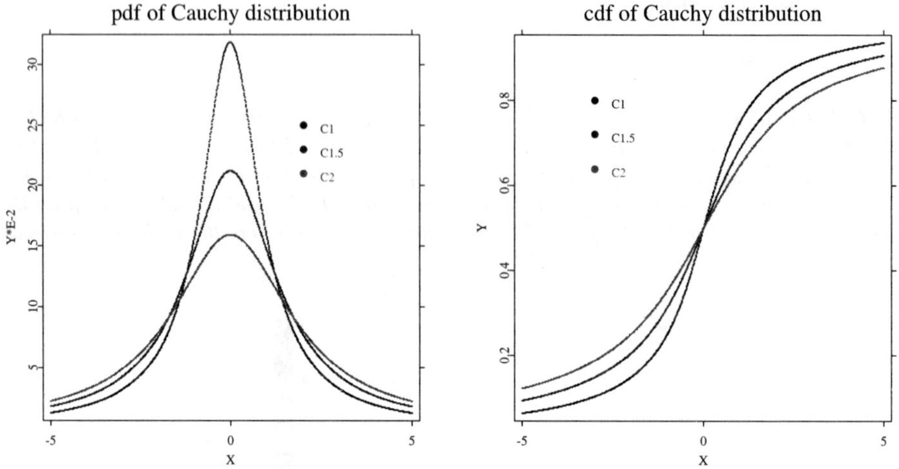

Figure 4.11. pdf (left) and cdf (right) of Cauchy distribution with $m = 0$ and different scale parameters (C1 stands for Cauchy distribution with $s = 1$) MVAcauchy

$$\begin{aligned} \int_{-\frac{\pi}{2}}^{\frac{\pi}{2}} \frac{1}{\pi}d\alpha &= 1 \\ \int_{-\infty}^{\infty} \frac{1}{s\pi}\frac{1}{1+(\frac{x}{s})^2}dx &= \frac{1}{\pi}\left\{ \arctan\left(\frac{x}{s}\right)\right\}_{-\infty}^{\infty} \\ &= \frac{1}{\pi}\left\{\frac{\pi}{2} - \left(-\frac{\pi}{2}\right)\right\} \\ &= 1 \end{aligned}$$

So the pdf of x can be written as:

$$f(x) = \frac{1}{s\pi}\frac{1}{1+(\frac{x}{s})^2}$$

The general formula for the pdf and cdf of the Cauchy distribution is

$$f_{Cauchy}(x; m, s) = \frac{1}{s\pi}\frac{1}{1+(\frac{x-m}{s})^2} \tag{4.71}$$

$$F_{Cauchy}(x; m, s) = \frac{1}{2} + \frac{1}{\pi}\arctan\left(\frac{x-m}{s}\right) \tag{4.72}$$

where m and s are location and scale parameter respectively. The case in the above example where $m = 0$ and $s = 1$ is called the standard Cauchy distribution with pdf and cdf as following,

$$f_{Cauchy}(x) = \frac{1}{\pi(1+x^2)} \tag{4.73}$$

$$F_{Cauchy}(x; m, s) = \frac{1}{2} + \frac{\arctan(x)}{\pi} \tag{4.74}$$

The mean, variance, skewness and kurtosis of Cauchy distribution are all undefined, since its moment generating function diverges. But it has mode and median, both equal the location parameter m.

Mixture Model

Mixture modeling concerns modeling a statistical distribution by a mixture (or weighted sum) of different distributions. For many choices of component density functions, the mixture model can approximate any continuous density to arbitrary accuracy, provided that the number of component density functions is sufficiently large and the parameters of the model are chosen correctly. The pdf of a mixture distribution consists of n distributions and can be written as:

$$f(x) = \sum_{l=1}^{n} w_l p_l(x) \tag{4.75}$$

under the constraints:

$$\begin{aligned} 0 \leq w_l &\leq 1 \\ \sum_{l=1}^{n} w_l &= 1 \\ \int p_l(x)dx &= 1 \end{aligned}$$

where $p_l(x)$ is the pdf of the l'th component density and w_l is a weight. The mean, variance, skewness and kurtosis of a mixture are

$$\mu = \sum_{l=1}^{n} w_l \mu_l \tag{4.76}$$

$$\sigma^2 = \sum_{l=1}^{n} w_l \{\sigma_l^2 + (\mu_l - \mu)^2\} \tag{4.77}$$

$$Skewness = \sum_{l=1}^{n} w_l \left\{ \left(\frac{\sigma_l}{\sigma}\right)^3 SK_l + \frac{3\sigma_l^2(\mu_l - \mu)}{\sigma^3} + \left(\frac{\mu_l - \mu}{\sigma}\right)^3 \right\} \tag{4.78}$$

$$Kurtosis = \sum_{l-1}^{n} w_l \left\{ \left(\frac{\sigma_l}{\sigma}\right)^4 K_l + \frac{6(\mu_l - \mu)^2\sigma_l^2}{\sigma^4} + \frac{4(\mu_l - \mu)\sigma_l^3}{\sigma^4} SK_l + \left(\frac{\mu_l - \mu}{\sigma}\right)^4 \right\}, \tag{4.79}$$

where μ_l, σ_l, SK_l and K_l are respectively mean, variance, skewness and kurtosis of l'th distribution.

Mixture models are ubiquitous in virtually every facet of statistical analysis, machine learning and data mining. For data sets comprising continuous variables, the most common approach involves mixture distributions having Gaussian components.

The pdf for a Gaussian mixture is:

$$f_{GM}(x) = \sum_{l=1}^{n} \frac{w_l}{\sqrt{2\pi}\sigma_l} e^{-\frac{(x-\mu_l)^2}{2\sigma_l^2}}. \tag{4.80}$$

For a Gaussian mixture consisting of Gaussian distributions with mean 0, this can be simplified to:

$$f_{GM}(x) = \sum_{l=1}^{n} \frac{w_l}{\sqrt{2\pi}\sigma_l} e^{-\frac{x^2}{2\sigma_l^2}}, \tag{4.81}$$

with variance, skewness and kurtosis

$$\sigma^2 = \sum_{l=1}^{n} w_l \sigma_l^2 \tag{4.82}$$

$$Skewness = 0 \tag{4.83}$$

$$Kurtosis = \sum_{l=1}^{n} w_l \left(\frac{\sigma_l}{\sigma}\right)^4 3 \tag{4.84}$$

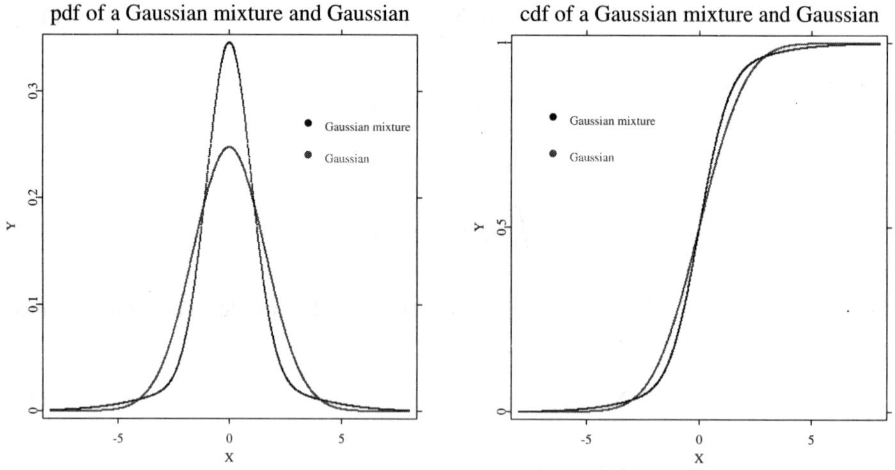

Figure 4.12. pdf (left) and cdf (right) of a Gaussian mixture (Example 4.23) MVAmixture

EXAMPLE 4.24 *Consider a Gaussian Mixture which is 80% N(0,1) and 20% N(0,9). The pdf of N(0,1) and N(0,9) are*

$$f_{N(0,1)}(x) = \frac{1}{\sqrt{2\pi}} e^{-\frac{x^2}{2}}$$

$$f_{N(0,9)}(x) = \frac{1}{3\sqrt{2\pi}} e^{-\frac{x^2}{18}}$$

so the pdf of the Gaussian Mixture is

$$f_{GM}(x) = \frac{1}{5\sqrt{2\pi}} \left(4e^{-\frac{x^2}{2}} + \frac{1}{3} e^{-\frac{x^2}{18}}\right)$$

Notice that the Gaussian Mixture is not a Gaussian distribution:

$$\begin{aligned}
\mu &= 0 \\
\sigma^2 &= 0.8 \times 1 + 0.2 \times 9 = 2.6 \\
Skewness &= 0 \\
Kurtosis &= 0.8 \times \left(\frac{1}{\sqrt{2.6}}\right)^4 \times 3 + 0.2 \times \left(\frac{\sqrt{9}}{\sqrt{2.6}}\right)^4 \times 3 = 7.54
\end{aligned}$$

The kurtosis of this Gaussian mixture is higher than 3.

A summary of the basic statistics is given in Table 4.2.

	t	Laplace	Cauchy
mean	0	μ	not defined
variance	$\frac{n}{n-2}$	$2\theta^2$	not defined
skewness	0	0	not defined
kurtosis	$3 + \frac{6}{n-4}$	6	not defined

Table 4.2. basic statistics of t, Laplace and Cauchy distribution

GH	
mean	$\mu + \frac{\delta\beta}{\sqrt{\alpha^2+\beta^2}} \frac{K_{\lambda+1}(\delta\sqrt{\alpha^2+\beta^2})}{K_\lambda(\delta\sqrt{\alpha^2+\beta^2})}$
variance	$\delta^2 \left[\frac{K_{\lambda+1}(\delta\sqrt{\alpha^2+\beta^2})}{\delta\sqrt{\alpha^2+\beta^2}K_\lambda(\delta\sqrt{\alpha^2+\beta^2})} + \frac{\beta^2}{\alpha^2+\beta^2} \left[\frac{K_{\lambda+2}(\delta\sqrt{\alpha^2+\beta^2})}{K_\lambda(\delta\sqrt{\alpha^2+\beta^2})} - \left\{ \frac{K_{\lambda+1}(\delta\sqrt{\alpha^2+\beta^2})}{K_\lambda(\delta\sqrt{\alpha^2+\beta^2})} \right\}^2 \right] \right]$
Mixture	
mean	$\sum_{l=1}^n w_l \mu_l$
variance	$\sum_{l=1}^n w_l \{\sigma_l^2 + (\mu_l - \mu)^2\}$
skewness	$\sum_{l=1}^n w_l \left\{ \left(\frac{\sigma_l}{\sigma}\right)^3 SK_l + \frac{3\sigma_l^2(\mu_l-\mu)}{\sigma^3} + \left(\frac{\mu_l-\mu}{\sigma}\right)^3 \right\}$
kurtosis	$\sum_{l=1}^n w_l \left\{ \left(\frac{\sigma_l}{\sigma}\right)^4 K_l + \frac{6(\mu_l-\mu)^2\sigma_l^2}{\sigma^4} + \frac{4(\mu_l-\mu)\sigma_l^3}{\sigma^4} SK_l + \left(\frac{\mu_l-\mu}{\sigma}\right)^4 \right\}$

Table 4.3. basic statistics of GH distribution and mixture model

Multivariate Generalized Hyperbolic Distribution

The multivariate Generalized Hyperbolic Distribution (GH_d) has the following pdf

$$f_{GH_d}(x; \lambda, \alpha, \beta, \delta, \Delta, \mu) = a_d \frac{K_{\lambda-\frac{d}{2}}\left\{\alpha\sqrt{\delta^2 + (x-\mu)^\top \Delta^{-1}(x-\mu)}\right\}}{\left\{\alpha^{-1}\sqrt{\delta^2 + (x-\mu)^\top \Delta^{-1}(x-\mu)}\right\}^{\frac{d}{2}-\lambda}} e^{\beta^\top(x-\mu)} \tag{4.85}$$

$$a_d = a_d(\lambda, \alpha, \beta, \delta, \Delta) = \frac{\left(\sqrt{\alpha^2 - \beta^\top \Delta\beta}/\delta\right)^\lambda}{(2\pi)^{\frac{d}{2}} K_\lambda(\delta\sqrt{\alpha^2 - \beta^\top\Delta\beta}}, \tag{4.86}$$

and characteristic function

$$\phi(t) = \left(\frac{\alpha^2 - \beta^\top \Delta \beta}{\alpha^2 - \beta^\top \Delta \beta + \frac{1}{2} t^\top \Delta t - i\beta^\top \Delta t}\right)^{\frac{\lambda}{2}} \times \frac{K_\lambda\left(\delta\sqrt{\alpha^2 - \beta^\top \Delta \beta^\top + \frac{1}{2} t^\top \Delta t - i\beta^\top \Delta t}\right)}{K_\lambda\left(\delta\sqrt{\alpha^2 - \beta^\top \Delta \beta^\top}\right)} \tag{4.87}$$

These parameters have the following domain of variation:

$$\begin{array}{ll} \lambda \in \mathbb{R}, & \beta, \mu \in \mathbb{R}^d \\ \delta > 0, & \alpha > \beta^\top \Delta \beta \\ \Delta \in \mathbb{R}^{d \times d} & \text{positive definite matrix} \\ |\Delta| = 1 & \end{array}$$

For $\lambda = \frac{d+1}{2}$ we obtain the multivariate hyperbolic (HYP) distribution; for $\lambda = -\frac{1}{2}$ we get the multivariate normal inverse Gaussian (NIG) distribution.

Blæsild and Jensen (1981) introduced a second parameterization (ζ, Π, Σ), where

$$\zeta = \delta\sqrt{\alpha^2 - \beta^\top \Delta \beta} \tag{4.88}$$

$$\Pi = \beta\sqrt{\frac{\Delta}{\alpha^2 - \beta^\top \Delta \beta}} \tag{4.89}$$

$$\Sigma = \delta^2 \Delta \tag{4.90}$$

The mean and variance of $X \sim GH_d$

$$\mathrm{E}[X] = \mu + \delta R_\lambda(\zeta)\Pi\Delta^{\frac{1}{2}} \tag{4.91}$$

$$\mathrm{Var}[X] = \delta^2\{\zeta^{-1} R_\lambda(\zeta)\Delta + S_\lambda(\zeta)(\Pi\Delta^{\frac{1}{2}})^\top(\Pi\Delta^{\frac{1}{2}})\} \tag{4.92}$$

where

$$R_\lambda(x) = \frac{K_{\lambda+1}(x)}{K_\lambda(x)} \tag{4.93}$$

$$S_\lambda(x) = \frac{K_{\lambda+2}(x)K_\lambda(x) - K^2_{\lambda+1}(x)}{K^2_\lambda(x)} \tag{4.94}$$

THEOREM 4.12 *Suppose that X is a d-dimensional variate distributed according to the generalized hyperbolic distribution GH_d. Let (X_1, X_2) be a partitioning of X, let r and k denote the dimensions of X_1 and X_2, respectively, and let (β_1, β_2) and (μ_1, μ_2) be similar partitions of β and μ, let*

$$\Delta = \begin{pmatrix} \Delta_{11} & \Delta_{12} \\ \Delta_{21} & \Delta_{22} \end{pmatrix} \tag{4.95}$$

be a partition of Δ such that Δ_{11} is a $r \times r$ matrix. Then one has the following

1. *The distribution of X_1 is the r-dimensional generalized hyperbolic distribution,*

$GH_r(\lambda^*, \alpha^*, \beta^*, \delta^*, \mu^*, \Delta^*)$, where

$$\begin{aligned}
\lambda^* &= \lambda \\
\alpha^* &= |\Delta_{11}|^{-\frac{1}{2r}}\{\alpha^2 - \beta_2(\Delta_{22} - \Delta_{21}\Delta_{11}^{-1}\Delta_{12})\beta_2^\top\}^{\frac{1}{2}} \\
\beta^* &= \beta_1 + \beta_2\Delta_{21}\Delta_{11}^{-1} \\
\delta^* &= \delta|\Delta_{11}|^{\frac{1}{2\rho}} \\
\mu^* &= \mu_1 \\
\Delta^* &= |\Delta|^{-\frac{1}{r}}\Delta_{11}
\end{aligned}$$

2. *The conditional distribution of X_2 given $X_1 = x_1$ is the k-dimensional generalized hyperbolic distribution $GH_k(\tilde{\lambda}, \tilde{\alpha}, \tilde{\beta}, \tilde{\delta}, \tilde{\mu}, \tilde{\Delta})$, where*

$$\begin{aligned}
\tilde{\lambda} &= \lambda - \frac{r}{2} \\
\tilde{\alpha} &= \alpha|\Delta_{11}|^{\frac{1}{2k}} \\
\tilde{\beta} &= \beta_2 \\
\tilde{\delta} &= |\Delta_{11}|^{-\frac{1}{2k}}\{\delta^2 + (x_1 - \mu_1)\Delta_{11}^{-1}(x_1 - \mu_1)^\top\}^{\frac{1}{2}} \\
\tilde{\mu} &= \mu_2 + (x_1 - \mu_1)\Delta_{11}^{-1}\Delta_{12} \\
\tilde{\Delta} &= |\Delta_{11}|^{\frac{1}{k}}(\Delta_{22} - \Delta_{21}\Delta_{11}^{-1}\Delta_{12})
\end{aligned}$$

3. *Let $Y = XA + B$ be a regular affine transformation of X and let $||A||$ denote the absolute value of the determinant of A. The distribution of Y is the d-dimensional generalized hyperbolic distribution $GH_d(\lambda^+, \alpha^+, \beta^+, \delta^+, \mu^+, \Delta^+)$, where*

$$\begin{aligned}
\lambda^+ &= \lambda \\
\alpha^+ &= \alpha||A||^{-\frac{1}{d}} \\
\beta^+ &= \beta(A^{-1})^\top \\
\delta^+ &= ||A||^{\frac{1}{d}} \\
\mu^+ &= \mu A + B \\
\Delta^+ &= ||A||^{-\frac{2}{d}}A^\top\Delta A
\end{aligned}$$

Multivariate t-distribution

If X and Y are independent and distributed as $N_p(\mu, \Sigma)$ and $\mathcal{X}_n^2$ respectively, and $X\sqrt{n/Y} = t - \mu$, then the pdf of t is given by

$$f_t(t; n, \Sigma, \mu) = \frac{\Gamma\{(n+p)/2\}}{\Gamma(n/2)n^{p/2}\pi^{p/2}|\Sigma|^{1/2}\{1 + \frac{1}{n}(t-\mu)^\top\Sigma^{-1}(t-\mu)\}^{(n+p)/2}} \tag{4.96}$$

The distribution of t is the noncentral t-distribution with n degrees of freedom and the noncentrality parameter μ, Giri (1996).

Multivariate Laplace Distribution

Let g and G be the pdf and cdf of a d-dimensional Gaussian distribution $N_d(0,\Sigma)$, the pdf and cdf of a multivariate Laplace distribution can be written as

$$f_{MLaplace_d}(x;m,\Sigma) = \int_0^\infty g(z^{-\frac{1}{2}}x - z^{\frac{1}{2}}m)z^{-\frac{d}{2}}e^{-z}dz \tag{4.97}$$

$$F_{MLaplace_d}(x,m,\Sigma) = \int_0^\infty G(z^{-\frac{1}{2}}x - z^{\frac{1}{2}}m)e^{-z}dz \tag{4.98}$$

the pdf can also be described as

$$\begin{aligned} f_{MLaplace_d}(x;m,\Sigma) &= \frac{2e^{x^\top\Sigma^{-1}m}}{(2\pi)^{\frac{d}{2}}|\Sigma|^{\frac{1}{2}}}\left(\frac{x^\top\Sigma^{-1}x}{2+m^\top\Sigma^{-1}m}\right)^{\frac{\lambda}{2}} \\ &\quad \times K_\lambda\left(\sqrt{(2+m^\top\Sigma^{-1}m)(x^\top\Sigma^{-1}x)}\right) \end{aligned} \tag{4.99}$$

where $\lambda = \frac{2-d}{2}$ and $K_\lambda(x)$ is the modified Bessel function of the third kind

$$K_\lambda(x) = \frac{1}{2}\left(\frac{x}{2}\right)^\lambda \int_0^\infty t^{-\lambda-1}e^{-t-\frac{x^2}{4t}}dt, \qquad x>0 \tag{4.100}$$

Multivariate Laplace distribution has mean and variance

$$\mathrm{E}[X] = m \tag{4.101}$$

$$\mathrm{Cov}[X] = \Sigma + mm^\top \tag{4.102}$$

Multivariate Mixture Model

A multivariate mixture model comprises multivariate distributions, e.g. the pdf of a multivariate Gaussian distribution can be written as

$$f(x) = \sum_{l=1}^{n} \frac{w_l}{|2\pi\Sigma_l|^{\frac{1}{2}}} e^{-\frac{1}{2}(x-\mu_l)^\top\Sigma^{-1}(x-\mu_l)} \tag{4.103}$$

Generalized Hyperbolic Distribution

The GH distribution has an exponential decaying speed

$$f_{GH}(x;\lambda,\alpha,\beta,\delta,\mu=0) \sim x^{\lambda-1}e^{-(\alpha-\beta)x} \qquad \text{as} \quad x\to\infty, \tag{4.104}$$

Figure 4.14 illustrates the tail behavior of GH distributions with different value of λ with $\alpha=1,\beta=0,\delta=1,\mu=0$. The left panel contains part of pdf curves of GH distribution and the right panel demonstrates the approximation by the function mentioned above. It is clear that among the four distributions, GH with $\lambda=1.5$ has the lowest decaying speed, while NIG decays fastest.

In Figure 4.15, Chen, Härdle and Jeong (2005), four distributions and especially their tail-behavior are compared. In order to keep the comparability of these distributions, we specified

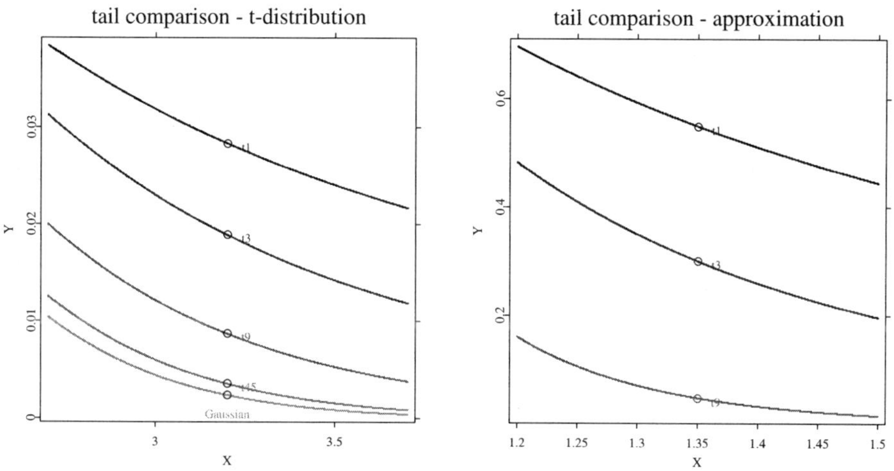

Figure 4.13. Tail comparison of t-distribution, pdf (left) and approximation (right) `MVAtdistail`

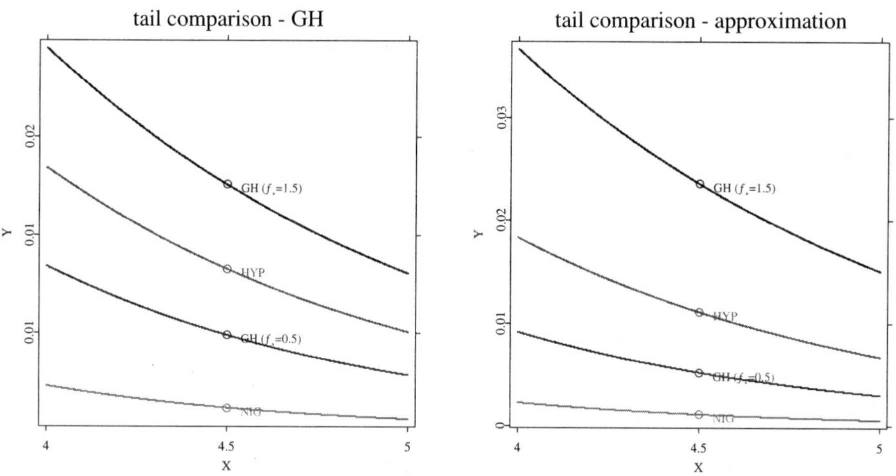

Figure 4.14. Tail comparison of GH distribution, pdf (left) and approximation (right) `MVAghdistail`

the means to 0 and standardized the variances to 1. Furthermore we used one important subclass of the GH distribution: the normal-inverse Gaussian (NIG) distribution with $\lambda = -\frac{1}{2}$ introduced above. On the left panel, the complete forms of these distributions are revealed. The Cauchy (dots) distribution has the lowest peak and the fattest tails. In other words, it has the flattest distribution. The NIG distribution decays second fast in the tails although it has the highest peak, which is more clearly displayed on the right panel.

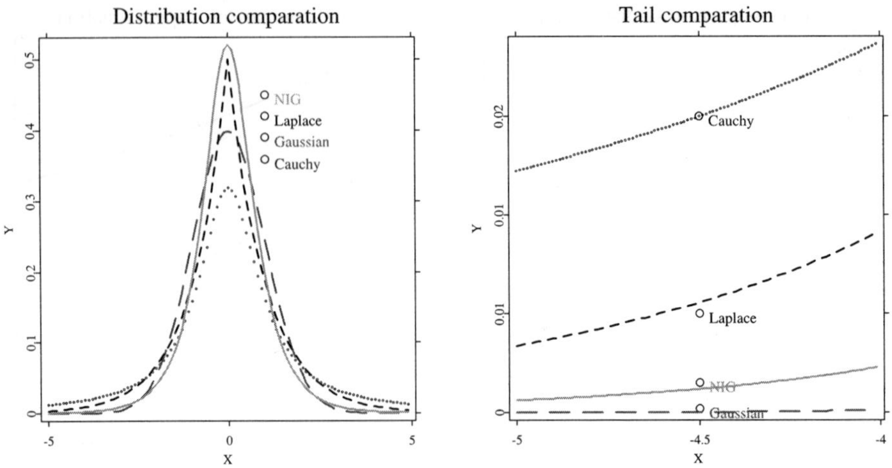

Figure 4.15. Graphical comparison of the NIG distribution (line), standard normal distribution MVAghadatail

4.7 Copulae

The cumulative distribution function (cdf) of a 2-dimensional vector (X_1, X_2) is given by

$$F(x_1, x_2) = P(X_1 \leq x_1, Y_1 \leq y_1) \tag{4.105}$$

For the case that X_1 and X_2 are independent, their joint cumulative distribution function $F(x_1, x_2)$ can be written as a product of their 1-dimensional marginals:

$$F(x_1, x_2) = F_{X_1}(x_1) F_{X_2}(x_2) = P(X_1 \leq x_1) P(X_2 \leq x_2) \tag{4.106}$$

But how can we model dependence of X_1 and X_2? Most people would suggest linear correlation. Correlation is though an appropriate measure of dependence only when the random variables have an eliptical or spherical distribution, which include the normal multivariate distribution. Although the terms "correlation" and "dependency" are often used interchangeably, correlation is actually a rather imperfect measure of dependency, and there are many circumstances where correlation should not be used.

Copulae represent an elegant concept of connecting marginals with joint cummulative distribution functions. Copulae are functions that join or "couple" multivariate distribution functions to their 1-dimensional marginal distribution functions. Let us consider a d-dimensional vector $X = (X_1, ..., X_d)^\top$. Using copulae, the marginal distribution functions $F_{X_i} (i = 1, ..., d)$ can be separately modelled from their dependence structure and then coupled together to form the multivariate distribution F_X. Copula functions have a long history in probability theory and statistics, since they date back to (Sklar, 1959). Their application in finance is very recent: the idea first appears in (Embrechts, Mc Neil and Straumann, 1999) in connection with correlation as a measure of dependence. Copulae are important in Value-at-Risk calculations and constitute an essential tool in quantitative finance (Härdle, Kleinow and Stahl, 2002).

First let us concentrate on the 2-dimensional case, then we will extend this concept to the d-dimensional case, for a random variable in $\mathbb{R}^d$ with $d \geq 1$. To be able to define a copula

function, first we need to represent a concept of the *volume of a rectangle, a 2-increading function* and *a grounded function.*
Let U_1 and U_2 be two sets in $\overline{\mathbb{R}} = \mathbb{R} \cup \{+\infty\} \cup \{-\infty\}$ and consider the function $F : U_1 \times U_2 \longrightarrow \overline{\mathbb{R}}$.

DEFINITION 4.2 *The F-volume of a rectangle $B = [x_1, x_2] \times [y_1, y_2] \subset U_1 \times U_2$ is defined as:*

$$V_F(B) = F(x_2, y_2) - F(x_1, y_2) - F(x_2, y_1) + F(x_1, y_1) \tag{4.107}$$

DEFINITION 4.3 *F is said to be a 2-increasing function if for every $B = [x_1, x_2] \times [y_1, y_2] \subset U_1 \times U_2$,*

$$V_F(B) \ \geq \ 0 \tag{4.108}$$

REMARK 4.2 *Note, that "to be 2-increasing function" neither implies nor is implied by "to be increasing in each argument".*

The following lemmas (Nelsen, 1999) will be very useful later for establishing the continuity of copulae.

LEMMA 4.1 *Let U_1 and U_2 be nonempty sets in $\overline{\mathbb{R}}$ and let $F : U_1 \times U_2 \longrightarrow \overline{\mathbb{R}}$ be a two-increasing function. Let x_1, x_2 be in U_1 with $x_1 \leq x_2$, and y_1, y_2 be in U_2 with $y_1 \leq y_2$. Then the function $t \mapsto F(t, y_2) - F(t, y_1)$ is nondecreasing on U_1 and the function $t \mapsto F(x_2, t) - F(x_1, t)$ is nondecreasing on U_2.*

DEFINITION 4.4 *If U_1 and U_2 have a smallest element $\min U_1$ and $\min U_2$ respectively, then we say, that a function $F : U_1 \times U_2 \longrightarrow \mathbb{R}$ is grounded if :*

$$\begin{aligned} \text{for all } x \in U_1 : \ \ F(x, \min U_2) &= 0 \text{ and} && (4.109)\\ \text{for all } y \in U_2 : \ \ F(\min U_1, y) &= 0 && (4.110)\end{aligned}$$

In the following, we will refer to this definition of a distribution function :

DEFINITION 4.5 *A distribution function is a function from $\overline{\mathbb{R}}^2 \mapsto [0, 1]$ which:*

- *is grounded*
- *is 2-increasing*
- *satisfies $F(\infty, \infty) = 1$*

LEMMA 4.2 *Let U_1 and U_2 be nonempty sets in $\overline{\mathbb{R}}$ and let $F : U_1 \times U_2 \longrightarrow \overline{\mathbb{R}}$ be a grounded two-increasing function. Then F is nondecreasing in each argument.*

DEFINITION 4.6 *If U_1 and U_2 have a greatest element $\max U_1$ and $\max U_2$ respectively, then we say, that a function $F : U_1 \times U_2 \longrightarrow \mathbb{R}$ has margins and that the margins of F are given by:*

$$\begin{aligned} F(x) &= F(x, \max U_2) \text{ for all } x \in U_1 && (4.111)\\ F(y) &= F(\max U_1, y) \text{ for all } y \in U_2 && (4.112)\end{aligned}$$

LEMMA 4.3 *Let U_1 and U_2 be nonempty sets in $\overline{\mathbb{R}}$ and let $F : U_1 \times U_2 \longrightarrow \overline{\mathbb{R}}$ be a grounded two-increasing function which has margins. Let (x_1, y_1), $(x_2, y_2) \in S_1 \times S_2$. Then*

$$|F(x_2, y_2) - F(x_1, y_1)| \leq |F(x_2) - F(x_1)| + |F(y_2) - F(y_1)| \tag{4.113}$$

DEFINITION 4.7 *A two-dimensional copula is a function C defined on the unit square $I^2 = I \times I$ with I the unit interval ($I = [0, 1]$) such that*

- *for every $u \in I$ holds: $C(u, 0) = C(0, v) = 0$, i.e. C is grounded.*
- *for every $u_1, u_2, v_1, v_2 \in I$ with $u_1 \leq u_2$ and $v_1 \leq v_2$ holds:*

$$C(u_2, v_2) - C(u_2, v_1) - C(u_1, v_2) + C(u_1, v_1) \geq 0, \tag{4.114}$$

  i.e. C is 2-increasing.
- *for every $u \in I$ holds $C(u, 1) = u$ and $C(1, v) = v$.*

Informally, a copula is a joint distribution function defined on the unit square $[0, 1]^2$ which has uniform marginals. That means that if $F_{X_1}(x_1)$ and $F_{X_2}(x_2)$ are univariate distribution functions, then $C\{F_{X_1}(x_1), F_{X_2}(x_2)\}$ is a 2-dimensional distribution function with marginals $F_{X_1}(x_1)$ and $F_{X_2}(x_2)$.

EXAMPLE 4.25 *The functions* $\max(u + v - 1, 0)$, uv, $\min(u, v)$ *can be easily checked to be copula functions. They are called respectively the minimum, product and maximum copula.*

EXAMPLE 4.26 *Consider the function*

$$\begin{aligned} C_\rho^{Gauss}(u, v) &= \Phi_\rho\left\{\Phi^{-1}(u), \Phi^{-1}(v)\right\} \\ &= \int_{-\infty}^{\Phi_1^{-1}(u)} \int_{-\infty}^{\Phi_2^{-1}(v)} f_\rho(x_1, x_2)dx_2dx_1 \end{aligned} \tag{4.115}$$

where Φ_ρ is the joint 2-dimensional standard normal distribution function with correlation coefficient ρ, while Φ_1 and Φ_2 refer to standard normal cdfs and

$$f_\rho(x_1, x_2) = \frac{1}{2\pi\sqrt{1-\rho^2}} \exp\left\{-\frac{x_1^2 - 2\rho x_1 x_2 + x_2^2}{2(1-\rho^2)}\right\} \tag{4.116}$$

denotes the bivariate normal pdf.

It is easy to see, that C^{Gauss} is a copula, the so called Gaussian or normal copula, since it is 2-increasing and

$$\Phi_\rho\left\{\Phi^{-1}(u), \Phi^{-1}(0)\right\} = \Phi_\rho\left\{\Phi^{-1}(0), \Phi^{-1}(v)\right\} = 0 \tag{4.117}$$

$$\Phi_\rho\left\{\Phi^{-1}(u), \Phi^{-1}(1)\right\} = u \text{ and } \Phi_\rho\left\{\Phi^{-1}(1), \Phi^{-1}(v)\right\} = v \tag{4.118}$$

EXAMPLE 4.27 *One can easily verify, that the function*

$$C_\theta^{GH}(u, v) = \exp\left[-\left\{(-\ln u)^\theta + (-\ln v)^\theta\right\}^{1/\theta}\right] \tag{4.119}$$

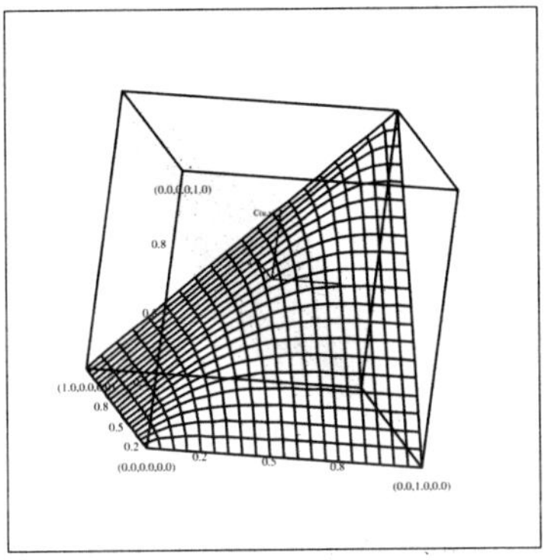

Figure 4.16. Surface plot of the Gumbel-Hougaard copula, $\theta = 3$ MVAghsurface

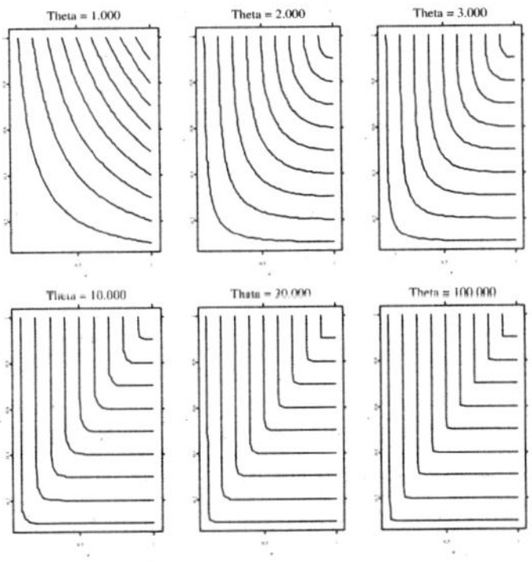

Figure 4.17. Contour plots of the Gumbel-Hougard copula MVAghcontour

is also a copula function, a so called Gumbel-Hougaard family of copulae. The parameter θ may take values in the interval $[1, \infty)$. Figure 4.16 represents a surface plot of the Gumbel-Hougaard copula in the case of $\theta = 3$.

A simple and useful way to represent the graph of a copula is the contour diagram, that is, graphs of its level sets - the sets in I^2 given by $C(u, v) =$ a constant. In the Figure 4.17 we present the countour diagrams of the Gumbel-Hougard copula for different values of the copula parameter θ.

For $\theta = 1$ the Gumbel-Hougaard copula reduces to the product copula, i.e.

$$C_1(u, v) = \Pi(u, v) = uv \tag{4.120}$$

For $\theta \to \infty$, one finds for the Gumbel-Hougaard copula:

$$C_\theta(u, v) \longrightarrow \min(u, v) = M(u, v) \tag{4.121}$$

where M is also a copula such that $C(u, v) \leq M(u, v)$ for an arbitrary copula C. The copula M is called the Fréchet-Hoeffding upper bound.
The two-dimensional function $W(u, v) = \max(u + v - 1, 0)$ defines a copula with $W(u, v) \leq C(u, v)$ for any other copula C. W is called the Fréchet-Hoeffding lower bound.

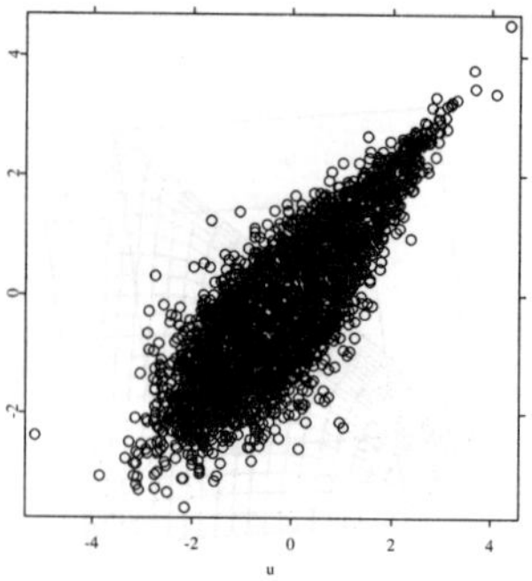

Figure 4.18. 10000-sample output for $\sigma_1 = 1$, $\sigma_2 = 1$, $\theta = 3$ MVAsample1000

In the next Figure 4.18 we show an example of Gumbel-Hougaard copula sampling for fixed parameters $\sigma_1 = 1$, $\sigma_2 = 1$ and $\theta = 3$.

It follows immediately from the definition that copulae are increasing in each argument and that their range is the unit interval. In addition, one can demonstrate the so-called Fréchet-Hoeffding inequality, which we have already used in Example 1.3. and which states that each copula function is bounded by the minimum and maximum one:

$$W(u,v) = \max(u+v-1,0) \le C(u,v) \le \min(u,v) = M(u,v) \tag{4.122}$$

The full relationship between copula functions and random variables however depends on the following theorem, due to Sklar, which says that not only copulae are joint distribution functions, but the converse also holds: joint distribution functions can be rewritten in terms of the marginals and a copula. It is very useful in modelling because we often only know the marginal distributions. This result implies that the construction of any multivariate distribution can be divided into the construction of the marginals and the copula separately.

THEOREM 4.13 **(Sklar's theorem)** *Let F be a two-dimensional distribution function with marginal distribution functions F_{X_1} and F_{X_2}. Then a copula C exists such that for all x_1, $x_2 \in \overline{\mathbb{R}}^2$:*

$$F(x_1, x_2) = C\{F_{X_1}(x_1), F_{X_2}(x_2)\} \tag{4.123}$$

Moreover, if F_{X_1} and F_{X_2} are continuous then C is unique. Otherwise C is uniquely determined on the Cartesian product $Im(F_{X_1}) \times Im(F_{X_2})$.
Conversely, if C is a copula and F_{X_1} and F_{X_2} are distribution functions then F defined by (24) is a two-dimensional distribution function with marginals F_{X_1} and F_{X_2}.

With the Sklar's theorem the use of the name "copula" becomes obvious. It was chosen to describe "a function that links a multidimensional distribution to its 1-dimensional marginals".

EXAMPLE 4.28 *Let us verify that the Gaussian copula satisfies Sklar's theorem in both directions. On the one side, let*

$$F(x_1, x_2) = \int_{-\infty}^{x_1}\int_{-\infty}^{x_2} \frac{1}{2\pi\sqrt{1-\rho^2}} \exp\left\{-\frac{x_1^2 - 2\rho x_1 x_2 + x_2^2}{2(1-\rho^2)}\right\} dx_2 dx_1 \tag{4.124}$$

be a 2-dimensional normal distribution function with standard normal cdf's $F_{X_1}(x_1)$ *and* $F_{X_2}(x_2)$. *Since* $F_{X_1}(x_1)$ *and* $F_{X_2}(x_2)$ *are continuous, a unique copula* C *exists such that for all* $x_1, x_2 \in \overline{\mathbb{R}}^2$ *a 2-dimensional distribution function can be written as a copula in* $F_{X_1}(x_1)$ *and* $F_{X_2}(x_2)$:

$$F(x_1, x_2) = C\{\Phi_{X_1}(x_1), \Phi_{X_2}(x_2)\} \tag{4.125}$$

The Gaussian copula satisfies the above equality, therefore it is the unique copula mentioned in Sklar's theorem. This proves that the Gaussian copula, together with Gaussian marginals, gives the two-dimensional normal distribution.

Conversely, if C *is a copula and* F_{X_1} *and* F_{X_2} *are standard normal distribution functions, then*

$$C\{F_{X_1}(x_1), F_{X_2}(x_2)\} \tag{4.126}$$

$$= \int_{-\infty}^{\phi_1^{-1}\{F_{X_1}(x_1)\}} \int_{-\infty}^{\phi_2^{-1}\{F_{X_2}(x_2)\}} \frac{1}{2\pi\sqrt{1-\rho^2}} \exp\left\{-\frac{x_1^2 - 2\rho x_1 x_2 + x_2^2}{2(1-\rho^2)}\right\} dx_2 dx_1$$

is evidently a joint (two-dimensional) distribution function. Its margins are

$$C\{F_{X_1}(x_1), F_{X_2}(+\infty)\} = \Phi_\rho\left[\Phi^{-1}\{F_{X_1}(x_1)\}, +\infty\right] = F_{X_1}(x_1) \tag{4.127}$$

$$C\{F_{X_1}(+\infty), F_{X_2}(x_2)\} = \Phi_\rho\left[+\infty, \Phi^{-1}\{F_{X_2}(x_2)\}\right] = F_{X_2}(x_2) \tag{4.128}$$

The following proposition shows one attractive feature of the copula representation of dependence, i.e. that the dependence structure described by a copula is invariant under increasing and continuous transformations of the marginal distributions.

THEOREM 4.14 *If* (X_1, X_2) *have copula* C *and set* g_1, g_2 *two continuous increasing functions, then* $\{g_1(X_1), g_2(X_2)\}$ *have the copula* C*, too.*

EXAMPLE 4.29 *Independence implies that the product of the cdf's* F_{X_1} *and* F_{X_2} *equals the joint distribution function* F*, i.e.:*

$$F(x_1, x_2) = F_{X_1}(x_1)F_{X_2}(x_2) \tag{4.129}$$

Thus, we obtain the independence or product copula $C = \Pi(u, v) = uv$.

While it is easily understood how a product copula describes an independence relationship, the converse is also true. Namely, the joint distribution function of two independent random variables can be interpreted as a product copula. This concept is formalized in the following theorem:

THEOREM 4.15 *Let* X_1 *and* X_2 *be random variables with continuous distribution functions* F_{X_1} *and* F_{X_2} *and the joint distribution function* F. *Then* X_1 *and* X_2 *are independent if and only if* $C_{X_1,X_2} = \Pi$.

EXAMPLE 4.30 *Let us consider the Gaussian copula for the case* $\rho = 0$*, i.e. vanishing correlation. In this case the Gaussian copula becomes*

$$\begin{aligned} C_0^{Gauss}(u,v) &= \int_{-\infty}^{\Phi_1^{-1}(u)} \varphi(x_1)dx_1 \int_{-\infty}^{\Phi_2^{-1}(v)} \varphi(x_2)dx_2 \\ &= uv \\ &= \Pi(u,v) \end{aligned} \tag{4.130}$$

The following theorem, which follows directly from Lemma 4.3, establishes the continuity of copulae .

THEOREM 4.16 *Let* C *be a copula. Then for any* $u_1, v_1, u_2, v_2 \in I$ *holds*

$$|C(u_2,v_2) - C(u_1,v_1)| \leq |u_2 - u_1| + |v_2 - v_1| \tag{4.131}$$

From (4.131) it follows that every copula C is uniformly continuous on its domain.

A further important property of copulae concerns the partial derivatives of a copula with respect to its variables:

THEOREM 4.17 *Let* $C(u,v)$ *be a copula. For any* $u \in I$*, the partial derivative* $\frac{\partial C(u,v)}{\partial v}$ *exists for almost all* $u \in I$*. For such* u *and* v *one has:*

$$\frac{\partial C(u,v)}{\partial v} \in I \tag{4.132}$$

The analogous statement is true for the partial derivative $\frac{\partial C(u,v)}{\partial u}$*:*

$$\frac{\partial C(u,v)}{\partial u} \in I \tag{4.133}$$

Moreover, the functions

$$u \mapsto C_v(u) \stackrel{\text{def}}{=} \partial C(u,v)/\partial v \text{ and}$$

$$v \mapsto C_u(v) \stackrel{\text{def}}{=} \partial C(u,v)/\partial u$$

are defined and nonincreasing almost everywhere on I*.*

Until now, we have considered copulae only in a 2-dimensional setting. Let us now extend this concept to the d-dimensional case, for a random variable in $\mathbb{R}^d$ with $d \geq 1$.

Let $U_1, U_2, ..., U_d$ be nonempty sets in $\overline{\mathbb{R}}$ and consider the function $F : U_1 \times U_2 \times ... \times U_d \longrightarrow \overline{\mathbb{R}}$. For $a = (a_1, a_2, ..., a_d)$ and $b = (b_1, b_2, ..., b_d)$ with $a \leq b$ (i.e. $a_k \leq b_k$ for all k) let $B = [a,b] = [a_1, b_1] \times [a_2, b_2] \times ... \times [a_n, b_n]$ be the d-box with vertices $c = (c_1, c_2, ..., c_d)$. It is obvious, that each c_k is either equal to a_k or to b_k.

DEFINITION 4.8 *The F-volume of a d-box* $B = [a,b] = [a_1,b_1] \times [a_2,b_2] \times ... \times [a_d,b_d] \subset U_1 \times U_2 \times ... \times U_d$ *is defined as follows:*

$$V_F(B) = \sum_{k=1}^{d} sign(c_k)F(c_k) \tag{4.134}$$

where $sign(c_k) = 1$, *if* $c_k = a_k$ *for even* k *and* $sign(c_k) = -1$, *if* $c_k = a_k$ *for odd* k.

EXAMPLE 4.31 *For the case* $d = 3$, *the F-volume of a 3-box* $B = [a,b] = [x_1,x_2] \times [y_1,y_2] \times [z_1,z_2]$ *is defined as:*

$$\begin{aligned} V_F(B) = & F(x_2,y_2,z_2) - F(x_2,y_2,z_1) - F(x_2,y_1,z_2) - F(x_1,y_2,z_2) \\ & + F(x_2,y_1,z_1) + F(x_1,y_2,z_1) + F(x_1,y_1,z_2) - F(x_1,y_1,z_1) \end{aligned}$$

DEFINITION 4.9 *F is said to be a d-increasing function if for all d-boxes B with vertices in* $U_1 \times U_2 \times ... \times U_d$ *holds:*

$$V_F(B) \ \geq \ 0 \tag{4.135}$$

DEFINITION 4.10 *If* $U_1, U_2, ..., U_d$ *have a smallest element* $\min U_1, \min U_2, ..., \min U_d$ *respectively, then we say, that a function* $F : U_1 \times U_2 \times ... \times U_d \longrightarrow \overline{\mathbb{R}}$ *is grounded if :*

$$F(x) = 0 \text{ for all } x \in U_1 \times U_2 \times ... \times U_d \tag{4.136}$$

such that $x_k = \min U_k$ *for at least one k.*

The lemmas, which we have presented for the 2-dimensional case, have analogous multivariate version, see Nelsen (1999).

DEFINITION 4.11 *A d-dimensional copula (or d-copula) is a function C defined on the unit d-cube* $I^d = I \times I \times ... \times I$ *such that*

- *for every* $u \in I^d$ *holds:* $C(u) = 0$, *if at least one coordinate of u is equal to 0; i.e. C is grounded.*
- *for every* $a, b \in I^d$ *with* $a \leq b$ *holds:*

  $$V_C([a,b]) \ \geq \ 0; \tag{4.137}$$

  i.e. C is 2-increasing.
- *for every* $u \in I^d$ *holds:* $C(u) = u_k$, *if all coordinates of u are 1 except* u_k.

Analogously to the 2-dimensional setting, let us state the Sklar's theorem for the d-dimensional case.

THEOREM 4.18 (Sklar's theorem in d-dimensional case) *Let F be a d-dimensional distribution function with marginal distribution functions* $F_{X_1}, F_{X_2}, ..., F_{X_d}$. *Then a d-copula C exists such that for all* $x_1, ..., x_2 \in \overline{\mathbb{R}}^n$:

$$F(x_1, x_2, ..., x_d) = C\left\{F_{X_1}(x_1), F_{X_2}(x_2), ..., F_{X_d}(x_d)\right\} \tag{4.138}$$

Moreover, if $F_{X_1}, F_{X_2}, ..., F_{X_n}$ *are continuous then C is unique. Otherwise C is uniquely determined on the Cartesian product* $Im(F_{X_1}) \times Im(F_{X_2}) \times ... \times Im(F_{X_d})$.
Conversely, if C is a copula and $F_{X_1}, F_{X_2}, ..., F_{X_d}$ *are distribution functions then F defined by (46) is a d-dimensional distribution function with marginals* $F_{X_1}, F_{X_2}, ..., F_{X_d}$.

In order to illustrate the d-copulae we present the following examples:

EXAMPLE 4.32 *Let Φ denote the univariate standard normal distribution function and $\Phi_{\Sigma,d}$ the d-dimensional standard normal distribution function with correlation matrix Σ. Then the function*

$$\begin{aligned} C_\rho^{Gauss}(u,\Sigma) &= \Phi_{\Sigma,d}\left\{\Phi^{-1}(u_1),...,\Phi^{-1}(u_d)\right\} \\ &= \int_{-\infty}^{\phi_1^{-1}(u_d)} \cdots \int_{-\infty}^{\phi_2^{-1}(u_1)} f_\Sigma(x_1,...,x_n)dx_1...dx_d \end{aligned} \tag{4.139}$$

is the d-dimensional Gaussian or normal copula with correlation matrix Σ. The function

$$f_\rho(x_1,...,x_d) = \frac{1}{\sqrt{\det(\Sigma)}} \times \tag{4.140}$$

$$\exp\left\{-\frac{(\Phi^{-1}(u_1),...,\Phi^{-1}(u_d))^\top(\Sigma^{-1}-\mathcal{I}_d)(\Phi^{-1}(u_1),...,\Phi^{-1}(u_d))}{2}\right\}$$

is copula density function.
The copula dependence parameter α is the collection of all unknown correlation coefficients in Σ. If $\alpha \neq 0$, then the corresponding normal copula allows to generate joint symmetric dependence. However, it is not possible to model a tail dependence, i.e. joint extreme events have a zero - probability.

EXAMPLE 4.33 *Let us consider the following function*

$$C_\theta^{GH}(u_1,...,u_d) = \exp\left[-\left\{\sum_{j=1}^{d}(-\ln u_j)^\theta\right\}^{1/\theta}\right] \tag{4.141}$$

One can easily see that this function is a d-dimensional Gumbel-Hougaard copula function. Unlike the Gaussian copula, the Gumbel-Hougaard copula can generate an upper tail dependence.

EXAMPLE 4.34 *As in the 2-dimensional setting, let us consider the d-dimentional Gumbel-Hougaard copula for the case $\theta = 1$. In this case the Gumbel-Hougaard copula reduces to the d-dimentional product copula, i.e.*

$$C_1(u_1,...,u_d) = \prod_{j=1}^{d} u_j = \Pi^d(u) \tag{4.142}$$

The extension of the 2-dimensional copula M, which one gets from the d-dimensional Gumbel-Hougaard copula for $\theta \to \infty$ is denoted $M^d(u)$:

$$C_\theta(u_1,...u_d) \longrightarrow \min(u_1,...,u_d) = M^d(u) \tag{4.143}$$

The d-dimensional function

$$W^d(u) = \max(u_1+u_2+...+u_d-d+1,0) \tag{4.144}$$

defines a copula with $W(u) \leq C(u)$ for any other d-dimensional copula function $C(u)$. $W^d(u)$ is the Fréchet-Hoeffding lower bound in the d-dimensional case.

The functions M^d and Π^d are d-copulas for all $d \geq 2$, whereas the function W^d fails to be a d-copula for any $d > 2$ (Nelsen, 1999). However, the d-dimensional version of the Fréchet-Hoeffding inequality can be written as follows:

$$W^d(u) \leq C(u) \leq M^d(u) \tag{4.145}$$

As we have already mentioned, copula functions have been widely applied in empirical finance. For instance, the Gaussian copula has been used for Value-at-Risk calculations for exchange rates and asset returns (Härdle et al., 2002) and the Gumbel-Hougaard copula for the joint tail dependence and risk management modelling for multivariate high frequency data Dias and Embrechts (2004).

4.8 Bootstrap

Recall that we need large sample sizes in order to sufficiently approximate the critical values computable by the CLT. Here large means n= 50 for one-dimensional data. How can we construct confidence intervals in the case of smaller sample sizes? One way is to use a method called the *Bootstrap*. The Bootstrap algorithm uses the data twice:

1. estimate the parameter of interest,
2. simulate from an estimated distribution to approximate the asymptotic distribution of the statistics of interest.

In detail, bootstrap works as follows. Consider the observations $x_1, \ldots, x_n$ of the sample $X_1, \ldots, X_n$ and estimate the empirical distribution function (edf) F_n. In the case of one-dimensional data

$$F_n(x) = \frac{1}{n} \sum_{i=1}^{n} \boldsymbol{I}(X_i \leq x). \tag{4.146}$$

This is a step function which is constant between neighboring data points.

EXAMPLE 4.35 *Suppose that we have $n = 100$ standard normal $N(0,1)$ data points X_i, $i = 1, \ldots, n$. The cdf of X is $\Phi(x) = \int_{-\infty}^{x} \varphi(u)du$ and is shown in Figure 4.19 as the thin, solid line. The empirical distribution function (edf) is displayed as a thick step function line. Figure 4.20 shows the same setup for $n = 1000$ observations.*

Now draw with replacement a new sample from this empirical distribution. That is we sample with replacement n^* observations $X_1^*, \ldots, X_{n^*}^*$ from the original sample. This is called a Bootstrap sample. Usually one takes $n^* = n$.

Since we sample with replacement, a single observation from the original sample may appear several times in the Bootstrap sample. For instance, if the original sample consists of the three observations x_1, x_2, x_3, then a Bootstrap sample might look like $X_1^* = x_3, X_2^* = x_2, X_3^* = x_3$. Computationally, we find the Bootstrap sample by using a uniform random number generator to draw from the indices $1, 2, \ldots, n$ of the original samples.

The Bootstrap observations are drawn randomly from the empirical distribution, i.e., the probability for each original observation to be selected into the Bootstrap sample is $1/n$ for each draw. It is easy to compute that

$$E_{F_n}(X_i^*) = \frac{1}{n} \sum_{i=1}^{n} x_i = \bar{x}.$$

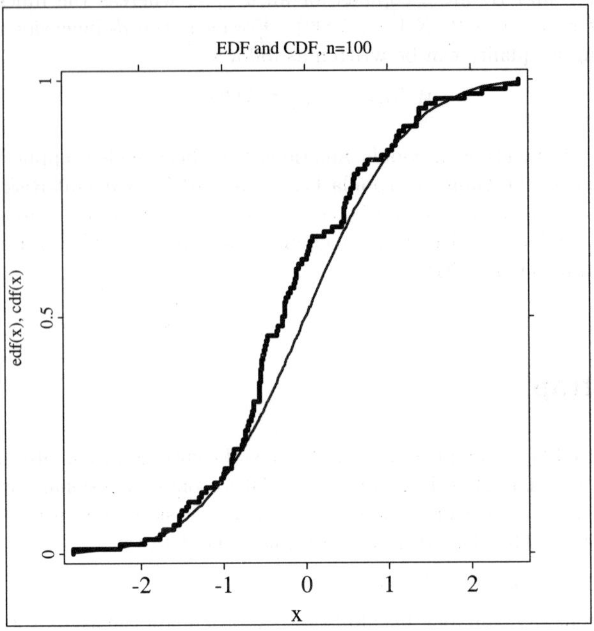

Figure 4.19. The standard normal cdf (thin line) and the empirical distribution function (thick line) for $n = 100$. `MVAedfnormal`

This is the expected value given that the cdf is the original mean of the sample $x_1 \dots, x_n$. The same holds for the variance, i.e.,

$$Var_{F_n}(X_i^*) = \widehat{\sigma}^2,$$

where $\widehat{\sigma}^2 = \frac{1}{n}\sum(x_i - \bar{x})^2$. The cdf of the bootstrap observations is defined as in (4.146). Figure 4.21 shows the cdf of the $n = 100$ original observations as a solid line and two bootstrap cdf's as thin lines.

The CLT holds for the bootstrap sample. Analogously to Corollary 4.1 we have the following corollary.

COROLLARY 4.2 *If $X_1^*, \dots, X_n^*$ is a bootstrap sample from $X_1, \dots, X_n$, then the distribution of*

$$\sqrt{n}\left(\frac{\bar{x}^* - \bar{x}}{\widehat{\sigma}^*}\right)$$

also becomes $N(0,1)$ asymptotically, where $\bar{x}^* = \frac{1}{n}\sum_{i=1}^n X_i^*$ and $(\widehat{\sigma}^*)^2 = \frac{1}{n}\sum_{i=1}^n (X_i^* - \bar{x}^*)^2$.

How do we find a confidence interval for μ using the Bootstrap method? Recall that the quantile $u_{1-\alpha/2}$ might be bad for small sample sizes because the true distribution of $\sqrt{n}\left(\frac{\bar{x}-\mu}{\widehat{\sigma}}\right)$ might be far away from the limit distribution $N(0,1)$. The Bootstrap idea enables us to "simulate" this distribution by computing $\sqrt{n}\left(\frac{\bar{x}^*-\bar{x}}{\widehat{\sigma}^*}\right)$ for **many** Bootstrap samples. In this way

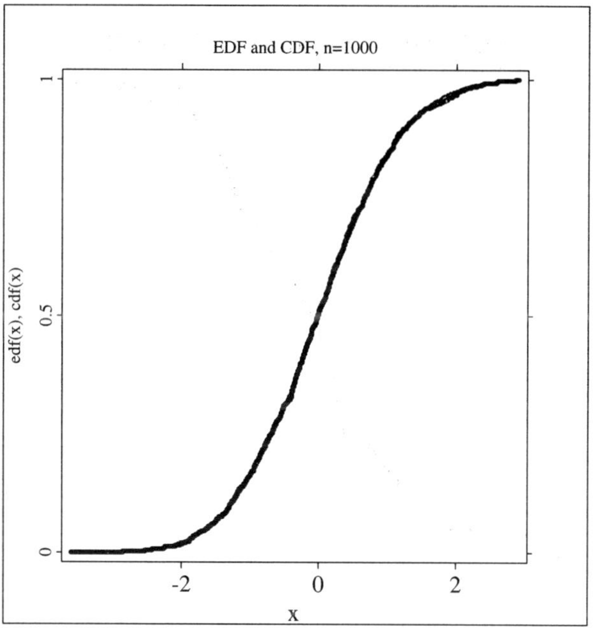

Figure 4.20. The standard normal cdf (thin line) and the empirical distribution function (thick line) for $n = 1000$. `MVAedfnormal`

we can estimate an empirical $(1-\alpha/2)$-quantile $u^*_{1-\alpha/2}$. The bootstrap improved confidence interval is then

$$C^*_{1-\alpha} = \left[\overline{x} - \frac{\widehat{\sigma}}{\sqrt{n}} u^*_{1-\alpha/2}, \; \overline{x} + \frac{\widehat{\sigma}}{\sqrt{n}} u^*_{1-\alpha/2}\right].$$

By Corollary 4.2 we have

$$P(\mu \in C^*_{1-\alpha}) \longrightarrow 1-\alpha \quad \text{as } n \to \infty,$$

but with an improved speed of convergence, see Hall (1992).

Summary

- $\hookrightarrow$ For small sample sizes the bootstrap improves the precision of the confidence interval.
- $\hookrightarrow$ The bootstrap distribution $\mathcal{L}\{\sqrt{n}(\overline{x}^* - \overline{x})/\hat{\sigma}^*\}$ converges to the same asymptotic limit as the distribution $\mathcal{L}\{\sqrt{n}(\overline{x}^* - \overline{x})/\hat{\sigma}\}$.

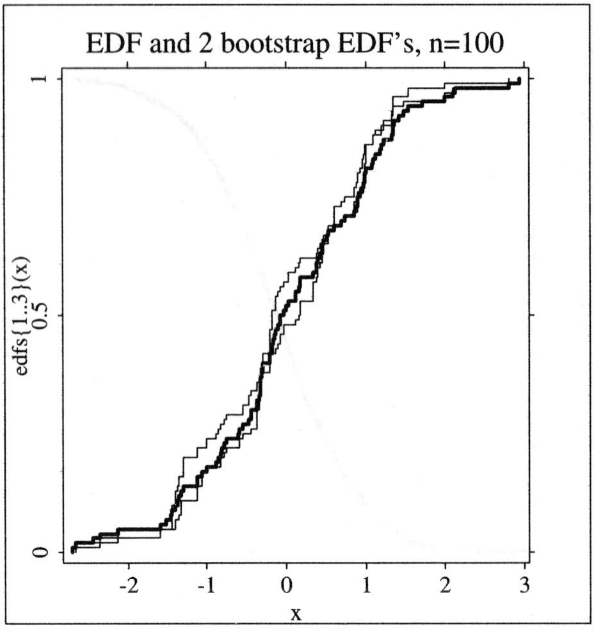

Figure 4.21. The cdf F_n (thick line) and two bootstrap cdf's F_n^* (thin lines). `MVAedfbootstrap`

4.9 Exercises

EXERCISE 4.1 *Assume that the random vector Y has the following normal distribution: $Y \sim N_p(0, \mathcal{I})$. Transform it according to (4.49) to create $X \sim N(\mu, \Sigma)$ with mean $\mu = (3, 2)^\top$ and $\Sigma = \begin{pmatrix} 1 & -1.5 \\ -1.5 & 4 \end{pmatrix}$. How would you implement the resulting formula on a computer?*

EXERCISE 4.2 *Prove Theorem 4.7 using Theorem 4.5.*

EXERCISE 4.3 *Suppose that X has mean zero and covariance $\Sigma = \begin{pmatrix} 1 & 0 \\ 0 & 2 \end{pmatrix}$. Let $Y = X_1 + X_2$. Write Y as a linear transformation, i.e., find the transformation matrix $\mathcal{A}$. Then compute $Var(Y)$ via (4.26). Can you obtain the result in another fashion?*

EXERCISE 4.4 *Calculate the mean and the variance of the estimate $\hat{\beta}$ in (3.50).*

EXERCISE 4.5 *Compute the conditional moments $E(X_2 \mid x_1)$ and $E(X_1 \mid x_2)$ for the pdf of Example 4.5.*

EXERCISE 4.6 *Prove the relation (4.28).*

EXERCISE 4.7 *Prove the relation (4.29). Hint: Note that $Var(E(X_2|X_1)) = E(E(X_2|X_1)\, E(X_2^\top|X_1)) - E(X_2)\, E(X_2^\top))$ and that $E(Var(X_2|X_1)) = E[E(X_2 X_2^\top|X_1) - E(X_2|X_1)\, E(X_2^\top|X_1)]$.*

EXERCISE 4.8 *Compute (4.46) for the pdf of Example 4.5.*

EXERCISE 4.9

$$\textit{Show that} \quad f_Y(y) = \begin{cases} \frac{1}{2}y_1 - \frac{1}{4}y_2 & 0 \le y_1 \le 2, \ |y_2| \le 1 - |1 - y_1| \\ 0 & \textit{otherwise} \end{cases} \quad \textit{is a pdf!}$$

EXERCISE 4.10 *Compute (4.46) for a two-dimensional standard normal distribution. Show that the transformed random variables Y_1 and Y_2 are independent. Give a geometrical interpretation of this result based on iso-distance curves.*

EXERCISE 4.11 *Consider the Cauchy distribution which has no moment, so that the CLT cannot be applied. Simulate the distribution of $\overline{x}$ (for different n's). What can you expect for $n \to \infty$?*
Hint: The Cauchy distribution can be simulated by the quotient of two independent standard normally distributed random variables.

EXERCISE 4.12 *A European car company has tested a new model and reports the consumption of gasoline (X_1) and oil (X_2). The expected consumption of gasoline is 8 liters per 100 km (μ_1) and the expected consumption of oil is 1 liter per 10.000 km (μ_2). The measured consumption of gasoline is 8.1 liters per 100 km ($\overline{x}_1$) and the measured consumption of oil is 1.1 liters per 10,000 km ($\overline{x}_2$). The asymptotic distribution of $\sqrt{n}\left\{\binom{\overline{x}_1}{\overline{x}_2} - \binom{\mu_1}{\mu_2}\right\}$ is $N\left(\binom{0}{0}, \begin{pmatrix} 0.1 & 0.05 \\ 0.05 & 0.1 \end{pmatrix}\right)$.*

For the American market the basic measuring units are miles (1 mile $\approx$ 1.6 km) and gallons (1 gallon $\approx$ 3.8 liter). The consumptions of gasoline (Y_1) and oil (Y_2) are usually reported in miles per gallon. Can you express $\overline{y}_1$ and $\overline{y}_2$ in terms of $\overline{x}_1$ and $\overline{x}_2$? Recompute the asymptotic distribution for the American market!

EXERCISE 4.13 *Consider the pdf $f(x_1, x_2) = e^{-(x_1+x_2)}, x_1, x_2 > 0$ and let $U_1 = X_1 + X_2$ and $U_2 = X_1 - X_2$. Compute $f(u_1, u_2)$.*

EXERCISE 4.14 *Consider the pdf's*

$$\begin{array}{rcll} f(x_1,x_2) & = & 4x_1x_2e^{-x_1^2} & x_1, x_2 > 0, \\ f(x_1,x_2) & = & 1 & 0 < x_1, x_2 < 1 \textit{ and } x_1 + x_2 < 1 \\ f(x_1,x_2) & = & \frac{1}{2}e^{-x_1} & x_1 > |x_2|. \end{array}$$

For each of these pdf's compute $E(X)$, $Var(X)$, $E(X_1|X_2)$, $E(X_2|X_1)$, $V(X_1|X_2)$ and $V(X_2|X_1)$.

EXERCISE 4.15 *Consider the pdf $f(x_1, x_2) = \frac{3}{2}x_1^{\frac{1}{2}}$, $0 < x_1 < x_2 < 1$. Compute $P(X_1 < 0.25)$, $P(X_2 < 0.25)$ and $P(X_2 < 0.25|X_1 < 0.25)$.*

EXERCISE 4.16 *Consider the pdf $f(x_1, x_2) = \frac{1}{2\pi}$, $0 < x_1 < 2\pi, 0 < x_2 < 1$. Let $U_1 = \sin X_1\sqrt{-2\log X_2}$ and $U_2 = \cos X_1\sqrt{-2\log X_2}$. Compute $f(u_1, u_2)$.*

EXERCISE 4.17 *Consider $f(x_1, x_2, x_3) = k(x_1 + x_2x_3)$; $0 < x_1, x_2, x_3 < 1$.*

a) *Determine k so that f is a valid pdf of $(X_1, X_2, X_3) = X$.*

b) *Compute the* (3×3) *matrix* Σ_X.

c) *Compute the* (2×2) *matrix of the conditional variance of* (X_2, X_3) *given* $X_1 = x_1$.

EXERCISE 4.18 *Let* $X \sim N_2\left(\binom{1}{2}, \begin{pmatrix} 2 & a \\ a & 2 \end{pmatrix}\right)$.

a) *Represent the contour ellipses for* $a = 0;\ -\frac{1}{2};\ +\frac{1}{2};\ 1$.

b) *For* $a = \frac{1}{2}$ *find the regions of* X *centered on* μ *which cover the area of the true parameter with probability* 0.90 *and* 0.95.

EXERCISE 4.19 *Consider the pdf*

$$f(x_1, x_2) = \frac{1}{8x_2} e^{-\left(\frac{x_1}{2x_2} + \frac{x_2}{4}\right)} \qquad x_1, x_2 > 0.$$

Compute $f(x_2)$ *and* $f(x_1|x_2)$. *Also give the best approximation of* X_1 *by a function of* X_2. *Compute the variance of the error of the approximation.*

EXERCISE 4.20 *Prove Theorem 4.6.*

5 Theory of the Multinormal

In the preceeding chapter we saw how the multivariate normal distribution comes into play in many applications. It is useful to know more about this distribution, since it is often a good approximate distribution in many situations. Another reason for considering the multinormal distribution relies on the fact that it has many appealing properties: it is stable under linear transforms, zero correlation corresponds to independence, the marginals and all the conditionals are also multivariate normal variates, etc. The mathematical properties of the multinormal make analyses much simpler.

In this chapter we will first concentrate on the probabilistic properties of the multinormal, then we will introduce two "companion" distributions of the multinormal which naturally appear when sampling from a multivariate normal population: the Wishart and the Hotelling distributions. The latter is particularly important for most of the testing procedures proposed in Chapter 7.

5.1 Elementary Properties of the Multinormal

Let us first summarize some properties which were already derived in the previous chapter.

- The pdf of $X \sim N_p(\mu, \Sigma)$ is

$$f(x) = |2\pi\Sigma|^{-1/2} \exp\left\{-\frac{1}{2}(x-\mu)^\top\Sigma^{-1}(x-\mu)\right\}. \tag{5.1}$$

  The expectation is $E(X) = \mu$, the covariance can be calculated as $Var(X) = E(X-\mu)(X-\mu)^\top = \Sigma$.

- Linear transformations turn normal random variables into normal random variables. If $X \sim N_p(\mu, \Sigma)$ and $\mathcal{A}(p \times p), c \in \mathbb{R}^p$, then $Y = \mathcal{A}X + c$ is p-variate Normal, i.e.,

$$Y \sim N_p(\mathcal{A}\mu + c, \mathcal{A}\Sigma\mathcal{A}^\top). \tag{5.2}$$

- If $X \sim N_p(\mu, \Sigma)$, then the Mahalanobis transformation is

$$Y = \Sigma^{-1/2}(X-\mu) \sim N_p(0, \mathcal{I}_p) \tag{5.3}$$

  and it holds that

$$Y^\top Y = (X-\mu)^\top \Sigma^{-1}(X-\mu) \sim \chi^2_p. \tag{5.4}$$

Often it is interesting to partition X into sub-vectors X_1 and X_2. The following theorem tells us how to correct X_2 to obtain a vector which is independent of X_1.

THEOREM 5.1 *Let* $X = \binom{X_1}{X_2} \sim N_p(\mu, \Sigma)$, $X_1 \in \mathbb{R}^r$, $X_2 \in \mathbb{R}^{p-r}$. *Define* $X_{2.1} = X_2 - \Sigma_{21}\Sigma_{11}^{-1}X_1$ *from the partitioned covariance matrix*

$$\Sigma = \begin{pmatrix} \Sigma_{11} & \Sigma_{12} \\ \Sigma_{21} & \Sigma_{22} \end{pmatrix}.$$

Then

$$X_1 \sim N_r(\mu_1, \Sigma_{11}), \tag{5.5}$$
$$X_{2.1} \sim N_{p-r}(\mu_{2.1}, \Sigma_{22.1}) \tag{5.6}$$

are independent with

$$\mu_{2.1} = \mu_2 - \Sigma_{21}\Sigma_{11}^{-1}\mu_1, \quad \Sigma_{22.1} = \Sigma_{22} - \Sigma_{21}\Sigma_{11}^{-1}\Sigma_{12}. \tag{5.7}$$

Proof:

$$\begin{aligned} X_1 &= \mathcal{A}X \quad \text{with} \quad \mathcal{A} = [\, \mathcal{I}_r \ , \ 0 \,] \\ X_{2.1} &= \mathcal{B}X \quad \text{with} \quad \mathcal{B} = [\, -\Sigma_{21}\Sigma_{11}^{-1} \ , \ \mathcal{I}_{p-r} \,]. \end{aligned}$$

Then, by (5.2) X_1 and $X_{2.1}$ are both normal. Note that

$$Cov(X_1, X_{2.1}) = \mathcal{A}\Sigma\mathcal{B}^\top = \begin{pmatrix} \boxed{\begin{matrix} 1 & & 0 \\ & \ddots & \\ 0 & & 1 \end{matrix}} & 0 \end{pmatrix} \begin{pmatrix} \Sigma_{11} & \Sigma_{12} \\ \Sigma_{21} & \Sigma_{22} \end{pmatrix} \begin{pmatrix} (-\Sigma_{21}\Sigma_{11}^{-1})^\top \\ \boxed{\begin{matrix} 1 & & 0 \\ & \ddots & \\ 0 & & 1 \end{matrix}} \end{pmatrix},$$

$$\begin{aligned} \mathcal{A}\Sigma &= (\mathcal{I}\ 0) \begin{pmatrix} \Sigma_{11} & \Sigma_{12} \\ \Sigma_{21} & \Sigma_{22} \end{pmatrix} = (\Sigma_{11} \quad \Sigma_{12}), \\ \Rightarrow \mathcal{A}\Sigma\mathcal{B}^\top &= (\Sigma_{11} \quad \Sigma_{12}) \begin{pmatrix} (-\Sigma_{21}\Sigma_{11}^{-1})^\top \\ \mathcal{I} \end{pmatrix} = \left(-\Sigma_{11}\left(\Sigma_{21}\Sigma_{11}^{-1}\right)^\top + \Sigma_{12}\right). \end{aligned}$$

Recall that $\Sigma_{21} = (\Sigma_{12})^\top$. Hence $\mathcal{A}\Sigma\mathcal{B}^\top = -\Sigma_{11}\Sigma_{11}^{-1}\Sigma_{12} + \Sigma_{12} \equiv 0$!

Using (5.2) again we also have the joint distribution of $(X_1, X_{2.1})$, namely

$$\begin{pmatrix} X_1 \\ X_{2.1} \end{pmatrix} = \begin{pmatrix} \mathcal{A} \\ \mathcal{B} \end{pmatrix} X \sim N_p\left(\begin{pmatrix} \mu_1 \\ \mu_{2.1} \end{pmatrix}, \begin{pmatrix} \Sigma_{11} & 0 \\ 0 & \Sigma_{22.1} \end{pmatrix}\right).$$

With this block diagonal structure of the covariance matrix, the joint pdf of $(X_1, X_{2.1})$ can easily be factorized into

$$\begin{aligned} f(x_1, x_{2.1}) &= |2\pi\Sigma_{11}|^{-\frac{1}{2}} \exp\left\{-\frac{1}{2}(x_1 - \mu_1)^\top \Sigma_{11}^{-1}(x_1 - \mu_1)\right\} \times \\ &\quad |2\pi\Sigma_{22.1}|^{-\frac{1}{2}} \exp\left\{-\frac{1}{2}(x_{2.1} - \mu_{2.1})^\top \Sigma_{22.1}^{-1}(x_{2.1} - \mu_{2.1})\right\} \end{aligned}$$

from which the independence between X_1 and $X_{2.1}$ follows. □

The next two corollaries are direct consequences of Theorem 5.1.

COROLLARY 5.1 *Let* $X = \begin{pmatrix} X_1 \\ X_2 \end{pmatrix} \sim N_p(\mu, \Sigma)$, $\Sigma = \begin{pmatrix} \Sigma_{11} & \Sigma_{12} \\ \Sigma_{21} & \Sigma_{22} \end{pmatrix}$. $\Sigma_{12} = 0$ *if and only if* X_1 *is independent of* X_2.

The independence of two linear transforms of a multinormal X can be shown via the following corollary.

COROLLARY 5.2 *If $X \sim N_p(\mu, \Sigma)$ and given some matrices $\mathcal{A}$ and $\mathcal{B}$, then $\mathcal{A}X$ and $\mathcal{B}X$ are independent if and only if $\mathcal{A}\Sigma\mathcal{B}^\top = 0$.*

The following theorem is also useful. It generalizes Theorem 4.6. The proof is left as an exercise.

THEOREM 5.2 *If $X \sim N_p(\mu, \Sigma)$, $\mathcal{A}(q \times p)$, $c \in \mathbb{R}^q$ and $q \le p$, then $Y = \mathcal{A}X + c$ is a q-variate Normal, i.e.,*

$$Y \sim N_q(\mathcal{A}\mu + c, \mathcal{A}\Sigma\mathcal{A}^\top).$$

The conditional distribution of X_2 given X_1 is given by the next theorem.

THEOREM 5.3 *The conditional distribution of X_2 given $X_1 = x_1$ is normal with mean $\mu_2 + \Sigma_{21}\Sigma_{11}^{-1}(x_1 - \mu_1)$ and covariance $\Sigma_{22.1}$, i.e.,*

$$(X_2 \mid X_1 = x_1) \sim N_{p-r}(\mu_2 + \Sigma_{21}\Sigma_{11}^{-1}(x_1 - \mu_1), \Sigma_{22.1}). \tag{5.8}$$

Proof:
Since $X_2 = X_{2.1} + \Sigma_{21}\Sigma_{11}^{-1}X_1$, for a fixed value of $X_1 = x_1$, X_2 is equivalent to $X_{2.1}$ plus a constant term:

$$(X_2|X_1 = x_1) = (X_{2.1} + \Sigma_{21}\Sigma_{11}^{-1}x_1),$$

which has the normal distribution $N(\mu_{2.1} + \Sigma_{21}\Sigma_{11}^{-1}x_1, \Sigma_{22.1})$. □

Note that the conditional mean of $(X_2 \mid X_1)$ is a linear function of X_1 and that the conditional variance does not depend on the particular value of X_1. In the following example we consider a specific distribution.

EXAMPLE 5.1 *Suppose that $p = 2$, $r = 1$, $\mu = \begin{pmatrix} 0 \\ 0 \end{pmatrix}$ and $\Sigma = \begin{pmatrix} 1 & -0.8 \\ -0.8 & 2 \end{pmatrix}$. Then $\Sigma_{11} = 1$, $\Sigma_{21} = -0.8$ and $\Sigma_{22.1} = \Sigma_{22} - \Sigma_{21}\Sigma_{11}^{-1}\Sigma_{12} = 2 - (0.8)^2 = 1.36$. Hence the marginal pdf of X_1 is*

$$f_{X_1}(x_1) = \frac{1}{\sqrt{2\pi}} \exp\left(-\frac{x_1^2}{2}\right)$$

and the conditional pdf of $(X_2 \mid X_1 = x_1)$ is given by

$$f(x_2 \mid x_1) = \frac{1}{\sqrt{2\pi(1.36)}} \exp\left\{-\frac{(x_2 + 0.8x_1)^2}{2 \times (1.36)}\right\}.$$

As mentioned above, the conditional mean of $(X_2 \mid X_1)$ is linear in X_1. The shift in the density of $(X_2 \mid X_1)$ can be seen in Figure 5.1.

Sometimes it will be useful to reconstruct a joint distribution from the marginal distribution of X_1 and the conditional distribution $(X_2|X_1)$. The following theorem shows under which conditions this can be easily done in the multinormal framework.

THEOREM 5.4 *If $X_1 \sim N_r(\mu_1, \Sigma_{11})$ and $(X_2|X_1 = x_1) \sim N_{p-r}(\mathcal{A}x_1 + b, \Omega)$ where Ω does not depend on x_1, then $X = \binom{X_1}{X_2} \sim N_p(\mu, \Sigma)$, where*

$$\mu = \begin{pmatrix} \mu_1 \\ \mathcal{A}\mu_1 + b \end{pmatrix}$$

$$\Sigma = \begin{pmatrix} \Sigma_{11} & \Sigma_{11}\mathcal{A}^\top \\ \mathcal{A}\Sigma_{11} & \Omega + \mathcal{A}\Sigma_{11}\mathcal{A}^\top \end{pmatrix}.$$

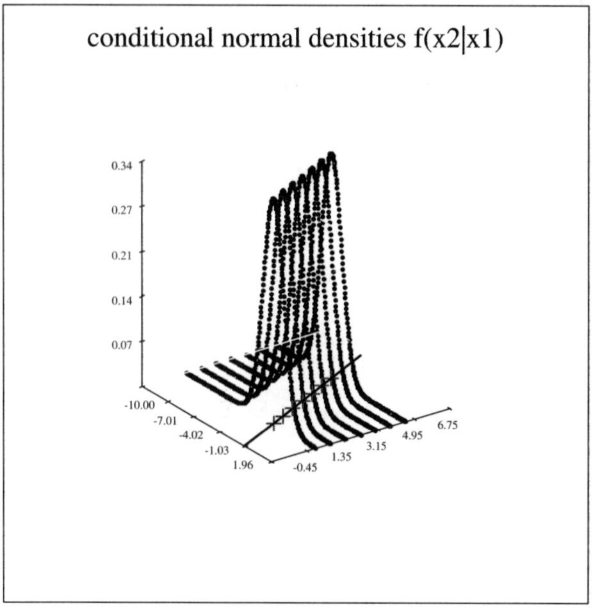

Figure 5.1. Shifts in the conditional density. MVAcondnorm

EXAMPLE 5.2 *Consider the following random variables*

$$X_1 \sim N_1(0,1),$$
$$X_2|X_1 = x_1 \sim N_2\left(\begin{pmatrix} 2x_1 \\ x_1+1 \end{pmatrix}, \begin{pmatrix} 1 & 0 \\ 0 & 1 \end{pmatrix}\right).$$

Using Theorem (5.4), where $\mathcal{A} = (2 \quad 1)^\top$, $b = (0 \quad 1)^\top$ *and* $\Omega = \mathcal{I}_2$, *we easily obtain the following result:*

$$X = \begin{pmatrix} X_1 \\ X_2 \end{pmatrix} \sim N_3\left(\begin{pmatrix} 0 \\ 0 \\ 1 \end{pmatrix}, \begin{pmatrix} 1 & 2 & 1 \\ 2 & 5 & 2 \\ 1 & 2 & 2 \end{pmatrix}\right).$$

In particular, the marginal distribution of X_2 *is*

$$X_2 \sim N_2\left(\begin{pmatrix} 0 \\ 1 \end{pmatrix}, \begin{pmatrix} 5 & 2 \\ 2 & 2 \end{pmatrix}\right),$$

thus conditional on X_1*, the two components of* X_2 *are independent but marginally they are not!*

Note that the marginal mean vector and covariance matrix of X_2 *could have also been computed directly by using (4.28)–(4.29). Using the derivation above, however, provides us with useful properties: we have multinormality!*

Conditional Approximations

As we saw in Chapter 4 (Theorem 4.3), the conditional expectation $E(X_2|X_1)$ is the mean squared error (MSE) best approximation of X_2 by a function of X_1. We have in this case that

$$X_2 = E(X_2|X_1) + U = \mu_2 + \Sigma_{21}\Sigma_{11}^{-1}(X_1 - \mu_1) + U. \tag{5.9}$$

Hence, the best approximation of $X_2 \in \mathbb{R}^{p-r}$ by $X_1 \in \mathbb{R}^r$ is the linear approximation that can be written as:

$$X_2 = \beta_0 + \mathcal{B}\, X_1 + U \tag{5.10}$$

with $\mathcal{B} = \Sigma_{21}\Sigma_{11}^{-1}$, $\beta_0 = \mu_2 - B\mu_1$ and $U \sim N(0, \Sigma_{22.1})$.

Consider now the particular case where $r = p - 1$. Now $X_2 \in \mathbb{R}$ and $\mathcal{B}$ is a row vector $\beta^\top$ of dimension $(1 \times r)$

$$X_2 = \beta_0 + \beta^\top X_1 + U. \tag{5.11}$$

This means, geometrically speaking, that the best MSE approximation of X_2 by a function of X_1 is hyperplane. The marginal variance of X_2 can be decomposed via (5.11):

$$\sigma_{22} = \beta^\top\Sigma_{11}\beta + \sigma_{22.1} = \sigma_{21}\Sigma_{11}^{-1}\sigma_{12} + \sigma_{22.1}. \tag{5.12}$$

The ratio

$$\rho^2_{2.1\ldots r} = \frac{\sigma_{21}\Sigma_{11}^{-1}\sigma_{12}}{\sigma_{22}} \tag{5.13}$$

is known as the square of the multiple correlation between X_2 and the r variables X_1. It is the percentage of the variance of X_2 which is explained by the linear approximation $\beta_0 + \beta^\top X_1$. The last term in (5.12) is the residual variance of X_2. The square of the multiple correlation corresponds to the coefficient of determination introduced in Section 3.4, see (3.39), but here it is defined in terms of the r.v. X_1 and X_2. It can be shown that $\rho_{2.1\ldots r}$ is also the maximum correlation attainable between X_2 and a linear combination of the elements of X_1, the optimal linear combination being precisely given by $\beta^\top X_1$. Note, that when $r = 1$, the multiple correlation $\rho_{2.1}$ coincides with the usual simple correlation $\rho_{X_2X_1}$ between X_2 and X_1.

EXAMPLE 5.3 *Consider the "classic blue" pullover example (Example 3.15) and suppose that X_1 (sales), X_2 (price), X_3 (advertisement) and X_4 (sales assistants) are normally distributed with*

$$\mu = \begin{pmatrix} 172.7 \\ 104.6 \\ 104.0 \\ 93.8 \end{pmatrix} \text{ and } \Sigma = \begin{pmatrix} 1037.21 & & & \\ -80.02 & 219.84 & & \\ 1430.70 & 92.10 & 2624.00 & \\ 271,44 & -91.58 & 210.30 & 177.36 \end{pmatrix}.$$

(These are in fact the sample mean and the sample covariance matrix but in this example we pretend that they are the true parameter values.)

The conditional distribution of X_1 given (X_2, X_3, X_4) is thus an univariate normal with mean

$$\mu_1 + \sigma_{12}\Sigma_{22}^{-1}\begin{pmatrix} X_2 - \mu_2 \\ X_3 - \mu_3 \\ X_4 - \mu_4 \end{pmatrix} = 65.670 - 0.216X_2 + 0.485X_3 + 0.844X_4$$

and variance

$$\sigma_{11.2} = \sigma_{11} - \sigma_{12}\Sigma_{22}^{-1}\sigma_{21} = 96.761$$

The linear approximation of the sales (X_1) *by the price* (X_2), *advertisement* (X_3) *and sales assistants* (X_4) *is provided by the conditional mean above.(Note that this coincides with the results of Example 3.15 due to the particular choice of* μ *and* Σ). *The quality of the approximation is given by the multiple correlation* $\rho^2_{1.234} = \frac{\sigma_{12}\Sigma_{22}^{-1}\sigma_{21}}{\sigma_{11}} = 0.907$. *(Note again that this coincides with the coefficient of determination* r^2 *found in Example 3.15).*

This example also illustrates the concept of partial correlation. The correlation matrix between the 4 variables is given by

$$P = \begin{pmatrix} 1 & -0.168 & 0.867 & 0.633 \\ -0.168 & 1 & 0.121 & -0.464 \\ 0.867 & 0.121 & 1 & 0.308 \\ 0.633 & -0.464 & 0.308 & 1 \end{pmatrix},$$

so that the correlation between X_1 *(sales) and* X_2 *(price) is* -0.168. *We can compute the conditional distribution of* (X_1, X_2) *given* (X_3, X_4), *which is a bivariate normal with mean:*

$$\begin{pmatrix} \mu_1 \\ \mu_2 \end{pmatrix} + \begin{pmatrix} \sigma_{13} & \sigma_{14} \\ \sigma_{23} & \sigma_{24} \end{pmatrix} \begin{pmatrix} \sigma_{33} & \sigma_{34} \\ \sigma_{43} & \sigma_{44} \end{pmatrix}^{-1} \begin{pmatrix} X_3 - \mu_3 \\ X_4 - \mu_4 \end{pmatrix} = \begin{pmatrix} 32.516 + 0.467X_3 + 0.977X_4 \\ 153.644 + 0.085X_3 - 0.617X_4 \end{pmatrix}$$

and covariance matrix:

$$\begin{pmatrix} \sigma_{11} & \sigma_{12} \\ \sigma_{21} & \sigma_{22} \end{pmatrix} - \begin{pmatrix} \sigma_{13} & \sigma_{14} \\ \sigma_{23} & \sigma_{24} \end{pmatrix} \begin{pmatrix} \sigma_{33} & \sigma_{34} \\ \sigma_{43} & \sigma_{44} \end{pmatrix}^{-1} \begin{pmatrix} \sigma_{31} & \sigma_{32} \\ \sigma_{41} & \sigma_{42} \end{pmatrix} = \begin{pmatrix} 104.006 & \\ -33.574 & 155.592 \end{pmatrix}.$$

In particular, the last covariance matrix allows the partial correlation between X_1 *and* X_2 *to be computed for a fixed level of* X_3 *and* X_4:

$$\rho_{X_1X_2|X_3X_4} = \frac{-33.574}{\sqrt{104.006 * 155.592}} = -0.264,$$

so that in this particular example with a fixed level of advertisement and sales assistance, the negative correlation between price and sales is more important than the marginal one.

Summary

- ↪ If $X \sim N_p(\mu, \Sigma)$, then a linear transformation $\mathcal{A}X + c$, $\mathcal{A}(q \times p)$, where $c \in \mathbb{R}^q$, has distribution $N_q(\mathcal{A}\mu + c, \mathcal{A}\Sigma\mathcal{A}^\top)$.
- ↪ Two linear transformations $\mathcal{A}X$ and $\mathcal{B}X$ with $X \sim N_p(\mu, \Sigma)$ are independent if and only if $\mathcal{A}\Sigma\mathcal{B}^\top = 0$.
- ↪ If X_1 and X_2 are partitions of $X \sim N_p(\mu, \Sigma)$, then the conditional distribution of X_2 given $X_1 = x_1$ is again normal.
- ↪ In the multivariate normal case, X_1 is independent of X_2 if and only if $\Sigma_{12} = 0$.
- ↪ The conditional expectation of $(X_2|X_1)$ is a linear function if $\binom{X_1}{X_2} \sim N_p(\mu, \Sigma)$.
- ↪ The multiple correlation coefficient is defined as $\rho^2_{2.1\ldots r} = \frac{\sigma_{21}\Sigma_{11}^{-1}\sigma_{12}}{\sigma_{22}}$.
- ↪ The multiple correlation coefficient is the percentage of the variance of X_2 explained by the linear approximation $\beta_0 + \beta^\top X_1$.

5.2 The Wishart Distribution

The Wishart distribution (named after its discoverer) plays a prominent role in the analysis of estimated covariance matrices. If the mean of $X \sim N_p(\mu, \Sigma)$ is known to be $\mu = 0$, then for a data matrix $\mathcal{X}(n \times p)$ the estimated covariance matrix is proportional to $\mathcal{X}^\top \mathcal{X}$. This is the point where the Wishart distribution comes in, because $\mathcal{M}(p \times p) = \mathcal{X}^\top \mathcal{X} = \sum_{i=1}^n x_i x_i^\top$ has a Wishart distribution $W_p(\Sigma, n)$.

EXAMPLE 5.4 *Set $p = 1$, then for $X \sim N_1(0, \sigma^2)$ the data matrix of the observations*

$$\mathcal{X} = (x_1, \dots, x_n)^\top \quad \textit{with} \quad \mathcal{M} = \mathcal{X}^\top \mathcal{X} = \sum_{i=1}^n x_i x_i$$

leads to the Wishart distribution $W_1(\sigma^2, n) = \sigma^2 \chi_n^2$. The one-dimensional Wishart distribution is thus in fact a χ^2 distribution.

When we talk about the distribution of a matrix, we mean of course the joint distribution of all its elements. More exactly: since $\mathcal{M} = \mathcal{X}^\top \mathcal{X}$ is symmetric we only need to consider the elements of the lower triangular matrix

$$\mathcal{M} = \begin{pmatrix} m_{11} & & & \\ m_{21} & m_{22} & & \\ \vdots & \vdots & \ddots & \\ m_{p1} & m_{p2} & \dots & m_{pp} \end{pmatrix}. \tag{5.14}$$

Hence the Wishart distribution is defined by the distribution of the vector

$$(m_{11}, \dots, m_{p1}, m_{22}, \dots, m_{p2}, \dots, m_{pp})^\top. \tag{5.15}$$

Linear transformations of the data matrix $\mathcal{X}$ also lead to Wishart matrices.

THEOREM 5.5 *If $\mathcal{M} \sim W_p(\Sigma, n)$ and $\mathcal{B}(p \times q)$, then the distribution of $\mathcal{B}^\top \mathcal{M} \mathcal{B}$ is Wishart $W_q(\mathcal{B}^\top \Sigma \mathcal{B}, n)$.*

With this theorem we can standardize Wishart matrices since with $\mathcal{B} = \Sigma^{-1/2}$ the distribution of $\Sigma^{-1/2} \mathcal{M} \Sigma^{-1/2}$ is $W_p(\mathcal{I}, n)$. Another connection to the χ^2-distribution is given by the following theorem.

THEOREM 5.6 *If $\mathcal{M} \sim W_p(\Sigma, m)$, and $a \in \mathbb{R}^p$ with $a^\top \Sigma a \neq 0$, then the distribution of $\dfrac{a^\top \mathcal{M} a}{a^\top \Sigma a}$ is χ_m^2.*

This theorem is an immediate consequence of Theorem 5.5 if we apply the linear transformation $x \mapsto a^\top x$. Central to the analysis of covariance matrices is the next theorem.

THEOREM 5.7 (Cochran) *Let $\mathcal{X}(n \times p)$ be a data matrix from a $N_p(0, \Sigma)$ distribution and let $\mathcal{C}(n \times n)$ be a symmetric matrix.*

(a) $\mathcal{X}^\top \mathcal{C} \mathcal{X}$ *has the distribution of weighted Wishart random variables, i.e.*

$$\mathcal{X}^\top \mathcal{C} \mathcal{X} = \sum_{i=1}^{n} \lambda_i W_p(\Sigma, 1),$$

where λ_i, $i = 1, \ldots, n$, *are the eigenvalues of* $\mathcal{C}$.

(b) $\mathcal{X}^\top \mathcal{C} \mathcal{X}$ *is Wishart if and only if* $\mathcal{C}^2 = \mathcal{C}$. *In this case*

$$\mathcal{X}^\top \mathcal{C} \mathcal{X} \sim W_p(\Sigma, r),$$

and $r = \operatorname{rank}(\mathcal{C}) = \operatorname{tr}(\mathcal{C})$.

(c) $n\mathcal{S} = \mathcal{X}^\top \mathcal{H} \mathcal{X}$ *is distributed as* $W_p(\Sigma, n-1)$ *(note that* $\mathcal{S}$ *is the sample covariance matrix).*

(d) $\bar{x}$ *and* $\mathcal{S}$ *are independent.*

The following properties are useful:

1. If $\mathcal{M} \sim W_p(\Sigma, n)$, then $E(\mathcal{M}) = n\Sigma$.
2. If $\mathcal{M}_i$ are independent Wishart $W_p(\Sigma, n_i)$ i $= 1, \cdots, k$, then $\mathcal{M} = \sum_{i=1}^{k} \mathcal{M}_i \sim W_p(\Sigma, n)$ where $n = \sum_{i=1}^{k} n_i$.
3. The density of $W_p(\Sigma, n-1)$ for a positive definite $\mathcal{M}$ is given by:

$$f_{\Sigma, n-1}(\mathcal{M}) = \frac{|\mathcal{M}|^{\frac{1}{2}(n-p-2)} e^{-\frac{1}{2}\operatorname{tr}(\mathcal{M}\Sigma^{-1})}}{2^{\frac{1}{2}p(n-1)} \pi^{\frac{1}{4}p(p-1)} |\Sigma|^{\frac{1}{2}(n-1)} \prod_{i=1}^{p} \Gamma\{\frac{n-i}{2}\}}, \tag{5.16}$$

where Γ is the gamma function, see Feller (1966): $\Gamma(z) = \int_0^\infty t^{z-1} e^{-t} dt$.

For further details on the Wishart distribution, see Mardia, Kent and Bibby (1979).

Summary

↪ The Wishart distribution is a generalization of the χ^2-distribution. In particular $W_1(\sigma^2, n) = \sigma^2 \chi^2_n$.

↪ The empirical covariance matrix $\mathcal{S}$ has a $\frac{1}{n} W_p(\Sigma, n-1)$ distribution.

↪ In the normal case, $\bar{x}$ and $\mathcal{S}$ are independent.

↪ For $\mathcal{M} \sim W_p(\Sigma, m)$, $\frac{a^\top \mathcal{M} a}{a^\top \Sigma a} \sim \chi^2_m$.

5.3 Hotelling's T^2-Distribution

Suppose that $Y \in \mathbb{R}^p$ is a standard normal random vector, i.e., $Y \sim N_p(0, \mathcal{I})$, independent of the random matrix $\mathcal{M} \sim W_p(\mathcal{I}, n)$. What is the distribution of $Y^\top \mathcal{M}^{-1} Y$? The answer is provided by the Hotelling T^2-distribution: $n\, Y^\top \mathcal{M}^{-1} Y$ is Hotelling T^2 (p, n) distributed.

The Hotelling T^2-distribution is a generalization of the Student t-distribution. The general multinormal distribution $N(\mu, \Sigma)$ is considered in Theorem 5.8. The Hotelling T^2-distribution will play a central role in hypothesis testing in Chapter 7.

THEOREM 5.8 *If $X \sim N_p(\mu, \Sigma)$ is independent of $\mathcal{M} \sim W_p(\Sigma, n)$, then*

$$n(X-\mu)^\top \mathcal{M}^{-1}(X-\mu) \sim T^2(p,n).$$

COROLLARY 5.3 *If $\overline{x}$ is the mean of a sample drawn from a normal population $N_p(\mu, \Sigma)$ and $\mathcal{S}$ is the sample covariance matrix, then*

$$(n-1)(\overline{x}-\mu)^\top \mathcal{S}^{-1}(\overline{x}-\mu) = n(\overline{x}-\mu)^\top \mathcal{S}_u^{-1}(\overline{x}-\mu) \sim T^2(p, n-1). \tag{5.17}$$

Recall that $\mathcal{S}_u = \frac{n}{n-1}\mathcal{S}$ is an unbiased estimator of the covariance matrix. A connection between the Hotelling T^2- and the F-distribution is given by the next theorem.

THEOREM 5.9

$$T^2(p,n) = \frac{np}{n-p+1} F_{p,n-p+1}.$$

EXAMPLE 5.5 *In the univariate case (p=1), this theorem boils down to the well known result:*

$$\left(\frac{\bar{x}-\mu}{\sqrt{\mathcal{S}_u}/\sqrt{n}}\right)^2 \sim T^2(1,n-1) = F_{1,n-1} = t^2_{n-1}$$

For further details on Hotelling T^2-distribution see Mardia et al. (1979). The next corollary follows immediately from (3.23),(3.24) and from Theorem 5.8. It will be useful for testing linear restrictions in multinormal populations.

COROLLARY 5.4 *Consider a linear transform of $X \sim N_p(\mu, \Sigma)$, $Y = \mathcal{A}X$ where $\mathcal{A}(q \times p)$ with $(q \le p)$. If $\overline{x}$ and $\mathcal{S}_X$ are the sample mean and the covariance matrix, we have*

$$\begin{aligned} \overline{y} &= \mathcal{A}\overline{x} \sim N_q(\mathcal{A}\mu, \frac{1}{n}\mathcal{A}\Sigma\mathcal{A}^\top) \\ n\mathcal{S}_Y &= n\mathcal{A}\mathcal{S}_X\mathcal{A}^\top \sim W_q(\mathcal{A}\Sigma\mathcal{A}^\top, n-1) \end{aligned}$$

$$(n-1)(\mathcal{A}\overline{x}-\mathcal{A}\mu)^\top(\mathcal{A}\mathcal{S}_X\mathcal{A}^\top)^{-1}(\mathcal{A}\overline{x}-\mathcal{A}\mu) \sim T^2(q, n-1)$$

The T^2 distribution is closely connected to the univariate t-statistic. In Example 5.4 we described the manner in which the Wishart distribution generalizes the χ^2-distribution. We can write (5.17) as:

$$T^2 = \sqrt{n}(\overline{x}-\mu)^\top \left(\frac{\sum_{j=1}^n (x_j-\overline{x})(x_j-\overline{x})^\top}{n-1}\right)^{-1} \sqrt{n}(\overline{x}-\mu)$$

which is of the form

$$\left(\begin{array}{c}\text{multivariate normal}\\ \text{random vector}\end{array}\right)^\top \left(\frac{\begin{array}{c}\text{Wishart random}\\ \text{matrix}\end{array}}{\text{degrees of freedom}}\right)^{-1} \left(\begin{array}{c}\text{multivariate normal}\\ \text{random vector}\end{array}\right).$$

This is analogous to

$$t^2 = \sqrt{n}(\overline{x} - \mu)(s^2)^{-1}\sqrt{n}(\overline{x} - \mu)$$

or

$$\begin{pmatrix} \text{normal} \\ \text{random variable} \end{pmatrix} \left(\frac{\chi^2\text{-random} \atop \text{variable}}{\text{degrees of freedom}} \right)^{-1} \begin{pmatrix} \text{normal} \\ \text{random variable} \end{pmatrix}$$

for the univariate case. Since the multivariate normal and Wishart random variables are independently distributed, their joint distribution is the product of the marginal normal and Wishart distributions. Using calculus, the distribution of T^2 as given above can be derived from this joint distribution.

Summary

- ↪ Hotelling's T^2-distribution is a generalization of the t-distribution. In particular $T(1, n) = t_n$.
- ↪ $(n-1)(\overline{x} - \mu)^\top \mathcal{S}^{-1}(\overline{x} - \mu)$ has a $T^2(p, n-1)$ distribution.
- ↪ The relation between Hotelling's T^2- and Fisher's F-distribution is given by $T^2(p, n) = \frac{np}{n-p+1} \; F_{p,n-p+1}$.

5.4 Spherical and Elliptical Distributions

The multinormal distribution belongs to the large family of elliptical distributions which has recently gained a lot of attention in financial mathematics. Elliptical distributions are often used, particularly in risk management.

DEFINITION 5.1 *A $(p \times 1)$ random vector Y is said to have a spherical distribution $S_p(\phi)$ if its characteristic function $\psi_Y(t)$ satisfies: $\psi_Y(t) = \phi(t^\top t)$ for some scalar function $\phi(.)$ which is then called the characteristic generator of the spherical distribution $S_p(\phi)$. We will write $Y \sim S_p(\phi)$.*

This is only one of several possible ways to define spherical distributions. We can see spherical distributions as an extension of the standard multinormal distribution $N_p(0, \mathcal{I}_p)$.

THEOREM 5.10 *Spherical random variables have the following properties:*

1. *All marginal distributions of a spherical distributed random vector are spherical.*
2. *All the marginal characteristic functions have the same generator.*
3. *Let $X \sim S_p(\phi)$, then X has the same distribution as $ru^{(p)}$ where $u^{(p)}$ is a random vector distributed uniformly on the unit sphere surface in $\mathbb{R}^p$ and $r \geq 0$ is a random variable independent of $u^{(p)}$. If $E(r^2) < \infty$, then*

$$E(X) = 0\,, \quad Cov(X) = \frac{E(r^2)}{p}\mathcal{I}_p.$$

The random radius r is related to the generator ϕ by a relation described in Fang, Kotz and Ng (1990, p.29). The moments of $X \sim S_p(\phi)$, provided that they exist, can be expressed in terms of one-dimensional integrals (Fang et al., 1990).

A spherically distributed random vector does not, in general, necessarily possess a density. However, if it does, the marginal densities of dimension smaller than $p-1$ are continuous and the marginal densities of dimension smaller than $p-2$ are differentiable (except possibly at the origin in both cases). Univariate marginal densities for p greater than 2 are nondecreasing on $(-\infty, 0)$ and nonincreasing on $(0, \infty)$.

DEFINITION 5.2 *A $(p \times 1)$ random vector X is said to have an elliptical distribution with parameters $\mu(p \times 1)$ and $\Sigma(p \times p)$ if X has the same distribution as $\mu + \mathcal{A}^\top Y$, where $Y \sim S_k(\phi)$ and $\mathcal{A}$ is a $(k \times p)$ matrix such that $\mathcal{A}^\top \mathcal{A} = \Sigma$ with $\operatorname{rank}(\Sigma) = k$. We shall write $X \sim EC_p(\mu, \Sigma, \phi)$.*

REMARK 5.1 *The elliptical distribution can be seen as an extension of $N_p(\mu, \Sigma)$.*

EXAMPLE 5.6 The multivariate t-distribution. *Let $Z \sim N_p(0, \mathcal{I}_p)$ and $s \sim \chi^2_m$ be independent. The random vector*

$$Y = \sqrt{m}\,\frac{Z}{s}$$

has a multivariate t-distribution with m degrees of freedom. Moreover the t-distribution belongs to the family of p-dimensioned spherical distributions.

EXAMPLE 5.7 The multinormal distribution. *Let $X \sim N_p(\mu, \Sigma)$. Then $X \sim EC_p(\mu, \Sigma, \phi)$ and $\phi(u) = \exp(-u/2)$. Figure 4.3 shows a density surface of the multivariate normal distribution: $f(x) = \det(2\pi\Sigma)^{-\frac{1}{2}} \exp\{-\frac{1}{2}(x-\mu)^\top \Sigma^{-1}(x-\mu)\}$ with $\Sigma = \begin{pmatrix} 1 & 0.6 \\ 0.6 & 1 \end{pmatrix}$ and $\mu = \begin{pmatrix} 0 \\ 0 \end{pmatrix}$. Notice that the density is constant on ellipses. This is the reason for calling this family of distributions "elliptical".*

THEOREM 5.11 *Elliptical random vectors X have the following properties:*

1. *Any linear combination of elliptically distributed variables are elliptical.*
2. *Marginal distributions of elliptically distributed variables are elliptical.*
3. *A scalar function $\phi(.)$ can determine an elliptical distribution $EC_p(\mu, \Sigma, \phi)$ for every $\mu \in \mathbb{R}^p$ and $\Sigma \geq 0$ with $\operatorname{rank}(\Sigma) = k$ iff $\phi(t^\top t)$ is a p-dimensional characteristic function.*
4. *Assume that X is nondegenerate. If $X \sim EC_p(\mu, \Sigma, \phi)$ and $X \sim EC_p(\mu^*, \Sigma^*, \phi^*)$, then there exists a constant $c > 0$ such that*

   $$\mu = \mu^*, \quad \Sigma = c\Sigma^*, \quad \phi^*(.) = \phi(c^{-1}.).$$

   In other words $\Sigma, \phi, \mathcal{A}$ are not unique, unless we impose the condition that $\det(\Sigma) = 1$.
5. *The characteristic function of $X, \psi(t) = E(e^{\mathbf{i}t^\top X})$ is of the form*

   $$\psi(t) = e^{\mathbf{i}t^\top \mu} \phi(t^\top \Sigma t)$$

   for a scalar function ϕ.

6. $X \sim EC_p(\mu, \Sigma, \phi)$ *with* $\text{rank}(\Sigma) = k$ iff X *has the same distribution as:*

$$\mu + r\mathcal{A}^\top u^{(k)} \tag{5.18}$$

where $r \geq 0$ *is independent of* $u^{(k)}$ *which is a random vector distributed uniformly on the unit sphere surface in* $\mathbb{R}^k$ *and* $\mathcal{A}$ *is a* $(k \times p)$ *matrix such that* $\mathcal{A}^\top \mathcal{A} = \Sigma$.

7. *Assume that* $X \sim EC_p(\mu, \Sigma, \phi)$ *and* $E(r^2) < \infty$. *Then*

$$E(X) = \mu \quad Cov(X) = \frac{E(r^2)}{\text{rank}(\Sigma)}\Sigma = -2\phi^\top(0)\Sigma.$$

8. *Assume that* $X \sim EC_p(\mu, \Sigma, \phi)$ *with* $\text{rank}(\Sigma) = k$. *Then*

$$Q(X) = (X - \mu)^\top \Sigma^-(X - \mu)$$

has the same distribution as r^2 *in equation (5.18).*

5.5 Exercises

EXERCISE 5.1 *Consider* $X \sim N_2(\mu, \Sigma)$ *with* $\mu = (2, 2)^\top$ *and* $\Sigma = \begin{pmatrix} 1 & 0 \\ 0 & 1 \end{pmatrix}$ *and the matrices* $\mathcal{A} = \begin{pmatrix} 1 \\ 1 \end{pmatrix}^\top$, $\mathcal{B} = \begin{pmatrix} 1 \\ -1 \end{pmatrix}^\top$. *Show that* $\mathcal{A}X$ *and* $\mathcal{B}X$ *are independent.*

EXERCISE 5.2 *Prove Theorem 5.4.*

EXERCISE 5.3 *Prove proposition (c) of Theorem 5.7.*

EXERCISE 5.4 *Let*

$$X \sim N_2\left(\begin{pmatrix} 1 \\ 2 \end{pmatrix}, \begin{pmatrix} 2 & 1 \\ 1 & 2 \end{pmatrix}\right)$$

and

$$Y \mid X \sim N_2\left(\begin{pmatrix} X_1 \\ X_1 + X_2 \end{pmatrix}, \begin{pmatrix} 1 & 0 \\ 0 & 1 \end{pmatrix}\right).$$

a) *Determine the distribution of* $Y_2 \mid Y_1$.
b) *Determine the distribution of* $W = X - Y$.

EXERCISE 5.5 *Consider* $\begin{pmatrix} X \\ Y \\ Z \end{pmatrix} \sim N_3(\mu, \Sigma)$. *Compute* μ *and* Σ *knowing that*

$$\begin{aligned} Y \mid Z &\sim N_1(-Z, 1) \\ \mu_{Z|Y} &= -\frac{1}{3} - \frac{1}{3}Y \\ X \mid Y, Z &\sim N_1(2 + 2Y + 3Z, 1). \end{aligned}$$

Determine the distributions of $X \mid Y$ *and of* $X \mid Y + Z$.

EXERCISE 5.6 *Knowing that*

$$\begin{aligned} Z &\sim N_1(0,1) \\ Y \mid Z &\sim N_1(1+Z,1) \\ X \mid Y,Z &\sim N_1(1-Y,1) \end{aligned}$$

a) *find the distribution of* $\begin{pmatrix} X \\ Y \\ Z \end{pmatrix}$ *and of* $Y \mid X, Z$.

b) *find the distribution of*

$$\begin{pmatrix} U \\ V \end{pmatrix} = \begin{pmatrix} 1+Z \\ 1-Y \end{pmatrix}.$$

c) *compute* $E(Y \mid U = 2)$.

EXERCISE 5.7 *Suppose* $\begin{pmatrix} X \\ Y \end{pmatrix} \sim N_2(\mu, \Sigma)$ *with* Σ *positive definite. Is it possible that*

a) $\mu_{X|Y} = 3Y^2$,

b) $\sigma_{XX|Y} = 2 + Y^2$,

c) $\mu_{X|Y} = 3 - Y$, *and*

d) $\sigma_{XX|Y} = 5$?

EXERCISE 5.8 *Let* $X \sim N_3 \left(\begin{pmatrix} 1 \\ 2 \\ 3 \end{pmatrix}, \begin{pmatrix} 11 & -6 & 2 \\ -6 & 10 & -4 \\ 2 & -4 & 6 \end{pmatrix} \right)$.

a) *Find the best linear approximation of* X_3 *by a linear function of* X_1 *and* X_2 *and compute the multiple correlation between* X_3 *and* (X_1, X_2).

b) *Let* $Z_1 = X_2 - X_3$, $Z_2 = X_2 + X_3$ *and* $(Z_3 \mid Z_1, Z_2) \sim N_1(Z_1 + Z_2, 10)$. *Compute the distribution of* $\begin{pmatrix} Z_1 \\ Z_2 \\ Z_3 \end{pmatrix}$.

EXERCISE 5.9 *Let* $(X, Y, Z)^\top$ *be a trivariate normal r.v. with*

$$\begin{aligned} Y \mid Z &\sim N_1(2Z, 24) \\ Z \mid X &\sim N_1(2X+3, 14) \\ X &\sim N_1(1, 4) \\ \text{and } \rho_{XY} &= 0.5. \end{aligned}$$

Find the distribution of $(X, Y, Z)^\top$ *and compute the partial correlation between* X *and* Y *for fixed* Z. *Do you think it is reasonable to approximate* X *by a linear function of* Y *and* Z?

EXERCISE 5.10 *Let* $X \sim N_4 \left(\begin{pmatrix} 1 \\ 2 \\ 3 \\ 4 \end{pmatrix}, \begin{pmatrix} 4 & 1 & 2 & 4 \\ 1 & 4 & 2 & 1 \\ 2 & 2 & 16 & 1 \\ 4 & 1 & 1 & 9 \end{pmatrix} \right)$.

a) *Give the best linear approximation of* X_2 *as a function of* (X_1, X_4) *and evaluate the quality of the approximation.*

b) *Give the best linear approximation of* X_2 *as a function of* (X_1, X_3, X_4) *and compare your answer with part a).*

EXERCISE 5.11 *Prove Theorem 5.2.*
(Hint: complete the linear transformation $Z = \begin{pmatrix} \mathcal{A} \\ \mathcal{I}_{p-q} \end{pmatrix} X + \begin{pmatrix} c \\ 0_{p-q} \end{pmatrix}$ *and then use Theorem 5.1 to get the marginal of the first* q *components of* Z*.)*

EXERCISE 5.12 *Prove Corollaries 5.1 and 5.2.*

6 Theory of Estimation

We know from our basic knowledge of statistics that one of the objectives in statistics is to better understand and model the underlying process which generates the data. This is known as statistical inference: we infer from information contained in a sample properties of the population from which the observations are taken. In multivariate statistical inference, we do exactly the same. The basic ideas were introduced in Section 4.5 on sampling theory: we observed the values of a multivariate random variable X and obtained a sample $\mathcal{X} = \{x_i\}_{i=1}^n$. Under random sampling, these observations are considered to be realizations of a sequence of i.i.d. random variables $X_1, \ldots, X_n$ where each X_i is a p-variate random variable which replicates the *parent* or *population* random variable X. In this chapter, for notational convenience, we will no longer differentiate between a random variable X_i and an observation of it, x_i, in our notation. We will simply write x_i and it should be clear from the context whether a random variable or an observed value is meant.

Statistical inference infers from the i.i.d. random sample $\mathcal{X}$ the properties of the population: typically, some unknown characteristic θ of its distribution. In parametric statistics, θ is a k-variate vector $\theta \in \mathbb{R}^k$ characterizing the unknown properties of the population pdf $f(x;\theta)$: this could be the mean, the covariance matrix, kurtosis, etc.

The aim will be to estimate θ from the sample $\mathcal{X}$ through estimators $\widehat{\theta}$ which are functions of the sample: $\widehat{\theta} = \widehat{\theta}(\mathcal{X})$. When an estimator $\widehat{\theta}$ is proposed, we must derive its sampling distribution to analyze its properties (is it related to the unknown quantity θ it is supposed to estimate?).

In this chapter the basic theoretical tools are developed which are needed to derive estimators and to determine their properties in general situations. We will basically rely on the maximum likelihood theory in our presentation. In many situations, the maximum likelihood estimators indeed share asymptotic optimal properties which make their use easy and appealing.

We will illustrate the multivariate normal population and also the linear regression model where the applications are numerous and the derivations are easy to do. In multivariate setups, the maximum likelihood estimator is at times too complicated to be derived analytically. In such cases, the estimators are obtained using numerical methods (nonlinear optimization). The general theory and the asymptotic properties of these estimators remain simple and valid. The following chapter, Chapter 7, concentrates on hypothesis testing and confidence interval issues.

6.1 The Likelihood Function

Suppose that $\{x_i\}_{i=1}^n$ is an i.i.d. sample from a population with pdf $f(x;\theta)$. The aim is to estimate $\theta \in \mathbb{R}^k$ which is a vector of unknown parameters. The *likelihood function* is defined as the joint density $L(\mathcal{X};\theta)$ of the observations x_i considered as a function of θ:

$$L(\mathcal{X};\theta) = \prod_{i=1}^{n} f(x_i;\theta), \tag{6.1}$$

where $\mathcal{X}$ denotes the sample of the data matrix with the observations $x_1^\top, \dots, x_n^\top$ in each row. The *maximum likelihood estimator* (MLE) of θ is defined as

$$\widehat{\theta} = \arg\max_{\theta} L(\mathcal{X};\theta).$$

Often it is easier to maximize the *log-likelihood function*

$$\ell(\mathcal{X};\theta) = \log L(\mathcal{X};\theta), \tag{6.2}$$

which is equivalent since the logarithm is a monotone one-to-one function. Hence

$$\widehat{\theta} = \arg\max_{\theta} L(\mathcal{X};\theta) = \arg\max_{\theta} \ell(\mathcal{X};\theta).$$

The following examples illustrate cases where the maximization process can be performed analytically, i.e., we will obtain an explicit analytical expression for $\widehat{\theta}$. Unfortunately, in other situations, the maximization process can be more intricate, involving nonlinear optimization techniques. In the latter case, given a sample $\mathcal{X}$ and the likelihood function, numerical methods will be used to determine the value of θ maximizing $L(\mathcal{X};\theta)$ or $\ell(\mathcal{X};\theta)$. These numerical methods are typically based on Newton-Raphson iterative techniques.

EXAMPLE 6.1 *Consider a sample $\{x_i\}_{i=1}^n$ from $N_p(\mu,\mathcal{I})$, i.e., from the pdf*

$$f(x;\theta) = (2\pi)^{-p/2}\exp\left\{-\frac{1}{2}(x-\theta)^\top(x-\theta)\right\}$$

where $\theta = \mu \in \mathbb{R}^p$ is the mean vector parameter. The log-likelihood is in this case

$$\ell(\mathcal{X};\theta) = \sum_{i=1}^{n}\log\{f(x_i;\theta)\} = \log(2\pi)^{-np/2} - \frac{1}{2}\sum_{i=1}^{n}(x_i-\theta)^\top(x_i-\theta). \tag{6.3}$$

The term $(x_i-\theta)^\top(x_i-\theta)$ equals

$$(x_i-\overline{x})^\top(x_i-\overline{x}) + (\overline{x}-\theta)^\top(\overline{x}-\theta) + 2(\overline{x}-\theta)^\top(x_i-\overline{x}).$$

Summing this term over $i = 1,\dots,n$ we see that

$$\sum_{i=1}^{n}(x_i-\theta)^\top(x_i-\theta) = \sum_{i=1}^{n}(x_i-\overline{x})^\top(x_i-\overline{x}) + n(\overline{x}-\theta)^\top(\overline{x}-\theta).$$

Hence

$$\ell(\mathcal{X};\theta) = \log(2\pi)^{-np/2} - \frac{1}{2}\sum_{i=1}^{n}(x_i-\overline{x})^\top(x_i-\overline{x}) - \frac{n}{2}(\overline{x}-\theta)^\top(\overline{x}-\theta).$$

Only the last term depends on θ and is obviously maximized for

$$\widehat{\theta} = \widehat{\mu} = \overline{x}.$$

Thus $\overline{x}$ is the MLE of θ for this family of pdfs $f(x,\theta)$.

A more complex example is the following one where we derive the MLE's for μ and Σ.

EXAMPLE 6.2 *Suppose $\{x_i\}_{i=1}^n$ is a sample from a normal distribution $N_p(\mu,\Sigma)$. Here $\theta = (\mu, \Sigma)$ with Σ interpreted as a vector. Due to the symmetry of Σ the unknown parameter θ is in fact $\{p + \frac{1}{2}p(p+1)\}$-dimensional. Then*

$$L(\mathcal{X};\theta) = |2\pi\Sigma|^{-n/2} \exp\left\{-\frac{1}{2}\sum_{i=1}^n (x_i-\mu)^\top \Sigma^{-1}(x_i-\mu)\right\} \tag{6.4}$$

and

$$\ell(\mathcal{X};\theta) = -\frac{n}{2}\log|2\pi\Sigma| - \frac{1}{2}\sum_{i=1}^n (x_i-\mu)^\top\Sigma^{-1}(x_i-\mu). \tag{6.5}$$

The term $(x_i-\mu)^\top\Sigma^{-1}(x_i-\mu)$ equals

$$(x_i-\overline{x})^\top\Sigma^{-1}(x_i-\overline{x}) + (\overline{x}-\mu)^\top\Sigma^{-1}(\bar{x}-\mu) + 2(\overline{x}-\mu)^\top\Sigma^{-1}(x_i-\overline{x}).$$

Summing this term over $i = 1,\ldots,n$ we see that

$$\sum_{i=1}^n (x_i-\mu)^\top\Sigma^{-1}(x_i-\mu) = \sum_{i=1}^n (x_i-\overline{x})^\top\Sigma^{-1}(x_i-\overline{x}) + n(\overline{x}-\mu)^\top\Sigma^{-1}(\overline{x}-\mu).$$

Note that from (2.14)

$$\begin{aligned}(x_i-\overline{x})^\top\Sigma^{-1}(x_i-\overline{x}) &= \operatorname{tr}\left\{(x_i-\overline{x})^\top\Sigma^{-1}(x_i-\overline{x})\right\}\\ &= \operatorname{tr}\left\{\Sigma^{-1}(x_i-\overline{x})(x_i-\overline{x})^\top\right\}.\end{aligned}$$

Therefore, by summing over the index i we finally arrive at

$$\begin{aligned}\sum_{i=1}^n (x_i-\mu)^\top\Sigma^{-1}(x_i-\mu) &= \operatorname{tr}\{\Sigma^{-1}\sum_{i=1}^n (x_i-\overline{x})(x_i-\overline{x})^\top\} + n(\overline{x}-\mu)^\top\Sigma^{-1}(\overline{x}-\mu)\\ &= \operatorname{tr}\{\Sigma^{-1}n\mathcal{S}\} + n(\overline{x}-\mu)^\top\Sigma^{-1}(\overline{x}-\mu).\end{aligned}$$

Thus the log-likelihood function for $N_p(\mu,\Sigma)$ is

$$\ell(\mathcal{X};\theta) = -\frac{n}{2}\log|2\pi\Sigma| - \frac{n}{2}\operatorname{tr}\{\Sigma^{-1}\mathcal{S}\} - \frac{n}{2}(\overline{x}-\mu)^\top\Sigma^{-1}(\overline{x}-\mu). \tag{6.6}$$

We can easily see that the third term is maximized by $\mu = \bar{x}$. In fact the MLE's are given by

$$\widehat{\mu} = \overline{x}, \quad \widehat{\Sigma} = \mathcal{S}.$$

The derivation of $\widehat{\Sigma}$ is a lot more complicated. It involves derivatives with respect to matrices with their notational complexities and will not be presented here: for a more elaborate proof see Mardia et al. (1979, p.103-104). Note that the unbiased covariance estimator $\mathcal{S}_u = \frac{n}{n-1}\mathcal{S}$ is not the MLE of Σ!

EXAMPLE 6.3 *Consider the linear regression model $y_i = \beta^\top x_i + \varepsilon_i$ for $i = 1,\ldots,n$, where ε_i is i.i.d. and $N(0,\sigma^2)$ and where $x_i \in \mathbb{R}^p$. Here $\theta = (\beta^\top, \sigma)$ is a $(p+1)$-dimensional parameter vector. Denote*

$$y = \begin{pmatrix} y_1\\ \vdots\\ y_n\end{pmatrix}, \quad \mathcal{X} = \begin{pmatrix} x_1^\top\\ \vdots\\ x_n^\top\end{pmatrix}.$$

Then

$$L(y, \mathcal{X}; \theta) = \prod_{i=1}^{n} \frac{1}{\sqrt{2\pi}\sigma} \exp\left\{-\frac{1}{2\sigma^2}(y_i - \beta^\top x_i)^2\right\}$$

and

$$\begin{aligned}
\ell(y, \mathcal{X}; \theta) &= \log\left(\frac{1}{(2\pi)^{n/2}\sigma^n}\right) - \frac{1}{2\sigma^2}\sum_{i=1}^{n}(y_i - \beta^\top x_i)^2 \\
&= -\frac{n}{2}\log(2\pi) - n\log\sigma - \frac{1}{2\sigma^2}(y - \mathcal{X}\beta)^\top(y - \mathcal{X}\beta) \\
&= -\frac{n}{2}\log(2\pi) - n\log\sigma - \frac{1}{2\sigma^2}(y^\top y + \beta^\top\mathcal{X}^\top\mathcal{X}\beta - 2\beta^\top\mathcal{X}^\top y).
\end{aligned}$$

Differentiating w.r.t. the parameters yields

$$\begin{aligned}
\frac{\partial}{\partial\beta}\ell &= -\frac{1}{2\sigma^2}(2\mathcal{X}^\top\mathcal{X}\beta - 2\mathcal{X}^\top y) \\
\frac{\partial}{\partial\sigma}\ell &= -\frac{n}{\sigma} + \frac{1}{\sigma^3}\{(y - \mathcal{X}\beta)^\top(y - \mathcal{X}\beta)\}.
\end{aligned}$$

Note that $\frac{\partial}{\partial\beta}\ell$ *denotes the vector of the derivatives w.r.t. all components of* β *(the gradient). Since the first equation only depends on* β*, we start with deriving* $\widehat{\beta}$*.*

$$\mathcal{X}^\top\mathcal{X}\widehat{\beta} = \mathcal{X}^\top y \quad \Longrightarrow \quad \widehat{\beta} = (\mathcal{X}^\top\mathcal{X})^{-1}\mathcal{X}^\top y$$

Plugging $\widehat{\beta}$ *into the second equation gives*

$$\frac{n}{\widehat{\sigma}} = \frac{1}{\widehat{\sigma}^3}(y - \mathcal{X}\widehat{\beta})^\top(y - \mathcal{X}\widehat{\beta}) \Longrightarrow \widehat{\sigma}^2 = \frac{1}{n}||y - \mathcal{X}\widehat{\beta}||^2,$$

where $|| \bullet ||^2$ *denotes the Euclidean vector norm from Section 2.6. We see that the MLE* $\widehat{\beta}$ *is identical with the least squares estimator (3.52). The variance estimator*

$$\widehat{\sigma}^2 = \frac{1}{n}\sum_{i=1}^{n}(y_i - \widehat{\beta}^\top x_i)^2$$

is nothing else than the residual sum of squares (RSS) from (3.37) generalized to the case of multivariate x_i*.*

Note that when the x_i *are considered to be fixed we have*

$$E(y) = \mathcal{X}\beta \text{ and } Var(y) = \sigma^2\mathcal{I}_n.$$

Then, using the properties of moments from Section 4.2 we have

$$E(\widehat{\beta}) = (\mathcal{X}^\top\mathcal{X})^{-1}\mathcal{X}^\top E(y) = \beta, \tag{6.7}$$

$$Var(\widehat{\beta}) = \sigma^2(\mathcal{X}^\top\mathcal{X})^{-1}. \tag{6.8}$$

Summary

- ↪ If $\{x_i\}_{i=1}^n$ is an i.i.d. sample from a distribution with pdf $f(x;\theta)$, then $L(\mathcal{X};\theta) = \prod_{i=1}^n f(x_i;\theta)$ is the likelihood function. The maximum likelihood estimator (MLE) is that value of θ which maximizes $L(\mathcal{X};\theta)$. Equivalently one can maximize the log-likelihood $\ell(\mathcal{X};\theta)$.
- ↪ The MLE's of μ and Σ from a $N_p(\mu,\Sigma)$ distribution are $\widehat{\mu} = \overline{x}$ and $\widehat{\Sigma} = \mathcal{S}$. Note that the MLE of Σ is not unbiased.
- ↪ The MLE's of β and σ in the linear model $y = \mathcal{X}\beta + \varepsilon,\ \varepsilon \sim N_n(0,\sigma^2\mathcal{I})$ are given by the least squares estimator $\widehat{\beta} = (\mathcal{X}^\top\mathcal{X})^{-1}\mathcal{X}^\top y$ and $\widehat{\sigma}^2 = \frac{1}{n}\|y - \mathcal{X}\widehat{\beta}\|^2$. $E(\widehat{\beta}) = \beta$ and $Var(\widehat{\beta}) = \sigma^2(\mathcal{X}^\top\mathcal{X})^{-1}$.

6.2 The Cramer-Rao Lower Bound

As pointed out above, an important question in estimation theory is whether an estimator $\widehat{\theta}$ has certain desired properties, in particular, if it converges to the unknown parameter θ it is supposed to estimate. One typical property we want for an estimator is unbiasedness, meaning that on the average, the estimator hits its target: $E(\widehat{\theta}) = \theta$. We have seen for instance (see Example 6.2) that $\bar{x}$ is an unbiased estimator of μ and $\mathcal{S}$ is a biased estimator of Σ in finite samples. If we restrict ourselves to unbiased estimation then the natural question is whether the estimator shares some optimality properties in terms of its sampling variance. Since we focus on unbiasedness, we look for an estimator with the smallest possible variance.

In this context, the Cramer-Rao lower bound will give the minimal achievable variance for any unbiased estimator. This result is valid under very general regularity conditions (discussed below). One of the most important applications of the Cramer-Rao lower bound is that it provides the asymptotic optimality property of maximum likelihood estimators. The Cramer-Rao theorem involves the *score function* and its properties which will be derived first.

The score function $s(\mathcal{X};\theta)$ is the derivative of the log likelihood function w.r.t. $\theta \in \mathbb{R}^k$

$$s(\mathcal{X};\theta) = \frac{\partial}{\partial\theta}\ell(\mathcal{X};\theta) = \frac{1}{L(\mathcal{X};\theta)}\frac{\partial}{\partial\theta}L(\mathcal{X};\theta). \tag{6.9}$$

The covariance matrix $\mathcal{F}_n = Var\{s(\mathcal{X};\theta)\}$ is called the *Fisher information matrix*. In what follows, we will give some interesting properties of score functions.

THEOREM 6.1 *If $s = s(\mathcal{X};\theta)$ is the score function and if $\hat{\theta} = t = t(\mathcal{X},\theta)$ is any function of $\mathcal{X}$ and θ, then under regularity conditions*

$$E(st^\top) = \frac{\partial}{\partial\theta}E(t^\top) - E\left(\frac{\partial t^\top}{\partial\theta}\right). \tag{6.10}$$

The proof is left as an exercise (see Exercise 6.9). The regularity conditions required for this theorem are rather technical and ensure that the expressions (expectations and derivations) appearing in (6.10) are well defined. In particular, the support of the density $f(x;\theta)$ should not depend on θ. The next corollary is a direct consequence.

COROLLARY 6.1 *If $s = s(\mathcal{X};\theta)$ is the score function, and $\hat{\theta} = t = t(\mathcal{X})$ is any unbiased estimator of θ (i.e., $E(t) = \theta$), then*

$$E(st^\top) = Cov(s,t) = \mathcal{I}_k. \tag{6.11}$$

Note that the score function has mean zero (see Exercise 6.10).

$$E\{s(\mathcal{X};\theta)\} = 0. \tag{6.12}$$

Hence, $E(ss^\top) = Var(s) = \mathcal{F}_n$ and by setting $s = t$ in Theorem 6.1 it follows that

$$\mathcal{F}_n = -E\left\{\frac{\partial^2}{\partial\theta\partial\theta^\top}\ell(\mathcal{X};\theta)\right\}.$$

REMARK 6.1 *If $x_1, \cdots, x_n$ are i.i.d., $\mathcal{F}_n = n\mathcal{F}_1$ where $\mathcal{F}_1$ is the Fisher information matrix for sample size n=1.*

EXAMPLE 6.4 *Consider an i.i.d. sample $\{x_i\}_{i=1}^n$ from $N_p(\theta, \mathcal{I})$. In this case the parameter θ is the mean μ. It follows from (6.3) that:*

$$\begin{aligned} s(\mathcal{X};\theta) &= \frac{\partial}{\partial\theta}\ell(\mathcal{X};\theta) \\ &= -\frac{1}{2}\frac{\partial}{\partial\theta}\left\{\sum_{i=1}^n (x_i-\theta)^\top(x_i-\theta)\right\} \\ &= n(\overline{x}-\theta). \end{aligned}$$

Hence, the information matrix is

$$\mathcal{F}_n = Var\{n(\overline{x}-\theta)\} = n\mathcal{I}_p.$$

How well can we estimate θ? The answer is given in the following theorem which is due to Cramer and Rao. As pointed out above, this theorem gives a lower bound for unbiased estimators. Hence, all estimators, which are unbiased and attain this lower bound, are *minimum variance estimators.*

THEOREM 6.2 (Cramer-Rao) *If $\hat{\theta} = t = t(\mathcal{X})$ is any unbiased estimator for θ, then under regularity conditions*

$$Var(t) \geq \mathcal{F}_n^{-1}, \tag{6.13}$$

where

$$\mathcal{F}_n = E\{s(\mathcal{X};\theta)s(\mathcal{X};\theta)^\top\} = Var\{s(\mathcal{X};\theta)\} \tag{6.14}$$

is the Fisher information matrix.

Proof:
Consider the correlation $\rho_{Y,Z}$ between Y and Z where $Y = a^\top t$, $Z = c^\top s$. Here s is the score and the vectors a, $c \in \mathbb{R}^p$. By Corollary 6.1 $Cov(s,t) = \mathcal{I}$ and thus

$$\begin{aligned} Cov(Y,Z) &= a^\top Cov(t,s)c = a^\top c \\ Var(Z) &= c^\top Var(s)c = c^\top \mathcal{F}_n c. \end{aligned}$$

Hence,

$$\rho^2_{Y,Z} = \frac{Cov^2(Y,Z)}{Var(Y)\ Var(Z)} = \frac{(a^\top c)^2}{a^\top\ Var(t)a \cdot c^\top \mathcal{F}_n c} \leq 1. \tag{6.15}$$

In particular, this holds for any $c \neq 0$. Therefore it holds also for the maximum of the left-hand side of (6.15) with respect to c. Since

$$\max_c \frac{c^\top aa^\top c}{c^\top \mathcal{F}_n c} = \max_{c^\top \mathcal{F}_n c=1} c^\top aa^\top c$$

and

$$\max_{c^\top \mathcal{F}_n c=1} c^\top aa^\top c = a^\top \mathcal{F}_n^{-1} a$$

by our maximization Theorem 2.5 we have

$$\frac{a^\top \mathcal{F}_n^{-1} a}{a^\top\ Var(t)a} \leq 1 \quad \forall\, a \in \mathbb{R}^p, \quad a \neq 0,$$

i.e.,

$$a^\top \{Var(t) - \mathcal{F}_n^{-1}\} a \geq 0 \quad \forall\, a \in \mathbb{R}^p, \quad a \neq 0,$$

which is equivalent to $Var(t) \geq \mathcal{F}_n^{-1}$. □

Maximum likelihood estimators (MLE's) attain the lower bound if the sample size n goes to infinity. The next Theorem 6.3 states this and, in addition, gives the asymptotic sampling distribution of the maximum likelihood estimation, which turns out to be multinormal.

THEOREM 6.3 *Suppose that the sample $\{x_i\}_{i=1}^n$ is i.i.d. If $\widehat{\theta}$ is the MLE for $\theta \in \mathbb{R}^k$, i.e., $\widehat{\theta} = \arg\max_\theta L(\mathcal{X};\theta)$, then under some regularity conditions, as $n \to \infty$:*

$$\sqrt{n}(\widehat{\theta} - \theta) \xrightarrow{\mathcal{L}} N_k(0, \mathcal{F}_1^{-1}) \tag{6.16}$$

where $\mathcal{F}_1$ denotes the Fisher information for sample size $n = 1$.

As a consequence of Theorem 6.3 we see that under regularity conditions the MLE is asymptotically unbiased, efficient (minimum variance) and normally distributed. Also it is a consistent estimator of θ.

Note that from property (5.4) of the multinormal it follows that asymptotically

$$n(\widehat{\theta} - \theta)^\top \mathcal{F}_1 (\widehat{\theta} - \theta) \xrightarrow{\mathcal{L}} \chi^2_p. \tag{6.17}$$

If $\widehat{\mathcal{F}}_1$ is a consistent estimator of $\mathcal{F}_1$ (*e.g.* $\widehat{\mathcal{F}}_1 = \mathcal{F}_1(\widehat{\theta})$), we have equivalently

$$n(\widehat{\theta} - \theta)^\top \widehat{\mathcal{F}}_1 (\widehat{\theta} - \theta) \xrightarrow{\mathcal{L}} \chi^2_p. \tag{6.18}$$

This expression is sometimes useful in testing hypotheses about θ and in constructing confidence regions for θ in a very general setup. These issues will be raised in more details in the next chapter but from (6.18) it can be seen, for instance, that when n is large,

$$P\left(n(\widehat{\theta} - \theta)^\top \widehat{\mathcal{F}}_1 (\widehat{\theta} - \theta) \leq \chi^2_{1-\alpha;p}\right) \approx 1 - \alpha,$$

where $\chi^2_{\nu;p}$ denotes the ν-quantile of a χ^2_p random variable. So, the ellipsoid $n(\widehat{\theta} - \theta)^\top \widehat{\mathcal{F}}_1 (\widehat{\theta} - \theta) \leq \chi^2_{1-\alpha;p}$ provides in $\mathbb{R}^p$ an asymptotic $(1-\alpha)$-confidence region for θ.

Summary

↪ The score function is the derivative $s(\mathcal{X};\theta) = \frac{\partial}{\partial\theta}\ell(\mathcal{X};\theta)$ of the log-likelihood with respect to θ. The covariance matrix of $s(\mathcal{X};\theta)$ is the Fisher information matrix.

↪ The score function has mean zero: $E\{s(\mathcal{X};\theta)\} = 0$.

↪ The Cramer-Rao bound says that any unbiased estimator $\hat{\theta} = t = t(\mathcal{X})$ has a variance that is bounded from below by the inverse of the Fisher information. Thus, an unbiased estimator, which attains this lower bound, is a minimum variance estimator.

↪ For i.i.d. data $\{x_i\}_{i=1}^n$ the Fisher information matrix is: $\mathcal{F}_n = n\mathcal{F}_1$.

↪ MLE's attain the lower bound in an asymptotic sense, i.e.,

$$\sqrt{n}(\widehat{\theta} - \theta) \xrightarrow{\mathcal{L}} N_k(0, \mathcal{F}_1^{-1})$$

if $\widehat{\theta}$ is the MLE for $\theta \in \mathbb{R}^k$, i.e., $\widehat{\theta} = \arg\max_\theta L(\mathcal{X};\theta)$.

6.3 Exercises

EXERCISE 6.1 *Consider a uniform distribution on the interval* $[0, \theta]$. *What is the MLE of* θ*? (Hint: the maximization here cannot be performed by means of derivatives. Here the support of* x *depends on* θ*!)*

EXERCISE 6.2 *Consider an i.i.d. sample of size n from the bivariate population with pdf* $f(x_1, x_2) = \frac{1}{\theta_1\theta_2} e^{-\left(\frac{x_1}{\theta_1} + \frac{x_2}{\theta_2}\right)}$, $x_1, x_2 > 0$. *Compute the MLE of* $\theta = (\theta_1, \theta_2)$. *Find the Cramer-Rao lower bound. Is it possible to derive a minimal variance unbiased estimator of* θ*?*

EXERCISE 6.3 *Show that the MLE of Example 6.1,* $\widehat{\mu} = \overline{x}$, *is a minimal variance estimator for any finite sample size n (i.e., without applying Theorem 6.3).*

EXERCISE 6.4 *We know from Example 6.4 that the MLE of Example 6.1 has* $\mathcal{F}_1 = \mathcal{I}_p$. *This leads to*

$$\sqrt{n}(\overline{x} - \mu) \xrightarrow{\mathcal{L}} N_p(0, \mathcal{I})$$

by Theorem 6.3. Can you give an analogous result for the square $\overline{x}^2$ *for the case* $p = 1$*?*

EXERCISE 6.5 *Consider an i.i.d. sample of size n from the bivariate population with pdf* $f(x_1, x_2) = \frac{1}{\theta_1^2\theta_2}\frac{1}{x_2} e^{-\left(\frac{x_1}{\theta_1 x_2} + \frac{x_2}{\theta_1\theta_2}\right)}$, $x_1, x_2 > 0$. *Compute the MLE of* $\theta = (\theta_1, \theta_2)$. *Find the Cramer-Rao lower bound and the asymptotic variance of* $\widehat{\theta}$.

EXERCISE 6.6 *Consider a sample* $\{x_i\}_{i=1}^n$ *from* $N_p(\mu, \Sigma_0)$ *where* Σ_0 *is known. Compute the Cramer-Rao lower bound for* μ. *Can you derive a minimal unbiased estimator for* μ*?*

EXERCISE 6.7 *Let* $X \sim N_p(\mu, \Sigma)$ *where* Σ *is unknown but we know* $\Sigma = \mathrm{diag}(\sigma_{11}, \sigma_{22}, \ldots, \sigma_{pp})$. *From an i.i.d. sample of size* n, *find the MLE of* μ *and of* Σ.

EXERCISE 6.8 *Reconsider the setup of the previous exercise. Suppose that*

$$\Sigma = \mathrm{diag}(\sigma_{11}, \sigma_{22}, \ldots, \sigma_{pp}).$$

Can you derive in this case the Cramer-Rao lower bound for $\theta^\top = (\mu_1 \ldots \mu_p, \sigma_{11} \ldots \sigma_{pp})$*?*

EXERCISE 6.9 *Prove Theorem 6.1. Hint: start from* $\frac{\partial}{\partial\theta} E(t^\top) = \frac{\partial}{\partial\theta} \int t^\top(\mathcal{X};\theta) L(\mathcal{X};\theta) d\mathcal{X}$, *then permute integral and derivatives and note that* $s(\mathcal{X};\theta) = \frac{1}{L(\mathcal{X};\theta)} \frac{\partial}{\partial\theta} L(\mathcal{X};\theta)$.

EXERCISE 6.10 *Prove expression (6.12).*
(Hint: start from $E(s(\mathcal{X};\theta)) = \int \frac{1}{L(\mathcal{X};\theta)} \frac{\partial}{\partial\theta} L(\mathcal{X};\theta) L(\mathcal{X};\theta) \partial\mathcal{X}$ *and then permute integral and derivative.)*

7 Hypothesis Testing

In the preceding chapter, the theoretical basis of estimation theory was presented. Now we turn our interest towards testing issues: we want to test the hypothesis H_0 that the unknown parameter θ belongs to some subspace of $\mathbb{R}^q$. This subspace is called the *null set* and will be denoted by $\Omega_0 \subset \mathbb{R}^q$.

In many cases, this null set corresponds to restrictions which are imposed on the parameter space: H_0 corresponds to a "reduced model". As we have already seen in Chapter 3, the solution to a testing problem is in terms of a *rejection region* R which is a set of values in the sample space which leads to the decision of rejecting the null hypothesis H_0 in favor of an alternative H_1, which is called the "full model".

In general, we want to construct a rejection region R which controls the size of the type I error, *i.e.* the probability of rejecting the null hypothesis when it is true. More formally, a solution to a testing problem is of predetermined size α if:

$$P(\text{Rejecting } H_0 \mid H_0 \text{ is true}) = \alpha.$$

In fact, since H_0 is often a composite hypothesis, it is achieved by finding R such that

$$\sup_{\theta \in \Omega_0} P(\mathcal{X} \in R \mid \theta) = \alpha.$$

In this chapter we will introduce a tool which allows us to build a rejection region in general situations: it is based on the likelihood ratio principle. This is a very useful technique because it allows us to derive a rejection region with an asymptotically appropriate size α. The technique will be illustrated through various testing problems and examples. We concentrate on multinormal populations and linear models where the size of the test will often be exact even for finite sample sizes n.

Section 7.1 gives the basic ideas and Section 7.2 presents the general problem of testing linear restrictions. This allows us to propose solutions to frequent types of analys (including comparisons of several means, repeated measurements and profile analysis). Each case can be viewed as a simple specific case of testing linear restrictions. Special attention is devoted to confidence intervals and confidence regions for means and for linear restrictions on means in a multinormal setup.

7.1 Likelihood Ratio Test

Suppose that the distribution of $\{x_i\}_{i=1}^n$, $x_i \in \mathbb{R}^p$, depends on a parameter vector θ. We will consider two hypotheses:

$$\begin{aligned} H_0 &: \theta \in \Omega_0 \\ H_1 &: \theta \in \Omega_1. \end{aligned}$$

The hypothesis H_0 corresponds to the "reduced model" and H_1 to the "full model". This notation was already used in Chapter 3.

EXAMPLE 7.1 *Consider a multinormal $N_p(\theta, \mathcal{I})$. To test if θ equals a certain fixed value θ_0 we construct the test problem:*

$$\begin{aligned} H_0 &: \theta = \theta_0 \\ H_1 &: \text{no constraints on } \theta \end{aligned}$$

or, equivalently, $\Omega_0 = \{\theta_0\}$, $\Omega_1 = \mathbb{R}^p$.

Define $L_j^* = \max_{\theta \in \Omega_j} L(\mathcal{X}; \theta)$, the maxima of the likelihood for each of the hypotheses. Consider the *likelihood ratio* (LR)

$$\lambda(\mathcal{X}) = \frac{L_0^*}{L_1^*} \tag{7.1}$$

One tends to favor H_0 if the LR is high and H_1 if the LR is low. The *likelihood ratio test* (LRT) tells us when exactly to favor H_0 over H_1. A likelihood ratio test of size α for testing H_0 against H_1 has the rejection region

$$R = \{\mathcal{X} : \lambda(\mathcal{X}) < c\}$$

where c is determined so that $\sup_{\theta \in \Omega_0} P_\theta(\mathcal{X} \in R) = \alpha$. The difficulty here is to express c as a function of α, because $\lambda(\mathcal{X})$ might be a complicated function of $\mathcal{X}$.

Instead of λ we may equivalently use the log-likelihood

$$-2 \log \lambda = 2(\ell_1^* - \ell_0^*).$$

In this case the rejection region will be $R = \{\mathcal{X} : -2 \log \lambda(\mathcal{X}) > k\}$. What is the distribution of λ or of $-2 \log \lambda$ from which we need to compute c or k?

THEOREM 7.1 *If $\Omega_1 \subset \mathbb{R}^q$ is a q-dimensional space and if $\Omega_0 \subset \Omega_1$ is an r-dimensional subspace, then under regularity conditions*

$$\forall\, \theta \in \Omega_0 : -2 \log \lambda \xrightarrow{\mathcal{L}} \chi^2_{q-r} \qquad \text{as} \quad n \to \infty.$$

An asymptotic rejection region can now be given by simply computing the $1 - \alpha$ quantile $k = \chi^2_{1-\alpha;q-r}$. The LRT rejection region is therefore

$$R = \{\mathcal{X} : -2 \log \lambda(\mathcal{X}) > \chi^2_{1-\alpha;q-r}\}.$$

The Theorem 7.1 is thus very helpful: it gives a general way of building rejection regions in many problems. Unfortunately, it is only an asymptotic result, meaning that the size of the test is only approximately equal to α, although the approximation becomes better when the sample size n increases. The question is "how large should n be?". There is no definite rule: we encounter here the same problem that was already discussed with respect to the Central Limit Theorem in Chapter 4.

Fortunately, in many standard circumstances, we can derive exact tests even for finite samples because the test statistic $-2 \log \lambda(\mathcal{X})$ or a simple transformation of it turns out to have a simple form. This is the case in most of the following standard testing problems. All of them can be viewed as an illustration of the likelihood ratio principle.

Test Problem 1 is an *amuse-bouche*: in testing the mean of a multinormal population with a known covariance matrix the likelihood ratio statistic has a very simple quadratic form with a known distribution under H_0.

TEST PROBLEM 1 Suppose that $X_1, \ldots, X_n$ is an i.i.d. random sample from a $N_p(\mu, \Sigma)$ population.

$$H_0 : \mu = \mu_0, \ \Sigma \text{ known versus } H_1 : \text{ no constraints.}$$

In this case H_0 is a simple hypothesis, i.e., $\Omega_0 = \{\mu_0\}$ and therefore the dimension r of Ω_0 equals 0. Since we have imposed no constraints in H_1, the space Ω_1 is the whole $\mathbb{R}^p$ which leads to $q = p$. From (6.6) we know that

$$\ell_0^* = \ell(\mu_0, \Sigma) = -\frac{n}{2}\log|2\pi\Sigma| - \frac{1}{2}n\,\mathrm{tr}(\Sigma^{-1}\mathcal{S}) - \frac{1}{2}n(\overline{x} - \mu_0)^\top\Sigma^{-1}(\overline{x} - \mu_0).$$

Under H_1 the maximum of $\ell(\mu, \Sigma)$ is

$$\ell_1^* = \ell(\overline{x}, \Sigma) = -\frac{n}{2}\log|2\pi\Sigma| - \frac{1}{2}n\,\mathrm{tr}(\Sigma^{-1}\mathcal{S}).$$

Therefore,

$$-2\log\lambda = 2(\ell_1^* - \ell_0^*) = n(\overline{x} - \mu_0)^\top\Sigma^{-1}(\overline{x} - \mu_0) \tag{7.2}$$

which, by Theorem 4.7, has a χ^2_p-distribution under H_0.

EXAMPLE 7.2 *Consider the bank data again. Let us test whether the population mean of the forged bank notes is equal to*

$$\mu_0 = (214.9, 129.9, 129.7, 8.3, 10.1, 141.5)^\top.$$

(This is in fact the sample mean of the genuine bank notes.) The sample mean of the forged bank notes is

$$\overline{x} = (214.8, 130.3, 130.2, 10.5, 11.1, 139.4)^\top.$$

Suppose for the moment that the estimated covariance matrix $\mathcal{S}_f$ given in (3.5) is the true covariance matrix Σ. We construct the likelihood ratio test and obtain

$$\begin{aligned} -2\log\lambda &= 2(\ell_1^* - \ell_0^*) = n(\overline{x} - \mu_0)^\top\Sigma^{-1}(\overline{x} - \mu_0) \\ &= 7362.32, \end{aligned}$$

the quantile $k = \chi^2_{0.95;6}$ equals 12.592. The rejection rejection consists of all values in the sample space which lead to values of the likelihood ratio test statistic larger than 12.592. Under H_0 the value of $-2\log\lambda$ is therefore highly significant. Hence, the true mean of the forged bank notes is significantly different from μ_0!

Test Problem 2 is the same as the preceding one but in a more realistic situation where the covariance matrix is unknown: here the Hotelling's T^2-distribution will be useful to determine an exact test and a confidence region for the unknown μ.

TEST PROBLEM 2 Suppose that $X_1, \ldots, X_n$ is an i.i.d. random sample from a $N_p(\mu, \Sigma)$ population.

$$H_0 : \mu = \mu_0, \ \Sigma \text{ unknown versus } H_1 : \text{ no constraints.}$$

Under H_0 it can be shown that

$$\begin{aligned}
\mathcal{S}_0 &= \frac{1}{n}\left[x - 1_n\mu_0^\top - 1_n\overline{x}^\top + 1_n\overline{x}^\top\right]^\top \left[x - 1_n\mu_0^\top - 1_n\overline{x}^\top + 1_n\overline{x}^\top\right] \\
&= \mathcal{S} + (\overline{x} - \mu_0)(\overline{x} - \mu_0)^\top \\
\ell_0^* &= \ell(\mu_0, \mathcal{S} + dd^\top), \quad d = (\overline{x} - \mu_0)
\end{aligned} \tag{7.3}$$

and under H_1 we have

$$\ell_1^* = \ell(\overline{x}, \mathcal{S}).$$

This leads after some calculation to

$$\begin{aligned}
-2\log\lambda &= 2(\ell_1^* - \ell_0^*) \\
&= -n\log|\mathcal{S}| - n\,\mathrm{tr}(\mathcal{S}^{-1}\mathcal{S}) - n\,(\overline{x} - \overline{x})^\top \mathcal{S}^{-1}(\overline{x} - \overline{x}) + n\log|\mathcal{S} + dd^\top| \\
&\quad + n\,\mathrm{tr}\left[\left(\mathcal{S} + dd^\top\right)^{-1}\mathcal{S}\right] + n\,(\overline{x} - \mu_0)^\top(\mathcal{S} + dd^\top)^{-1}(\overline{x} - \mu_0) \\
&= n\log|\frac{\mathcal{S} + dd^\top}{\mathcal{S}}| + n\,\mathrm{tr}\left[(\mathcal{S} + dd^\top)^{-1}\mathcal{S}\right] + nd^\top(\mathcal{S} + dd^\top)^{-1}d - np \\
&= n\log|\frac{\mathcal{S} + dd^\top}{\mathcal{S}}| + n\,\mathrm{tr}\left[(\mathcal{S} + dd^\top)^{-1}(dd^\top + \mathcal{S})\right] - np \\
&= n\log|\frac{\mathcal{S} + dd^\top}{\mathcal{S}}| \\
&= n\log|1 + \mathcal{S}^{-1/2}dd^\top\mathcal{S}^{-1/2}|
\end{aligned}$$

By using the result for the determinant of partitioned matrix, it equals to

$$n\log\begin{vmatrix} 1 & -d^\top\mathcal{S}^{-1/2} \\ \mathcal{S}^{-1/2}d & I \end{vmatrix} = n\log\begin{vmatrix} 1 & -d^\top\mathcal{S}^{-1/2}{}_1 & -d^\top\mathcal{S}^{-1/2}{}_2 & \ldots & -d^\top\mathcal{S}^{-1/2}{}_p \\ \mathcal{S}^{-1/2}d_1 & 1 & 0 & \ldots & 0 \\ \mathcal{S}^{-1/2}d_2 & 0 & 1 & & 0 \\ \vdots & \vdots & & \ddots & \\ \mathcal{S}^{-1/2}d_p & 0 & 0 & \ldots & 1 \end{vmatrix}$$

$$= \quad n\log + n\log \sum_{i=1}^{p} -d^\top \mathcal{S}^{-1/2}{}_i(-1)^{1+(i+1)} \begin{vmatrix} \mathcal{S}^{-1/2}d_1 & 1 & 0 & \dots & 0 \\ \mathcal{S}^{-1/2}d_2 & 0 & 1 & \dots & 0 \\ \vdots & & & \ddots & \\ \mathcal{S}^{-1/2}d_i & 0 & 0 & \dots & 0 \\ \vdots & & & & \\ \mathcal{S}^{-1/2}d_p & 0 & 0 & \dots & 1 \end{vmatrix}$$

$$= \quad n\log 1 + \sum_{i=1}^{p} -d^\top \mathcal{S}^{-1/2}{}_i(-1)^{2+i}\mathcal{S}^{-1/2}d_i(-1)^{i+1}$$

$$= \quad n\log(1 + d^\top \mathcal{S}^{-1}d). \tag{7.4}$$

This statistic is a monotone function of $(n-1)d^\top \mathcal{S}^{-1}d$. This means that $-2\log\lambda > k$ if and only if $(n-1)d^\top \mathcal{S}^{-1}d > k'$. The latter statistic has by Corollary 5.3, under H_0, a Hotelling's T^2-distribution. Therefore,

$$(n-1)(\bar{x}-\mu_0)^\top \mathcal{S}^{-1}(\bar{x}-\mu_0) \sim T^2(p, n-1), \tag{7.5}$$

or equivalently

$$\left(\frac{n-p}{p}\right)(\bar{x}-\mu_0)^\top \mathcal{S}^{-1}(\bar{x}-\mu_0) \sim F_{p,n-p}. \tag{7.6}$$

In this case an exact rejection region may be defined as

$$\left(\frac{n-p}{p}\right)(\bar{x}-\mu_0)^\top \mathcal{S}^{-1}(\bar{x}-\mu_0) > F_{1-\alpha;p,n-p}.$$

Alternatively, we have from Theorem 7.1 that under H_0 the asymptotic distribution of the test statistic is

$$-2\log\lambda \xrightarrow{\mathcal{L}} \chi^2_p, \quad \text{as } n\to\infty$$

which leads to the (asymptotically valid) rejection region

$$n\log\{1 + (\bar{x}-\mu_0)^\top \mathcal{S}^{-1}(\bar{x}-\mu_0)\} > \chi^2_{1-\alpha;p},$$

but of course, in this case, we would prefer to use the exact F-test provided just above.

EXAMPLE 7.3 *Consider the problem of Example 7.2 again. We know that $\mathcal{S}_f$ is the empirical analogue for Σ_f, the covariance matrix for the forged banknotes. The test statistic (7.5) has the value 1153.4 or its equivalent for the F distribution in (7.6) is 182.5 which is highly significant ($F_{0.95;6,94} = 2.1966$) so that we conclude that $\mu_f \neq \mu_0$.*

Confidence Region for μ

When estimating a multidimensional parameter $\theta \in \mathbb{R}^k$ from a sample, we saw in Chapter 6 how to determine the estimator $\widehat{\theta} = \widehat{\theta}(\mathcal{X})$. For the observed data we end up with a point estimate, which is the corresponding observed value of $\widehat{\theta}$. We know $\widehat{\theta}(\mathcal{X})$ is a random variable and we often prefer to determine a *confidence region* for θ. A confidence region (CR) is a random subset of $\mathbb{R}^k$ (determined by appropriate statistics) such that we are "confident", at a certain given level $1-\alpha$, that this region contains θ:

$$P(\theta \in \text{CR}) = 1-\alpha.$$

This is just a multidimensional generalization of the basic univariate confidence interval. Confidence regions are particularly useful when a hypothesis H_0 on θ is rejected, because they eventually help in identifying which component of θ is responsible for the rejection.

There are only a few cases where confidence regions can be easily assessed, and include most of the testing problems on mean presented in this section.

Corollary 5.3 provides a pivotal quantity which allows confidence regions for μ to be constructed. Since $\left(\frac{n-p}{p}\right)(\bar{x}-\mu)^\top \mathcal{S}^{-1}(\bar{x}-\mu) \sim F_{p,n-p}$, we have

$$P\left(\left(\frac{n-p}{p}\right)(\mu-\bar{x})^\top \mathcal{S}^{-1}(\mu-\bar{x}) < F_{1-\alpha;p,n-p}\right) = 1-\alpha.$$

Then,

$$\mathrm{CR} = \left\{\mu \in \mathbb{R}^p \mid (\mu-\bar{x})^\top \mathcal{S}^{-1}(\mu-\bar{x}) \le \frac{p}{n-p} F_{1-\alpha;p,n-p}\right\}$$

is a confidence region at level (1-α) for μ. It is the interior of an iso-distance ellipsoid in $\mathbb{R}^p$ centered at $\bar{x}$, with a scaling matrix $\mathcal{S}^{-1}$ and a distance constant $\left(\frac{p}{n-p}\right) F_{1-\alpha;p,n-p}$. When p is large, ellipsoids are not easy to handle for practical purposes. One is thus interested in finding confidence intervals for $\mu_1, \mu_2, \dots, \mu_p$ so that simultaneous confidence on all the intervals reaches the desired level of say, $1-\alpha$.

In the following, we consider a more general problem. We construct *simultaneous confidence intervals* for all possible linear combinations $a^\top\mu$, $a \in \mathbb{R}^p$ of the elements of μ.

Suppose for a moment that we fix a particular projection vector a. We are back to a standard univariate problem of finding a confidence interval for the mean $a^\top\mu$ of a univariate random variable $a^\top X$. We can use the t-statistics and an obvious confidence interval for $a^\top\mu$ is given by the values $a^\top\mu$ such that

$$\left|\frac{\sqrt{n-1}(a^\top\mu - a^\top\bar{x})}{\sqrt{a^\top \mathcal{S} a}}\right| \le t_{1-\frac{\alpha}{2};n-1}$$

or equivalently

$$t^2(a) = \frac{(n-1)\left\{a^\top(\mu-\bar{x})\right\}^2}{a^\top \mathcal{S} a} \le F_{1-\alpha;1,n-1}.$$

This provides the $(1-\alpha)$ confidence interval for $a^\top\mu$:

$$\left(a^\top\bar{x} - \sqrt{F_{1-\alpha;1,n-1}\frac{a^\top \mathcal{S} a}{n-1}} \le a^\top\mu \le a^\top\bar{x} + \sqrt{F_{1-\alpha;1,n-1}\frac{a^\top \mathcal{S} a}{n-1}}\right).$$

Now it is easy to prove (using Theorem 2.5) that:

$$\max_a t^2(a) = (n-1)(\bar{x}-\mu)^\top \mathcal{S}^{-1}(\bar{x}-\mu) \sim T^2(p, n-1).$$

Therefore, simultaneously for all $a \in \mathbb{R}^p$, the interval

$$\left(a^\top\bar{x} - \sqrt{K_\alpha a^\top \mathcal{S} a},\ a^\top\bar{x} + \sqrt{K_\alpha a^\top \mathcal{S} a}\right) \tag{7.7}$$

where $K_\alpha = \frac{p}{n-p} F_{1-\alpha;p,n-p}$, will contain $a^\top\mu$ with probability $(1-\alpha)$.

A particular choice of a are the columns of the identity matrix $\mathcal{I}_p$, providing simultaneous confidence intervals for $\mu_1, \dots, \mu_p$. We have therefore with probability $(1-\alpha)$ for $j = 1, \dots, p$

$$\bar{x}_j - \sqrt{\frac{p}{n-p} F_{1-\alpha;p,n-p} s_{jj}} \le \mu_j \le \bar{x}_j + \sqrt{\frac{p}{n-p} F_{1-\alpha;p,n-p} s_{jj}}. \tag{7.8}$$

It should be noted that these intervals define a rectangle inscribing the confidence ellipsoid for μ given above. They are particularly useful when a null hypothesis H_0 of the type described above is rejected and one would like to see which component(s) are mainly responsible for the rejection.

EXAMPLE 7.4 *The 95% confidence region for μ_f, the mean of the forged banknotes, is given by the ellipsoid:*

$$\left\{\mu \in \mathbb{R}^6 \left| (\mu - \bar{x}_f)^\top S_f^{-1}(\mu - \bar{x}_f) \le \frac{6}{94} F_{0.95;6,94} \right.\right\}.$$

The 95% simultaneous confidence intervals are given by (we use $F_{0.95;6,94} = 2.1966$)

$$\begin{array}{rcccl} 214.692 & \le & \mu_1 & \le & 214.954 \\ 130.205 & \le & \mu_2 & \le & 130.395 \\ 130.082 & \le & \mu_3 & \le & 130.304 \\ 10.108 & \le & \mu_4 & \le & 10.952 \\ 10.896 & \le & \mu_5 & \le & 11.370 \\ 139.242 & \le & \mu_6 & \le & 139.658. \end{array}$$

Comparing the inequalities with $\mu_0 = (214.9, 129.9, 129.7, 8.3, 10.1, 141.5)^\top$ shows that almost all components (except the first one) are responsible for the rejection of μ_0 in Example 7.2 and 7.3.

In addition, the method can provide other confidence intervals. We have at the same level of confidence (choosing $a^\top = (0,\ 0,\ 0,\ 1,\ -1,\ 0)$)

$$-1.211 \le \mu_4 - \mu_5 \le 0.005$$

showing that for the forged bills, the lower border is essentially smaller than the upper border.

REMARK 7.1 *It should be noted that the confidence region is an ellipsoid whose characteristics depend on the whole matrix S. In particular, the slope of the axis depends on the eigenvectors of S and therefore on the covariances s_{ij}. However, the rectangle inscribing the confidence ellipsoid provides the simultaneous confidence intervals for μ_j, $j = 1, \ldots, p$. They do not depend on the covariances s_{ij}, but only on the variances s_{jj} (see (7.8)). In particular, it may happen that a tested value μ_0 is covered by the confidence ellipsoid but not covered by the intervals (7.8). In this case, μ_0 is rejected by a test based on the simultaneous confidence intervals but not rejected by a test based on the confidence ellipsoid. The simultaneous confidence intervals are easier to handle than the full ellipsoid but we have lost some information, namely the covariance between the components (see Exercise 7.14).*

The following Problem concerns the covariance matrix in a multinormal population: in this situation the test statistic has a slightly more complicated distribution. We will therefore invoke the approximation of Theorem 7.1 in order to derive a test of approximate size α.

TEST PROBLEM 3 Suppose that $X_1, \ldots, X_n$ is an i.i.d. random sample from a $N_p(\mu, \Sigma)$ population.

$$H_0 : \Sigma = \Sigma_0,\ \mu \text{ unknown versus } H_1 : \text{ no constraints.}$$

Under H_0 we have $\widehat{\mu} = \overline{x}$, and $\Sigma = \Sigma_0$, whereas under H_1 we have $\widehat{\mu} = \overline{x}$, and $\widehat{\Sigma} = \mathcal{S}$. Hence

$$\begin{aligned} \ell_0^* &= \ell(\overline{x}, \Sigma_0) = -\frac{1}{2} n \log |2\pi\Sigma_0| - \frac{1}{2} n \operatorname{tr}(\Sigma_0^{-1}\mathcal{S}) \\ \ell_1^* &= \ell(\overline{x}, \mathcal{S}) = -\frac{1}{2} n \log |2\pi\mathcal{S}| - \frac{1}{2} np \end{aligned}$$

and thus

$$\begin{aligned} -2\log\lambda &= 2(\ell_1^* - \ell_0^*) \\ &= n \operatorname{tr}(\Sigma_0^{-1}\mathcal{S}) - n \log |\Sigma_0^{-1}\mathcal{S}| - np. \end{aligned}$$

Note that this statistic is a function of the eigenvalues of $\Sigma_0^{-1}\mathcal{S}$! Unfortunately, the exact finite sample distribution of $-2\log\lambda$ is very complicated. Asymptotically, we have under H_0

$$-2\log\lambda \xrightarrow{\mathcal{L}} \chi^2_m \qquad \text{as} \quad n \to \infty$$

with $m = \frac{1}{2}\{p(p+1)\}$, since a $(p \times p)$ covariance matrix has only these m parameters as a consequence of its symmetry.

EXAMPLE 7.5 *Consider the US companies data set (Table B.5) and suppose we are interested in the companies of the energy sector, analyzing their assets* (X_1) *and sales* (X_2). *The sample is of size 15 and provides the value of* $S = 10^7 \times \begin{bmatrix} 1.6635 & 1.2410 \\ 1.2410 & 1.3747 \end{bmatrix}$. *We want to test if* $Var\binom{X_1}{X_2} = 10^7 \times \begin{bmatrix} 1.2248 & 1.1425 \\ 1.1425 & 1.5112 \end{bmatrix} = \Sigma_0$. *($\Sigma_0$ is in fact the empirical variance matrix for X_1 and X_2 for the manufacturing sector). The test statistic (* `MVAusenergy`*) turns out to be* $-2\log\lambda = 5.4046$ *which is not significant for* χ^2_3 *(p-value=0.1445). So we can not conclude that* $\Sigma \neq \Sigma_0$.

In the next testing problem, we address a question that was already stated in Chapter 3, Section 3.6: testing a particular value of the coefficients β in a linear model. The presentation is done in general terms so that it can be built on in the next section where we will test linear restrictions on β.

TEST PROBLEM 4 Suppose that $Y_1, \ldots, Y_n$ are independent r.v.'s with $Y_i \sim N_1(\beta^\top x_i, \sigma^2), x_i \in \mathbb{R}^p$.

$$H_0 : \beta = \beta_0,\ \sigma^2 \text{ unknown versus } H_1 : \text{ no constraints.}$$

Under H_0 we have $\beta = \beta_0, \widehat{\sigma}_0^2 = \frac{1}{n}||y - \mathcal{X}\beta_0||^2$ and under H_1 we have $\hat{\beta} = (\mathcal{X}^\top\mathcal{X})^{-1}\mathcal{X}^\top y, \hat{\sigma}^2 = \frac{1}{n}||y - \mathcal{X}\beta||^2$ (see Example 6.3). Hence by Theorem 7.1

$$\begin{aligned} -2\log\lambda &= 2(\ell_1^* - \ell_0^*) \\ &= n \log\left(\frac{||y - \mathcal{X}\beta_0||^2}{||y - \mathcal{X}\hat{\beta}||^2}\right) \\ &\xrightarrow{\mathcal{L}} \chi^2_p. \end{aligned}$$

We draw upon the result (3.45) which gives us:

$$F = \frac{(n-p)}{p}\left(\frac{||y - \mathcal{X}\beta_0||^2}{||y - \mathcal{X}\hat{\beta}||^2} - 1\right) \sim F_{p,n-p},$$

so that in this case we again have an exact distribution.

EXAMPLE 7.6 *Let us consider our "classic blue" pullovers again. In Example 3.11 we tried to model the dependency of sales on prices. As we have seen in Figure 3.5 the slope of the regression curve is rather small, hence we might ask if* $\binom{\alpha}{\beta} = \binom{211}{0}$. *Here*

$$y = \begin{pmatrix} y_1 \\ \vdots \\ y_{10} \end{pmatrix} = \begin{pmatrix} x_{1,1} \\ \vdots \\ x_{10,1} \end{pmatrix}, \qquad \mathcal{X} = \begin{pmatrix} 1 & x_{1,2} \\ \vdots & \vdots \\ 1 & x_{10,2} \end{pmatrix}.$$

The test statistic for the LR test is

$$-2 \log \lambda = 9.10$$

which under the χ^2_2 *distribution is significant. The exact* F*-test statistic*

$$F = 5.93$$

is also significant under the $F_{2,8}$ *distribution* ($F_{2,8;0.95} = 4.46$).

Summary

- ↪ The hypotheses $H_0 : \theta \in \Omega_0$ against $H_1 : \theta \in \Omega_1$ can be tested using the likelihood ratio test (LRT). The likelihood ratio (LR) is the quotient $\lambda(\mathcal{X}) = L_0^*/L_1^*$ where the L_j^* are the maxima of the likelihood for each of the hypotheses.
- ↪ The test statistic in the LRT is $\lambda(\mathcal{X})$ or equivalently its logarithm $\log \lambda(\mathcal{X})$. If Ω_1 is q-dimensional and $\Omega_0 \subset \Omega_1$ r-dimensional, then the asymptotic distribution of $-2\log\lambda$ is χ^2_{q-r}. This allows H_0 to be tested against H_1 by calculating the test statistic $-2\log\lambda = 2(\ell_1^* - \ell_0^*)$ where $\ell_j^* = \log L_j^*$.
- ↪ The hypothesis $H_0 : \mu = \mu_0$ for $X \sim N_p(\mu, \Sigma)$, where Σ is known, leads to $-2\log\lambda = n(\overline{x} - \mu_0)^\top \Sigma^{-1}(\overline{x} - \mu_0) \sim \chi^2_p$.
- ↪ The hypothesis $H_0 : \mu = \mu_0$ for $X \sim N_p(\mu, \Sigma)$, where Σ is unknown, leads to $-2\log\lambda = n \log\{1 + (\overline{x} - \mu_0)^\top \mathcal{S}^{-1}(\overline{x} - \mu_0)\} \longrightarrow \chi^2_p$, and $(n-1)(\bar{x} - \mu_0)^\top \mathcal{S}^{-1}(\bar{x} - \mu_0) \sim T^2(p, n-1)$.
- ↪ The hypothesis $H_0 : \Sigma = \Sigma_0$ for $X \sim N_p(\mu, \Sigma)$, where μ is unknown, leads to $-2\log\lambda = n \operatorname{tr}\left(\Sigma_0^{-1}\mathcal{S}\right) - n\log|\Sigma_0^{-1}\mathcal{S}| - np \longrightarrow \chi^2_m$, $m = \frac{1}{2}p(p+1)$.
- ↪ The hypothesis $H_0 : \beta = \beta_0$ for $Y_i \sim N_1(\beta^\top x_i, \sigma^2)$, where σ^2 is unknown, leads to $\;2\log\lambda = n\log\left(\frac{\|y - \mathcal{X}\beta_0\|^2}{\|y - \mathcal{X}\hat{\beta}\|^2}\right) \longrightarrow \chi^2_p$.

7.2 Linear Hypothesis

In this section, we present a very general procedure which allows a linear hypothesis to be tested, i.e., a linear restriction, either on a vector mean μ or on the coefficient β of a linear model. The presented technique covers many of the practical testing problems on means or regression coefficients.

Linear hypotheses are of the form $\mathcal{A}\mu = a$ with known matrices $\mathcal{A}(q \times p)$ and $a(q \times 1)$ with $q \le p$.

EXAMPLE 7.7 *Let $\mu = (\mu_1, \mu_2)^\top$. The hypothesis that $\mu_1 = \mu_2$ can be equivalently written as:*

$$\mathcal{A}\mu = \begin{pmatrix} 1 & -1 \end{pmatrix} \begin{pmatrix} \mu_1 \\ \mu_2 \end{pmatrix} = 0 = a.$$

The general idea is to test a normal population $H_0: \mathcal{A}\mu = a$ (restricted model) against the full model H_1 where no restrictions are put on μ. Due to the properties of the multinormal, we can easily adapt the Test Problems 1 and 2 to this new situation. Indeed we know, from Theorem 5.2, that $y_i = \mathcal{A}x_i \sim N_q(\mu_y, \Sigma_y)$, where $\mu_y = \mathcal{A}\mu$ and $\Sigma_y = \mathcal{A}\Sigma\mathcal{A}^\top$.

Testing the null $H_0: \mathcal{A}\mu = a$, is the same as testing $H_0: \mu_y = a$. The appropriate statistics are $\bar{y}$ and $\mathcal{S}_y$ which can be derived from the original statistics $\bar{x}$ and $\mathcal{S}$ available from $\mathcal{X}$:

$$\bar{y} = \mathcal{A}\bar{x}, \quad \mathcal{S}_y = \mathcal{A}\mathcal{S}\mathcal{A}^\top.$$

Here the difference between the translated sample mean and the tested value is $d = \mathcal{A}\bar{x} - a$. We are now in the situation to proceed to Test Problem 5 and 6.

TEST PROBLEM 5 Suppose $X_1, \ldots, X_n$ is an i.i.d. random sample from a $N_p(\mu, \Sigma)$ population.

$$H_0: \mathcal{A}\mu = a, \ \Sigma \text{ known versus } H_1: \text{ no constraints.}$$

By (7.2) we have that, under H_0:

$$n(\mathcal{A}\bar{x} - a)^\top (\mathcal{A}\Sigma\mathcal{A}^\top)^{-1}(\mathcal{A}\bar{x} - a) \sim \mathcal{X}_q^2,$$

and we reject H_0 if this test statistic is too large at the desired significance level.

EXAMPLE 7.8 *We consider hypotheses on partitioned mean vectors $\mu = \binom{\mu_1}{\mu_2}$. Let us first look at*

$$H_0: \mu_1 = \mu_2, \ \textit{versus } H_1: \textit{no constraints},$$

for $N_{2p}(\binom{\mu_1}{\mu_2}, \begin{pmatrix} \Sigma & 0 \\ 0 & \Sigma \end{pmatrix})$ with known Σ. This is equivalent to $\mathcal{A} = (\mathcal{I}, -\mathcal{I})$, $a = (0, \ldots, 0)^\top \in \mathbb{R}^p$ and leads to:

$$-2\log\lambda = n(\overline{x}_1 - \overline{x}_2)(2\Sigma)^{-1}(\overline{x}_1 - \overline{x}_2) \sim \chi_p^2.$$

Another example is the test whether $\mu_1 = 0$, i.e.,

$$H_0: \mu_1 = 0, \ \textit{versus } H_1: \textit{no constraints},$$

for $N_{2p}(\binom{\mu_1}{\mu_2}, \begin{pmatrix} \Sigma & 0 \\ 0 & \Sigma \end{pmatrix})$ with known Σ. This is equivalent to $\mathcal{A}\mu = a$ with $\mathcal{A} = (\mathcal{I}, 0)$, and $a = (0, \ldots, 0)^\top \in \mathbb{R}^p$. Hence:

$$-2\log\lambda = n\overline{x}_1\Sigma^{-1}\overline{x}_1 \sim \chi_p^2.$$

TEST PROBLEM 6 Suppose $X_1, \ldots, X_n$ is an i.i.d. random sample from a $N_p(\mu, \Sigma)$ population.

$$H_0 : \mathcal{A}\mu = a, \ \Sigma \text{ unknown versus } H_1 : \quad \text{no constraints.}$$

From Corollary (5.4) and under H_0 it follows immediately that

$$(n-1)(\mathcal{A}\overline{x} - a)^\top(\mathcal{A}\mathcal{S}\mathcal{A}^\top)^{-1}(\mathcal{A}\overline{x} - a) \sim \ T^2(q, n-1) \tag{7.9}$$

since indeed under H_0,

$$\mathcal{A}\overline{x} \sim N_q(a, n^{-1}\mathcal{A}\Sigma\mathcal{A}^\top)$$

is independent of

$$n\mathcal{A}\mathcal{S}\mathcal{A}^\top \sim W_q(\mathcal{A}\Sigma\mathcal{A}^\top, n-1).$$

EXAMPLE 7.9 *Let's come back again to the bank data set and suppose that we want to test if* $\mu_4 = \mu_5$, *i.e., the hypothesis that the lower border mean equals the larger border mean for the forged bills. In this case:*

$$\begin{aligned} \mathcal{A} &= (0\ 0\ 0\ 1 - 1\ 0) \\ a &= 0. \end{aligned}$$

The test statistic is:

$$99(\mathcal{A}\bar{x})^\top(\mathcal{A}\mathcal{S}_f\mathcal{A}^\top)^{-1}(\mathcal{A}\bar{x}) \sim T^2(1, 99) = F_{1,99}.$$

The observed value is 13.638 *which is significant at the* 5% *level.*

Repeated Measurements

In many situations, n independent sampling units are observed at p different times or under p different experimental conditions (different treatments,...). So here we repeat p one-dimensional measurements on n different subjects. For instance, we observe the results from n students taking p different exams. We end up with a $(n \times p)$ matrix. We can thus consider the situation where we have $X_1, \ldots, X_n$ i.i.d. from a normal distribution $N_p(\mu, \Sigma)$ when there are p repeated measurements. The hypothesis of interest in this case is that there are no treatment effects, $H_0 : \mu_1 = \mu_2 = \ldots = \mu_p$. This hypothesis is a direct application of Test Problem 6. Indeed, introducing an appropriate matrix transform on μ we have:

$$H_0 : \ \mathcal{C}\mu = 0 \text{ where } \mathcal{C}((p-1) \times p) = \begin{pmatrix} 1 & -1 & 0 & \cdots & 0 \\ 0 & 1 & -1 & \cdots & 0 \\ \vdots & \vdots & \vdots & \vdots & \vdots \\ 0 & \cdots & 0 & 1 & -1 \end{pmatrix}. \tag{7.10}$$

Note that in many cases one of the experimental conditions is the "control" (a placebo, standard drug or reference condition). Suppose it is the first component. In that case one is interested in studying differences to the control variable. The matrix $\mathcal{C}$ has therefore a different form

$$\mathcal{C}((p-1) \times p) = \begin{pmatrix} 1 & -1 & 0 & \cdots & 0 \\ 1 & 0 & -1 & \cdots & 0 \\ \vdots & \vdots & \vdots & \vdots & \vdots \\ 1 & 0 & 0 & \cdots & -1 \end{pmatrix}.$$

By (7.9) the null hypothesis will be rejected if:

$$\frac{(n-p+1)}{p-1}\bar{x}^\top \mathcal{C}^\top(\mathcal{C}\mathcal{S}\mathcal{C}^\top)^{-1}\mathcal{C}\bar{x} > F_{1-\alpha;p-1,n-p+1}.$$

As a matter of fact, $\mathcal{C}\mu$ is the mean of the random variable $y_i = \mathcal{C}x_i$

$$y_i \sim N_{p-1}(\mathcal{C}\mu, \mathcal{C}\Sigma\mathcal{C}^\top).$$

Simultaneous confidence intervals for linear combinations of the mean of y_i have been derived above in (7.7). For all $a \in \mathbb{R}^{p-1}$, with probability $(1-\alpha)$ we have:

$$a^\top \mathcal{C}\mu \in a^\top \mathcal{C}\bar{x} \pm \sqrt{\frac{(p-1)}{n-p+1}F_{1-\alpha;p-1,n-p+1}a^\top \mathcal{C}\mathcal{S}\mathcal{C}^\top a}.$$

Due to the nature of the problem here, the row sums of the elements in $\mathcal{C}$ are zero: $\mathcal{C}1_p = 0$, therefore $a^\top\mathcal{C}$ is a vector having sum of elements equals to 0 . This is called a *contrast*. Let $b = \mathcal{C}^\top a$. We have $b^\top 1_p = \sum_{j=1}^{p} b_j = 0$. The result above thus provides for all contrasts of μ, and $b^\top\mu$ simultaneous confidence intervals at level $(1-\alpha)$

$$b^\top \mu \in b^\top \bar{x} \pm \sqrt{\frac{(p-1)}{n-p+1}F_{1-\alpha;p-1,n-p+1}b^\top \mathcal{S}b}.$$

Examples of contrasts for $p=4$ are $b^\top = (1\ -1\ 0\ 0)$ or $(1\ 0\ 0\ -1)$ or even $(1\ -\frac{1}{3}\ -\frac{1}{3}\ -\frac{1}{3})$ when the control is to be compared with the mean of 3 different treatments.

EXAMPLE 7.10 *Bock (1975) considers the evolution of the vocabulary of children from the eighth through eleventh grade. The data set contains the scores of a vocabulary test of 40 randomly chosen children that are observed from grades 8 to 11. This is a repeated measurement situation, $(n = 40, p = 4)$, since the same children were observed from grades 8 to 11. The statistics of interest are:*

$$\bar{x} = (1.086, 2.544, 2.851, 3.420)^\top$$
$$\mathcal{S} = \begin{pmatrix} 2.902 & 2.438 & 2.963 & 2.183 \\ 2.438 & 3.049 & 2.775 & 2.319 \\ 2.963 & 2.775 & 4.281 & 2.939 \\ 2.183 & 2.319 & 2.939 & 3.162 \end{pmatrix}.$$

Suppose we are interested in the yearly evolution of the children. Then the matrix $\mathcal{C}$ providing successive differences of μ_j is:

$$\mathcal{C} = \begin{pmatrix} 1 & -1 & 0 & 0 \\ 0 & 1 & -1 & 0 \\ 0 & 0 & 1 & -1 \end{pmatrix}.$$

The value of the test statistic is $F_{obs} = 53.134$ which is highly significant for $F_{3.37}$. There are significant differences between the successive means. However, the analysis of the contrasts shows the following simultaneous 95% confidence intervals

$$\begin{array}{rcccr} -1.958 & \le & \mu_1 - \mu_2 & \le & -0.959 \\ -0.949 & \le & \mu_2 - \mu_3 & \le & 0.335 \\ -1.171 & \le & \mu_3 - \mu_4 & \le & 0.036. \end{array}$$

Thus, the rejection of H_0 is mainly due to the difference between the childrens' performances in the first and second year. The confidence intervals for the following contrasts may also be of interest:

$$\begin{array}{rcccl} -2.283 & \le & \mu_1 - \frac{1}{3}(\mu_2+\mu_3+\mu_4) & \le & -1.423 \\ -1.777 & \le & \frac{1}{3}(\mu_1+\mu_2+\mu_3) - \mu_4 & \le & -0.742 \\ -1.479 & \le & \mu_2 - \mu_4 & \le & -0.272. \end{array}$$

They show that μ_1 is different from the average of the 3 other years (the same being true for μ_4) and μ_4 turns out to be higher than μ_2 (and of course higher than μ_1).

Test Problem 7 illustrates how the likelihood ratio can be applied to testing a linear restriction on the coefficient β of a linear model. It is also shown how a transformation of the test statistic leads to an exact F test as presented in Chapter 3.

TEST PROBLEM 7 Suppose $Y_1, \ldots, Y_n$, are independent with $Y_i \sim N_1(\beta^\top x_i, \sigma^2)$, and $x_i \in \mathbb{R}^p$.

$$H_0 : \mathcal{A}\beta = a,\ \sigma^2 \text{ unknown versus } H_1 : \text{ no constraints.}$$

To get the constrained maximum likelihood estimators under H_0, let $f(\beta, \lambda) = (y - x\beta)^\top(y - x\beta) - \lambda^\top(\mathcal{A}\beta - a)$ where $\lambda \in \mathbb{R}^q$ and solve $\frac{\partial f(\beta,\lambda)}{\partial \beta} = 0$ and $\frac{\partial f(\beta,\lambda)}{\partial \lambda} = 0$ (Exercise 3.24), thus we obtain:

$$\tilde{\beta} = \hat{\beta} - (\mathcal{X}^\top\mathcal{X})^{-1}\mathcal{A}^\top\{\mathcal{A}(\mathcal{X}^\top\mathcal{X})^{-1}\mathcal{A}^\top\}^{-1}(\mathcal{A}\hat{\beta} - a)$$

for β and $\tilde{\sigma}^2 = \frac{1}{n}(y - \mathcal{X}\tilde{\beta})^\top(y - \mathcal{X}\tilde{\beta})$. The estimate $\hat{\beta}$ denotes the unconstrained MLE as before. Hence, the LR statistic is

$$\begin{aligned} -2\log\lambda &= 2(\ell_1^* - \ell_0^*) \\ &= n\log\left(\frac{||y - \mathcal{X}\tilde{\beta}||^2}{||y - \mathcal{X}\hat{\beta}||^2}\right) \\ &\xrightarrow{\mathcal{L}} \chi_q^2 \end{aligned}$$

where q is the number of elements of a. This problem also has an exact F-test since

$$\frac{n-p}{q}\left(\frac{||y - \mathcal{X}\tilde{\beta}||^2}{||y - \mathcal{X}\hat{\beta}||^2} - 1\right) = \frac{n-p}{q}\frac{(\mathcal{A}\hat{\beta} - a)^\top\{\mathcal{A}(\mathcal{X}^\top\mathcal{X})^{-1}\mathcal{A}^\top\}^{-1}(\mathcal{A}\hat{\beta} - a)}{(y - \mathcal{X}\hat{\beta})^\top(y - \mathcal{X}\hat{\beta})} \sim F_{q,n-p}.$$

EXAMPLE 7.11 *Let us continue with the "classic blue" pullovers. We can once more test if $\beta = 0$ in the regression of sales on prices. It holds that*

$$\beta = 0 \quad \textit{iff} \quad (0\ 1)\begin{pmatrix}\alpha\\ \beta\end{pmatrix} = 0.$$

The LR statistic here is

$$-2\log\lambda = 0.284$$

which is not significant for the χ_1^2 distribution. The F-test statistic

$$F = 0.231$$

is also not significant. Hence, we can assume independence of sales and prices (alone). Recall that this conclusion has to be revised if we consider the prices together with advertisement costs and hours of sales managers.

Recall the different conclusion that was made in Example 7.6 when we rejected $H_0 : \alpha = 211$ and $\beta = 0$. The rejection there came from the fact that the *pair of values* was rejected. Indeed, if $\beta = 0$ the estimator of α would be $\bar{y} = 172.70$ and this is too far from 211.

EXAMPLE 7.12 *Let us now consider the multivariate regression in the "classic blue" pullovers example. From Example 3.15 we know that the estimated parameters in the model*

$$X_1 = \alpha + \beta_1 X_2 + \beta_2 X_3 + \beta_3 X_4 + \varepsilon$$

are

$$\hat{\alpha} = 65.670, \quad \hat{\beta}_1 = -0.216, \quad \hat{\beta}_2 = 0.485, \quad \hat{\beta}_3 = 0.844.$$

Hence, we could postulate the approximate relation:

$$\beta_1 \approx -\frac{1}{2}\beta_2,$$

which means in practice that augmenting the price by 20 EUR requires the advertisement costs to increase by 10 EUR in order to keep the number of pullovers sold constant. Vice versa, reducing the price by 20 EUR yields the same result as before if we reduced the advertisement costs by 10 EUR. Let us now test whether the hypothesis

$$H_0: \quad \beta_1 = -\frac{1}{2}\beta_2$$

is valid. This is equivalent to

$$\left(0 \;\; 1 \;\; \frac{1}{2} \;\; 0\right) \begin{pmatrix} \alpha \\ \beta_1 \\ \beta_2 \\ \beta_3 \end{pmatrix} = 0.$$

The LR statistic in this case is equal to (`MVAlrtest`*)*

$$-2 \log \lambda = 0.012,$$

the F statistic is

$$F = 0.007.$$

Hence, in both cases we will not reject the null hypothesis.

Comparison of Two Mean Vectors

In many situations, we want to compare two groups of individuals for whom a set of p characteristics has been observed. We have two random samples $\{x_{i1}\}_{i=1}^{n_1}$ and $\{x_{j2}\}_{j=1}^{n_2}$ from two distinct p-variate normal populations. Several testing issues can be addressed in this framework. In Test Problem 8 we will first test the hypothesis of equal mean vectors in the two groups under the assumption of equality of the two covariance matrices. This task can be solved by adapting Test Problem 2.

In Test Problem 9 a procedure for testing the equality of the two covariance matrices is presented. If the covariance matrices differ, the procedure of Test Problem 8 is no longer valid. If the equality of the covariance matrices is rejected, an easy rule for comparing two means with no restrictions on the covariance matrices is provided in Test Problem 10.

TEST PROBLEM 8 Assume that $X_{i1} \sim N_p(\mu_1, \Sigma)$, with $i = 1, \cdots, n_1$ and $X_{j2} \sim N_p(\mu_2, \Sigma)$, with $j = 1, \cdots, n_2$, where all the variables are independent.

$$H_0 : \mu_1 = \mu_2, \text{ versus } H_1 : \text{ no constraints.}$$

Both samples provide the statistics $\bar{x}_k$ and $\mathcal{S}_k$, $k = 1, 2$. Let $\delta = \mu_1 - \mu_2$. We have

$$(\bar{x}_1 - \bar{x}_2) \sim N_p\left(\delta, \frac{n_1 + n_2}{n_1 n_2}\Sigma\right) \tag{7.11}$$

$$n_1 S_1 + n_2 S_2 \sim W_p(\Sigma, n_1 + n_2 - 2). \tag{7.12}$$

Let $\mathcal{S}=(n_1+n_2)^{-1}(n_1 S_1+n_2 S_2)$ be the weighted mean of $\mathcal{S}_1$ and $\mathcal{S}_2$. Since the two samples are independent and since $\mathcal{S}_k$ is independent of $\bar{x}_k$ (for $k = 1, 2$) it follows that $\mathcal{S}$ is independent of $(\bar{x}_1 - \bar{x}_2)$. Hence, Theorem 5.8 applies and leads to a T^2-distribution:

$$\frac{n_1 n_2 (n_1 + n_2 - 2)}{(n_1 + n_2)^2} \left\{(\bar{x}_1 - \bar{x}_2) - \delta\right\}^\top \mathcal{S}^{-1} \left\{(\bar{x}_1 - \bar{x}_2) - \delta\right\}) \sim T^2(p, n_1 + n_2 - 2) \tag{7.13}$$

or

$$\left\{(\bar{x}_1 - \bar{x}_2) - \delta\right\}^\top \mathcal{S}^{-1} \left\{(\bar{x}_1 - \bar{x}_2) - \delta\right\} \sim \frac{p(n_1 + n_2)^2}{(n_1 + n_2 - p - 1) n_1 n_2} F_{p, n_1 + n_2 - p - 1}.$$

This result, as in Test Problem 2, can be used to test H_0: δ=0 or to construct a confidence region for $\delta \in \mathbb{R}^p$. The rejection region is given by:

$$\frac{n_1 n_2 (n_1 + n_2 - p - 1)}{p(n_1 + n_2)^2} (\bar{x}_1 - \bar{x}_2)^\top \mathcal{S}^{-1} (\bar{x}_1 - \bar{x}_2) \geq F_{1-\alpha; p, n_1 + n_2 - p - 1}. \tag{7.14}$$

A $(1 - \alpha)$ confidence region for δ is given by the ellipsoid centered at $(\bar{x}_1 - \bar{x}_2)$

$$\left\{\delta - (\bar{x}_1 - \bar{x}_2)\right\}^\top \mathcal{S}^{-1} \left\{\delta - (\bar{x}_1 - \bar{x}_2)\right\} \leq \frac{p(n_1 + n_2)^2}{(n_1 + n_2 - p - 1)(n_1 n_2)} F_{1-\alpha; p, n_1 + n_2 - p - 1},$$

and the simultaneous confidence intervals for all linear combinations $a^\top \delta$ of the elements of δ are given by

$$a^\top \delta \in a^\top (\bar{x}_1 - \bar{x}_2) \pm \sqrt{\frac{p(n_1 + n_2)^2}{(n_1 + n_2 - p - 1)(n_1 n_2)} F_{1-\alpha; p, n_1 + n_2 - p - 1} a^\top \mathcal{S} a}.$$

In particular we have at the $(1 - \alpha)$ level, for $j = 1, \ldots, p$,

$$\delta_j \in (\bar{x}_{1j} - \bar{x}_{2j}) + \sqrt{\frac{p(n_1 + n_2)^2}{(n_1 + n_2 - p - 1)(n_1 n_2)} F_{1-\alpha; p, n_1 + n_2 - p - 1} s_{jj}}. \tag{7.15}$$

EXAMPLE 7.13 *Let us come back to the questions raised in Example 7.5. We compare the means of assets (X_1) and of sales (X_2) for two sectors, energy (group 1) and manufacturing (group 2). With $n_1 = 15$, $n_2 = 10$, and $p = 2$ we obtain the statistics:*

$$\bar{x}_1 = \begin{pmatrix} 4084 \\ 2580.5 \end{pmatrix}, \; \bar{x}_2 = \begin{pmatrix} 4307.2 \\ 4925.2 \end{pmatrix}$$

and

$$\mathcal{S}_1 = 10^7 \begin{pmatrix} 1.6635 & 1.2410 \\ 1.2410 & 1.3747 \end{pmatrix}, \mathcal{S}_2 = 10^7 \begin{pmatrix} 1.2248 & 1.1425 \\ 1.1425 & 1.5112 \end{pmatrix},$$

so that

$$\mathcal{S} = 10^7 \begin{pmatrix} 1.4880 & 1.2016 \\ 1.2016 & 1.4293 \end{pmatrix}.$$

The observed value of the test statistic (7.14) is $F = 2.7036$. *Since* $F_{0.95;2,22} = 3.4434$ *the hypothesis of equal means of the two groups is not rejected although it would be rejected at a less severe level* ($F > F_{0.90;2,22} = 2.5613$). *By directly applying (7.15), the 95% simultaneous confidence intervals for the differences (* `MVAsimcidif`*) are obtained as:*

$$\begin{array}{rcccl} -4628.6 & \le & \mu_{1a} - \mu_{2a} & \le & 4182.2 \\ -6662.4 & \le & \mu_{1s} - \mu_{2s} & \le & 1973.0. \end{array}$$

EXAMPLE 7.14 *In order to illustrate the presented test procedures it is interesting to analyze some simulated data. This simulation will point out the importance of the covariances in testing means. We created 2 independent normal samples in* $\mathbb{R}^4$ *of sizes* $n_1 = 30$ *and* $n_2 = 20$ *with:*

$$\mu_1 = (8, 6, 10, 10)^\top$$

$$\mu_2 = (6, 6, 10, 13)^\top.$$

One may consider this as an example of $X = (X_1, \ldots, X_n)^\top$ *being the students' scores from 4 tests, where the 2 groups of students were subjected to two different methods of teaching. First we simulate the two samples with* $\Sigma = \mathcal{I}_4$ *and obtain the statistics:*

$$\begin{array}{rcl} \bar{x}_1 & = & (7.607, 5.945, 10.213, 9.635)^\top \\ \bar{x}_2 & = & (6.222, 6.444, 9.560, 13.041)^\top \\ \mathcal{S}_1 & = & \begin{pmatrix} 0.812 & -0.229 & -0.034 & 0.073 \\ -0.229 & 1.001 & 0.010 & -0.059 \\ -0.034 & 0.010 & 1.078 & -0.098 \\ 0.073 & -0.059 & -0.098 & 0.823 \end{pmatrix} \\ \mathcal{S}_2 & = & \begin{pmatrix} 0.559 & -0.057 & -0.271 & 0.306 \\ -0.057 & 1.237 & 0.181 & 0.021 \\ -0.271 & 0.181 & 1.159 & -0.130 \\ 0.306 & 0.021 & -0.130 & 0.683 \end{pmatrix}. \end{array}$$

The test statistic (7.14) takes the value $F = 60.65$ *which is highly significant: the small variance allows the difference to be detected even with these relatively moderate sample sizes. We conclude (at the 95% level) that:*

$$\begin{array}{rcccr} 0.6213 & \le & \delta_1 & \le & 2.2691 \\ -1.5217 & \le & \delta_2 & \le & 0.5241 \\ -0.3766 & \le & \delta_3 & \le & 1.6830 \\ -4.2614 & \le & \delta_4 & \le & -2.5494 \end{array}$$

which confirms that the means for X_1 *and* X_4 *are different.*

Consider now a different simulation scenario where the standard deviations are 4 times larger: $\Sigma = 16\mathcal{I}_4$. *Here we obtain:*

$$\begin{aligned}
\bar{x}_1 &= (7.312, 6.304, 10.840, 10.902)^\top \\
\bar{x}_2 &= (6.353, 5.890, 8.604, 11.283)^\top \\
\mathcal{S}_1 &= \begin{pmatrix} 21.907 & 1.415 & -2.050 & 2.379 \\ 1.415 & 11.853 & 2.104 & -1.864 \\ -2.050 & 2.104 & 17.230 & 0.905 \\ 2.379 & -1.864 & 0.905 & 9.037 \end{pmatrix} \\
\mathcal{S}_2 &= \begin{pmatrix} 20.349 & -9.463 & 0.958 & -6.507 \\ -9.463 & 15.502 & -3.383 & -2.551 \\ 0.958 & -3.383 & 14.470 & -0.323 \\ -6.507 & -2.551 & -0.323 & 10.311 \end{pmatrix}.
\end{aligned}$$

Now the test statistic takes the value 1.54 which is no longer significant ($F_{0.95,4,45} = 2.58$). *Now we cannot reject the null hypothesis (which we know to be false!) since the increase in variances prohibits the detection of differences of such magnitude.*

The following situation illustrates once more the role of the covariances between covariates. Suppose that $\Sigma = 16\mathcal{I}_4$ *as above but with* $\sigma_{14} = \sigma_{41} = -3.999$ *(this corresponds to a negative correlation* $r_{41} = -0.9997$*). We have:*

$$\begin{aligned}
\bar{x}_1 &= (8.484, 5.908, 9.024, 10.459)^\top \\
\bar{x}_2 &= (4.959, 7.307, 9.057, 13.803)^\top \\
\mathcal{S}_1 &= \begin{pmatrix} 14.649 & -0.024 & 1.248 & -3.961 \\ -0.024 & 15.825 & 0.746 & 4.301 \\ 1.248 & 0.746 & 9.446 & 1.241 \\ -3.961 & 4.301 & 1.241 & 20.002 \end{pmatrix} \\
\mathcal{S}_2 &= \begin{pmatrix} 14.035 & -2.372 & 5.596 & -1.601 \\ -2.372 & 9.173 & -2.027 & -2.954 \\ 5.596 & -2.027 & 9.021 & -1.301 \\ -1.601 & -2.954 & -1.301 & 9.593 \end{pmatrix}.
\end{aligned}$$

The value of F *is* 3.853 *which is significant at the 5% level (p-value = 0.0089). So the null hypothesis* $\delta = \mu_1 - \mu_2 = 0$ *is outside the 95% confidence ellipsoid. However, the simultaneous confidence intervals, which do not take the covariances into account are given by:*

$$\begin{aligned}
-0.1837 \le &\ \delta_1 \le 7.2343 \\
-4.9452 \le &\ \delta_2 \le 2.1466 \\
-3.0091 \le &\ \delta_3 \le 2.9438 \\
-7.2336 \le &\ \delta_4 \le 0.5450.
\end{aligned}$$

They contain the null value (see Remark 7.1 above) although they are very asymmetric for δ_1 *and* δ_4.

EXAMPLE 7.15 *Let us compare the vectors of means of the forged and the genuine bank notes. The matrices* $\mathcal{S}_f$ *and* $\mathcal{S}_g$ *were given in Example 3.1 and since here* $n_f = n_g = 100$, $\mathcal{S}$ *is the simple average of* $\mathcal{S}_f$ *and* $\mathcal{S}_g : \mathcal{S} = \frac{1}{2}(\mathcal{S}_f + \mathcal{S}_g)$.

$$\begin{aligned}
\bar{x}_g &= (214.97, 129.94, 129.72, 8.305, 10.168, 141.52)^\top \\
\bar{x}_f &= (214.82, 130.3, 130.19, 10.53, 11.133, 139.45)^\top.
\end{aligned}$$

The test statistic is given by (7.14) and turns out to be $F = 391.92$ which is highly significant for $F_{6,193}$. The 95% simultaneous confidence intervals for the differences $\delta_j = \mu_{gj} - \mu_{fj}$, $j = 1, \dots, p$ are:

$$\begin{array}{rcccr} -0.0443 & \le & \delta_1 & \le & 0.3363 \\ -0.5186 & \le & \delta_2 & \le & -0.1954 \\ -0.6416 & \le & \delta_3 & \le & -0.3044 \\ -2.6981 & \le & \delta_4 & \le & -1.7519 \\ -1.2952 & \le & \delta_5 & \le & -0.6348 \\ 1.8072 & \le & \delta_6 & \le & 2.3268. \end{array}$$

All of the components (except for the first one) show significant differences in the means. The main effects are taken by the lower border (X_4) and the diagonal (X_6).

The preceding test implicitly uses the fact that the two samples are extracted from two different populations with common variance Σ. In this case, the test statistic (7.14) measures the distance between the two centers of gravity of the two groups w.r.t. the common metric given by the pooled variance matrix $\mathcal{S}$. If $\Sigma_1 \neq \Sigma_2$ no such matrix exists. There are no satisfactory test procedures for testing the equality of variance matrices which are robust with respect to normality assumptions of the populations. The following test extends Bartlett's test for equality of variances in the univariate case. But this test is known to be very sensitive to departures from normality.

TEST PROBLEM 9 (Comparison of Covariance Matrices)
Let $X_{ih} \sim N_p(\mu_h, \Sigma_h)$, $i = 1, \dots, n_h$, $h = 1, \dots, k$ be independent random variables,

$$H_0 : \Sigma_1 = \Sigma_2 = \cdots = \Sigma_k \text{ versus } H_1 : \quad \text{no constraints.}$$

Each subsample provides $\mathcal{S}_h$, an estimator of Σ_h, with

$$n_h \mathcal{S}_h \sim W_p(\Sigma_h, n_h - 1).$$

Under H_0, $\sum_{h=1}^{k} n_h \mathcal{S}_h \sim W_p(\Sigma, n-k)$ (Section 5.2), where Σ is the common covariance matrix X_{ih} and $n = \sum_{h=1}^{k} n_h$. Let $\mathcal{S} = \frac{n_1\mathcal{S}_1 + \cdots + n_k\mathcal{S}_k}{n}$ be the weighted average of the $\mathcal{S}_h$ (this is in fact the MLE of Σ when H_0 is true). The likelihood ratio test leads to the statistic

$$-2\log\lambda = n \log \mid \mathcal{S} \mid - \sum_{h=1}^{k} n_h \log \mid \mathcal{S}_h \mid \tag{7.16}$$

which under H_0 is approximately distributed as a $\mathcal{X}^2_m$ where $m = \frac{1}{2}(k-1)p(p+1)$.

EXAMPLE 7.16 *Let's come back to Example 7.13, where the mean of assets and sales have been compared for companies from the energy and manufacturing sector assuming that $\Sigma_1 = \Sigma_2$. The test of $\Sigma_1 = \Sigma_2$ leads to the value of the test statistic*

$$-2\log\lambda = 0.9076 \tag{7.17}$$

which is not significant (p-value for a $\chi^2_3 = 0.82$). We cannot reject H_0 and the comparison of the means performed above is valid.

EXAMPLE 7.17 *Let us compare the covariance matrices of the forged and the genuine bank notes (the matrices S_f and S_g are shown in Example 3.1). A first look seems to suggest that $\Sigma_1 \neq \Sigma_2$. The pooled variance S is given by $\mathcal{S} = \frac{1}{2}(\mathcal{S}_f + \mathcal{S}_g)$ since here $n_f = n_g$. The test statistic here is $-2\log\lambda = 127.21$, which is highly significant χ^2 with 21 degrees of freedom. As expected, we reject the hypothesis of equal covariance matrices, and as a result the procedure for comparing the two means in Example 7.15 is not valid.*

What can we do with unequal covariance matrices? When both n_1 and n_2 are large, we have a simple solution:

> **TEST PROBLEM 10** (Comparison of two means, unequal covariance matrices, large samples)
> Assume that $X_{i1} \sim N_p(\mu_1, \Sigma_1)$, with $i = 1, \cdots, n_1$ and $X_{j2} \sim N_p(\mu_2, \Sigma_2)$, with $j = 1, \cdots, n_2$ are independent random variables.
>
> $$H_0 : \mu_1 = \mu_2 \text{ versus } H_1 : \text{ no constraints.}$$

Letting $\delta = \mu_1 - \mu_2$, we have

$$(x_1 - x_2) \sim N_p\left(\delta, \frac{\Sigma_1}{n_1} + \frac{\Sigma_2}{n_2}\right).$$

Therefore, by (5.4)

$$(\bar{x}_1 - \bar{x}_2)^\top \left(\frac{\Sigma_1}{n_1} + \frac{\Sigma_2}{n_2}\right)^{-1} (\bar{x}_1 - \bar{x}_2) \sim \chi^2_p.$$

Since $\mathcal{S}_i$ is a consistent estimator of Σ_i for $i = 1, 2$, we have

$$(\bar{x}_1 - \bar{x}_2)^\top \left(\frac{\mathcal{S}_1}{n_1} + \frac{\mathcal{S}_2}{n_2}\right)^{-1} (\bar{x}_1 - \bar{x}_2) \xrightarrow{\mathcal{L}} \chi^2_p. \tag{7.18}$$

This can be used in place of (7.13) for testing H_0, defining a confidence region for δ or constructing simultaneous confidence intervals for $\delta_j, j = 1, \ldots, p$.

For instance, the rejection region at the level α will be

$$(\bar{x}_1 - \bar{x}_2)^\top \left(\frac{\mathcal{S}_1}{n_1} + \frac{\mathcal{S}_2}{n_2}\right)^{-1} (\bar{x}_1 - \bar{x}_2) > \chi^2_{1-\alpha;p} \tag{7.19}$$

and the $(1 - \alpha)$ simultaneous confidence intervals for δ_j, $j = 1, \ldots, p$ are:

$$\delta_j \in (\bar{x}_1 - \bar{x}_2) \pm \sqrt{\chi^2_{1-\alpha;p}\left(\frac{s^{(1)}_{jj}}{n_1} + \frac{s^{(2)}_{jj}}{n_2}\right)} \tag{7.20}$$

where $s^{(i)}_{jj}$ is the (j, j) element of the matrix $\mathcal{S}_i$. This may be compared to (7.15) where the pooled variance was used.

REMARK 7.2 *We see, by comparing the statistics (7.19) with (7.14), that we measure here the distance between $\bar{x}_1$ and $\bar{x}_2$ using the metric $\left(\frac{\mathcal{S}_1}{n_1} + \frac{\mathcal{S}_2}{n_2}\right)$. It should be noticed that when $n_1 = n_2$, the two methods are essentially the same since then $\mathcal{S} = \frac{1}{2}(\mathcal{S}_1 + \mathcal{S}_2)$. If the covariances are different but have the same eigenvectors (different eigenvalues), one can apply the common principal component (CPC) technique, see Chapter 9.*

EXAMPLE 7.18 *Let us use the last test to compare the forged and the genuine bank notes again (n_1 and n_2 are both large). The test statistic (7.19) turns out to be 2436.8 which is again highly significant. The 95% simultaneous confidence intervals are:*

$$\begin{array}{rcr} -0.0389 & \le \delta_1 \le & 0.3309 \\ -0.5140 & \le \delta_2 \le & -0.2000 \\ -0.6368 & \le \delta_3 \le & -0.3092 \\ -2.6846 & \le \delta_4 \le & -1.7654 \\ -1.2858 & \le \delta_5 \le & -0.6442 \\ 1.8146 & \le \delta_6 \le & 2.3194 \end{array}$$

showing that all the components except the first are different from zero, the largest difference coming from X_6 (length of the diagonal) and X_4 (lower border). The results are very similar to those obtained in Example (7.15). This is due to the fact that here $n_1 = n_2$ as we already mentioned in the remark above.

Profile Analysis

Another useful application of Test Problem 6 is the repeated measurements problem applied to two independent groups. This problem arises in practice when we observe repeated measurements of characteristics (or measures of the same type under different experimental conditions) on the different groups which have to be compared. It is important that the p measures (the "profile") are comparable and in particular are reported in the same units. For instance, they may be measures of blood pressure at p different points in time, one group being the control group and the other the group receiving a new treatment. The observations may be the scores obtained from p different tests of two different experimental groups. One is then interested in comparing the profiles of each group: the profile being just the vectors of the means of the p responses (the comparison may be visualized in a two dimensional graph using the parallel coordinate plot introduced in Section 1.7).

We are thus in the same statistical situation as for the comparison of two means:

$$X_{i1} \sim N_p(\mu_1, \Sigma) \quad i = 1, \dots, n_1$$

$$X_{i2} \sim N_p(\mu_2, \Sigma) \quad i = 1, \dots, n_2$$

where all variables are independent. Suppose the two population profiles look like in Figure 7.1.

The following questions are of interest:

1. Are the profiles similar in the sense of being parallel (which means no interaction between the treatments and the groups)?
2. If the profiles are parallel, are they at the same level?
3. If the profiles are parallel, is there any treatment effect, i.e., are the profiles horizontal (profiles keep the same no matter which treatment received)?

The above questions are easily translated into linear constraints on the means and a test statistic can be obtained accordingly.

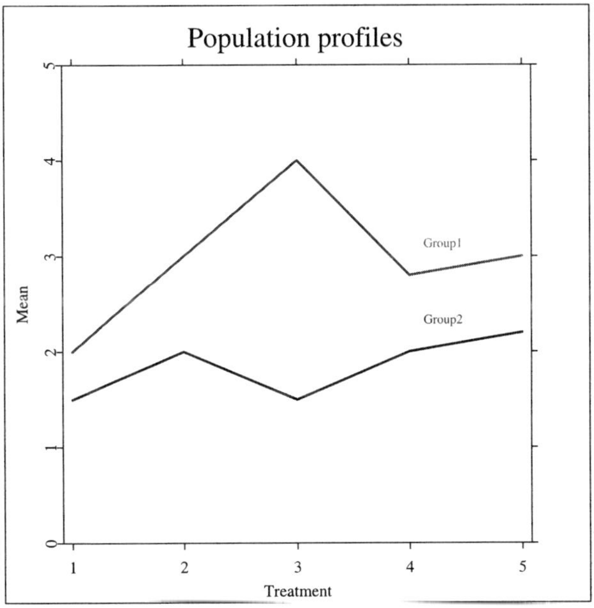

Figure 7.1. Example of population profiles MVAprofil

Parallel Profiles

Let $\mathcal{C}$ be a $(p-1) \times p$ matrix defined as $\mathcal{C} = \begin{pmatrix} 1 & -1 & 0 & \cdots & 0 \\ 0 & 1 & -1 & \cdots & 0 \\ \vdots & \vdots & \vdots & \vdots & \vdots \\ 0 & \cdots & 0 & 1 & -1 \end{pmatrix}$.

The hypothesis to be tested is

$$H_0^{(1)} : \mathcal{C}(\mu_1 - \mu_2) = 0.$$

From (7.11), (7.12) and Corollary 5.4 we know that under H_0:

$$\frac{n_1 n_2}{(n_1+n_2)^2}(n_1+n_2-2)\left\{\mathcal{C}(\bar{x}_1-\bar{x}_2)\right\}^{\top}(\mathcal{C}\mathcal{S}\mathcal{C}^{\top})^{-1}\mathcal{C}(\bar{x}_1-\bar{x}_2) \sim T^2(p-1, n_1+n_2-2) \quad (7.21)$$

where $\mathcal{S}$ is the pooled covariance matrix. The hypothesis is rejected if

$$\frac{n_1 n_2(n_1+n_1-p)}{(n_1+n_2)^2(p-1)}\left(\mathcal{C}\bar{x}\right)^{\top}\left(\mathcal{C}\mathcal{S}\mathcal{C}^{\top}\right)^{-1}\mathcal{C}\bar{x} > F_{1-\alpha;p-1,n_1+n_2-p}.$$

Equality of Two Levels

The question of equality of the two levels is meaningful only if the two profiles are parallel. In the case of interactions (rejection of $H_0^{(1)}$), the two populations react differently to the treatments and the question of the level has no meaning.

The equality of the two levels can be formalized as

$$H_0^{(2)} : 1_p^\top(\mu_1 - \mu_2) = 0$$

since

$$1_p^\top(\bar{x}_1 - \bar{x}_2) \sim N_1\left(1_p^\top(\mu_1 - \mu_2), \frac{n_1 + n_2}{n_1 n_2} 1_p^\top \Sigma 1_p\right)$$

and

$$(n_1 + n_2) 1_p^\top \mathcal{S} 1_p \sim W_1(1_p^\top \Sigma 1_p, n_1 + n_2 - 2).$$

Using Corollary 5.4 we have that:

$$\begin{aligned} \frac{n_1 n_2}{(n_1+n_2)^2}(n_1+n_2-2)\frac{\left\{1_p^\top(\bar{x}_1 - \bar{x}_2)\right\}^2}{1_p^\top \mathcal{S} 1_p} &\sim T^2(1, n_1+n_2-2) \\ &= F_{1,n_1+n_2-2}. \end{aligned} \tag{7.22}$$

The rejection region is

$$\frac{n_1 n_2 (n_1+n_2-2)}{(n_1+n_2)^2} \frac{\left\{1_p^\top(\bar{x}_1 - \bar{x}_2)\right\}^2}{1_p^\top \mathcal{S} 1_p} > F_{1-\alpha;1,n_1+n_2-2}.$$

Treatment Effect

If it is rejected that the profiles are parallel, then two independent analyses should be done on the two groups using the repeated measurement approach. But if it is accepted that they are parallel, then we can exploit the information contained in both groups (eventually at different levels) to test a treatment effect, i.e., if the two profiles are horizontal. This may be written as:

$$H_0^{(3)} : \mathcal{C}(\mu_1 + \mu_2) = 0.$$

Consider the average profile $\bar{x}$:

$$\bar{x} = \frac{n_1 \bar{x}_1 + n_2 \bar{x}_2}{n_1 + n_2}.$$

Clearly,

$$\bar{x} \sim N_p\left(\frac{n_1\mu_1 + n_2\mu_2}{n_1 + n_2}, \frac{1}{n_1 + n_2}\Sigma\right).$$

Now it is not hard to prove that $H_0^{(3)}$ with $H_0^{(1)}$ implies that

$$\mathcal{C}\left(\frac{n_1\mu_1 + n_2\mu_2}{n_1 + n_2}\right) = 0.$$

So under parallel, horizontal profiles we have

$$\sqrt{n_1 + n_2}\,\mathcal{C}\bar{x} \sim N_p(0, \mathcal{C}\Sigma\mathcal{C}^\top).$$

From Corollary 5.4 we again obtain

$$(n_1 + n_2 - 2)(\mathcal{C}\bar{x})^\top(\mathcal{C}\mathcal{S}\mathcal{C}^\top)^{-1}\mathcal{C}\bar{x} \sim T^2(p-1, n_1+n_2-2). \tag{7.23}$$

This leads to the rejection region of $H_0^{(3)}$, namely

$$\frac{n_1 + n_2 - p}{p - 1}(\mathcal{C}\bar{x})^\top(\mathcal{C}\mathcal{S}\mathcal{C}^\top)^{-1}\mathcal{C}\bar{x} > F_{1-\alpha;p-1,n_1+n_2-p}.$$

EXAMPLE 7.19 *Morrison (1990) proposed a test in which the results of 4 sub-tests of the Wechsler Adult Intelligence Scale (WAIS) are compared for 2 categories of people: group 1 contains $n_1 = 37$ people who do not have a senile factor and group 2 contains $n_2 = 12$ people who have a senile factor. The four WAIS sub-tests are X_1 (information), X_2 (similarities), X_3 (arithmetic) and X_4 (picture completion). The relevant statistics are*

$$\begin{aligned}
\bar{x}_1 &= (12.57, 9.57, 11.49, 7.97)^\top \\
\bar{x}_2 &= (8.75, 5.33, 8.50, 4.75)^\top \\
\mathcal{S}_1 &= \begin{pmatrix} 11.164 & 8.840 & 6.210 & 2.020 \\ 8.840 & 11.759 & 5.778 & 0.529 \\ 6.210 & 5.778 & 10.790 & 1.743 \\ 2.020 & 0.529 & 1.743 & 3.594 \end{pmatrix} \\
\mathcal{S}_2 &= \begin{pmatrix} 9.688 & 9.583 & 8.875 & 7.021 \\ 9.583 & 16.722 & 11.083 & 8.167 \\ 8.875 & 11.083 & 12.083 & 4.875 \\ 7.021 & 8.167 & 4.875 & 11.688 \end{pmatrix}.
\end{aligned}$$

The test statistic for testing if the two profiles are parallel is $F = 0.4634$, which is not significant (p-value $= 0.71$). Thus it is accepted that the two are parallel. The second test statistic (testing the equality of the levels of the 2 profiles) is $F = 17.21$, which is highly significant (p-value $\approx 10^{-4}$). The global level of the test for the non-senile people is superior to the senile group. The final test (testing the horizontality of the average profile) has the test statistic $F = 53.32$, which is also highly significant (p-value $\approx 10^{-14}$). This implies that there are substantial differences among the means of the different subtests.

Summary

- ↪ Hypotheses about μ can often be written as $\mathcal{A}\mu = a$, with matrix $\mathcal{A}$, and vector a.
- ↪ The hypothesis $H_0 : \mathcal{A}\mu = a$ for $X \sim N_p(\mu, \Sigma)$ with Σ known leads to $-2\log\lambda = n(\mathcal{A}\bar{x} - a)^\top(\mathcal{A}\Sigma\mathcal{A}^\top)^{-1}(\mathcal{A}\bar{x} - a) \sim \chi^2_q$, where q is the number of elements in a.
- ↪ The hypothesis $H_0 : \mathcal{A}\mu = a$ for $X \sim N_p(\mu, \Sigma)$ with Σ unknown leads to $-2\log\lambda = n\log\{1 + (\mathcal{A}\bar{x} - a)^\top(\mathcal{A}\mathcal{S}\mathcal{A}^\top)^{-1}(\mathcal{A}\bar{x} - a)\} \longrightarrow \chi^2_q$, where q is the number of elements in a and we have an exact test $(n-1)(\mathcal{A}\bar{x} - a)^\top(\mathcal{A}\mathcal{S}\mathcal{A}^\top)^{-1}(\mathcal{A}\bar{x} - a) \sim T^2(q, n-1)$.
- ↪ The hypothesis $H_0 : \mathcal{A}\beta = a$ for $Y_i \sim N_1(\beta^\top x_i, \sigma^2)$ with σ^2 unknown leads to $-2\log\lambda = \frac{n}{2}\log\left(\frac{\|y - \mathcal{X}\tilde{\beta}\|^2}{\|y - \mathcal{X}\hat{\beta}\|^2} - 1\right) \longrightarrow \chi^2_q$, with q being the length of a and with

$$\frac{n-p}{q}\frac{\left(\mathcal{A}\hat{\beta} - a\right)\left\{\mathcal{A}\left(\mathcal{X}^\top\mathcal{X}\right)^{-1}\mathcal{A}^\top\right\}^{-1}\left(\mathcal{A}\hat{\beta} - a\right)}{\left(y - \mathcal{X}\hat{\beta}\right)^\top\left(y - \mathcal{X}\hat{\beta}\right)} \sim F_{q,n-p}.$$

7.3 Boston Housing

Returning to the Boston housing data set, we are now in a position to test if the means of the variables vary according to their location, for example, when they are located in a district with high valued houses. In Chapter 1, we built 2 groups of observations according to the value of X_{14} being less than or equal to the median of X_{14} (a group of 256 districts) and greater than the median (a group of 250 districts). In what follows, we use the transformed variables motivated in Section 1.8.

Testing the equality of the means from the two groups was proposed in a multivariate setup, so we restrict the analysis to the variables X_1, X_5, X_8, X_{11}, and X_{13} to see if the differences between the two groups that were identified in Chapter 1 can be confirmed by a formal test. As in Test Problem 8, the hypothesis to be tested is

$$H_0: \ \mu_1 = \mu_2, \text{ where } \mu_1 \in \mathbb{R}^5, n_1 = 256, \text{ and } n_2 = 250.$$

Σ is not known. The F-statistic given in (7.13) is equal to 126.30, which is much higher than the critical value $F_{0.95;5,500} = 2.23$. Therefore, we reject the hypothesis of equal means.

To see which component, X_1, X_5, X_8, X_{11}, or X_{13}, is responsible for this rejection, take a look at the simultaneous confidence intervals defined in (7.14):

$$\begin{aligned} \delta_1 &\in (1.4020, 2.5499) \\ \delta_5 &\in (0.1315, 0.2383) \\ \delta_8 &\in (-0.5344, -0.2222) \\ \delta_{11} &\in (1.0375, 1.7384) \\ \delta_{13} &\in (1.1577, 1.5818). \end{aligned}$$

These confidence intervals confirm that all of the δ_j are significantly different from zero (note there is a negative effect for X_8: weighted distances to employment centers) MVAsimcibh.

We could also check if the factor "being bounded by the river" (variable X_4) has some effect on the other variables. To do this compare the means of $(X_5, X_8, X_9, X_{12}, X_{13}, X_{14})^\top$. There are two groups: $n_1 = 35$ districts bounded by the river and $n_2 = 471$ districts not bounded by the river. Test Problem 8 ($H_0 : \mu_1 = \mu_2$) is applied again with $p = 6$. The resulting test statistic, $F = 5.81$, is highly significant ($F_{0.95;6,499} = 2.12$). The simultaneous confidence intervals indicate that only X_{14} (the value of the houses) is responsible for the hypothesis being rejected! At a significance level of 0.95

$$\begin{aligned} \delta_5 &\in (-0.0603, 0.1919) \\ \delta_8 &\in (-0.5225, 0.1527) \\ \delta_9 &\in (-0.5051, 0.5938) \\ \delta_{12} &\in (-0.3974, 0.7481) \\ \delta_{13} &\in (-0.8595, 0.3782) \\ \delta_{14} &\in (0.0014, 0.5084). \end{aligned}$$

Testing Linear Restrictions

In Chapter 3 a linear model was proposed that explained the variations of the price X_{14} by the variations of the other variables. Using the same procedure that was shown in Testing Problem 7, we are in a position to test a set of linear restrictions on the vector of regression coefficients β.

The model we estimated in Section 3.7 provides the following (MVAlinregbh):

Variable	$\hat{\beta}_j$	$SE(\hat{\beta}_j)$	t	p-value
constant	4.1769	0.3790	11.020	0.0000
X_1	−0.0146	0.0117	−1.254	0.2105
X_2	0.0014	0.0056	0.247	0.8051
X_3	−0.0127	0.0223	−0.570	0.5692
X_4	0.1100	0.0366	3.002	0.0028
X_5	−0.2831	0.1053	−2.688	0.0074
X_6	0.4211	0.1102	3.822	0.0001
X_7	0.0064	0.0049	1.317	0.1885
X_8	−0.1832	0.0368	−4.977	0.0000
X_9	0.0684	0.0225	3.042	0.0025
X_{10}	−0.2018	0.0484	−4.167	0.0000
X_{11}	−0.0400	0.0081	−4.946	0.0000
X_{12}	0.0445	0.0115	3.882	0.0001
X_{13}	−0.2626	0.0161	−16.320	0.0000

Recall that the estimated residuals $Y - \mathcal{X}\widehat{\beta}$ did not show a big departure from normality, which means that the testing procedure developed above can be used.

1. First a global test of significance for the regression coefficients is performed,

$$H_0: \ (\beta_1, \ldots, \beta_{13}) = 0.$$

This is obtained by defining $\mathcal{A} = (0_{13}, \mathcal{I}_{13})$ and $a = 0_{13}$ so that H_0 is equivalent to $\mathcal{A}\beta = a$ where $\beta = (\beta_0, \beta_1, \ldots, \beta_{13})^\top$. Based on the observed values $F = 123.20$. This is highly significant ($F_{0.95;13,492} = 1.7401$), thus we reject H_0. Note that under H_0 $\widehat{\beta}_{H_0} = (3.0345, 0, \ldots, 0)$ where $3.0345 = \overline{y}$.

2. Since we are interested in the effect that being located close to the river has on the value of the houses, the second test is $H_0: \ \beta_4 = 0$. This is done by fixing

$$\mathcal{A} = (0, 0, 0, 0, 1, 0, 0, 0, 0, 0, 0, 0, 0, 0)^\top$$

and $a = 0$ to obtain the equivalent hypothesis $H_0: \mathcal{A}\beta = a$. The result is again significant: $F = 9.0125$ ($F_{0.95;1,492} = 3.8604$) with a p-value of 0.0028. Note that this is the same p-value obtained in the individual test $\beta_4 = 0$ in Chapter 3, computed using a different setup.

3. A third test notices the fact that some of the regressors in the full model (3.57) appear to be insignificant (that is they have high individual p-values). It can be confirmed from a joint test if the corresponding reduced model, formulated by deleting the insignificant variables, is rejected by the data. We want to test $H_0: \ \beta_1 = \beta_2 = \beta_3 = \beta_7 = 0$. Hence,

$$\mathcal{A} = \begin{pmatrix} 0 & 1 & 0 & 0 & 0 & 0 & 0 & 0 & 0 & 0 & 0 & 0 & 0 & 0 \\ 0 & 0 & 1 & 0 & 0 & 0 & 0 & 0 & 0 & 0 & 0 & 0 & 0 & 0 \\ 0 & 1 & 0 & 1 & 0 & 0 & 0 & 0 & 0 & 0 & 0 & 0 & 0 & 0 \\ 0 & 1 & 0 & 0 & 0 & 0 & 0 & 1 & 0 & 0 & 0 & 0 & 0 & 0 \end{pmatrix}$$

and $a = 0_4$. The test statistic is 0.9344, which is not significant for $F_{4,492}$. Given that the p-value is equal to 0.44, we cannot reject the null hypothesis nor the corresponding reduced model. The value of $\widehat{\beta}$ under the null hypothesis is

$$\widehat{\beta}_{H_0} = (4.16, 0, 0, 0, 0.11, -0.31, 0.47, 0, -0.19, 0.05, -0.20, -0.04, 0.05, -0.26)^\top.$$

Variable	$\hat{\beta}_j$	SE	t	p-value
const	4.1582	0.3628	11.462	0.0000
X_4	0.1087	0.0362	2.999	0.0028
X_5	−0.3055	0.0973	−3.140	0.0018
X_6	0.4668	0.1059	4.407	0.0000
X_8	−0.1855	0.0327	−5.679	0.0000
X_9	0.0492	0.0183	2.690	0.0074
X_{10}	−0.2096	0.0446	−4.705	0.0000
X_{11}	−0.0410	0.0078	−5.280	0.0000
X_{12}	0.0481	0.0112	4.306	0.0000
X_{13}	−0.2588	0.0149	−17.396	0.0000

Table 7.1. Linear Regression for Boston Housing Data Set. `MVAlinreg2bh`

A possible reduced model is

$$X_{14} = \beta_0 + \beta_4 X_4 + \beta_5 X_5 + \beta_6 X_6 + \beta_8 X_8 + \cdots + \beta_{13} X_{13} + \varepsilon.$$

Estimating this reduced model using OLS, as was done in Chapter 3, provides the results shown in Table 7.1.

Note that the reduced model has $r^2 = 0.763$ which is very close to $r^2 = 0.765$ obtained from the full model. Clearly, including variables X_1, X_2, X_3, and X_7 does not provide valuable information in explaining the variation of X_{14}, the price of the houses.

7.4 Exercises

EXERCISE 7.1 *Use Theorem 7.1 to derive a test for testing the hypothesis that a dice is balanced, based on n tosses of that dice. (Hint: use the multinomial probability function.)*

EXERCISE 7.2 *Consider $N_3(\mu, \Sigma)$. Formulate the hypothesis $H_0 : \mu_1 = \mu_2 = \mu_3$ in terms of $\mathcal{A}\mu = a$.*

EXERCISE 7.3 *Simulate a normal sample with $\mu = \binom{1}{2}$ and $\Sigma = \begin{pmatrix} 1 & 0.5 \\ 0.5 & 2 \end{pmatrix}$ and test $H_0 : 2\mu_1 - \mu_2 = 0.2$ first with Σ known and then with Σ unknown. Compare the results.*

EXERCISE 7.4 *Derive expression (7.3) for the likelihood ratio test statistic in Test Problem 2.*

EXERCISE 7.5 *With the simulated data set of Example 7.14, test the hypothesis of equality of the covariance matrices.*

EXERCISE 7.6 *In the U.S. companies data set, test the equality of means between the energy and manufacturing sectors, taking the full vector of observations X_1 to X_6. Derive the simultaneous confidence intervals for the differences.*

EXERCISE 7.7 *Let* $X \sim N_2(\mu, \Sigma)$ *where* Σ *is known to be* $\Sigma = \begin{pmatrix} 2 & -1 \\ -1 & 2 \end{pmatrix}$. *We have an i.i.d. sample of size* $n = 6$ *providing* $\bar{x}^\top = \left(1\ \frac{1}{2}\right)$. *Solve the following test problems* ($\alpha = 0.05$):

$$\begin{array}{llll} a)\ H_0\colon & \mu = \left(2, \frac{2}{3}\right)^\top & H_1\colon & \mu \neq \left(2, \frac{2}{3}\right)^\top \\ b)\ H_0\colon & \mu_1 + \mu_2 = \frac{7}{2} & H_1\colon & \mu_1 + \mu_2 \neq \frac{7}{2} \\ c)\ H_0\colon & \mu_1 - \mu_2 = \frac{1}{2} & H_1\colon & \mu_1 - \mu_2 \neq \frac{1}{2} \\ d)\ H_0\colon & \mu_1 = 2 & H_1\colon & \mu_1 \neq 2 \end{array}$$

For each case, represent the rejection region graphically (comment!).

EXERCISE 7.8 *Repeat the preceeding exercise with* Σ *unknown and* $S = \begin{pmatrix} 2 & -1 \\ -1 & 2 \end{pmatrix}$. *Compare the results.*

EXERCISE 7.9 *Consider* $X \sim N_3(\mu, \Sigma)$. *An i.i.d. sample of size* $n = 10$ *provides:*

$$\begin{aligned} \bar{x} &= (1, 0, 2)^\top \\ S &= \begin{pmatrix} 3 & 2 & 1 \\ 2 & 3 & 1 \\ 1 & 1 & 4 \end{pmatrix}. \end{aligned}$$

a) *Knowing that the eigenvalues of* S *are integers, describe a 95% confidence region for* μ. *(Hint: to compute eigenvalues use* $|S| = \prod_{j=1}^{3} \lambda_j$ *and* $tr(S) = \sum_{j=1}^{3} \lambda_j$*).*

b) *Calculate the simultaneous confidence intervals for* μ_1, μ_2 *and* μ_3.

c) *Can we assert that* μ_1 *is an average of* μ_2 *and* μ_3*?*

EXERCISE 7.10 *Consider two independent i.i.d. samples, each of size 10, from two bivariate normal populations. The results are summarized below:*

$$\bar{x}_1 = (3, 1)^\top;\ \bar{x}_2 = (1, 1)^\top$$

$$S_1 = \begin{pmatrix} 4 & -1 \\ -1 & 2 \end{pmatrix};\ S_2 = \begin{pmatrix} 2 & -2 \\ -2 & 4 \end{pmatrix}.$$

Provide a solution to the following tests:

$$\begin{array}{llll} a)\ H_0\colon & \mu_1 = \mu_2 & H_1\colon & \mu_1 \neq \mu_2 \\ b)\ H_0\colon & \mu_{11} = \mu_{21} & H_1\colon & \mu_{11} \neq \mu_{21} \\ c)\ H_0\colon & \mu_{12} = \mu_{22} & H_1\colon & \mu_{12} \neq \mu_{22} \end{array}$$

Compare the solutions and comment.

EXERCISE 7.11 *Prove expression (7.4) in the Test Problem 2 with log-likelihoods* ℓ_0^* *and* ℓ_1^*. *(Hint: use (2.29)).*

EXERCISE 7.12 *Assume that* $X \sim N_p(\mu, \Sigma)$ *where* Σ *is unknown.*

a) *Derive the log likelihood ratio test for testing the independence of the* p *components, that is* $H_0 : \Sigma$ *is a diagonal matrix. (Solution:* $-2 \log \lambda = -n \log |R|$ *where* R *is the correlation matrix, which is asymptotically a* $\chi^2_{\frac{1}{2}p(p-1)}$ *under* H_0*).*

b) *Assume that* Σ *is a diagonal matrix (all the variables are independent). Can an asymptotic test for* $H_0: \mu = \mu_o$ *against* $H_1: \mu \neq \mu_o$ *be derived? How would this compare to* p *independent univariate* $t-$*tests on each* μ_j*?*

c) *Show an easy derivation of an asymptotic test for testing the equality of the* p *means (Hint: use* $(C\bar{X})^\top(CSC^\top)^{-1}C\bar{X} \to \chi^2_{p-1}$ *where* $S = diag(s_{11}, \ldots, s_{pp})$ *and* C *is defined as in (7.10)). Compare this to the simple ANOVA procedure used in Section 3.5.*

EXERCISE 7.13 *The yields of wheat have been measured in 30 parcels that have been randomly attributed to 3 lots prepared by one of 3 different fertilizer A B and C. The data are*

Fertilizer Yield	A	B	C
1	4	6	2
2	3	7	1
3	2	7	1
4	5	5	1
5	4	5	3
6	4	5	4
7	3	8	3
8	3	9	3
9	3	9	2
10	1	6	2

Using Exercise 7.12,

a) *test the independence between the 3 variables.*

b) *test whether* $\mu^\top = [2\ 6\ 4]$ *and compare this to the 3 univariate* $t-$*tests.*

c) *test whether* $\mu_1 = \mu_2 = \mu_3$ *using simple ANOVA and the* χ^2 *approximation.*

EXERCISE 7.14 *Consider an i.i.d. sample of size* $n = 5$ *from a bivariate normal distribution*

$$X \sim N_2\left(\mu, \begin{pmatrix} 3 & \rho \\ \rho & 1 \end{pmatrix}\right)$$

where ρ *is a known parameter. Suppose* $\bar{x}^\top = (1\ 0)$*. For what value of* ρ *would the hypothesis* $H_0: \mu^\top = (0\ 0)$ *be rejected in favor of* $H_1: \mu^\top \neq (0\ 0)$ *(at the 5% level)?*

EXERCISE 7.15 *Using Example 7.14, test the last two cases described there and test the sample number one* ($n_1 = 30$)*, to see if they are from a normal population with* $\Sigma = 4I_4$ *(the sample covariance matrix to be used is given by* S_1*).*

EXERCISE 7.16 *Consider the bank data set. For the counterfeit bank notes, we want to know if the length of the diagonal* (X_6) *can be predicted by a linear model in* X_1 *to* X_5*. Estimate the linear model and test if the coefficients are significantly different from zero.*

EXERCISE 7.17 *In Example 7.10, can you predict the vocabulary score of the children in eleventh grade, by knowing the results from grades 8–9 and 10? Estimate a linear model and test its significance.*

EXERCISE 7.18 *Test the equality of the covariance matrices from the two groups in the WAIS subtest (Example 7.19).*

EXERCISE 7.19 *Prove expressions (7.21), (7.22) and (7.23).*

EXERCISE 7.20 *Using Theorem 6.3 and expression (7.16), construct an asymptotic rejection region of size α for testing, in a general model $f(x, \theta)$, with $\theta \in \mathbb{R}^k$, $H_0 : \theta = \theta_0$ against $H_1 : \theta \neq \theta_0$.*

EXERCISE 7.21 *Exercise 6.5 considered the pdf $f(x_1, x_2) = \frac{1}{\theta_1^2 \theta_2^2 x_2} e^{-\left(\frac{x_1}{\theta_1 x_2} + \frac{x_2}{\theta_1 \theta_2}\right)}$ $x_1, x_2 > 0$. Solve the problem of testing $H_0 : \theta^\top = (\theta_{01}, \theta_{02})$ from an iid sample of size n on $x = (x_1, x_2)^\top$, where n is large.*

EXERCISE 7.22 *In Olkin and Veath (1980), the evolution of citrate concentrations in plasma is observed at 3 different times of day, X_1 (8 am), X_2 (11 am) and X_3 (3 pm), for two groups of patients who follow a different diets. (The patients were randomly attributed to each group under a balanced design $n_1 = n_2 = 5$).*
The data are:

Group	X_1(8 am)	X_2(11 am)	X_3(3 pm)
	125	137	121
	144	173	147
I	105	119	125
	151	149	128
	137	139	109
	93	121	107
	116	135	106
II	109	83	100
	89	95	83
	116	128	100

Test if the profiles of the groups are parallel, if they are at the same level and if they are horizontal.

Part III

Multivariate Techniques

8 Decomposition of Data Matrices by Factors

In Chapter 1 basic descriptive techniques were developed which provided tools for "looking" at multivariate data. They were based on adaptations of bivariate or univariate devices used to reduce the dimensions of the observations. In the following three chapters, issues of reducing the dimension of a multivariate data set will be discussed. The perspectives will be different but the tools will be related.

In this chapter, we take a descriptive perspective and show how using a geometrical approach provides a "best" way of reducing the dimension of a data matrix. It is derived with respect to a least-squares criterion. The result will be low dimensional graphical pictures of the data matrix. This involves the decomposition of the data matrix into "factors". These "factors" will be sorted in decreasing order of importance. The approach is very general and is the core idea of many multivariate techniques. We deliberately use the word "factor" here as a tool or transformation for structural interpretation in an exploratory analysis. In practice, the matrix to be decomposed will be some transformation of the original data matrix and as shown in the following chapters, these transformations provide easier interpretations of the obtained graphs in lower dimensional spaces.

Chapter 9 addresses the issue of reducing the dimensionality of a multivariate random variable by using linear combinations (the principal components). The identified principal components are ordered in decreasing order of importance. When applied in practice to a data matrix, the principal components will turn out to be the factors of a transformed data matrix (the data will be centered and eventually standardized).

Factor analysis is discussed in Chapter 10. The same problem of reducing the dimension of a multivariate random variable is addressed but in this case the number of factors is fixed from the start. Each factor is interpreted as a latent characteristic of the individuals revealed by the original variables. The non-uniqueness of the solutions is dealt with by searching for the representation with the easiest interpretation for the analysis.

Summarizing, this chapter can be seen as a foundation since it develops a basic tool for reducing the dimension of a multivariate data matrix.

8.1 The Geometric Point of View

As a matter of introducing certain ideas, assume that the data matrix $\mathcal{X}(n \times p)$ is composed of n observations (or individuals) of p variables.

There are in fact two ways of looking at $\mathcal{X}$, row by row or column by column:

(1) Each row (observation) is a vector $x_i^\top = (x_{i1}, \dots, x_{ip}) \in \mathbb{R}^p$.

From this point of view our data matrix $\mathcal{X}$ is representable as a cloud of n points in $\mathbb{R}^p$ as shown in Figure 8.1.

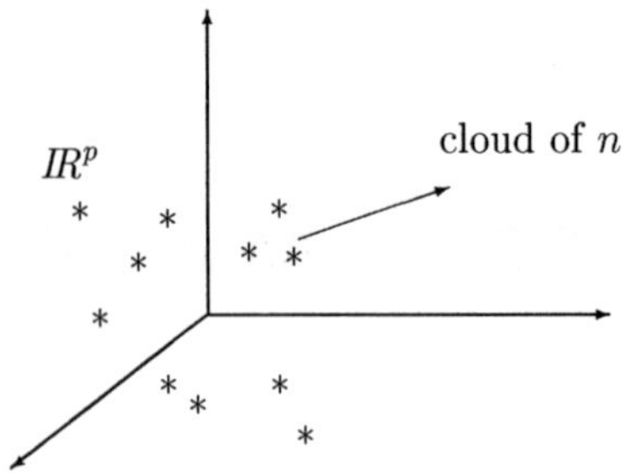

Figure 8.1.

(2) Each column (variable) is a vector $x_{[j]} = (x_{1j} \dots x_{nj})^\top \in \mathbb{R}^n$.
From this point of view the data matrix $\mathcal{X}$ is a cloud of p points in $\mathbb{R}^n$ as shown in Figure 8.2.

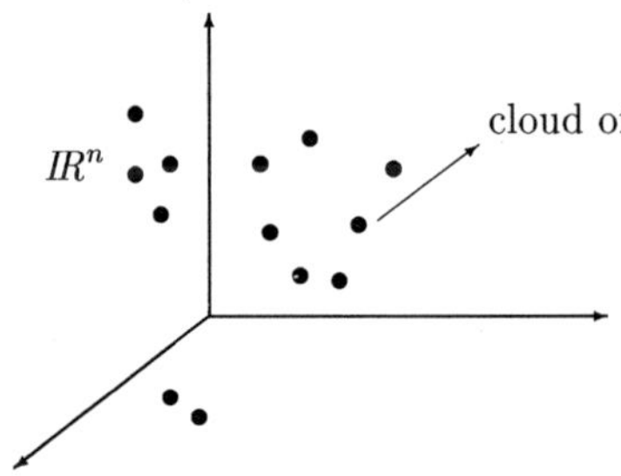

Figure 8.2.

When n and/or p are large (larger than 2 or 3), we cannot produce interpretable graphs of these clouds of points. Therefore, the aim of the factorial methods to be developed here is two-fold. We shall try to simultaneously approximate the column space $C(\mathcal{X})$ and the row space $C(\mathcal{X}^\top)$ with smaller subspaces. The hope is of course that this can be done without loosing too much information about the variation and structure of the point clouds in both spaces. Ideally, this will provide insights into the structure of $\mathcal{X}$ through graphs in $\mathbb{R}$, $\mathbb{R}^2$ or $\mathbb{R}^3$. The main focus then is to find the dimension reducing factors.

Summary

↪ Each row (individual) of $\mathcal{X}$ is a p-dimensional vector. From this point of view $\mathcal{X}$ can be considered as a cloud of n points in $\mathbb{R}^p$.

↪ Each column (variable) of $\mathcal{X}$ is a n-dimensional vector. From this point of view $\mathcal{X}$ can be considered as a cloud of p points in $\mathbb{R}^n$.

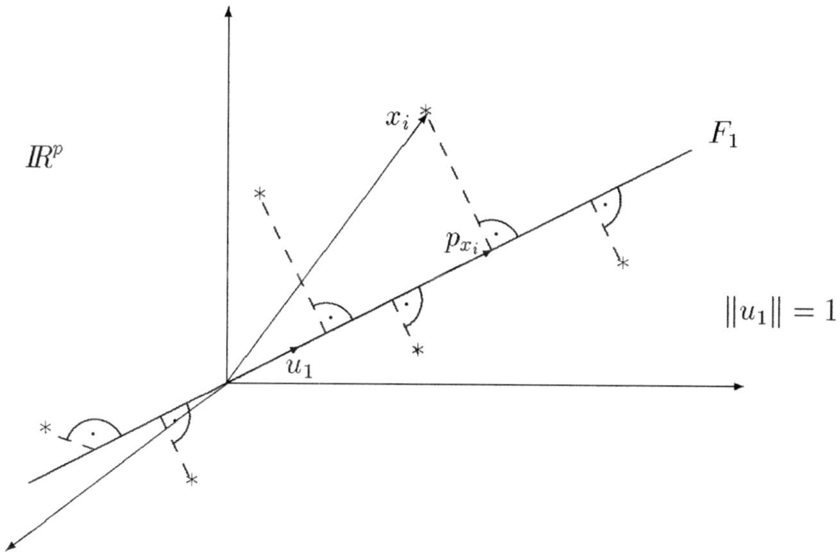

Figure 8.3.

8.2 Fitting the p-dimensional Point Cloud

Subspaces of Dimension 1

In this section $\mathcal{X}$ is represented by a cloud of n points in $\mathbb{R}^p$ (considering each row). The question is how to project this point cloud onto a space of lower dimension. To begin consider the simplest problem, namely finding a subspace of dimension 1. The problem boils down to finding a straight line F_1 through the origin. The direction of this line can be defined by a unit vector $u_1 \in \mathbb{R}^p$. Hence, we are searching for the vector u_1 which gives the "best" fit of the initial cloud of n points. The situation is depicted in Figure 8.3.

The representation of the i-th individual $x_i \in \mathbb{R}^p$ on this line is obtained by the projection of the corresponding point onto u_1, i.e., the projection point p_{x_i}. We know from (2.42) that the coordinate of x_i on F_1 is given by

$$p_{x_i} = x_i^\top \frac{u_1}{\|u_1\|} = x_i^\top u_1. \tag{8.1}$$

We define the *best line* F_1 in the following "least-squares" sense: Find $u_1 \in \mathbb{R}^p$ which minimizes

$$\sum_{i=1}^{n} \|x_i - p_{x_i}\|^2. \tag{8.2}$$

Since $\|x_i - p_{x_i}\|^2 = \|x_i\|^2 - \|p_{x_i}\|^2$ by Pythagoras's theorem, the problem of minimizing (8.2) is equivalent to maximizing $\sum_{i=1}^{n} \|p_{x_i}\|^2$. Thus the problem is to find $u_1 \in \mathbb{R}^p$ that maximizes

$\sum_{i=1}^{n} \|p_{x_i}\|^2$ under the constraint $\|u_1\| = 1$. With (8.1) we can write

$$\begin{pmatrix} p_{x_1} \\ p_{x_2} \\ \vdots \\ p_{x_n} \end{pmatrix} = \begin{pmatrix} x_1^\top u_1 \\ x_2^\top u_1 \\ \vdots \\ x_n^\top u_1 \end{pmatrix} = \mathcal{X} u_1$$

and the problem can finally be reformulated as: find $u_1 \in \mathbb{R}^p$ with $\|u_1\| = 1$ that maximizes the quadratic form $(\mathcal{X}u_1)^\top(\mathcal{X}u_1)$ or

$$\max_{u_1^\top u_1 = 1} u_1^\top (\mathcal{X}^\top \mathcal{X}) u_1. \tag{8.3}$$

The solution is given by Theorem 2.5 (using $\mathcal{A} = \mathcal{X}^\top \mathcal{X}$ and $\mathcal{B} = \mathcal{I}$ in the theorem).

THEOREM 8.1 *The vector u_1 which minimizes (8.2) is the eigenvector of $\mathcal{X}^\top \mathcal{X}$ associated with the largest eigenvalue λ_1 of $\mathcal{X}^\top \mathcal{X}$.*

Note that if the data have been centered, i.e., $\overline{x} = 0$, then $\mathcal{X} = \mathcal{X}_c$, where $\mathcal{X}_c$ is the centered data matrix, and $\frac{1}{n}\mathcal{X}^\top \mathcal{X}$ is the covariance matrix. Thus Theorem 8.1 says that we are searching for a maximum of the quadratic form (8.3) w.r.t. the covariance matrix $\mathcal{S}_\mathcal{X} = n^{-1}\mathcal{X}^\top \mathcal{X}$.

Representation of the Cloud on F_1

The coordinates of the n individuals on F_1 are given by $\mathcal{X}u_1$. $\mathcal{X}u_1$ is called the *first factorial variable* or the *first factor* and u_1 the *first factorial axis.* The n individuals, x_i, are now represented by a new factorial variable $z_1 = \mathcal{X}u_1$. This factorial variable is a linear combination of the original variables $(x_{[1]}, \ldots, x_{[p]})$ whose coefficients are given by the vector u_1, i.e.,

$$z_1 = u_{11}x_{[1]} + \ldots + u_{p1}x_{[p]}. \tag{8.4}$$

Subspaces of Dimension 2

If we approximate the n individuals by a plane (dimension 2), it can be shown via Theorem 2.5 that this space contains u_1. The plane is determined by the best linear fit (u_1) and a unit vector u_2 orthogonal to u_1 which maximizes the quadratic form $u_2^\top(\mathcal{X}^\top \mathcal{X})u_2$ under the constraints

$$\|u_2\| = 1, \text{ and } u_1^\top u_2 = 0.$$

THEOREM 8.2 *The second factorial axis, u_2, is the eigenvector of $\mathcal{X}^\top \mathcal{X}$ corresponding to the second largest eigenvalue λ_2 of $\mathcal{X}^\top \mathcal{X}$.*

The unit vector u_2 characterizes a second line, F_2, on which the points are projected. The coordinates of the n individuals on F_2 are given by $z_2 = \mathcal{X}u_2$. The variable z_2 is called the *second factorial variable* or *the second factor.* The representation of the n individuals in two-dimensional space ($z_1 = \mathcal{X}u_1$ vs. $z_2 = \mathcal{X}u_2$) is shown in Figure 8.4.

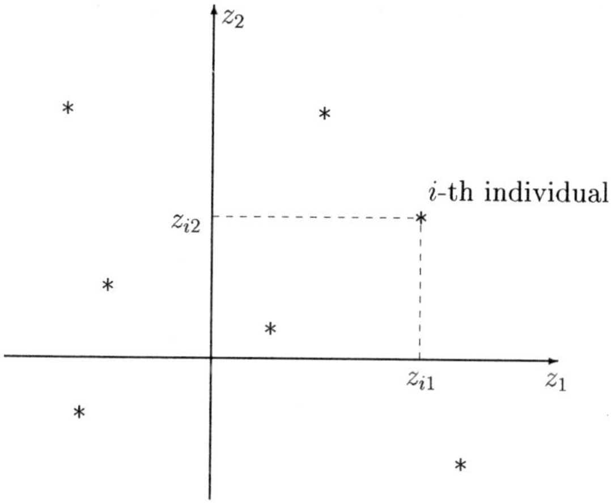

Figure 8.4. Representation of the individuals $x_1, \dots, x_n$ as a two-dimensional point cloud.

Subspaces of Dimension q ($q \le p$)

In the case of q dimensions the task is again to minimize (8.2) but with projection points in a q-dimensional subspace. Following the same argument as above, it can be shown via Theorem 2.5 that this best subspace is generated by $u_1, u_2, \dots, u_q$, the orthonormal eigenvectors of $\mathcal{X}^\top \mathcal{X}$ associated with the corresponding eigenvalues $\lambda_1 \ge \lambda_2 \ge \dots \ge \lambda_q$. The coordinates of the n individuals on the k-th factorial axis, u_k, are given by the k-th factorial variable $z_k = \mathcal{X} u_k$ for $k = 1, \dots, q$. Each factorial variable $z_k = (z_{1k}, z_{2k}, \dots, z_{nk})^\top$ is a linear combination of the original variables $x_{[1]}, x_{[2]}, \dots, x_{[p]}$ whose coefficients are given by the elements of the k-th vector $u_k : z_{ik} = \sum_{m=1}^{p} x_{im} u_{mk}$.

Summary

- ↪ The p-dimensional point cloud of individuals can be graphically represented by projecting each element into spaces of smaller dimensions.
- ↪ The first factorial axis is u_1 and defines a line F_1 through the origin. This line is found by minimizing the orthogonal distances (8.2). The factor u_1 equals the eigenvector of $\mathcal{X}^\top \mathcal{X}$ corresponding to its largest eigenvalue. The coordinates for representing the point cloud on a straight line are given by $z_1 = \mathcal{X} u_1$.
- ↪ The second factorial axis is u_2, where u_2 denotes the eigenvector of $\mathcal{X}^\top \mathcal{X}$ corresponding to its second largest eigenvalue. The coordinates for representing the point cloud on a plane are given by $z_1 = \mathcal{X} u_1$ and $z_2 = \mathcal{X} u_2$.
- ↪ The factor directions $1, \dots, q$ are $u_1, \dots, u_q$, which denote the eigenvectors of $\mathcal{X}^\top \mathcal{X}$ corresponding to the q largest eigenvalues. The coordinates for representing the point cloud of individuals on a q-dimensional subspace are given by $z_1 = \mathcal{X} u_1, \dots, z_q = \mathcal{X} u_q$.

8.3 Fitting the n-dimensional Point Cloud

Subspaces of Dimension 1

Suppose that $\mathcal{X}$ is represented by a cloud of p points (variables) in $\mathbb{R}^n$ (considering each column). How can this cloud be projected into a lower dimensional space? We start as before with one dimension. In other words, we have to find a straight line G_1, which is defined by the unit vector $v_1 \in \mathbb{R}^n$, and which gives the best fit of the initial cloud of p points.

Algebraically, this is the same problem as above (replace $\mathcal{X}$ by $\mathcal{X}^\top$ and follow Section 8.2): the representation of the j-th variable $x_{[j]} \in \mathbb{R}^n$ is obtained by the projection of the corresponding point onto the straight line G_1 or the direction v_1. Hence we have to find v_1 such that $\sum_{j=1}^{p} \|p_{x_{[j]}}\|^2$ is maximized, or equivalently, we have to find the unit vector v_1 which maximizes $(\mathcal{X}^\top v_1)^\top(\mathcal{X} v_1) = v_1^\top(\mathcal{X}\mathcal{X}^\top)v_1$. The solution is given by Theorem 2.5.

THEOREM 8.3 *v_1 is the eigenvector of $\mathcal{X}\mathcal{X}^\top$ corresponding to the largest eigenvalue μ_1 of $\mathcal{X}\mathcal{X}^\top$.*

Representation of the Cloud on G_1

The coordinates of the p variables on G_1 are given by $w_1 = \mathcal{X}^\top v_1$, the first factorial axis. The p variables are now represented by a linear combination of the original individuals $x_1, \ldots, x_n$, whose coefficients are given by the vector v_1, i.e., for $j = 1, \ldots, p$

$$w_{1j} = v_{11}x_{1j} + \ldots + v_{1n}x_{nj}. \tag{8.5}$$

Subspaces of Dimension q $(q \leq n)$

The representation of the p variables in a subspace of dimension q is done in the same manner as for the n individuals above. The best subspace is generated by the orthonormal eigenvectors $v_1, v_2, \ldots, v_q$ of $\mathcal{X}\mathcal{X}^\top$ associated with the eigenvalues $\mu_1 \geq \mu_2 \geq \ldots \geq \mu_q$. The coordinates of the p variables on the k-th factorial axis are given by the factorial variables $w_k = \mathcal{X}^\top v_k$, $k = 1, \ldots, q$. Each factorial variable $w_k = (w_{k1}, w_{k2}, \ldots, w_{kp})^\top$ is a linear combination of the original individuals $x_1, x_2, \ldots, x_n$ whose coefficients are given by the elements of the k-th vector v_k : $w_{kj} = \sum_{m=1}^{n} v_{km}x_{mj}$. The representation in a subspace of dimension $q = 2$ is depicted in Figure 8.5.

Summary

- ↪ The n-dimensional point cloud of variables can be graphically represented by projecting each element into spaces of smaller dimensions.
- ↪ The first factor direction is v_1 and defines a line G_1 through the origin. The vector v_1 equals the eigenvector of $\mathcal{X}\mathcal{X}^\top$ corresponding to the largest eigenvalue of $\mathcal{X}\mathcal{X}^\top$. The coordinates for representing the point cloud on a straight line are $w_1 = \mathcal{X}^\top v_1$.

Summary (continued)

↪ The second factor direction is v_2, where v_2 denotes the eigenvector of $\mathcal{X}\mathcal{X}^\top$ corresponding to its second largest eigenvalue. The coordinates for representing the point cloud on a plane are given by $w_1 = \mathcal{X}^\top v_1$ and $w_2 = \mathcal{X}^\top v_2$.

↪ The factor directions $1, \ldots, q$ are $v_1, \ldots, v_q$, which denote the eigenvectors of $\mathcal{X}\mathcal{X}^\top$ corresponding to the q largest eigenvalues. The coordinates for representing the point cloud of variables on a q-dimensional subspace are given by $w_1 = \mathcal{X}^\top v_1, \ldots, w_q = \mathcal{X}^\top v_q$.

8.4 Relations between Subspaces

The aim of this section is to present a duality relationship between the two approaches shown in Sections 8.2 and 8.3. Consider the eigenvector equations in $\mathbb{R}^n$

$$(\mathcal{X}\mathcal{X}^\top)v_k = \mu_k v_k \tag{8.6}$$

for $k \le r$, where $r = \text{rank}(\mathcal{X}\mathcal{X}^\top) = \text{rank}(\mathcal{X}) \le \min(p, n)$. Multiplying by $\mathcal{X}^\top$, we have

$$\mathcal{X}^\top(\mathcal{X}\mathcal{X}^\top)v_k = \mu_k \mathcal{X}^\top v_k \tag{8.7}$$

$$\text{or} \quad (\mathcal{X}^\top\mathcal{X})(\mathcal{X}^\top v_k) = \mu_k(\mathcal{X}^\top v_k) \tag{8.8}$$

so that each eigenvector v_k of $\mathcal{X}\mathcal{X}^\top$ corresponds to an eigenvector $(\mathcal{X}^\top v_k)$ of $\mathcal{X}^\top\mathcal{X}$ associated with the same eigenvalue μ_k. This means that every non-zero eigenvalue of $\mathcal{X}\mathcal{X}^\top$ is an eigenvalue of $\mathcal{X}^\top\mathcal{X}$. The corresponding eigenvectors are related by

$$u_k = c_k \mathcal{X}^\top v_k,$$

where c_k is some constant.

Now consider the eigenvector equations in $\mathbb{R}^p$:

$$(\mathcal{X}^\top\mathcal{X})u_k = \lambda_k u_k \tag{8.9}$$

for $k \le r$. Multiplying by $\mathcal{X}$, we have

$$(\mathcal{X}\mathcal{X}^\top)(\mathcal{X}u_k) = \lambda_k(\mathcal{X}u_k), \tag{8.10}$$

i.e., each eigenvector u_k of $\mathcal{X}^\top\mathcal{X}$ corresponds to an eigenvector $\mathcal{X}u_k$ of $\mathcal{X}\mathcal{X}^\top$ associated with the same eigenvalue λ_k. Therefore, every non-zero eigenvalue of $(\mathcal{X}^\top\mathcal{X})$ is an eigenvalue of $\mathcal{X}\mathcal{X}^\top$. The corresponding eigenvectors are related by

$$v_k = d_k \mathcal{X} u_k,$$

where d_k is some constant. Now, since $u_k^\top u_k = v_k^\top v_k = 1$ we have $c_k = d_k = \frac{1}{\sqrt{\lambda_k}}$. This lead to the following result:

THEOREM 8.4 (Duality Relations) *Let r be the rank of $\mathcal{X}$. For $k \le r$, the eigenvalues λ_k of $\mathcal{X}^\top\mathcal{X}$ and $\mathcal{X}\mathcal{X}^\top$ are the same and the eigenvectors (u_k and v_k, respectively) are related by*

$$u_k = \frac{1}{\sqrt{\lambda_k}}\mathcal{X}^\top v_k \tag{8.11}$$

$$v_k = \frac{1}{\sqrt{\lambda_k}}\mathcal{X} u_k. \tag{8.12}$$

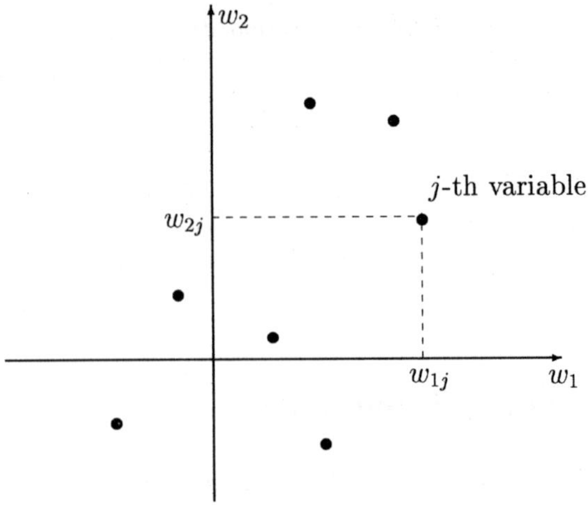

Figure 8.5. Representation of the variables $x_{[1]}, \ldots, x_{[p]}$ as a two-dimensional point cloud.

Note that the projection of the p variables on the factorial axis v_k is given by

$$w_k = \mathcal{X}^\top v_k = \frac{1}{\sqrt{\lambda_k}} \mathcal{X}^\top \mathcal{X} u_k = \sqrt{\lambda_k}\, u_k. \tag{8.13}$$

Therefore, the eigenvectors v_k do not have to be explicitly recomputed to get w_k.

Note that u_k and v_k provide the SVD of $\mathcal{X}$ (see Theorem 2.2). Letting $U = [u_1\ u_2\ \ldots\ u_r]$, $V = [v_1\ v_2\ \ldots\ v_r]$ and $\Lambda = \mathrm{diag}(\lambda_1, \ldots, \lambda_r)$ we have

$$\mathcal{X} = V\, \Lambda^{1/2}\, U^\top$$

so that

$$x_{ij} = \sum_{k=1}^{r} \lambda_k^{1/2}\, v_{ik}\, u_{jk}. \tag{8.14}$$

In the following section this method is applied in analysing consumption behavior across different household types.

Summary

↪ The matrices $\mathcal{X}^\top \mathcal{X}$ and $\mathcal{X}\mathcal{X}^\top$ have the same non-zero eigenvalues $\lambda_1, \ldots, \lambda_r$, where $r = \mathrm{rank}(\mathcal{X})$.

↪ The eigenvectors of $\mathcal{X}^\top \mathcal{X}$ can be calculated from the eigenvectors of $\mathcal{X}\mathcal{X}^\top$ and vice versa:

$$u_k = \frac{1}{\sqrt{\lambda_k}} \mathcal{X}^\top v_k \quad \text{and} \quad v_k = \frac{1}{\sqrt{\lambda_k}} \mathcal{X} u_k.$$

↪ The coordinates representing the variables (columns) of $\mathcal{X}$ in a q-dimensional subspace can be easily calculated by $w_k = \sqrt{\lambda_k} u_k$.

8.5 Practical Computation

The practical implementation of the techniques introduced begins with the computation of the eigenvalues $\lambda_1 \geq \lambda_2 \geq \ldots \geq \lambda_p$ and the corresponding eigenvectors $u_1, \ldots, u_p$ of $\mathcal{X}^\top \mathcal{X}$. (Since p is usually less than n, this is numerically less involved than computing v_k directly for $k = 1, \ldots, p$). The representation of the n individuals on a plane is then obtained by plotting $z_1 = \mathcal{X} u_1$ versus $z_2 = \mathcal{X} u_2$ ($z_3 = \mathcal{X} u_3$ may eventually be added if a third dimension is helpful). Using the Duality Relation (8.13) representations for the p variables can easily be obtained. These representations can be visualized in a scatterplot of $w_1 = \sqrt{\lambda_1}\, u_1$ against $w_2 = \sqrt{\lambda_2} u_2$ (and eventually against $w_3 = \sqrt{\lambda_3}\, u_3$). Higher dimensional factorial resolutions can be obtained (by computing z_k and w_k for $k > 3$) but, of course, cannot be plotted.

A standard way of evaluating the quality of the factorial representations in a subspace of dimension q is given by the ratio

$$\tau_q = \frac{\lambda_1 + \lambda_2 + \ldots + \lambda_q}{\lambda_1 + \lambda_2 + \ldots + \lambda_p}, \tag{8.15}$$

where $0 \leq \tau_q \leq 1$. In general, the scalar product $y^\top y$ is called the inertia of $y \in \mathbb{R}^n$ w.r.t. the origin. Therefore, the ratio τ_q is usually interpreted as the percentage of the inertia explained by the first q factors. Note that $\lambda_j = (\mathcal{X} u_j)^\top (\mathcal{X} u_j) = z_j^\top z_j$. Thus, λ_j is the inertia of the j-th factorial variable w.r.t. the origin. The denominator in (8.15) is a measure of the total inertia of the p variables, $x_{[j]}$. Indeed, by (2.3)

$$\sum_{j=1}^{p} \lambda_j = \mathrm{tr}(\mathcal{X}^\top \mathcal{X}) = \sum_{j=1}^{p} \sum_{i=1}^{n} x_{ij}^2 = \sum_{j=1}^{p} x_{[j]}^\top x_{[j]}.$$

REMARK 8.1 *It is clear that the sum $\sum_{j=1}^{q} \lambda_j$ is the sum of the inertia of the first q factorial variables $z_1, z_2, \ldots, z_q$.*

EXAMPLE 8.1 *We consider the data set in Table B.6 which gives the food expenditures of various French families (manual workers = MA, employees = EM, managers = CA) with varying numbers of children (2, 3, 4 or 5 children). We are interested in investigating whether certain household types prefer certain food types. We can answer this question using the factorial approximations developed here.*

The correlation matrix corresponding to the data is

$$\mathcal{R} = \begin{pmatrix} 1.00 & 0.59 & 0.20 & 0.32 & 0.25 & 0.86 & 0.30 \\ 0.59 & 1.00 & 0.86 & 0.88 & 0.83 & 0.66 & -0.36 \\ 0.20 & 0.86 & 1.00 & 0.96 & 0.93 & 0.33 & -0.49 \\ 0.32 & 0.88 & 0.96 & 1.00 & 0.98 & 0.37 & -0.44 \\ 0.25 & 0.83 & 0.93 & 0.98 & 1.00 & 0.23 & -0.40 \\ 0.86 & 0.66 & 0.33 & 0.37 & 0.23 & 1.00 & 0.01 \\ 0.30 & -0.36 & -0.49 & -0.44 & -0.40 & 0.01 & 1.00 \end{pmatrix}.$$

We observe a rather high correlation between meat and poultry, whereas the expenditure for milk and wine is rather small. Are there household types that prefer, say, meat over bread?

We shall now represent food expenditures and households simultaneously using two factors. First, note that in this particular problem the origin has no specific meaning (it represents a "zero" consumer). So it makes sense to compare the consumption of any family to that

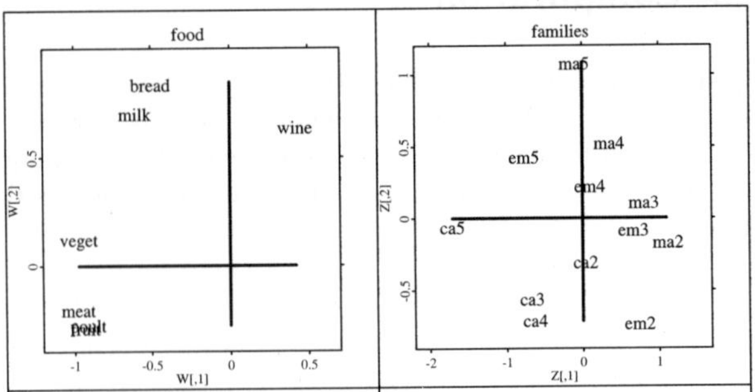

bread	-0.499	0.842
veget	-0.970	0.133
fruit	-0.929	-0.278
meat	-0.962	-0.191
poult	-0.911	-0.266
milk	-0.584	0.707
wine	0.428	0.648

ma2	1.129	-0.144
em2	0.746	-0.708
ca2	0.047	-0.286
ma3	0.806	0.128
em3	0.669	-0.064
ca3	-0.669	-0.535
ma4	0.368	0.542
em4	0.100	0.250
ca4	-0.632	-0.685
ma5	-0.087	1.096
em5	-0.770	0.447
ca5	-1.705	-0.040

Figure 8.6. Representation of food expenditures and family types in two dimensions. `MVAdecofood`

of an "average family" rather than to the origin. Therefore, the data is first centered (the origin is translated to the center of gravity, $\overline{x}$). Furthermore, since the dispersions of the 7 variables are quite different each variable is standardized so that each has the same weight in the analysis (mean 0 and variance 1). Finally, for convenience, we divide each element in the matrix by $\sqrt{n} = \sqrt{12}$. (This will only change the scaling of the plots in the graphical representation.)

The data matrix to be analyzed is

$$\mathcal{X}_* = \frac{1}{\sqrt{n}} \mathcal{H} \mathcal{X} \mathcal{D}^{-1/2},$$

where $\mathcal{H}$ is the centering matrix and $\mathcal{D} = \operatorname{diag}(s_{X_i X_i})$ (see Section 3.3). Note that by standardizing by $\sqrt{n}$, it follows that $\mathcal{X}_*^\top \mathcal{X}_* = \mathcal{R}$ where $\mathcal{R}$ is the correlation matrix of the original data. Calculating

$$\lambda = (4.33, 1.83, 0.63, 0.13, 0.06, 0.02, 0.00)^\top$$

shows that the directions of the first two eigenvectors play a dominant role ($\tau_2 = 88\%$), whereas the other directions contribute less than 15% of inertia. A two-dimensional plot should suffice for interpreting this data set.

The coordinates of the projected data points are given in the two lower windows of Figure 8.6. Let us first examine the food expenditure window. In this window we see the representation of the $p = 7$ variables given by the first two factors. The plot shows the factorial variables w_1 and w_2 in the same fashion as Figure 8.4. We see that the points for meat, poultry, vegetables and fruits are close to each other in the lower left of the graph. The expenditures for bread and milk can be found in the upper left whereas wine stands alone in the upper *right. The first factor, w_1, may be interpreted as the meat/fruit factor of consumption, the second factor, w_2, as the bread/wine component.*

In the lower window on the right-hand side, we show the factorial variables z_1 and z_2 from the fit of the $n = 12$ household types. Note that by the Duality Relations of Theorem 8.4, the factorial variables z_j are linear combinations of the factors w_k from the left window. The points displayed in the consumer window (graph on the right) are plotted relative to an average consumer represented by the origin. The manager families are located in the lower left corner of the graph whereas the manual workers and employees tend to be in the upper right. The factorial variables for CA5 (managers with five children) lie close to the meat/fruit factor. Relative to the average consumer this household type is a large consumer of meat/poultry and fruits/vegetables. In Chapter 9, we will return to these plots interpreting them in a much deeper way. At this stage, it suffices to notice that the plots provide a graphical representation in $\mathbb{R}^2$ of the information contained in the original, high-dimensional (12×7) data matrix.

Summary

↪ The practical implementation of factor decomposition of matrices consists of computing the eigenvalues $\lambda_1, \dots, \lambda_p$ and the eigenvectors $u_1, \dots, u_p$ of $\mathcal{X}^\top \mathcal{X}$. The representation of the n individuals is obtained by plotting $z_1 = \mathcal{X}u_1$ vs. $z_2 = \mathcal{X}u_2$ (and, if necessary, vs. $z_3 = \mathcal{X}u_3$). The representation of the p variables is obtained by plotting $w_1 = \sqrt{\lambda_1}u_1$ vs. $w_2 = \sqrt{\lambda_2}u_2$ (and, if necessary, vs. $w_3 = \sqrt{\lambda_3}u_3$).

↪ The quality of the factorial representation can be evaluated using τ_q which is the percentage of inertia explained by the first q factors.

8.6 Exercises

EXERCISE 8.1 *Prove that $n^{-1}\mathcal{Z}^\top\mathcal{Z}$ is the covariance of the centered data matrix, where $\mathcal{Z}$ is the matrix formed by the columns $z_k = \mathcal{X}u_k$.*

EXERCISE 8.2 *Compute the SVD of the French food data (Table B.6).*

EXERCISE 8.3 *Compute $\tau_3, \tau_4, \dots$ for the French food data (Table B.6).*

EXERCISE 8.4 *Apply the factorial techniques to the Swiss bank notes (Table B.2).*

EXERCISE 8.5 *Apply the factorial techniques to the time budget data (Table B.14).*

EXERCISE 8.6 *Assume that you wish to analyze p independent identically distributed random variables. What is the percentage of the inertia explained by the first factor? What is the percentage of the inertia explained by the first q factors?*

EXERCISE 8.7 *Assume that you have p i.i.d. r.v.'s. What does the eigenvector, corresponding to the first factor, look like.*

EXERCISE 8.8 *Assume that you have two random variables,* X_1 *and* $X_2 = 2X_1$*. What do the eigenvalues and eigenvectors of their correlation matrix look like? How many eigenvalues are nonzero?*

EXERCISE 8.9 *What percentage of inertia is explained by the first factor in the previous exercise?*

EXERCISE 8.10 *How do the eigenvalues and eigenvectors in Example 8.1 change if we take the prices in \$ instead of in EUR? Does it make a difference if some of the prices are in EUR and others in \$?*

9 Principal Components Analysis

Chapter 8 presented the basic geometric tools needed to produce a lower dimensional description of the rows and columns of a multivariate data matrix. Principal components analysis has the same objective with the exception that the rows of the data matrix $\mathcal{X}$ will now be considered as observations from a p-variate random variable X. The principle idea of reducing the dimension of X is achieved through linear combinations. Low dimensional linear combinations are often easier to interpret and serve as an intermediate step in a more complex data analysis. More precisely one looks for linear combinations which create the largest spread among the values of X. In other words, one is searching for linear combinations with the largest variances.

Section 9.1 introduces the basic ideas and technical elements behind principal components. No particular assumption will be made on X except that the mean vector and the covariance matrix exist. When reference is made to a data matrix $\mathcal{X}$ in Section 9.2, the empirical mean and covariance matrix will be used. Section 9.3 shows how to interpret the principal components by studying their correlations with the original components of X. Often analyses are performed in practice by looking at two-dimensional scatterplots. Section 9.4 develops inference techniques on principal components. This is particularly helpful in establishing the appropriate dimension reduction and thus in determining the quality of the resulting lower dimensional representations. Since principal component analysis is performed on covariance matrices, it is not scale invariant. Often, the measurement units of the components of X are quite different, so it is reasonable to standardize the measurement units. The normalized version of principal components is defined in Section 9.5. In Section 9.6 it is discovered that the empirical principal components are the factors of appropriate transformations of the data matrix. The classical way of defining principal components through linear combinations with respect to the largest variance is described here in geometric terms, i.e., in terms of the optimal fit within subspaces generated by the columns and/or the rows of $\mathcal{X}$ as was discussed in Chapter 8. Section 9.9 concludes with additional examples.

9.1 Standardized Linear Combination

The main objective of principal components analysis (PC) is to reduce the dimension of the observations. The simplest way of dimension reduction is to take just one element of the observed vector and to discard all others. This is not a very reasonable approach, as we have seen in the earlier chapters, since strength may be lost in interpreting the data. In the bank notes example we have seen that just one variable (e.g. X_1 = length) had no discriminatory power in distinguishing counterfeit from genuine bank notes. An alternative method is to weight all variables equally, i.e., to consider the simple average $p^{-1}\sum_{j=1}^{p} X_j$ of all the elements in the vector $X = (X_1, \ldots, X_p)^{\top}$. This again is undesirable, since all of the elements of X are considered with equal importance (weight).

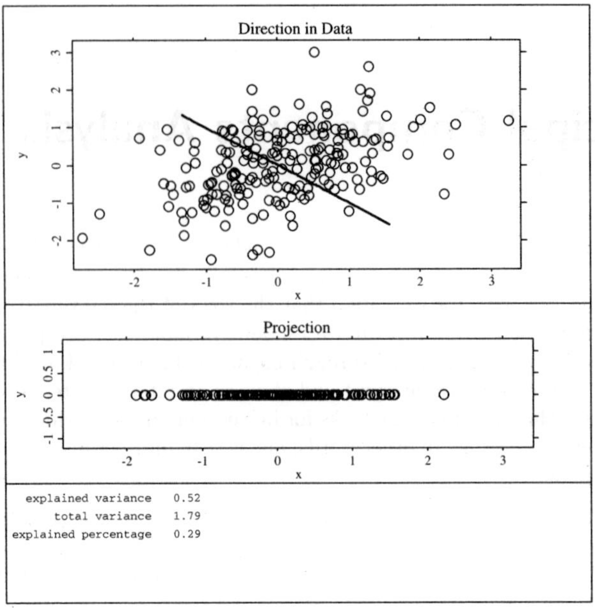

Figure 9.1. An arbitrary SLC. MVApcasimu

A more flexible approach is to study a weighted average, namely

$$\delta^\top X = \sum_{j=1}^{p} \delta_j X_j \quad \text{so that} \quad \sum_{j=1}^{p} \delta_j^2 = 1. \tag{9.1}$$

The weighting vector $\delta = (\delta_1, \ldots, \delta_p)^\top$ can then be optimized to investigate and to detect specific features. We call (9.1) a standardized linear combination (SLC). Which SLC should we choose? One aim is to maximize the variance of the projection $\delta^\top X$, i.e., to choose δ according to

$$\max_{\{\delta:\|\delta\|=1\}} Var(\delta^\top X) = \max_{\{\delta:\|\delta\|=1\}} \delta^\top Var(X)\delta. \tag{9.2}$$

The interesting "directions" of δ are found through the spectral decomposition of the covariance matrix. Indeed, from Theorem 2.5, the direction δ is given by the eigenvector γ_1 corresponding to the largest eigenvalue λ_1 of the covariance matrix $\Sigma = Var(X)$.

Figures 9.1 and 9.2 show two such projections (SLCs) of the same data set with zero mean. In Figure 9.1 an arbitrary projection is displayed. The upper window shows the data point cloud and the line onto which the data are projected. The middle window shows the projected values in the selected direction. The lower window shows the variance of the actual projection and the percentage of the total variance that is explained.

Figure 9.2 shows the projection that captures the majority of the variance in the data. This direction is of interest and is located along the main direction of the point cloud. The same line of thought can be applied to all data orthogonal to this direction leading to the second eigenvector. The SLC with the highest variance obtained from maximizing (9.2) is the first

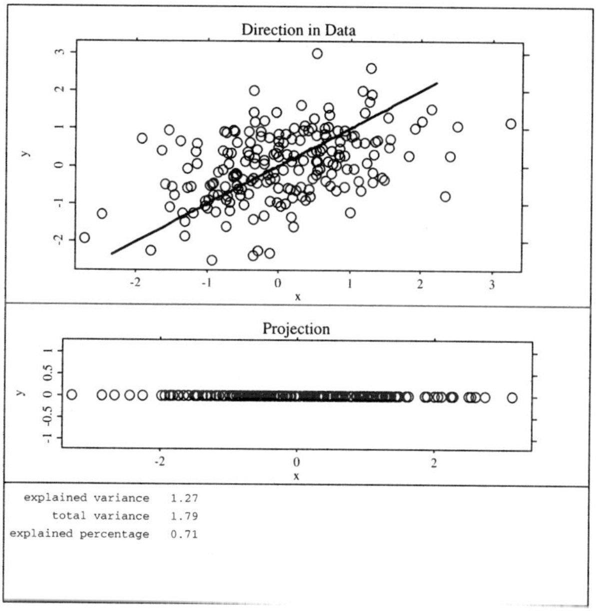

Figure 9.2. The most interesting SLC. MVApcasimu

principal component (PC) $y_1 = \gamma_1^\top X$. Orthogonal to the direction γ_1 we find the SLC with the second highest variance: $y_2 = \gamma_2^\top X$, the second PC.

Proceeding in this way and writing in matrix notation, the result for a random variable X with $E(X) = \mu$ and $Var(X) = \Sigma = \Gamma\Lambda\Gamma^\top$ is the PC transformation which is defined as

$$Y = \Gamma^\top (X - \mu). \tag{9.3}$$

Here we have centered the variable X in order to obtain a zero mean PC variable Y.

EXAMPLE 9.1 *Consider a bivariate normal distribution $N(0, \Sigma)$ with $\Sigma = \begin{pmatrix} 1 & \rho \\ \rho & 1 \end{pmatrix}$ and $\rho > 0$ (see Example 3.13). Recall that the eigenvalues of this matrix are $\lambda_1 = 1 + \rho$ and $\lambda_2 = 1 - \rho$ with corresponding eigenvectors*

$$\gamma_1 = \frac{1}{\sqrt{2}} \begin{pmatrix} 1 \\ 1 \end{pmatrix}, \quad \gamma_2 = \frac{1}{\sqrt{2}} \begin{pmatrix} 1 \\ -1 \end{pmatrix}.$$

The PC transformation is thus

$$Y = \Gamma^\top (X - \mu) = \frac{1}{\sqrt{2}} \begin{pmatrix} 1 & 1 \\ 1 & -1 \end{pmatrix} X$$

or

$$\begin{pmatrix} Y_1 \\ Y_2 \end{pmatrix} = \frac{1}{\sqrt{2}} \begin{pmatrix} X_1 + X_2 \\ X_1 - X_2 \end{pmatrix}.$$

So the first principal component is

$$Y_1 = \frac{1}{\sqrt{2}}(X_1 + X_2)$$

and the second is

$$Y_2 = \frac{1}{\sqrt{2}}(X_1 - X_2).$$

Let us compute the variances of these PCs using formulas (4.22)-(4.26):

$$\begin{aligned} Var(Y_1) &= Var\left\{\frac{1}{\sqrt{2}}(X_1 + X_2)\right\} = \frac{1}{2}\,Var(X_1 + X_2) \\ &= \frac{1}{2}\{\,Var(X_1) + Var(X_2) + 2\,Cov(X_1, X_2)\} \\ &= \frac{1}{2}(1 + 1 + 2\rho) = 1 + \rho \\ &= \lambda_1. \end{aligned}$$

Similarly we find that

$$Var(Y_2) = \lambda_2.$$

This can be expressed more generally and is given in the next theorem.

THEOREM 9.1 *For a given $X \sim (\mu, \Sigma)$ let $Y = \Gamma^\top(X - \mu)$ be the PC transformation. Then*

$$EY_j = 0, \quad j = 1, \ldots, p \tag{9.4}$$

$$Var(Y_j) = \lambda_j, \qquad j = 1, \ldots, p \tag{9.5}$$

$$Cov(Y_i, Y_j) = 0, \qquad i \neq j \tag{9.6}$$

$$Var(Y_1) \geq Var(Y_2) \geq \cdots \geq Var(Y_p) \geq 0 \tag{9.7}$$

$$\sum_{j=1}^{p} Var(Y_j) = \mathrm{tr}(\Sigma) \tag{9.8}$$

$$\prod_{j=1}^{p} Var(Y_j) = |\Sigma|. \tag{9.9}$$

Proof:
To prove (9.6), we use γ_i to denote the ith column of Γ. Then

$$Cov(Y_i, Y_j) = \gamma_i^\top Var(X - \mu)\gamma_j = \gamma_i^\top Var(X)\gamma_j.$$

As $Var(X) = \Sigma = \Gamma\Lambda\Gamma^\top$, $\Gamma^\top\Gamma = \mathcal{I}$, we obtain via the orthogonality of Γ:

$$\gamma_i^\top \Gamma\Lambda\Gamma^\top \gamma_j = \begin{cases} 0 & i \neq j \\ \lambda_i & i = j \end{cases}$$

In fact, as $Y_i = \gamma_i^\top(X - \mu)$ lies in the eigenvector space corresponding to γ_i, and eigenvector spaces corresponding to different eigenvalues are orthogonal to each other, we can directly see Y_i and Y_j are orthogonal to each other, so their covariance is 0. □

The connection between the PC transformation and the search for the best SLC is made in the following theorem, which follows directly from (9.2) and Theorem 2.5.

THEOREM 9.2 *There exists no SLC that has larger variance than $\lambda_1 = Var(Y_1)$.*

THEOREM 9.3 *If $Y = a^\top X$ is a SLC that is not correlated with the first k PCs of X, then the variance of Y is maximized by choosing it to be the $(k + 1)$-st PC.*

Summary

- ↪ A standardized linear combination (SLC) is a weighted average $\delta^\top X = \sum_{j=1}^{p} \delta_j X_j$ where δ is a vector of length 1.
- ↪ Maximizing the variance of $\delta^\top X$ leads to the choice $\delta = \gamma_1$, the eigenvector corresponding to the largest eigenvalue λ_1 of $\Sigma = Var(X)$.
  This is a projection of X into the one-dimensional space, where the components of X are weighted by the elements of γ_1. $Y_1 = \gamma_1^\top (X - \mu)$ is called the first principal component (PC).
- ↪ This projection can be generalized for higher dimensions. The PC transformation is the linear transformation $Y = \Gamma^\top (X - \mu)$, where $\Sigma = Var(X) = \Gamma \Lambda \Gamma^\top$ and $\mu = EX$.
  $Y_1, Y_2, \ldots, Y_p$ are called the first, second,..., and p-th PCs.
- ↪ The PCs have zero means, variance $Var(Y_j) = \lambda_j$, and zero covariances. From $\lambda_1 \geq \ldots \geq \lambda_p$ it follows that $Var(Y_1) \geq \ldots \geq Var(Y_p)$. It holds that $\sum_{j=1}^{p} Var(Y_j) = \operatorname{tr}(\Sigma)$ and $\prod_{j=1}^{p} Var(Y_j) = |\Sigma|$.
- ↪ If $Y = a^\top X$ is a SLC which is not correlated with the first k PCs of X then the variance of Y is maximized by choosing it to be the $(k+1)$-st PC.

9.2 Principal Components in Practice

In practice the PC transformation has to be replaced by the respective estimators: μ becomes $\overline{x}$, Σ is replaced by $\mathcal{S}$, etc. If g_1 denotes the first eigenvector of $\mathcal{S}$, the first principal component is given by $y_1 = (\mathcal{X} - 1_n \overline{x}^\top) g_1$. More generally if $\mathcal{S} = \mathcal{G} \mathcal{L} \mathcal{G}^\top$ is the spectral decomposition of $\mathcal{S}$, then the PCs are obtained by

$$\mathcal{Y} = (\mathcal{X} - 1_n \overline{x}^\top) \mathcal{G}. \tag{9.10}$$

Note that with the centering matrix $\mathcal{H} = \mathcal{I} - (n^{-1} 1_n 1_n^\top)$ and $\mathcal{H} 1_n \overline{x}^\top = 0$ we can write

$$\begin{aligned} \mathcal{S}_\mathcal{Y} &= n^{-1} \mathcal{Y}^\top \mathcal{H} \mathcal{Y} = n^{-1} \mathcal{G}^\top (\mathcal{X} - 1_n \overline{x}^\top)^\top \mathcal{H} (\mathcal{X} - 1_n \overline{x}^\top) \mathcal{G} \\ &= n^{-1} \mathcal{G}^\top \mathcal{X}^\top \mathcal{H} \mathcal{X} \mathcal{G} = \mathcal{G}^\top \mathcal{S} \mathcal{G} = \mathcal{L} \end{aligned} \tag{9.11}$$

where $\mathcal{L} = \operatorname{diag}(\ell_1, \ldots, \ell_p)$ is the matrix of eigenvalues of $\mathcal{S}$. Hence the variance of y_i equals the eigenvalue ℓ_i!

The PC technique is sensitive to scale changes. If we multiply one variable by a scalar we obtain different eigenvalues and eigenvectors. This is due to the fact that an eigenvalue decomposition is performed on of the covariance matrix and not on the correlation matrix (see Section 9.5). The following warning is therefore important:

⚠ The PC transformation should be applied to data that have approximately the same scale in each variable.

EXAMPLE 9.2 *Let us apply this technique to the bank data set. In this example we do not standardize the data. Figure 9.3 shows some PC plots of the bank data set. The genuine and counterfeit bank notes are marked by "o" and "+" respectively.*

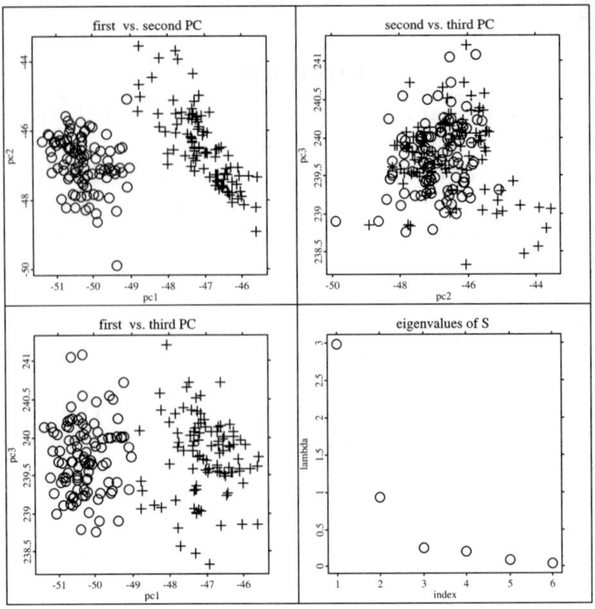

Figure 9.3. Principal components of the bank data. MVApcabank

Recall that the mean vector of $\mathcal{X}$ *is*

$$\overline{x} = (214.9, 130.1, 129.9, 9.4, 10.6, 140.5)^{\top}.$$

The vector of eigenvalues of $\mathcal{S}$ *is*

$$\ell = (2.985, 0.931, 0.242, 0.194, 0.085, 0.035)^{\top}.$$

The eigenvectors g_j *are given by the columns of the matrix*

$$\mathcal{G} = \begin{pmatrix} -0.044 & 0.011 & 0.326 & 0.562 & -0.753 & 0.098 \\ 0.112 & 0.071 & 0.259 & 0.455 & 0.347 & -0.767 \\ 0.139 & 0.066 & 0.345 & 0.415 & 0.535 & 0.632 \\ 0.768 & -0.563 & 0.218 & -0.186 & -0.100 & -0.022 \\ 0.202 & 0.659 & 0.557 & -0.451 & -0.102 & -0.035 \\ -0.579 & -0.489 & 0.592 & -0.258 & 0.085 & -0.046 \end{pmatrix}.$$

The first column of $\mathcal{G}$ *is the first eigenvector and gives the weights used in the linear combination of the original data in the first PC.*

EXAMPLE 9.3 *To see how sensitive the PCs are to a change in the scale of the variables, assume that* X_1, X_2, X_3 *and* X_6 *are measured in cm and that* X_4 *and* X_5 *remain in mm in the bank data set. This leads to:*

$$\bar{x} = (21.49,\ 13.01,\ 12.99,\ 9.41,\ 10.65,\ 14.05)^{\top}.$$

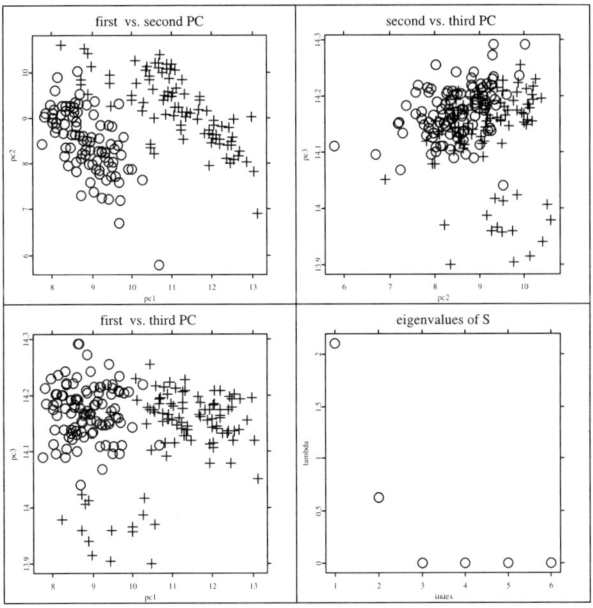

Figure 9.4. Principal components of the rescaled bank data. MVApcabankr

The covariance matrix can be obtained from S in (3.4) by dividing rows 1, 2, 3, 6 and columns 1, 2, 3, 6 by 10. We obtain:

$$\ell = (2.101,\ 0.623,\ 0.005,\ 0.002,\ 0.001,\ 0.0004)^{\top}$$

which clearly differs from Example 9.2. Only the first two eigenvectors are given:

$$g_1 = (-0.005,\ 0.011,\ 0.014,\ 0.992,\ 0.113,\ -0.052)^{\top}$$

$$g_2 = (-0.001,\ 0.013,\ 0.016,\ -0.117,\ 0.991,\ -0.069)^{\top}.$$

Comparing these results to the first two columns of $\mathcal{G}$ from Example 9.2, a completely different story is revealed. Here the first component is dominated by X_4 (lower margin) and the second by X_5 (upper margin), while all of the other variables have much less weight. The results are shown in Figure 9.4. Section 9.5 will show how to select a reasonable standardization of the variables when the scales are too different.

Summary

↪ The scale of the variables should be roughly the same for PC transformations.

Summary (continued)

↪ For the practical implementation of principal components analysis (PCA) we replace μ by the mean $\overline{x}$ and Σ by the empirical covariance $\mathcal{S}$. Then we compute the eigenvalues $\ell_1, \ldots, \ell_p$ and the eigenvectors $g_1, \ldots, g_p$ of $\mathcal{S}$. The graphical representation of the PCs is obtained by plotting the first PC vs. the second (and eventually vs. the third).

↪ The components of the eigenvectors g_i are the weights of the original variables in the PCs.

9.3 Interpretation of the PCs

Recall that the main idea of PC transformations is to find the most informative projections that maximize variances. The most informative SLC is given by the first eigenvector. In Section 9.2 the eigenvectors were calculated for the bank data. In particular, with centered x's, we had:

$$\begin{array}{rcl} y_1 &=& -0.044x_1 + 0.112x_2 + 0.139x_3 + 0.768x_4 + 0.202x_5 - 0.579x_6 \\ y_2 &=& 0.011x_1 + 0.071x_2 + 0.066x_3 - 0.563x_4 + 0.659x_5 - 0.489x_6 \end{array}$$

and

$$\begin{array}{rcl} x_1 &=& \text{length} \\ x_2 &=& \text{left height} \\ x_3 &=& \text{right height} \\ x_4 &=& \text{bottom frame} \\ x_5 &=& \text{top frame} \\ x_6 &=& \text{diagonal.} \end{array}$$

Hence, the first PC is essentially the difference between the bottom frame variable and the diagonal. The second PC is best described by the difference between the top frame variable and the sum of bottom frame and diagonal variables.

The weighting of the PCs tells us in which directions, expressed in original coordinates, the best variance explanation is obtained. A measure of how well the first q PCs explain variation is given by the relative proportion:

$$\psi_q = \frac{\sum_{j=1}^{q} \lambda_j}{\sum_{j=1}^{p} \lambda_j} = \frac{\sum_{j=1}^{q} Var(Y_j)}{\sum_{j=1}^{p} Var(Y_j)}. \tag{9.12}$$

Referring to the bank data example 9.2, the (cumulative) proportions of explained variance are given in Table 9.1. The first PC ($q = 1$) already explains 67% of the variation. The first three ($q = 3$) PCs explain 93% of the variation. Once again it should be noted that PCs are not scale invariant, e.g., the PCs derived from the correlation matrix give different results than the PCs derived from the covariance matrix (see Section 9.5).

A good graphical representation of the ability of the PCs to explain the variation in the data is given by the scree plot shown in the lower righthand window of Figure 9.3. The scree plot

eigenvalue	proportion of variance	cumulated proportion
2.985	0.67	0.67
0.931	0.21	0.88
0.242	0.05	0.93
0.194	0.04	0.97
0.085	0.02	0.99
0.035	0.01	1.00

Table 9.1. Proportion of variance of PC's

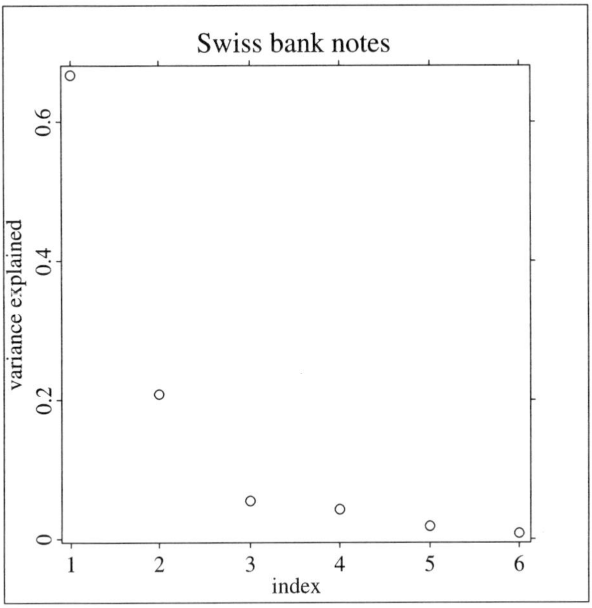

Figure 9.5. Relative proportion of variance explained by PCs. MVApcabanki

can be modified by using the relative proportions on the y-axis, as is shown in Figure 9.5 for the bank data set.

The covariance between the PC vector Y and the original vector X is calculated with the help of (9.4) as follows:

$$\begin{aligned} Cov(X,Y) &= E(XY^\top) - EXEY^\top = E(XY^\top) \\ &= E(XX^\top\Gamma) - \mu\mu^\top\Gamma = Var(X)\Gamma \\ &= \Sigma\Gamma \\ &= \Gamma\Lambda\Gamma^\top\Gamma \\ &= \Gamma\Lambda. \end{aligned} \tag{9.13}$$

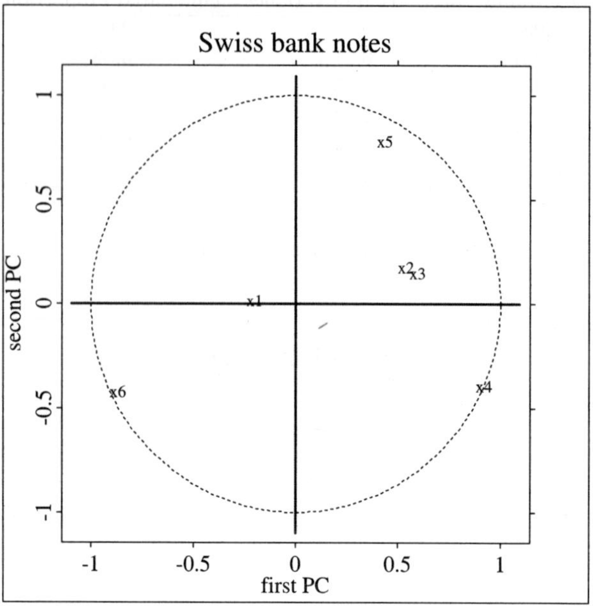

Figure 9.6. The correlation of the original variable with the PCs. MVApcabanki

Hence, the correlation, $\rho_{X_iY_j}$, between variable X_i and the PC Y_j is

$$\rho_{X_iY_j} = \frac{\gamma_{ij}\lambda_j}{(\sigma_{X_iX_i}\lambda_j)^{1/2}} = \gamma_{ij}\left(\frac{\lambda_j}{\sigma_{X_iX_i}}\right)^{1/2}. \tag{9.14}$$

Using actual data, this of course translates into

$$r_{X_iY_j} = g_{ij}\left(\frac{\ell_j}{s_{X_iX_i}}\right)^{1/2}. \tag{9.15}$$

The correlations can be used to evaluate the relations between the PCs Y_j where $j = 1, \dots, q$, and the original variables X_i where $i = 1, \dots, p$. Note that

$$\sum_{j=1}^{p} r^2_{X_iY_j} = \frac{\sum_{j=1}^{p} \ell_j g^2_{ij}}{s_{X_iX_i}} = \frac{s_{X_iX_i}}{s_{X_iX_i}} = 1. \tag{9.16}$$

Indeed, $\sum_{j=1}^{p} \ell_j g^2_{ij} = g_i^\top \mathcal{L} g_i$ is the (i,i)-element of the matrix $\mathcal{GLG}^\top = \mathcal{S}$, so that $r^2_{X_iY_j}$ may be seen as the proportion of variance of X_i explained by Y_j.

In the space of the first two PCs we plot these proportions, i.e., $r_{X_iY_1}$ versus $r_{X_iY_2}$. Figure 9.6 shows this for the bank notes example. This plot shows which of the original variables are most strongly correlated with PC Y_1 and Y_2.

From (9.16) it obviously follows that $r^2_{X_iY_1} + r^2_{X_iY_2} \le 1$ so that the points are always inside the circle of radius 1. In the bank notes example, the variables X_4, X_5 and X_6 correspond to correlations near the periphery of the circle and are thus well explained by the first two PCs.

	$r_{X_iY_1}$	$r_{X_iY_2}$	$r^2_{X_iY_1} + r^2_{X_iY_2}$
X_1 length	−0.201	0.028	0.041
X_2 left h.	0.538	0.191	0.326
X_3 right h.	0.597	0.159	0.381
X_4 lower	0.921	−0.377	0.991
X_5 upper	0.435	0.794	0.820
X_6 diagonal	−0.870	−0.410	0.926

Table 9.2. Correlation between the original variables and the PCs

Recall that we have interpreted the first PC as being essentially the difference between X_4 and X_6. This is also reflected in Figure 9.6 since the points corresponding to these variables lie on different sides of the vertical axis. An analogous remark applies to the second PC. We had seen that the second PC is well described by the difference between X_5 and the sum of X_4 and X_6. Now we are able to see this result again from Figure 9.6 since the point corresponding to X_5 lies above the horizontal axis and the points corresponding to X_4 and X_6 lie below.

The correlations of the original variables X_i and the first two PCs are given in Table 9.2 along with the cumulated percentage of variance of each variable explained by Y_1 and Y_2. This table confirms the above results. In particular, it confirms that the percentage of variance of X_1 (and X_2, X_3) explained by the first two PCs is relatively small and so are their weights in the graphical representation of the individual bank notes in the space of the first two PCs (as can be seen in the upper left plot in Figure 9.3). Looking simultaneously at Figure 9.6 and the upper left plot of Figure 9.3 shows that the genuine bank notes are roughly characterized by large values of X_6 and smaller values of X_4. The counterfeit bank notes show larger values of X_5 (see Example 7.15).

Summary

↪ The weighting of the PCs tells us in which directions, expressed in original coordinates, the best explanation of the variance is obtained. Note that the PCs are not scale invariant.

↪ A measure of how well the first q PCs explain variation is given by the relative proportion $\psi_q = \sum_{j=1}^{q} \lambda_j / \sum_{j=1}^{p} \lambda_j$. A good graphical representation of the ability of the PCs to explain the variation in the data is the scree plot of these proportions.

↪ The correlation between PC Y_j and an original variable X_i is $\rho_{X_iY_j} = \gamma_{ij} \left(\frac{\lambda_j}{\sigma_{X_iX_i}} \right)^{1/2}$. For a data matrix this translates into $r^2_{X_iY_j} = \frac{\ell_j g^2_{ij}}{s_{X_iX_i}}$. $r^2_{X_iY_j}$ can be interpreted as the proportion of variance of X_i explained by Y_j. A plot of $r_{X_iY_1}$ vs. $r_{X_iY_2}$ shows which of the original variables are most strongly correlated with the PCs, namely those that are close to the periphery of the circle of radius 1.

9.4 Asymptotic Properties of the PCs

In practice, PCs are computed from sample data. The following theorem yields results on the asymptotic distribution of the sample PCs.

THEOREM 9.4 *Let $\Sigma > 0$ with distinct eigenvalues, and let $\mathcal{U} \sim m^{-1}W_p(\Sigma, m)$ with spectral decompositions $\Sigma = \Gamma\Lambda\Gamma^\top$, and $\mathcal{U} = \mathcal{G}\mathcal{L}\mathcal{G}^\top$. Then*

(a) $\sqrt{m}(\ell - \lambda) \xrightarrow{\mathcal{L}} N_p(0, 2\Lambda^2)$,
where $\ell = (\ell_1, \dots, \ell_p)^\top$ and $\lambda = (\lambda_1, \dots, \lambda_p)^\top$ are the diagonals of $\mathcal{L}$ and Λ,

(b) $\sqrt{m}(g_j - \gamma_j) \xrightarrow{\mathcal{L}} N_p(0, \mathcal{V}_j)$,
with $\mathcal{V}_j = \lambda_j \sum_{k \neq j} \frac{\lambda_k}{(\lambda_k - \lambda_j)^2} \gamma_k \gamma_k^\top$,

(c) $Cov(g_j, g_k) = \mathcal{V}_{jk}$,
where the (r, s)-element of the matrix $\mathcal{V}_{jk}(p \times p)$ is $-\frac{\lambda_j \lambda_k \gamma_{rk} \gamma_{sj}}{m(\lambda_j - \lambda_k)^2}$,

(d) the elements in ℓ are asymptotically independent of the elements in $\mathcal{G}$.

EXAMPLE 9.4 *Since $n\mathcal{S} \sim W_p(\Sigma, n-1)$ if $X_1, \dots, X_n$ are drawn from $N(\mu, \Sigma)$, we have that*

$$\sqrt{n-1}(\ell_j - \lambda_j) \xrightarrow{\mathcal{L}} N(0, 2\lambda_j^2), \quad j = 1, \dots, p. \tag{9.17}$$

Since the variance of (9.17) depends on the true mean λ_j a log transformation is useful. Consider $f(\ell_j) = \log(\ell_j)$. Then $\frac{d}{d\ell_j} f|_{\ell_j = \lambda_j} = \frac{1}{\lambda_j}$ and by the Transformation Theorem 4.11 we have from (9.17) that

$$\sqrt{n-1}(\log \ell_j - \log \lambda_j) \longrightarrow N(0, 2). \tag{9.18}$$

Hence,

$$\sqrt{\frac{n-1}{2}} (\log \ell_j - \log \lambda_j) \xrightarrow{\mathcal{L}} N(0, 1)$$

and a two-sided confidence interval at the $1 - \alpha = 0.95$ significance level is given by

$$\log(\ell_j) - 1.96\sqrt{\frac{2}{n-1}} \leq \log \lambda_j \leq \log(\ell_j) + 1.96\sqrt{\frac{2}{n-1}}.$$

In the bank data example we have that

$$\ell_1 = 2.98.$$

Therefore,

$$\log(2.98) \pm 1.96\sqrt{\frac{2}{199}} = \log(2.98) \pm 0.1965.$$

It can be concluded for the true eigenvalue that

$$P\{\lambda_1 \in (2.448, 3.62)\} \approx 0.95.$$

Variance explained by the first q PCs.

The variance explained by the first q PCs is given by

$$\psi = \frac{\lambda_1 + \cdots + \lambda_q}{\sum_{j=1}^{p} \lambda_j}.$$

In practice this is estimated by

$$\widehat{\psi} = \frac{\ell_1 + \cdots + \ell_q}{\sum_{j=1}^{p} \ell_j}.$$

From Theorem 9.4 we know the distribution of $\sqrt{n-1}(\ell-\lambda)$. Since ψ is a nonlinear function of λ, we can again apply the Transformation Theorem 4.11 to obtain that

$$\sqrt{n-1}(\widehat{\psi} - \psi) \xrightarrow{\mathcal{L}} N(0, \mathcal{D}^\top \mathcal{V} \mathcal{D})$$

where $\mathcal{V} = 2\Lambda^2$ (from Theorem 9.4) and $\mathcal{D} = (d_1, \ldots, d_p)^\top$ with

$$d_j = \frac{\partial \psi}{\partial \lambda_j} = \begin{cases} \dfrac{1-\psi}{\operatorname{tr}(\Sigma)} & \text{for } 1 \le j \le q, \\ \dfrac{-\psi}{\operatorname{tr}(\Sigma)} & \text{for } q+1 \le j \le p. \end{cases}$$

Given this result, the following theorem can be derived.

THEOREM 9.5

$$\sqrt{n-1}(\widehat{\psi} - \psi) \xrightarrow{\mathcal{L}} N(0, \omega^2),$$

where

$$\begin{aligned} \omega^2 &= \mathcal{D}^\top \mathcal{V} \mathcal{D} = \frac{2}{\{\operatorname{tr}(\Sigma)\}^2} \left\{(1-\psi)^2(\lambda_1^2 + \cdots + \lambda_q^2) + \psi^2(\lambda_{q+1}^2 + \cdots + \lambda_p^2)\right\} \\ &= \frac{2\operatorname{tr}(\Sigma^2)}{\{\operatorname{tr}(\Sigma)\}^2}(\psi^2 - 2\beta\psi + \beta) \end{aligned}$$

and

$$\beta = \frac{\lambda_1^2 + \cdots + \lambda_q^2}{\lambda_1^2 + \cdots + \lambda_p^2}.$$

EXAMPLE 9.5 *From Section 9.3 it is known that the first PC for the Swiss bank notes resolves 67% of the variation. It can be tested whether the true proportion is actually 75%. Computing*

$$\begin{aligned} \widehat{\beta} &= \frac{\ell_1^2}{\ell_1^2 + \cdots + \ell_p^2} = \frac{(2.985)^2}{(2.985)^2 + (0.931)^2 + \cdots (0.035)^2} = 0.902 \\ \operatorname{tr}(\mathcal{S}) &= 4.472 \\ \operatorname{tr}(\mathcal{S}^2) &= \sum_{j=1}^{p} \ell_j^2 = 9.883 \\ \widehat{\omega}^2 &= \frac{2\operatorname{tr}(\mathcal{S}^2)}{\{\operatorname{tr}(\mathcal{S})\}^2}(\widehat{\psi}^2 - 2\widehat{\beta}\widehat{\psi} + \widehat{\beta}) \end{aligned}$$

$$= \frac{2 \cdot 9.883}{(4.472)^2}\{(0.668)^2 - 2(0.902)(0.668) + 0.902\} = 0.142.$$

Hence, a confidence interval at a significance of level $1 - \alpha = 0.95$ *is given by*

$$0.668 \pm 1.96\sqrt{\frac{0.142}{199}} = (0.615, 0.720).$$

Clearly the hypothesis that $\psi = 75\%$ *can be rejected!*

Summary

↪ The eigenvalues ℓ_j and eigenvectors g_j are asymptotically, normally distributed, in particular $\sqrt{n-1}(\ell - \lambda) \xrightarrow{\mathcal{L}} N_p(0, 2\Lambda^2)$.

↪ For the eigenvalues it holds that $\sqrt{\frac{n-1}{2}}\,(\log \ell_j - \log \lambda_j) \xrightarrow{\mathcal{L}} N(0,1)$.

↪ Given an asymptotic, normal distribution approximate confidence intervals and tests can be constructed for the proportion of variance which is explained by the first q PCs. The two-sided confidence interval at the $1-\alpha = 0.95$ level is given by $\log(\ell_j) - 1.96\sqrt{\frac{2}{n-1}} \le \log \lambda_j \le \log(\ell_j) + 1.96\sqrt{\frac{2}{n-1}}$.

↪ It holds for $\widehat{\psi}$, the estimate of ψ (the proportion of the variance explained by the first q PCs) that $\sqrt{n-1}(\widehat{\psi} - \psi) \xrightarrow{\mathcal{L}} N(0, \omega^2)$, where ω is given in Theorem 9.5.

9.5 Normalized Principal Components Analysis

In certain situations the original variables can be heterogeneous w.r.t. their variances. This is particularly true when the variables are measured on heterogeneous scales (such as years, kilograms, dollars, ...). In this case a description of the information contained in the data needs to be provided which is robust w.r.t. the choice of scale. This can be achieved through a standardization of the variables, namely

$$\mathcal{X}_S = \mathcal{H}\mathcal{X}\mathcal{D}^{-1/2} \tag{9.19}$$

where $\mathcal{D} = \text{diag}(s_{X_1X_1}, \ldots, s_{X_pX_p})$. Note that $\overline{x}_S = 0$ and $\mathcal{S}_{\mathcal{X}_S} = \mathcal{R}$, the correlation matrix of $\mathcal{X}$. The PC transformations of the matrix $\mathcal{X}_S$ are refereed to as the *Normalized Principal Components* (NPCs). The spectral decomposition of $\mathcal{R}$ is

$$\mathcal{R} = \mathcal{G}_{\mathcal{R}}\mathcal{L}_{\mathcal{R}}\mathcal{G}_{\mathcal{R}}^{\top}, \tag{9.20}$$

where $\mathcal{L}_{\mathcal{R}} = \text{diag}(\ell_1^{\mathcal{R}}, \ldots, \ell_p^{\mathcal{R}})$ and $\ell_1^{\mathcal{R}} \ge \ldots \ge \ell_p^{\mathcal{R}}$ are the eigenvalues of $\mathcal{R}$ with corresponding eigenvectors $g_1^{\mathcal{R}}, \ldots, g_p^{\mathcal{R}}$ (note that here $\sum_{j=1}^{p} \ell_j^{\mathcal{R}} = \text{tr}(\mathcal{R}) = p$).

The NPCs, Z_j, provide a representation of each individual, and is given by

$$\mathcal{Z} = \mathcal{X}_S\mathcal{G}_{\mathcal{R}} = (z_1, \ldots, z_p). \tag{9.21}$$

After transforming the variables, once again, we have that

$$\overline{z} = 0, \tag{9.22}$$

$$\mathcal{S}_{\mathcal{Z}} = \mathcal{G}_{\mathcal{R}}^{\top}\mathcal{S}_{\mathcal{X}_S}\mathcal{G}_{\mathcal{R}} = \mathcal{G}_{\mathcal{R}}^{\top}\mathcal{R}\mathcal{G}_{\mathcal{R}} = \mathcal{L}_{\mathcal{R}}. \tag{9.23}$$

⚠ The NPCs provide a perspective similar to that of the PCs, but in terms of the relative position of individuals, NPC gives each variable the same weight (with the PCs the variable with the largest variance received the largest weight).

Computing the covariance and correlation between X_i and Z_j is straightforward:

$$\mathcal{S}_{X_S,Z} = \frac{1}{n}\mathcal{X}_S^{\top}\mathcal{Z} = \mathcal{G}_{\mathcal{R}}\mathcal{L}_{\mathcal{R}}, \tag{9.24}$$

$$\mathcal{R}_{X_S,Z} = \mathcal{G}_{\mathcal{R}}\mathcal{L}_{\mathcal{R}}\mathcal{L}_{\mathcal{R}}^{-1/2} = \mathcal{G}_{\mathcal{R}}\mathcal{L}_{\mathcal{R}}^{1/2}. \tag{9.25}$$

The correlations between the original variables X_i and the NPCs Z_j are:

$$r_{X_iZ_j} = \sqrt{\ell_j}\,g_{R,ij} \tag{9.26}$$

$$\sum_{j=1}^{p} r_{X_iZ_j}^2 = 1 \tag{9.27}$$

(compare this to (9.15) and (9.16)). The resulting NPCs, the Z_j, can be interpreted in terms of the original variables and the role of each PC in explaining the variation in variable X_i can be evaluated.

9.6 Principal Components as a Factorial Method

The empirical PCs (normalized or not) turn out to be equivalent to the factors that one would obtain by decomposing the appropriate data matrix into its factors (see Chapter 8). It will be shown that the PCs are the factors representing the rows of the centered data matrix and that the NPCs correspond to the factors of the standardized data matrix. The representation of the columns of the standardized data matrix provides (at a scale factor) the correlations between the NPCs and the original variables. The derivation of the (N)PCs presented above will have a nice geometric justification here since they are the best fit in subspaces generated by the columns of the (transformed) data matrix $\mathcal{X}$. This analogy provides complementary interpretations of the graphical representations shown above.

Assume, as in Chapter 8, that we want to obtain representations of the individuals (the rows of $\mathcal{X}$) and of the variables (the columns of $\mathcal{X}$) in spaces of smaller dimension. To keep the representations simple, some prior transformations are performed. Since the origin has no particular statistical meaning in the space of individuals, we will first shift the origin to the center of gravity, $\overline{x}$, of the point cloud. This is the same as analyzing the centered data matrix $\mathcal{X}_C = \mathcal{H}\mathcal{X}$. Now all of the variables have zero means, thus the technique used in Chapter 8 can be applied to the matrix $\mathcal{X}_C$. Note that the spectral decomposition of $\mathcal{X}_C^{\top}\mathcal{X}_C$ is related to that of $\mathcal{S}_X$, namely

$$\mathcal{X}_C^{\top}\mathcal{X}_C = \mathcal{X}^{\top}\mathcal{H}^{\top}\mathcal{H}\mathcal{X} = n\mathcal{S}_X = n\mathcal{G}\mathcal{L}\mathcal{G}^{\top}. \tag{9.28}$$

The factorial variables are obtained by projecting $\mathcal{X}_C$ on $\mathcal{G}$,

$$\mathcal{Y} = \mathcal{X}_C\mathcal{G} = (y_1, \ldots, y_p). \tag{9.29}$$

These are the same principal components obtained above, see formula (9.10). (Note that the y's here correspond to the z's in Section 8.2.) Since $\mathcal{H}\mathcal{X}_C = \mathcal{X}_C$, it immediately follows that

$$\overline{y} = 0, \tag{9.30}$$
$$\mathcal{S}_Y = \mathcal{G}^\top \mathcal{S}_X \mathcal{G} = \mathcal{L} = \operatorname{diag}(\ell_1, \dots, \ell_p). \tag{9.31}$$

The scatterplot of the individuals on the factorial axes are thus centered around the origin and are more spread out in the first direction (first PC has variance ℓ_1) than in the second direction (second PC has variance ℓ_2).

The representation of the variables can be obtained using the Duality Relations (8.11), and (8.12). The projections of the columns of $\mathcal{X}_C$ onto the eigenvectors v_k of $\mathcal{X}_C\mathcal{X}_C^\top$ are

$$\mathcal{X}_C^\top v_k = \frac{1}{\sqrt{n\ell_k}} \mathcal{X}_C^\top \mathcal{X}_C g_k = \sqrt{n\ell_k} g_k. \tag{9.32}$$

Thus the projections of the variables on the first p axes are the columns of the matrix

$$\mathcal{X}_C^\top \mathcal{V} = \sqrt{n}\mathcal{G}\mathcal{L}^{1/2}. \tag{9.33}$$

Considering the geometric representation, there is a nice statistical interpretation of the angle between two columns of $\mathcal{X}_C$. Given that

$$x_{C[j]}^\top x_{C[k]} = n s_{X_j X_k}, \tag{9.34}$$
$$||x_{C[j]}||^2 = n s_{X_j X_j}, \tag{9.35}$$

where $x_{C[j]}$ and $x_{C[k]}$ denote the j-th and k-th column of $\mathcal{X}_C$, it holds that in the full space of the variables, if θ_{jk} is the angle between two variables, $x_{C[j]}$ and $x_{C[k]}$, then

$$\cos\theta_{jk} = \frac{x_{C[j]}^\top x_{C[k]}}{||x_{C[j]}|| \; ||x_{C[k]}||} = r_{X_j X_k} \tag{9.36}$$

(Example 2.11 shows the general connection that exists between the angle and correlation of two variables). As a result, the relative positions of the variables in the scatterplot of the first columns of $\mathcal{X}_C^\top \mathcal{V}$ may be interpreted in terms of their correlations; the plot provides a picture of the correlation structure of the original data set. Clearly, one should take into account the percentage of variance explained by the chosen axes when evaluating the correlation.

The NPCs can also be viewed as a factorial method for reducing the dimension. The variables are again standardized so that each one has mean zero and unit variance and is independent of the scale of the variables. The factorial analysis of $\mathcal{X}_S$ provides the NPCs. The spectral decomposition of $\mathcal{X}_S^\top \mathcal{X}_S$ is related to that of $\mathcal{R}$, namely

$$\mathcal{X}_S^\top \mathcal{X}_S = \mathcal{D}^{-1/2}\mathcal{X}^\top \mathcal{H}\mathcal{X}\mathcal{D}^{-1/2} = n\mathcal{R} = n\mathcal{G}_\mathcal{R}\mathcal{L}_\mathcal{R}\mathcal{G}_\mathcal{R}^\top.$$

The NPCs Z_j, given by (9.21), may be viewed as the projections of the rows of $\mathcal{X}_S$ onto $\mathcal{G}_\mathcal{R}$.

The representation of the variables are again given by the columns of

$$\mathcal{X}_S^\top \mathcal{V}_\mathcal{R} = \sqrt{n}\mathcal{G}_\mathcal{R}\mathcal{L}_\mathcal{R}^{1/2}. \tag{9.37}$$

Comparing (9.37) and (9.25) we see that the projections of the variables in the factorial analysis provide the correlation between the NPCs $\mathcal{Z}_k$ and the original variables $x_{[j]}$ (up to the factor $\sqrt{n}$ which could be the scale of the axes).

This implies that a deeper interpretation of the representation of the individuals can be obtained by looking simultaneously at the graphs plotting the variables. Note that

$$x_{S[j]}^{\top} x_{S[k]} = n r_{X_j X_k}, \tag{9.38}$$

$$\|x_{S[j]}\|^2 = n, \tag{9.39}$$

where $x_{S[j]}$ and $x_{S[k]}$ denote the j-th and k-th column of $\mathcal{X}_S$. Hence, in the full space, all the standardized variables (columns of $\mathcal{X}_S$) are contained within the "sphere" in $\mathbb{R}^n$, which is centered at the origin and has radius $\sqrt{n}$ (the scale of the graph). As in (9.36), given the angle θ_{jk} between two columns $x_{S[j]}$ and $x_{S[k]}$, it holds that

$$\cos\theta_{jk} = r_{X_j X_k}. \tag{9.40}$$

Therefore, when looking at the representation of the variables in the spaces of reduced dimension (for instance the first two factors), we have a picture of the correlation structure between the original X_i's in terms of their angles. Of course, the quality of the representation in those subspaces has to be taken into account, which is presented in the next section.

Quality of the Representations

As said before, an overall measure of the quality of the representation is given by

$$\psi = \frac{\ell_1 + \ell_2 + \ldots + \ell_q}{\sum_{j=1}^{p} \ell_j}.$$

In practice, q is chosen to be equal to 1, 2 or 3. Suppose for instance that $\psi = 0.93$ for $q = 2$. This means that the graphical representation in two dimensions captures 93% of the total variance. In other words, there is minimal dispersion in a third direction (no more than 7%).

It can be useful to check if each individual is well represented by the PCs. Clearly, the proximity of two individuals on the projected space may not necessarily coincide with the proximity in the full original space $\mathbb{R}^p$, which may lead to erroneous interpretations of the graphs. In this respect, it is worth computing the angle ϑ_{ik} between the representation of an individual i and the k-th PC or NPC axis. This can be done using (2.40), i.e.,

$$\cos\vartheta_{ik} = \frac{y_i^{\top} e_k}{\|y_i\|\|e_k\|} = \frac{y_{ik}}{\|x_{Ci}\|}$$

for the PCs or analogously

$$\cos\zeta_{ik} = \frac{z_i^{\top} e_k}{\|z_i\|\|e_k\|} = \frac{z_{ik}}{\|x_{Si}\|}$$

for the NPCs, where e_k denotes the k-th unit vector $e_k = (0, \ldots, 1, \ldots, 0)^{\top}$. An individual i will be represented on the k-th PC axis if its corresponding angle is small, i.e., if $\cos^2\vartheta_{ik}$ for $k = 1, \ldots, p$ is close to one. Note that for each individual i,

$$\sum_{k=1}^{p} \cos^2\vartheta_{ik} = \frac{y_i^{\top} y_i}{x_{Ci}^{\top} x_{Ci}} = \frac{x_{Ci}^{\top} \mathcal{G}\mathcal{G}^{\top} x_{Ci}}{x_{Ci}^{\top} x_{Ci}} = 1$$

The values $\cos^2\vartheta_{ik}$ are sometimes called the relative contributions of the k-th axis to the representation of the i-th individual, e.g., if $\cos^2\vartheta_{i1} + \cos^2\vartheta_{i2}$ is large (near one), we know that the individual i is well represented on the plane of the first two principal axes since its corresponding angle with the plane is close to zero.

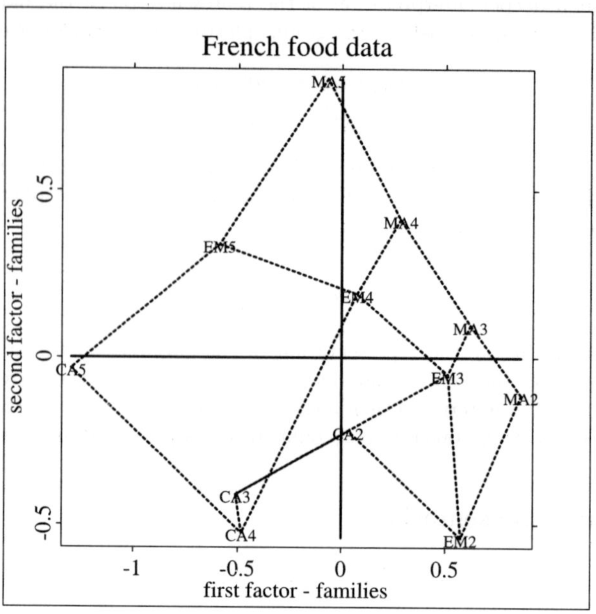

Figure 9.7. Representation of the individuals. MVAnpcafood

We already know that the quality of the representation of the variables can be evaluated by the percentage of X_i's variance that is explained by a PC, which is given by $r^2_{X_iY_j}$ or $r^2_{X_iZ_j}$ according to (9.16) and (9.27) respectively.

EXAMPLE 9.6 *Let us return to the French food expenditure example, see Appendix B.6. This yields a two-dimensional representation of the individuals as shown in Figure 9.7.*

Calculating the matrix $\mathcal{G}_\mathcal{R}$ we have

$$\mathcal{G}_\mathcal{R} = \begin{pmatrix} -0.240 & 0.622 & -0.011 & -0.544 & 0.036 & 0.508 \\ -0.466 & 0.098 & -0.062 & -0.023 & -0.809 & -0.301 \\ -0.446 & -0.205 & 0.145 & 0.548 & -0.067 & 0.625 \\ -0.462 & -0.141 & 0.207 & -0.053 & 0.411 & -0.093 \\ -0.438 & -0.197 & 0.356 & -0.324 & 0.224 & -0.350 \\ -0.281 & 0.523 & -0.444 & 0.450 & 0.341 & -0.332 \\ 0.206 & 0.479 & 0.780 & 0.306 & -0.069 & -0.138 \end{pmatrix},$$

which gives the weights of the variables (milk, vegetables, etc.). The eigenvalues ℓ_j and the proportions of explained variance are given in Table 9.3.

The interpretation of the principal components are best understood when looking at the correlations between the original X_i's and the PCs. Since the first two PCs explain 88.1% of the variance, we limit ourselves to the first two PCs. The results are shown in Table 9.4. The two-dimensional graphical representation of the variables in Figure 9.8 is based on the first two columns of Table 9.4.

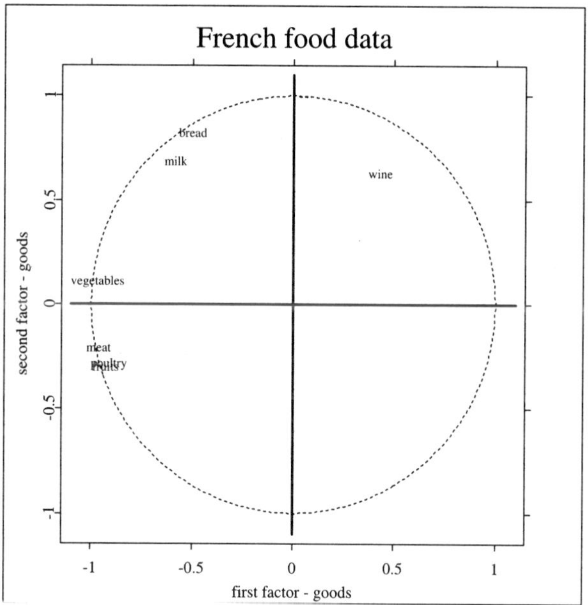

Figure 9.8. Representation of the variables. MVAnpcafood

eigenvalues	proportion of variance	cumulated proportion
4.333	0.6190	61.9
1.830	0.2620	88.1
0.631	0.0900	97.1
0.128	0.0180	98.9
0.058	0.0080	99.7
0.019	0.0030	99.9
0.001	0.0001	100.0

Table 9.3. Eigenvalues and explained variance

	$r_{X_iZ_1}$	$r_{X_iZ_2}$	$r^2_{X_iZ_1} + r^2_{X_iZ_2}$
X_1: bread	−0.499	0.842	0.957
X_2: vegetables	−0.970	0.133	0.958
X_3: fruits	−0.929	−0.278	0.941
X_4: meat	−0.962	−0.191	0.962
X_5: poultry	−0.911	−0.266	0.901
X_6: milk	−0.584	0.707	0.841
X_7: wine	0.428	0.648	0.604

Table 9.4. Correlations with PCs

The plots are the projections of the variables into $\mathbb{R}^2$. Since the quality of the representation is good for all the variables (except maybe X_7), their relative angles give a picture of their original correlation: wine is negatively correlated with the vegetables, fruits, meat and poultry groups ($\theta > 90^o$), whereas taken individually this latter grouping of variables are highly positively correlated with each other ($\theta \approx 0$). Bread and milk are positively correlated but poorly correlated with meat, fruits and poultry ($\theta \approx 90^o$).

Now the representation of the individuals in Figure 9.7 can be interpreted better. From Figure 9.8 and Table 9.4 we can see that the the first factor Z_1 is a vegetable–meat–poultry–fruit factor (with a negative sign), whereas the second factor is a milk–bread–wine factor (with a positive sign). Note that this corresponds to the most important weights in the first columns of $\mathcal{G}_\mathcal{R}$. In Figure 9.7 lines were drawn to connect families of the same size and families of the same professional types. A grid can clearly be seen (with a slight deformation by the manager families) that shows the families with higher expenditures (higher number of children) on the left.

Considering both figures together explains what types of expenditures are responsible for similarities in food expenditures. Bread, milk and wine expenditures are similar for manual workers and employees. Families of managers are characterized by higher expenditures on vegetables, fruits, meat and poultry. Very often when analyzing NPCs (and PCs), it is illuminating to use such a device to introduce qualitative aspects of individuals in order to enrich the interpretations of the graphs.

Summary

↪ NPCs are PCs applied to the standardized (normalized) data matrix $\mathcal{X}_S$.

↪ The graphical representation of NPCs provides a similar type of picture as that of PCs, the difference being in the relative position of individuals, i.e., each variable in NPCs has the same weight (in PCs, the variable with the largest variance has the largest weight).

↪ The quality of the representation is evaluated by $\psi = (\sum_{j=1}^{p} \ell_j)^{-1}(\ell_1 + \ell_2 + \ldots + \ell_q)$.

↪ The quality of the representation of a variable can be evaluated by the percentage of X_i's variance that is explained by a PC, i.e., $r^2_{X_i Y_j}$.

9.7 Common Principal Components

In many applications a statistical analysis is simultaneously done for groups of data. In this section a technique is presented that allows us to analyze group elements that have common PCs. From a statistical point of view, estimating PCs simultaneously in different groups will result in a joint dimension reducing transformation. This multi-group PCA, the so called common principle components analysis (CPCA), yields the joint eigenstructure across groups.

In addition to traditional PCA, the basic assumption of CPCA is that the space spanned by the eigenvectors is identical *across* several groups, whereas variances associated with the components are allowed to vary.

More formally, the hypothesis of common principle components can be stated in the following way (Flury, 1988):

$$H_{CPC}: \Sigma_i = \Gamma\Lambda_i\Gamma^\top, \qquad i = 1, ..., k$$

where Σ_i is a positive definite $p \times p$ population covariance matrix for every i, $\Gamma = (\gamma_1, ..., \gamma_p)$ is an orthogonal $p \times p$ transformation matrix and $\Lambda_i = \operatorname{diag}(\lambda_{i1}, ..., \lambda_{ip})$ is the matrix of eigenvalues. Moreover, assume that all λ_i are distinct.

Let $\mathcal{S}$ be the (unbiased) sample covariance matrix of an underlying p-variate normal distribution $N_p(\mu, \Sigma)$ with sample size n. Then the distribution of $n\mathcal{S}$ has $n-1$ degrees of freedom and is known as the Wishart distribution (Muirhead, 1982, p. 86):

$$n\mathcal{S} \sim \mathcal{W}_p(\Sigma, n-1).$$

The density is given in (5.16). Hence, for a given Wishart matrix $\mathcal{S}_i$ with sample size n_i, the likelihood function can be written as

$$L(\Sigma_1, ..., \Sigma_k) = C\prod_{i=1}^{k}\exp\left\{\operatorname{tr}\left(-\frac{1}{2}(n_i-1)\Sigma_i^{-1}\mathcal{S}_i\right)\right\}|\Sigma_i|^{-\frac{1}{2}(n_i-1)} \tag{9.41}$$

where C is a constant independent of the parameters Σ_i. Maximizing the likelihood is equivalent to minimizing the function

$$g(\Sigma_1, ..., \Sigma_k) = \sum_{i=1}^{k}(n_i-1)\left\{\ln|\Sigma_i| + \operatorname{tr}(\Sigma_i^{-1}\mathcal{S}_i)\right\}.$$

Assuming that H_{CPC} holds, i.e., in replacing Σ_i by $\Gamma\Lambda_i\Gamma^\top$, after some manipulations one obtains

$$g(\Gamma, \Lambda_1, ..., \Lambda_k) = \sum_{i=1}^{k}(n_i-1)\sum_{j=1}^{p}\left(\ln\lambda_{ij} + \frac{\gamma_j^\top\mathcal{S}_i\gamma_j}{\lambda_{ij}}\right).$$

As we know from Section 2.2, the vectors γ_j in Γ have to be orthogonal. Orthogonality of the vectors γ_j is achieved using the Lagrange method, i.e., we impose the p constraints $\gamma_j^\top\gamma_j = 1$ using the Lagrange multipliers μ_j, and the remaining $p(p-1)/2$ constraints $\gamma_h^\top\gamma_j = 0$ for $h \neq j$ using the multiplier $2\mu_{hj}$ (Flury, 1988). This yields

$$g^*(\Gamma, \Lambda_1, ..., \Lambda_k) = g(\cdot) - \sum_{j=1}^{p}\mu_j(\gamma_j^\top\gamma_j - 1) - 2\sum_{h=1}^{p}\sum_{j=h+1}^{p}\mu_{hj}\gamma_h^\top\gamma_j.$$

Taking partial derivatives with respect to all λ_{im} and γ_m, it can be shown that the solution of the CPC model is given by the generalized system of characteristic equations

$$\gamma_m^\top\left(\sum_{i=1}^{k}(n_i-1)\frac{\lambda_{im}-\lambda_{ij}}{\lambda_{im}\lambda_{ij}}\mathcal{S}_i\right)\gamma_j = 0, \qquad m, j = 1, ..., p, \quad m \neq j. \tag{9.42}$$

This system can be solved using

$$\lambda_{im} = \gamma_m^\top\mathcal{S}\gamma_m, \qquad i = 1, ..., k, \quad m = 1, ..., p$$

under the constraints

$$\gamma_m^\top\gamma_j = \begin{cases} 0 & m \neq j \\ 1 & m = j \end{cases}.$$

Flury (1988) proves existence and uniqueness of the maximum of the likelihood function, and Flury and Gautschi (1986) provide a numerical algorithm.

EXAMPLE 9.7 *As an example we provide the data sets XFGvolsurf01, XFGvolsurf02 and XFGvolsurf03 that have been used in Fengler, Härdle and Villa (2003) to estimate common principle components for the implied volatility surfaces of the DAX 1999. The data has been generated by smoothing an implied volatility surface day by day. Next, the estimated grid points have been grouped into maturities of* $\tau = 1$, $\tau = 2$ *and* $\tau = 3$ *months and transformed into a vector of time series of the "smile", i.e., each element of the vector belongs to a distinct moneyness ranging from 0.85 to 1.10.*

Figure 9.9 shows the first three eigenvectors in a parallel coordinate plot. The basic structure of the first three eigenvectors is not altered. We find a shift, a slope and a twist structure. This structure is common to all maturity groups, i.e., when exploiting PCA as a dimension reducing tool, the same transformation applies to each group! However, by comparing the size of eigenvalues among groups we find that variability is decreasing across groups as we move from the short term contracts to long term contracts.

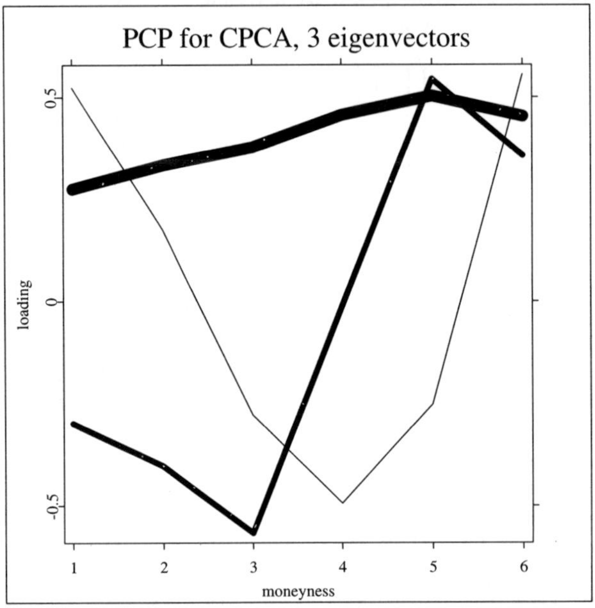

Figure 9.9. Factor loadings of the first (thick), the second (medium), and the third (thin) PC `MVAcpcaiv`

Before drawing conclusions we should convince ourselves that the CPC model is truly a good description of the data. This can be done by using a likelihood ratio test. The likelihood ratio statistic for comparing a restricted (the CPC) model against the unrestricted model (the model where all covariances are treated separately) is given by

$$T_{(n_1,n_2,\ldots,n_k)} = -2\ln\frac{L(\widehat{\Sigma}_1,\ldots,\widehat{\Sigma}_k)}{L(\mathcal{S}_1,\ldots,\mathcal{S}_k)}.$$

Inserting the likelihood function, we find that this is equivalent to

$$T_{(n_1,n_2,\ldots,n_k)} = \sum_{i=1}^{k}(n_i-1)\frac{det\ (\widehat{\Sigma}_i)}{det\ (\mathcal{S}_i)},$$

which has a χ^2 *distribution as* $\min(n_i)$ *tends to infinity with*

$$k\Big\{\frac{1}{2}p(p-1)+1\Big\}-\Big\{\frac{1}{2}p(p-1)+kp\Big\}=\frac{1}{2}(k-1)p(p-1)$$

degrees of freedom. This test is included in the quantlet `MVAcpcaiv`.

The calculations yield $T_{(n_1,n_2,\ldots,n_k)} = 31.836$, *which corresponds to the p-value* $p = 0.37512$ *for the* $\chi^2(30)$ *distribution. Hence we cannot reject the CPC model against the unrestricted model, where PCA is applied to each maturity separately.*

Using the methods in Section 9.3, we can estimate the amount of variability, ζ_l, *explained by the first* l *principle components: (only a few factors, three at the most, are needed to capture a large amount of the total variability present in the data). Since the model now captures the variability in both the strike and maturity dimensions, this is a suitable starting point for a simplified VaR calculation for delta-gamma neutral option portfolios using Monte Carlo methods, and is hence a valuable insight in risk management.*

9.8 Boston Housing

A set of transformations were defined in Chapter 1 for the Boston Housing data set that resulted in "regular" marginal distributions. The usefulness of principal component analysis with respect to such high-dimensional data sets will now be shown. The variable X_4 is dropped because it is a discrete 0–1 variable. It will be used later, however, in the graphical representations. The scale difference of the remaining 13 variables motivates a NPCA based on the correlation matrix.

The eigenvalues and the percentage of explained variance are given in Table 9.5.

eigenvalue	percentages	cumulated percentages
7.2852	0.5604	0.5604
1.3517	0.1040	0.6644
1.1266	0.0867	0.7510
0.7802	0.0600	0.8111
0.6359	0.0489	0.8600
0.5290	0.0407	0.9007
0.3397	0.0261	0.9268
0.2628	0.0202	0.9470
0.1936	0.0149	0.9619
0.1547	0.0119	0.9738
0.1405	0.0108	0.9846
0.1100	0.0085	0.9931
0.0900	0.0069	1.0000

Table 9.5. Eigenvalues and percentage of explained variance for Boston Housing data. `MVAnpcahous`

The first principal component explains 56% of the total variance and the first three components together explain more than 75%. These results imply that it is sufficient to look at 2, maximum 3, principal components.

Table 9.6 provides the correlations between the first three PC's and the original variables. These can be seen in Figure 9.10.

	PC_1	PC_2	PC_3
X_1	−0.9076	0.2247	0.1457
X_2	0.6399	−0.0292	0.5058
X_3	−0.8580	0.0409	−0.1845
X_5	−0.8737	0.2391	−0.1780
X_6	0.5104	0.7037	0.0869
X_7	−0.7999	0.1556	−0.2949
X_8	0.8259	−0.2904	0.2982
X_9	−0.7531	0.2857	0.3804
X_{10}	−0.8114	0.1645	0.3672
X_{11}	−0.5674	−0.2667	0.1498
X_{12}	0.4906	−0.1041	−0.5170
X_{13}	−0.7996	−0.4253	−0.0251
X_{14}	0.7366	0.5160	−0.1747

Table 9.6. Correlations of the first three PC's with the original variables. MVAnpcahous

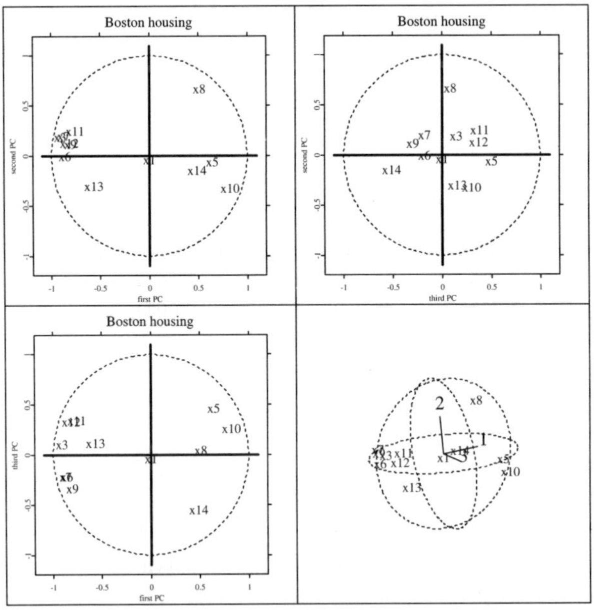

Figure 9.10. NPCA for the Boston housing data, correlations of first three PCs with the original variables. MVAnpcahousi

The correlations with the first PC show a very clear pattern. The variables X_2, X_6, X_8, X_{12}, and X_{14} are strongly positively correlated with the first PC, whereas the remaining variables are highly negatively correlated. The minimal correlation in the absolute value is 0.5. The first PC axis could be interpreted as a quality of life and house indicator. The second axis, given the polarities of X_{11} and X_{13} and of X_6 and X_{14}, can be interpreted as a social factor explaining only 10% of the total variance. The third axis is dominated by a polarity between

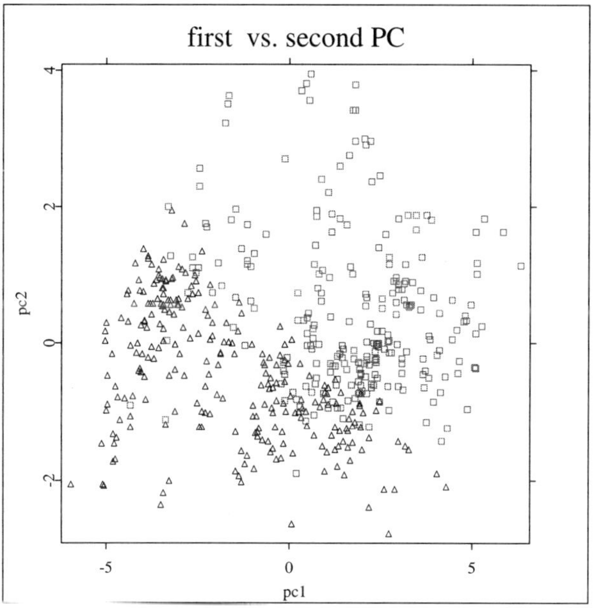

Figure 9.11. NPC analysis for the Boston housing data, scatterplot of the first two PCs. More expensive houses are marked with red color. MVAnpcahous

X_2 and X_{12}.

The set of individuals from the first two PCs can be graphically interpreted if the plots are color coded with respect to some particular variable of interest. Figure 9.11 color codes $X_{14} >$ median as red points. Clearly the first and second PCs are related to house value. The situation is less clear in Figure 9.12 where the color code corresponds to X_4, the Charles River indicator, i.e., houses near the river are colored red.

9.9 More Examples

EXAMPLE 9.8 *Let us now apply the PCA to the* standardized *bank data set (Table B.2). Figure 9.13 shows some PC plots of the bank data set. The genuine and counterfeit bank notes are marked by "o" and "+" respectively.*

The vector of eigenvalues of $\mathcal{R}$ is

$$\ell = (2.946, 1.278, 0.869, 0.450, 0.269, 0.189)^{\top}.$$

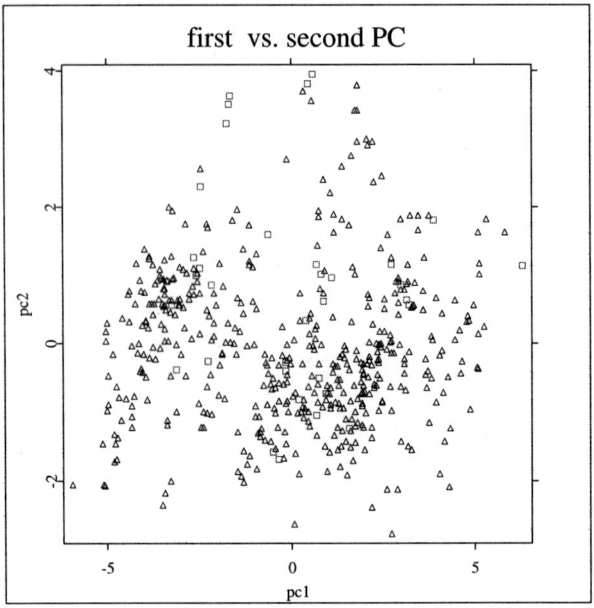

Figure 9.12. NPC analysis for the Boston housing data, scatterplot of the first two PCs. Houses close to the Charles River are indicated with red squares. `MVAnpcahous`

ℓ_j	proportion of variances	cumulated proportion
2.946	0.491	49.1
1.278	0.213	70.4
0.869	0.145	84.9
0.450	0.075	92.4
0.264	0.045	96.9
0.189	0.032	100.0

Table 9.7. Eigenvalues and proportions of explained variance

The eigenvectors g_j are given by the columns of the matrix

$$\mathcal{G} = \begin{pmatrix} -0.007 & -0.815 & 0.018 & 0.575 & 0.059 & 0.031 \\ 0.468 & -0.342 & -0.103 & -0.395 & -0.639 & -0.298 \\ 0.487 & -0.252 & -0.123 & -0.430 & 0.614 & 0.349 \\ 0.407 & 0.266 & -0.584 & 0.404 & 0.215 & -0.462 \\ 0.368 & 0.091 & 0.788 & 0.110 & 0.220 & -0.419 \\ -0.493 & -0.274 & -0.114 & -0.392 & 0.340 & -0.632 \end{pmatrix}.$$

Each original variable has the same weight in the analysis and the results are independent of the scale of each variable.

The proportions of explained variance are given in Table 9.7. It can be concluded that the representation in two dimensions should be sufficient. The correlations leading to Figure 9.14

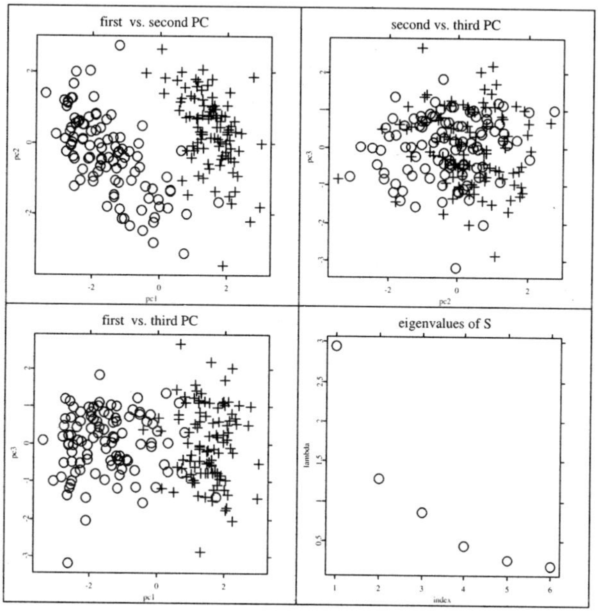

Figure 9.13. Principal components of the *standardized* bank data. MVAnpcabank

	$r_{X_iZ_1}$	$r_{X_iZ_2}$	$r^2_{X_iZ_1} + r^2_{X_iZ_2}$
X_1: length	−0.012	−0.922	0.85
X_2: left height	0.803	−0.387	0.79
X_3: right height	0.835	−0.285	0.78
X_4: lower	0.698	0.301	0.58
X_5: upper	0.631	0.104	0.41
X_6: diagonal	−0.847	−0.310	0.81

Table 9.8. Correlations with PCs

are given in Table 9.8. The picture is different from the one obtained in Section 9.3 (see Table 9.2). Here, the first factor is mainly a left–right vs. diagonal factor and the second one is a length factor (with negative weight). Take another look at Figure 9.13, where the individual bank notes are displayed. In the upper left graph it can be seen that the genuine bank notes are for the most part in the south-eastern portion of the graph featuring a larger diagonal, smaller height ($Z_1 < 0$) and also a larger length ($Z_2 < 0$). Note also that Figure 9.14 gives an idea of the correlation structure of the original data matrix.

EXAMPLE 9.9 *Consider the data of 79 U.S. companies given in Table B.5. The data is first standardized by subtracting the mean and dividing by the standard deviation. Note that the data set contains six variables: assets (X_1), sales (X_2), market value (X_3), profits (X_4), cash flow (X_5), number of employees (X_6).*

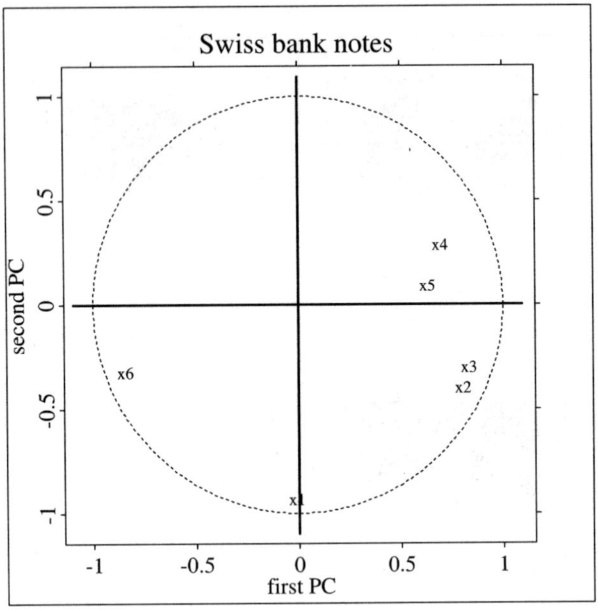

Figure 9.14. The correlations of the original variable with the PCs. MVAnpcabanki

Calculating the corresponding vector of eigenvalues gives

$$\ell = (5.039, 0.517, 0.359, 0.050, 0.029, 0.007)^{\top}$$

and the matrix of eigenvectors is

$$\mathcal{G} = \begin{pmatrix} 0.340 & -0.849 & -0.339 & 0.205 & 0.077 & -0.006 \\ 0.423 & -0.170 & 0.379 & -0.783 & -0.006 & -0.186 \\ 0.434 & 0.190 & -0.192 & 0.071 & -0.844 & 0.149 \\ 0.420 & 0.364 & -0.324 & 0.156 & 0.261 & -0.703 \\ 0.428 & 0.285 & -0.267 & -0.121 & 0.452 & 0.667 \\ 0.397 & 0.010 & 0.726 & 0.548 & 0.098 & 0.065 \end{pmatrix}.$$

Using this information the graphical representations of the first two principal components are given in Figure 9.15. The different sectors are marked by the following symbols:

H ... *Hi Tech and Communication*
E ... *Energy*
F ... *Finance*
M ... *Manufacturing*
R ... *Retail*
$\star$ ... *all other sectors.*

The two outliers in the right-hand side of the graph are IBM and General Electric (GE), which differ from the other companies with their high market values. As can be seen in the

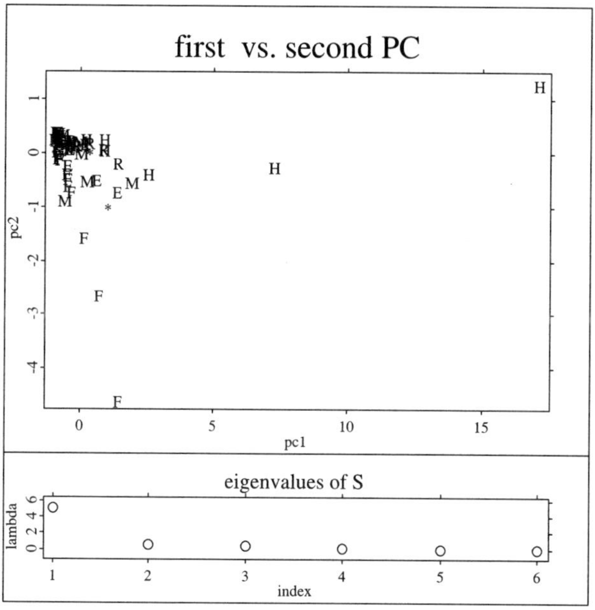

Figure 9.15. Principal components of the U.S. company data. `MVAnpcausco`

first column of $\mathcal{G}$, market value has the largest weight in the first PC, adding to the isolation of these two companies. If IBM and GE were to be excluded from the data set, a completely different picture would emerge, as shown in Figure 9.16. In this case the vector of eigenvalues becomes

$$\ell = (3.191, 1.535, 0.791, 0.292, 0.149, 0.041)^\top ,$$

and the corresponding matrix of eigenvectors is

$$\mathcal{G} = \begin{pmatrix} 0.263 & -0.408 & -0.800 & -0.067 & 0.333 & 0.099 \\ 0.438 & -0.407 & 0.162 & -0.509 & -0.441 & -0.403 \\ 0.500 & -0.003 & -0.035 & 0.801 & -0.264 & -0.190 \\ 0.331 & 0.623 & -0.080 & -0.192 & 0.426 & -0.526 \\ 0.443 & 0.450 & -0.123 & -0.238 & -0.335 & 0.646 \\ 0.427 & -0.277 & 0.558 & 0.021 & 0.575 & 0.313 \end{pmatrix}.$$

The percentage of variation explained by each component is given in Table 9.9. The first two components explain almost 79% of the variance. The interpretation of the factors (the axes of Figure 9.16) is given in the table of correlations (Table 9.10). The first two columns of this table are plotted in Figure 9.17.

From Figure 9.17 (and Table 9.10) it appears that the first factor is a "size effect", it is positively correlated with all the variables describing the size of the activity of the companies. It is also a measure of the economic strength of the firms. The second factor describes the "shape" of the companies ("profit-cash flow" vs. "assets-sales" factor), which is more difficult to interpret from an economic point of view.

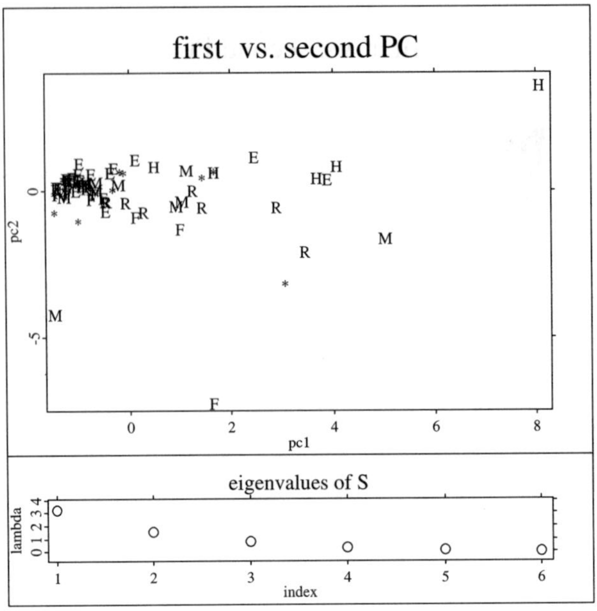

Figure 9.16. Principal components of the U.S. company data (without IBM and General Electric). MVAnpcausco2

ℓ_j	proportion of variance	cumulated proportion
3.191	0.532	0.532
1.535	0.256	0.788
0.791	0.132	0.920
0.292	0.049	0.968
0.149	0.025	0.993
0.041	0.007	1.000

Table 9.9. Eigenvalues and proportions of explained variance.

	$r_{X_iZ_1}$	$r_{X_iZ_2}$	$r^2_{X_iZ_1} + r^2_{X_iZ_2}$
X_1: assets	0.47	−0.510	0.48
X_2: sales	0.78	−0.500	0.87
X_3: market value	0.89	−0.003	0.80
X_4: profits	0.59	0.770	0.95
X_5: cash flow	0.79	0.560	0.94
X_6: employees	0.76	−0.340	0.70

Table 9.10. Correlations with PCs.

EXAMPLE 9.10 *Volle (1985) analyzes data on 28 individuals (Table B.14). For each individual, the time spent (in hours) on 10 different activities has been recorded over 100 days, as well as informative statistics such as the individual's sex, country of residence, professional*

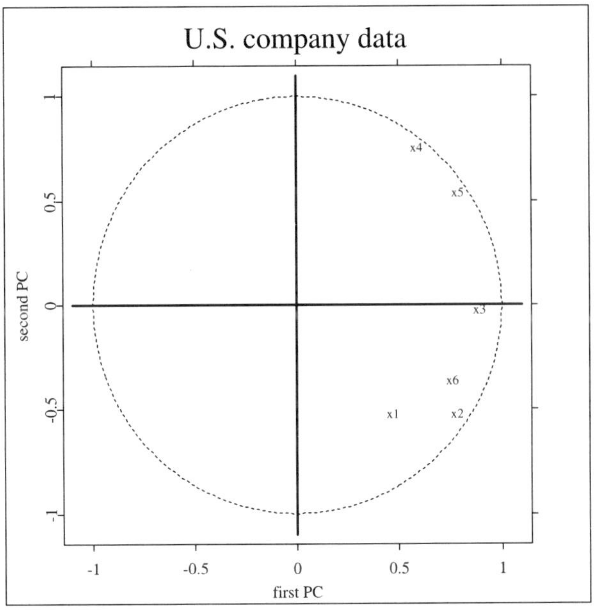

Figure 9.17. The correlation of the original variables with the PCs. MVAnpcausco2i

ℓ_j	proportion of variance	cumulated proportion
4.59	0.459	0.460
2.12	0.212	0.670
1.32	0.132	0.800
1.20	0.120	0.920
0.47	0.047	0.970
0.20	0.020	0.990
0.05	0.005	0.990
0.04	0.004	0.999
0.02	0.002	1.000
0.00	0.000	1.000

Table 9.11. Eigenvalues of correlation matrix for the time budget data.

activity and matrimonial status. The results of a NPCA are given below.

The eigenvalues of the correlation matrix are given in Table 9.11. Note that the last eigenvalue is exactly zero since the correlation matrix is singular (the sum of all the variables is always equal to $2400 = 24 \times 100$*). The results of the 4 first PCs are given in Tables 9.12 and 9.13.*

From these tables (and Figures 9.18 and 9.19), it appears that the professional and household activities are strongly contrasted in the first factor. Indeed on the horizontal axis of Figure 9.18 it can be seen that all the active men are on the right and all the inactive women

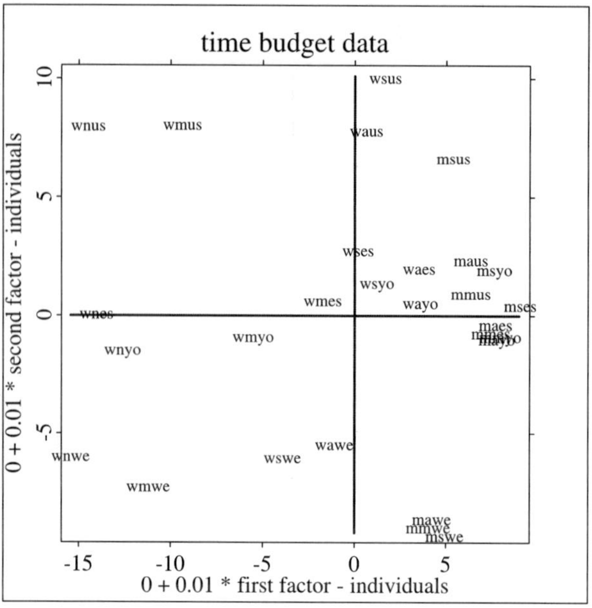

Figure 9.18. Representation of the individuals. MVAnpcatime

		$r_{X_iW_1}$	$r_{X_iW_2}$	$r_{X_iW_3}$	$r_{X_iW_4}$
X_1:	prof	0.9772	−0.1210	−0.0846	0.0669
X_2:	tran	0.9798	0.0581	−0.0084	0.4555
X_3:	hous	−0.8999	0.0227	0.3624	0.2142
X_4:	kids	−0.8721	0.1786	0.0837	0.2944
X_5:	shop	−0.5636	0.7606	−0.0046	−0.1210
X_6:	pers	−0.0795	0.8181	−0.3022	−0.0636
X_7:	eati	−0.5883	−0.6694	−0.4263	0.0141
X_8:	slee	−0.6442	−0.5693	−0.1908	−0.3125
X_9:	tele	−0.0994	0.1931	−0.9300	0.1512
X_{10}:	leis	−0.0922	0.1103	0.0302	−0.9574

Table 9.12. Correlation of variables with PCs.

are on the left. Active women and/or single women are inbetween. The second factor contrasts meal/sleeping vs. toilet/shopping (note the high correlation between meal and sleeping). Along the vertical axis of Figure 9.18 we see near the bottom of the graph the people from Western-European countries, who spend more time on meals and sleeping than people from the U. S. (who can be found close to the top of the graph). The other categories are inbetween.

In Figure 9.19 the variables television and other leisure activities hardly play any role (look at Table 9.12). The variable television appears in Z_3 (negatively correlated). Table 9.13 shows that this factor contrasts people from Eastern countries and Yugoslavia with men living in the U.S. The variable other leisure activities is the factor Z_4. It merely distinguishes between men

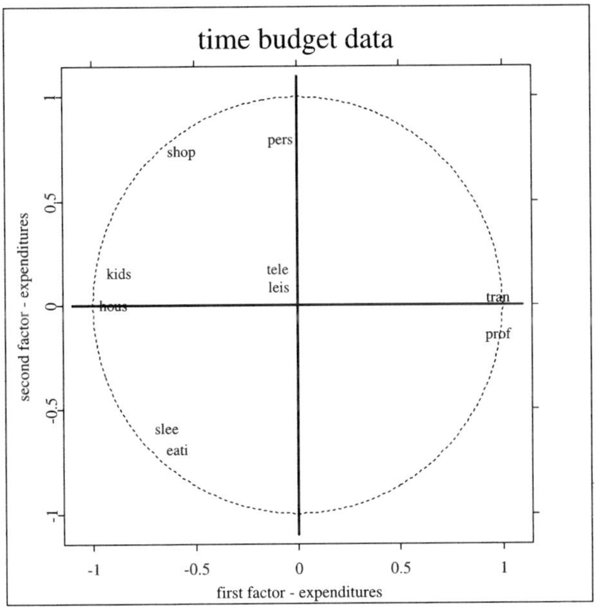

Figure 9.19. Representation of the variables. MVAnpcatime

and women in Eastern countries and in Yugoslavia. These last two factors are orthogonal to the preceeding axes and of course their contribution to the total variation is less important.

9.10 Exercises

EXERCISE 9.1 *Prove Theorem 9.1. (Hint: use (4.23).)*

EXERCISE 9.2 *Interpret the results of the PCA of the U.S. companies. Use the analysis of the bank notes in Section 9.3 as a guide. Compare your results with those in Example 9.9.*

EXERCISE 9.3 *Test the hypothesis that the proportion of variance explained by the first two PCs for the U.S. companies is* $\psi = 0.75$.

EXERCISE 9.4 *Apply the PCA to the car data (Table B.7). Interpret the first two PCs. Would it be necessary to look at the third PC?*

EXERCISE 9.5 *Take the athletic records for 55 countries (Appendix B.18) and apply the NPCA. Interpret your results.*

EXERCISE 9.6 *Apply a PCA to* $\Sigma = \begin{pmatrix} 1 & \rho \\ \rho & 1 \end{pmatrix}$, *where* $\rho > 0$. *Now change the scale of* X_1, *i.e., consider the covariance of* cX_1 *and* X_2. *How do the PC directions change with the screeplot?*

	Z_1	Z_2	Z_3	Z_4
maus	0.0633	0.0245	−0.0668	0.0205
waus	0.0061	0.0791	−0.0236	0.0156
wnus	−0.1448	0.0813	−0.0379	−0.0186
mmus	0.0635	0.0105	−0.0673	0.0262
wmus	−0.0934	0.0816	−0.0285	0.0038
msus	0.0537	0.0676	−0.0487	−0.0279
wsus	0.0166	0.1016	−0.0463	−0.0053
mawe	0.0420	−0.0846	−0.0399	−0.0016
wawe	−0.0111	−0.0534	−0.0097	0.0337
wnwe	−0.1544	−0.0583	−0.0318	−0.0051
mmwe	0.0402	−0.0880	−0.0459	0.0054
wmwe	−0.1118	−0.0710	−0.0210	0.0262
mswe	0.0489	−0.0919	−0.0188	−0.0365
wswe	−0.0393	−0.0591	−0.0194	−0.0534
mayo	0.0772	−0.0086	0.0253	−0.0085
wayo	0.0359	0.0064	0.0577	0.0762
wnyo	−0.1263	−0.0135	0.0584	−0.0189
mmyo	0.0793	−0.0076	0.0173	−0.0039
wmyo	−0.0550	−0.0077	0.0579	0.0416
msyo	0.0763	0.0207	0.0575	−0.0778
wsyo	0.0120	0.0149	0.0532	−0.0366
maes	0.0767	−0.0025	0.0047	0.0115
waes	0.0353	0.0209	0.0488	0.0729
wnes	−0.1399	0.0016	0.0240	−0.0348
mmes	0.0742	−0.0061	−0.0152	0.0283
wmes	−0.0175	0.0073	0.0429	0.0719
mses	0.0903	0.0052	0.0379	−0.0701
fses	0.0020	0.0287	0.0358	−0.0346

Table 9.13. PCs for time budget data.

EXERCISE 9.7 *Suppose that we have standardized some data using the Mahalanobis transformation. Would it be reasonable to apply a PCA?*

EXERCISE 9.8 *Apply a NPCA to the U.S. CRIME data set (Table B.10). Interpret the results. Would it be necessary to look at the third PC? Can you see any difference between the four regions? Redo the analysis excluding the variable "area of the state."*

EXERCISE 9.9 *Repeat Exercise 9.8 using the U.S. HEALTH data set (Table B.16).*

EXERCISE 9.10 *Do a NPCA on the GEOPOL data set (see Table B.15) which compares 41 countries w.r.t. different aspects of their development. Why or why not would a PCA be reasonable here?*

EXERCISE 9.11 *Let U be an uniform r.v. on $[0,1]$. Let $a \in \mathbb{R}^3$ be a vector of constants. Suppose that $X = Ua^\top = (X_1, X_2, X_3)$. What do you expect the NPCs of X to be?*

EXERCISE 9.12 *Let U_1 and U_2 be two independent uniform random variables on $[0, 1]$. Suppose that $X = (X_1, X_2, X_3, X_4)^\top$ where $X_1 = U_1$, $X_2 = U_2$, $X_3 = U_1 + U_2$ and $X_4 = U_1 - U_2$. Compute the correlation matrix $\mathcal{P}$ of X. How many PCs are of interest? Show that $\gamma_1 = \left(\frac{1}{\sqrt{2}}, \frac{1}{\sqrt{2}}, 1, 0\right)^\top$ and $\gamma_2 = \left(\frac{1}{\sqrt{2}}, \frac{-1}{\sqrt{2}}, 0, 1\right)^\top$ are eigenvectors of $\mathcal{P}$ corresponding to the non trivial λ's. Interpret the first two NPCs obtained.*

EXERCISE 9.13 *Simulate a sample of size $n = 50$ for the r.v. X in Exercise 9.12 and analyze the results of a NPCA.*

EXERCISE 9.14 *Bouroche (1980) reported the data on the state expenses of France from the period 1872 to 1971 (24 selected years) by noting the percentage of 11 categories of expenses. Do a NPCA of this data set. Do the three main periods (before WWI, between WWI and WWII, and after WWII) indicate a change in behavior w.r.t. to state expenses?*

10 Factor Analysis

A frequently applied paradigm in analyzing data from multivariate observations is to model the relevant information (represented in a multivariate variable X) as coming from a limited number of latent factors. In a survey on household consumption, for example, the consumption levels, X, of p different goods during one month could be observed. The variations and covariations of the p components of X throughout the survey might in fact be explained by two or three main social behavior factors of the household. For instance, a basic desire of comfort or the willingness to achieve a certain social level or other social latent concepts might explain most of the consumption behavior. These unobserved factors are much more interesting to the social scientist than the observed quantitative measures (X) themselves, because they give a better understanding of the behavior of households. As shown in the examples below, the same kind of factor analysis is of interest in many fields such as psychology, marketing, economics, politic sciences, etc.

How can we provide a statistical model addressing these issues and how can we interpret the obtained model? This is the aim of factor analysis. As in Chapter 8 and Chapter 9, the driving statistical theme of this chapter is to reduce the dimension of the observed data. The perspective used, however, is different: we assume that there is a model (it will be called the "Factor Model") stating that most of the covariances between the p elements of X can be explained by a limited number of latent factors. Section 10.1 defines the basic concepts and notations of the orthogonal factor model, stressing the non-uniqueness of the solutions. We show how to take advantage of this non-uniqueness to derive techniques which lead to easier interpretations. This will involve (geometric) rotations of the factors. Section 10.2 presents an empirical approach to factor analysis. Various estimation procedures are proposed and an optimal rotation procedure is defined. Many examples are used to illustrate the method.

10.1 The Orthogonal Factor Model

The aim of factor analysis is to explain the outcome of p variables in the data matrix $\mathcal{X}$ using fewer variables, the so-called *factors*. Ideally all the information in $\mathcal{X}$ can be reproduced by a smaller number of factors. These factors are interpreted as latent (unobserved) common characteristics of the observed $x \in \mathbb{R}^p$. The case just described occurs when every observed $x = (x_1, \dots, x_p)^\top$ can be written as

$$x_j = \sum_{\ell=1}^{k} q_{j\ell} f_\ell + \mu_j,\ j = 1, ..., p. \tag{10.1}$$

Here f_ℓ, for $\ell = 1, \dots, k$ denotes the factors. The number of factors, k, should always be much smaller than p. For instance, in psychology x may represent p results of a test measuring intelligence scores. One common latent factor explaining $x \in \mathbb{R}^p$ could be the overall level of "intelligence". In marketing studies, x may consist of p answers to a survey on the levels of satisfaction of the customers. These p measures could be explained by common latent factors like the attraction level of the product or the image of the brand, and so on. Indeed it is possible to create a representation of the observations that is similar to the one in (10.1) by means of principal components, but only if the last $p-k$ eigenvalues corresponding to the covariance matrix are equal to zero. Consider a p-dimensional random vector X with mean μ and covariance matrix $\text{Var}(X)=\Sigma$. A model similar to (10.1) can be written for X in matrix notation, namely

$$X = \mathcal{Q}F + \mu, \tag{10.2}$$

where F is the k-dimensional vector of the k factors. When using the factor model (10.2) it is often assumed that the factors F are centered, uncorrelated and standardized: $E(F)=0$ and $\text{Var}(F)=\mathcal{I}_k$. We will now show that if the last $p-k$ eigenvalues of Σ are equal to zero, we can easily express X by the factor model (10.2).

The spectral decomposition of Σ is given by $\Gamma\Lambda\Gamma^\top$. Suppose that only the first k eigenvalues are positive, i.e., $\lambda_{k+1}=\ldots=\lambda_p=0$. Then the (singular) covariance matrix can be written as

$$\Sigma = \sum_{\ell=1}^{k} \lambda_\ell \gamma_\ell \gamma_\ell^\top = (\Gamma_1 \Gamma_2) \begin{pmatrix} \Lambda_1 & 0 \\ 0 & 0 \end{pmatrix} \begin{pmatrix} \Gamma_1^\top \\ \Gamma_2^\top \end{pmatrix}.$$

In order to show the connection to the factor model (10.2), recall that the PCs are given by $Y=\Gamma^\top(X-\mu)$. Rearranging we have $X-\mu=\Gamma Y=\Gamma_1 Y_1+\Gamma_2 Y_2$, where the components of Y are partitioned according to the partition of Γ above, namely

$$Y = \begin{pmatrix} Y_1 \\ Y_2 \end{pmatrix} = \begin{pmatrix} \Gamma_1^\top \\ \Gamma_2^\top \end{pmatrix} (X-\mu), \quad \text{where} \quad \begin{pmatrix} \Gamma_1^\top \\ \Gamma_2^\top \end{pmatrix} (X-\mu) \sim \left(0, \begin{pmatrix} \Lambda_1 & 0 \\ 0 & 0 \end{pmatrix}\right).$$

In other words, Y_2 has a singular distribution with mean and covariance matrix equal to zero. Therefore, $X-\mu=\Gamma_1 Y_1+\Gamma_2 Y_2$ implies that $X-\mu$ is equivalent to $\Gamma_1 Y_1$, which can be written as

$$X = \Gamma_1 \Lambda_1^{1/2} \Lambda_1^{-1/2} Y_1 + \mu.$$

Defining $\mathcal{Q}=\Gamma_1\Lambda_1^{1/2}$ and $F=\Lambda_1^{-1/2}Y_1$, we obtain the factor model (10.2).

Note that the covariance matrix of model (10.2) can be written as

$$\Sigma = E(X-\mu)(X-\mu)^\top = \mathcal{Q}E(FF^\top)\mathcal{Q}^\top = \mathcal{Q}\mathcal{Q}^\top = \sum_{j=1}^{k} \lambda_j \gamma_j \gamma_j^\top. \tag{10.3}$$

We have just shown how the variable X can be completely determined by a weighted sum of k (where $k<p$) uncorrelated factors. The situation used in the derivation, however, is too idealistic. In practice the covariance matrix is rarely singular.

It is common praxis in factor analysis to split the influences of the factors into common and specific ones. There are, for example, highly informative factors that are common to all of the components of X and factors that are specific to certain components. The factor analysis model used in praxis is a generalization of (10.2):

$$X = \mathcal{Q}F + U + \mu, \tag{10.4}$$

where $\mathcal{Q}$ is a $(p \times k)$ matrix of the (non-random) *loadings* of the *common factors* $F(k \times 1)$ and U is a $(p \times 1)$ matrix of the (random) *specific factors*. It is assumed that the factor variables F are uncorrelated random vectors and that the specific factors are uncorrelated and have zero covariance with the common factors. More precisely, it is assumed that:

$$\begin{aligned} EF &= 0, \\ Var(F) &= \mathcal{I}_k, \\ EU &= 0, \\ Cov(U_i, U_j) &= 0, \quad i \neq j \\ Cov(F, U) &= 0. \end{aligned} \tag{10.5}$$

Define

$$Var(U) = \Psi = \text{diag}(\psi_{11}, \ldots, \psi_{pp}).$$

The generalized factor model (10.4) together with the assumptions given in (10.5) constitute the *orthogonal factor model.*

Orthogonal Factor Model

$$\underset{(p\times 1)}{X} = \underset{(p\times k)}{\mathcal{Q}} \; \underset{(k\times 1)}{F} + \underset{(p\times 1)}{U} + \underset{(p\times 1)}{\mu}$$

μ_j	=	mean of variable j
U_j	=	j-th specific factor
F_ℓ	=	ℓ-th common factor
$q_{j\ell}$	=	loading of the j-th variable on the ℓ-th factor

The random vectors F and U are unobservable and uncorrelated.

Note that (10.4) implies for the components of $X = (X_1, \ldots, X_p)^\top$ that

$$X_j = \sum_{\ell=1}^{k} q_{j\ell} F_\ell + U_j + \mu_j, \quad j = 1, \ldots, p. \tag{10.6}$$

Using (10.5) we obtain $\sigma_{X_j X_j} = Var(X_j) = \sum_{\ell=1}^{k} q_{j\ell}^2 + \psi_{jj}$. The quantity $h_j^2 = \sum_{\ell=1}^{k} q_{j\ell}^2$ is called the *communality* and ψ_{jj} the *specific variance.* Thus the covariance of X can be rewritten as

$$\begin{aligned}\Sigma &= E(X-\mu)(X-\mu)^\top = E(\mathcal{Q}F+U)(\mathcal{Q}F+U)^\top \\ &= \mathcal{Q}E(FF^\top)\mathcal{Q}^\top + E(UU^\top) = \mathcal{Q}\, Var(F)\mathcal{Q}^\top + Var(U) \\ &= \mathcal{Q}\mathcal{Q}^\top + \Psi. \end{aligned} \tag{10.7}$$

In a sense, the factor model explains the variations of X for the most part by a small number of latent factors F common to its p components and entirely explains all the correlation structure between its components, plus some "noise" U which allows specific variations of each component to enter. The specific factors adjust to capture the individual variance of each component. Factor analysis relies on the assumptions presented above. If the assumptions are not met, the analysis could be spurious. Although principal components analysis and factor analysis might be related (this was hinted at in the derivation of the factor model), they are quite different in nature. PCs are linear transformations of X arranged in decreasing order of variance and used to reduce the dimension of the data set, whereas in factor analysis, we try to model the variations of X using a linear transformation of a fixed, limited number of latent factors. The objective of factor analysis is to find the loadings $\mathcal{Q}$ and the specific variance Ψ. Estimates of $\mathcal{Q}$ and Ψ are deduced from the covariance structure (10.7).

Interpretation of the Factors

Assume that a factor model with k factors was found to be reasonable, i.e., most of the (co)variations of the p measures in X were explained by the k fixed latent factors. The next natural step is to try to understand what these factors represent. To interpret F_ℓ, it makes sense to compute its correlations with the original variables X_j first. This is done for

$\ell = 1, \ldots, k$ and for $j = 1, \ldots, p$ to obtain the matrix P_{XF}. The sequence of calculations used here are in fact the same that were used to interprete the PCs in the principal components analysis.

The following covariance between X and F is obtained via (10.5),

$$\Sigma_{XF} = E\{(\mathcal{Q}F + U)F^\top\} = \mathcal{Q}.$$

The correlation is

$$P_{XF} = D^{-1/2}\mathcal{Q}, \tag{10.8}$$

where $D = \mathrm{diag}(\sigma_{X_1X_1}, \ldots, \sigma_{X_pX_p})$. Using (10.8) it is possible to construct a figure analogous to Figure 9.6 and thus to consider which of the original variables $X_1, \ldots, X_p$ play a role in the unobserved common factors $F_1, \ldots, F_k$.

Returning to the psychology example where X are the observed scores to p different intelligence tests (the WAIS data set in Table B.12 provides an example), we would expect a model with one factor to produce a factor that is positively correlated with all of the components in X. For this example the factor represents the overall level of intelligence of an individual. A model with two factors could produce a refinement in explaining the variations of the p scores. For example, the first factor could be the same as before (overall level of intelligence), whereas the second factor could be positively correlated with some of the tests, X_j, that are related to the individual's ability to think abstractly and negatively correlated with other tests, X_i, that are related to the individual's practical ability. The second factor would then concern a particular dimension of the intelligence stressing the distinctions between the "theoretical" and "practical" abilities of the individual. If the model is true, most of the information coming from the p scores can be summarized by these two latent factors. Other practical examples are given below.

Invariance of Scale

What happens if we change the scale of X to $Y = \mathcal{C}X$ with $\mathcal{C} = \mathrm{diag}(c_1, \ldots, c_p)$? If the k-factor model (10.6) is true for X with $\mathcal{Q} = \mathcal{Q}_X$, $\Psi = \Psi_X$, then, since

$$Var(Y) = \mathcal{C}\Sigma\mathcal{C}^\top = \mathcal{C}\mathcal{Q}_X\mathcal{Q}_X^\top\mathcal{C}^\top + \mathcal{C}\Psi_X\mathcal{C}^\top,$$

the same k-factor model is also true for Y with $\mathcal{Q}_Y = \mathcal{C}\mathcal{Q}_X$ and $\Psi_Y = \mathcal{C}\Psi_X\mathcal{C}^\top$. In many applications, the search for the loadings $\mathcal{Q}$ and for the specific variance Ψ will be done by the decomposition of the correlation matrix of X rather than the covariance matrix Σ. This corresponds to a factor analysis of a linear transformation of X (i.e., $Y = D^{-1/2}(X - \mu)$). The goal is to try to find the loadings $\mathcal{Q}_Y$ and the specific variance Ψ_Y such that

$$P = \mathcal{Q}_Y\,\mathcal{Q}_Y^\top + \Psi_Y. \tag{10.9}$$

In this case the interpretation of the factors F immediately follows from (10.8) given the following correlation matrix:

$$P_{XF} = P_{YF} = \mathcal{Q}_Y. \tag{10.10}$$

Because of the scale invariance of the factors, the loadings and the specific variance of the model, where X is expressed in its original units of measure, are given by

$$\begin{aligned} \mathcal{Q}_X &= D^{1/2}\mathcal{Q}_Y \\ \Psi_X &= D^{1/2}\Psi_Y D^{1/2}. \end{aligned}$$

It should be noted that although the factor analysis model (10.4) enjoys the scale invariance property, the actual estimated factors could be scale dependent. We will come back to this point later when we discuss the method of principal factors.

Non-Uniqueness of Factor Loadings

The factor loadings are not unique! Suppose that $\mathcal{G}$ is an orthogonal matrix. Then X in (10.4) can also be written as

$$X = (\mathcal{Q}\mathcal{G})(\mathcal{G}^\top F) + U + \mu.$$

This implies that, if a k-factor of X with factors F and loadings $\mathcal{Q}$ is true, then the k-factor model with factors $\mathcal{G}^\top F$ and loadings $\mathcal{Q}\mathcal{G}$ is also true. In practice, we will take advantage of this non-uniqueness. Indeed, referring back to Section 2.6 we can conclude that premultiplying a vector F by an orthogonal matrix corresponds to a rotation of the system of axis, the direction of the first new axis being given by the first row of the orthogonal matrix. It will be shown that choosing an appropriate rotation will result in a matrix of loadings $\mathcal{Q}\mathcal{G}$ that will be easier to interpret. We have seen that the loadings provide the correlations between the factors and the original variables, therefore, it makes sense to search for rotations that give factors that are maximally correlated with various groups of variables.

From a numerical point of view, the non-uniqueness is a drawback. We have to find loadings $\mathcal{Q}$ and specific variances Ψ satisfying the decomposition $\Sigma = \mathcal{Q}\mathcal{Q}^\top + \Psi$, but no straightforward numerical algorithm can solve this problem due to the multiplicity of the solutions. An acceptable technique is to impose some chosen constraints in order to get—in the best case—an unique solution to the decomposition. Then, as suggested above, once we have a solution we will take advantage of the rotations in order to obtain a solution that is easier to interprete.

An obvious question is: what kind of constraints should we impose in order to eliminate the non-uniqueness problem? Usually, we impose additional constraints where

$$\mathcal{Q}^\top \Psi^{-1} \mathcal{Q} \quad \text{is diagonal} \tag{10.11}$$

or

$$\mathcal{Q}^\top \mathcal{D}^{-1} \mathcal{Q} \quad \text{is diagonal.} \tag{10.12}$$

How many parameters does the model (10.7) have without constraints?

$$\begin{array}{lll} \mathcal{Q}(p \times k) & \text{has} & p \cdot k \ \text{ parameters, and} \\ \Psi(p \times p) & \text{has} & p \ \text{ parameters.} \end{array}$$

Hence we have to determine $pk + p$ parameters! Conditions (10.11) respectively (10.12) introduce $\frac{1}{2}\{k(k-1)\}$ constraints, since we require the matrices to be diagonal. Therefore, the degrees of freedom of a model with k factors is:

$$\begin{aligned} d &= (\#\text{ parameters for } \Sigma \text{ unconstrained}) - (\#\text{ parameters for } \Sigma \text{ constrained}) \\ &= \tfrac{1}{2}p(p+1) - (pk + p - \tfrac{1}{2}k(k-1)) \\ &= \tfrac{1}{2}(p-k)^2 - \tfrac{1}{2}(p+k). \end{aligned}$$

If $d < 0$, then the model is undetermined: there are infinitly many solutions to (10.7). This means that the number of parameters of the factorial model is larger than the number of parameters of the original model, or that the number of factors k is "too large" relative to p. In some cases $d = 0$: there is a unique solution to the problem (except for rotation). In practice we usually have that $d > 0$: there are more equations than parameters, thus an exact solution does not exist. In this case approximate solutions are used. An approximation of Σ, for example, is $\mathcal{Q}\mathcal{Q}^\top + \Psi$. The last case is the most interesting since the factorial model has less parameters than the original one. Estimation methods are introduced in the next section.

Evaluating the degrees of freedom, d, is particularly important, because it already gives an idea of the upper bound on the number of factors we can hope to identify in a factor model. For instance, if $p = 4$, we could not identify a factor model with 2 factors (this results in $d = -1$ which has infinitly many solutions). With $p = 4$, only a one factor model gives an approximate solution ($d = 2$). When $p = 6$, models with 1 and 2 factors provide approximate solutions and a model with 3 factors results in an unique solution (up to the rotations) since $d = 0$. A model with 4 or more factors would not be allowed, but of course, the aim of factor analysis is to find suitable models with a small number of factors, i.e., smaller than p. The next two examples give more insights into the notion of degrees of freedom.

EXAMPLE 10.1 *Let $p = 3$ and $k = 1$, then $d = 0$ and*

$$\Sigma = \begin{pmatrix} \sigma_{11} & \sigma_{12} & \sigma_{13} \\ \sigma_{21} & \sigma_{22} & \sigma_{23} \\ \sigma_{31} & \sigma_{32} & \sigma_{33} \end{pmatrix} = \begin{pmatrix} q_1^2 + \psi_{11} & q_1 q_2 & q_1 q_3 \\ q_1 q_2 & q_2^2 + \psi_{22} & q_2 q_3 \\ q_1 q_3 & q_2 q_3 & q_3^2 + \psi_{33} \end{pmatrix}$$

with $\mathcal{Q} = \begin{pmatrix} q_1 \\ q_2 \\ q_3 \end{pmatrix}$ *and* $\Psi = \begin{pmatrix} \psi_{11} & 0 & 0 \\ 0 & \psi_{22} & 0 \\ 0 & 0 & \psi_{33} \end{pmatrix}$. *Note that here the constraint (10.8) is automatically verified since $k = 1$. We have*

$$q_1^2 = \frac{\sigma_{12}\sigma_{13}}{\sigma_{23}};\ q_2^2 = \frac{\sigma_{12}\sigma_{23}}{\sigma_{13}};\ q_3^2 = \frac{\sigma_{13}\sigma_{23}}{\sigma_{12}}$$

and

$$\psi_{11} = \sigma_{11} - q_1^2;\ \psi_{22} = \sigma_{22} - q_2^2;\ \psi_{33} = \sigma_{33} - q_3^2.$$

In this particular case ($k = 1$), the only rotation is defined by $\mathcal{G} = -1$, so the other solution for the loadings is provided by $-\mathcal{Q}$.

EXAMPLE 10.2 *Suppose now $p = 2$ and $k = 1$, then $d < 0$ and*

$$\Sigma = \begin{pmatrix} 1 & \rho \\ \rho & 1 \end{pmatrix} = \begin{pmatrix} q_1^2 + \psi_{11} & q_1 q_2 \\ q_1 q_2 & q_2^2 + \psi_{22} \end{pmatrix}.$$

We have infinitely many solutions: for any α ($\rho < \alpha < 1$), a solution is provided by

$$q_1 = \alpha;\ q_2 = \rho/\alpha;\ \psi_{11} = 1 - \alpha^2;\ \psi_{22} = 1 - (\rho/\alpha)^2.$$

The solution in Example 10.1 may be unique (up to a rotation), but it is not proper in the sense that it cannot be interpreted statistically. Exercise 10.5 gives an example where the specific variance ψ_{11} is negative.

⚠ Even in the case of a unique solution ($d = 0$), the solution may be inconsistent with statistical interpretations.

Summary

↪ The factor analysis model aims to describe how the original p variables in a data set depend on a small number of latent factors $k < p$, i.e., it assumes that $X = \mathcal{Q}F + U + \mu$. The ($k$-dimensional) random vector F contains the common factors, the (p-dimensional) U contains the specific factors and $\mathcal{Q}(p \times k)$ contains the factor loadings.

↪ It is assumed that F and U are uncorrelated and have zero means, i.e., $F \sim (0, \mathcal{I})$, $U \sim (0, \Psi)$ where Ψ is diagonal matrix and $Cov(F, U) = 0$. This leads to the covariance structure $\Sigma = \mathcal{Q}\mathcal{Q}^\top + \Psi$.

↪ The interpretation of the factor F is obtained through the correlation $P_{XF} = D^{-1/2}\mathcal{Q}$.

↪ A normalized analysis is obtained by the model $P = \mathcal{Q}\mathcal{Q}^\top + \Psi$. The interpretation of the factors is given directly by the loadings $\mathcal{Q}: \ P_{XF} = \mathcal{Q}$.

↪ The factor analysis model is scale invariant. The loadings are not unique (only up to multiplication by an orthogonal matrix).

↪ Whether a model has an unique solution or not is determined by the degrees of freedom $d = 1/2(p-k)^2 - 1/2(p+k)$.

10.2 Estimation of the Factor Model

In practice, we have to find estimates $\widehat{\mathcal{Q}}$ of the loadings $\mathcal{Q}$ and estimates $\widehat{\Psi}$ of the specific variances Ψ such that analogously to (10.7)

$$\mathcal{S} = \widehat{\mathcal{Q}}\widehat{\mathcal{Q}}^\top + \widehat{\Psi},$$

where $\mathcal{S}$ denotes the empirical covariance of $\mathcal{X}$. Given an estimate $\widehat{\mathcal{Q}}$ of $\mathcal{Q}$, it is natural to set

$$\widehat{\psi}_{jj} = s_{X_jX_j} - \sum_{\ell=1}^{k} \widehat{q}_{j\ell}^2.$$

We have that $\widehat{h}_j^2 = \sum_{\ell=1}^{k} \widehat{q}_{j\ell}^2$ is an estimate for the communality h_j^2.

In the ideal case $d = 0$, there is an exact solution. However, d is usually greater than zero, therefore we have to find $\widehat{\mathcal{Q}}$ and $\widehat{\Psi}$ such that S is approximated by $\widehat{\mathcal{Q}}\widehat{\mathcal{Q}}^\top + \widehat{\Psi}$. As mentioned above, it is often easier to compute the loadings and the specific variances of the standardized model.

Define $\mathcal{Y} = \mathcal{H}\mathcal{X}\mathcal{D}^{-1/2}$, the standardization of the data matrix $\mathcal{X}$, where, as usual, $\mathcal{D} = \operatorname{diag}(s_{X_1X_1}, \ldots, s_{X_pX_p})$ and the centering matrix $\mathcal{H} = \mathcal{I} - n^{-1}1_n1_n^\top$ (recall from Chapter 2 that $\mathcal{S} = \frac{1}{n}\mathcal{X}^\top\mathcal{H}\mathcal{X}$). The estimated factor loading matrix $\widehat{\mathcal{Q}}_Y$ and the estimated specific variance $\widehat{\Psi}_Y$ of $\mathcal{Y}$ are

$$\widehat{\mathcal{Q}}_Y = \mathcal{D}^{-1/2}\widehat{\mathcal{Q}}_X \quad \text{and} \quad \widehat{\Psi}_Y = \mathcal{D}^{-1}\widehat{\Psi}_X.$$

For the correlation matrix $\mathcal{R}$ of $\mathcal{X}$, we have that

$$\mathcal{R} = \widehat{\mathcal{Q}}_Y\widehat{\mathcal{Q}}_Y^\top + \widehat{\Psi}_Y.$$

The interpretations of the factors are formulated from the analysis of the loadings $\widehat{\mathcal{Q}}_Y$.

EXAMPLE 10.3 *Let us calculate the matrices just defined for the car data given in Table B.7. This data set consists of the averaged marks (from 1 =low to 6 =high) for 24 car types. Considering the three variables price, security and easy handling, we get the following correlation matrix:*

$$\mathcal{R} = \begin{pmatrix} 1 & 0.975 & 0.613 \\ 0.975 & 1 & 0.620 \\ 0.613 & 0.620 & 1 \end{pmatrix}.$$

We will first look for one factor, i.e., $k = 1$. Note that (# number of parameters of Σ unconstrained – # parameters of Σ constrained) is equal to $\frac{1}{2}(p-k)^2 - \frac{1}{2}(p+k) = \frac{1}{2}(3-1)^2 - \frac{1}{2}(3+1) = 0$. This implies that there is an exact solution! The equation

$$\begin{pmatrix} 1 & r_{X_1X_2} & r_{X_1X_3} \\ r_{X_1X_2} & 1 & r_{X_2X_3} \\ r_{X_1X_3} & r_{X_2X_3} & 1 \end{pmatrix} = \mathcal{R} = \begin{pmatrix} \widehat{q}_1^2 + \widehat{\psi}_{11} & \widehat{q}_1\widehat{q}_2 & \widehat{q}_1\widehat{q}_3 \\ \widehat{q}_1\widehat{q}_2 & \widehat{q}_2^2 + \widehat{\psi}_{22} & \widehat{q}_2\widehat{q}_3 \\ \widehat{q}_1\widehat{q}_3 & \widehat{q}_2\widehat{q}_3 & \widehat{q}_3^2 + \widehat{\psi}_{33} \end{pmatrix}$$

yields the communalities $\widehat{h}_i^2 = \widehat{q}_i^2$, where

$$\widehat{q}_1^2 = \frac{r_{X_1X_2} r_{X_1X_3}}{r_{X_2X_3}}, \qquad \widehat{q}_2^2 = \frac{r_{X_1X_2} r_{X_2X_3}}{r_{X_1X_3}} \quad \text{and} \quad \widehat{q}_3^2 = \frac{r_{X_1X_3} r_{X_2X_3}}{r_{X_1X_2}}.$$

Combining this with the specific variances $\widehat{\psi}_{11} = 1 - \widehat{q}_1^2$, $\widehat{\psi}_{22} = 1 - \widehat{q}_2^2$ and $\widehat{\psi}_{33} = 1 - \widehat{q}_3^2$, we obtain the following solution

$$\begin{aligned} \widehat{q}_1 &= 0.982 & \widehat{q}_2 &= 0.993 & \widehat{q}_3 &= 0.624 \\ \widehat{\psi}_{11} &= 0.035 & \widehat{\psi}_{22} &= 0.014 & \widehat{\psi}_{33} &= 0.610. \end{aligned}$$

Since the first two communalities ($\widehat{h}_i^2 = \widehat{q}_i^2$) are close to one, we can conclude that the first two variables, namely price and security, are explained by the single factor quite well. This factor can be interpreted as a "price+security" factor.

The Maximum Likelihood Method

Recall from Chapter 6 the log-likelihood function ℓ for a data matrix $\mathcal{X}$ of observations of $X \sim N_p(\mu, \Sigma)$:

$$\begin{aligned} \ell(\mathcal{X}; \mu, \Sigma) &= -\frac{n}{2} \log \mid 2\pi\Sigma \mid - \frac{1}{2} \sum_{i=1}^{n} (x_i - \mu)\Sigma^{-1}(x_i - \mu)^\top \\ &= -\frac{n}{2} \log \mid 2\pi\Sigma \mid - \frac{n}{2} \operatorname{tr}(\Sigma^{-1}\mathcal{S}) - \frac{n}{2}(\overline{x} - \mu)\Sigma^{-1}(\overline{x} - \mu)^\top. \end{aligned}$$

This can be rewritten as

$$\ell(\mathcal{X}; \widehat{\mu}, \Sigma) = -\frac{n}{2} \left\{ \log \mid 2\pi\Sigma \mid + \operatorname{tr}(\Sigma^{-1}\mathcal{S}) \right\}.$$

Replacing μ by $\widehat{\mu} = \overline{x}$ and substituting $\Sigma = \mathcal{Q}\mathcal{Q}^\top + \Psi$ this becomes

$$\ell(\mathcal{X}; \widehat{\mu}, \mathcal{Q}, \Psi) = -\frac{n}{2} \left[\log\{\mid 2\pi(\mathcal{Q}\mathcal{Q}^\top + \Psi) \mid\} + \operatorname{tr}\{(\mathcal{Q}\mathcal{Q}^\top + \Psi)^{-1}\mathcal{S}\} \right]. \tag{10.13}$$

Even in the case of a single factor ($k = 1$), these equations are rather complicated and iterative numerical algorithms have to be used (for more details see Mardia et al. (1979, p. 263ff)). A practical computation scheme is also given in Supplement 9A of Johnson and Wichern (1998).

Likelihood Ratio Test for the Number of Common Factors

Using the methodology of Chapter 7, it is easy to test the adequacy of the factor analysis model by comparing the likelihood under the null (factor analysis) and alternative (no constraints on covariance matrix) hypotheses.

Assuming that $\widehat{\mathcal{Q}}$ and $\widehat{\Psi}$ are the maximum likelihood estimates corresponding to (10.13), we obtain the following LR test statistic:

$$-2\log\left(\frac{\text{maximized likelihood under } H_0}{\text{maximized likelihood}}\right) = n\log\left(\frac{|\widehat{\mathcal{Q}}\widehat{\mathcal{Q}}^\top + \widehat{\Psi}|}{|\mathcal{S}|}\right), \tag{10.14}$$

which asymptotically has the $\chi^2_{\frac{1}{2}\{(p-k)^2-p-k\}}$ distribution.

The χ^2 approximation can be improved if we replace n by $n-1-(2p+4k+5)/6$ in (10.14) (Bartlett, 1954). Using Bartlett's correction, we reject the factor analysis model at the α level if

$$\{n-1-(2p+4k+5)/6\}\log\left(\frac{|\widehat{\mathcal{Q}}\widehat{\mathcal{Q}}^\top + \widehat{\Psi}|}{|\mathcal{S}|}\right) > \chi^2_{1-\alpha;\{(p-k)^2-p-k\}/2}, \tag{10.15}$$

and if the number of observations n is large and the number of common factors k is such that the χ^2 statistic has a positive number of degrees of freedom.

The Method of Principal Factors

The *method of principal factors* concentrates on the decomposition of the correlation matrix $\mathcal{R}$ or the covariance matrix $\mathcal{S}$. For simplicity, only the method for the correlation matrix $\mathcal{R}$ will be discussed. As pointed out in Chapter 9, the spectral decompositions of $\mathcal{R}$ and $\mathcal{S}$ yield different results and therefore, the method of principal factors may result in different estimators. The method can be motivated as follows: Suppose we know the exact Ψ, then the constraint (10.12) implies that the columns of $\mathcal{Q}$ are orthogonal since $\mathcal{D} = \mathcal{I}$ and it implies that they are eigenvectors of $\mathcal{Q}\mathcal{Q}^\top = \mathcal{R} - \Psi$. Furthermore, assume that the first k eigenvalues are positive. In this case we could calculate $\mathcal{Q}$ by means of a spectral decomposition of $\mathcal{Q}\mathcal{Q}^\top$ and k would be the number of factors.

The principal factors algorithm is based on good preliminary estimators $\widetilde{h}_j^2$ of the communalities h_j^2, for $j = 1, \dots, p$. There are two traditional proposals:

- $\widetilde{h}_j^2$, defined as the square of the multiple correlation coefficient of X_j with (X_l), for $l \neq j$, i.e., $\rho^2(V, W\widehat{\beta})$ with $V = X_j$, $W = (X_\ell)_{\ell\neq j}$ and where $\widehat{\beta}$ is the least squares regression parameter of a regression of V on W.
- $\widetilde{h}_j^2 = \max\limits_{\ell\neq j} |r_{X_j X_\ell}|$, where $\mathcal{R} = (r_{X_j X_\ell})$ is the correlation matrix of $\mathcal{X}$.

Given $\widetilde{\psi}_{jj} = 1 - \widetilde{h}_j^2$ we can construct the *reduced correlation matrix*, $\mathcal{R} - \widetilde{\Psi}$. The Spectral Decomposition Theorem says that

$$\mathcal{R} - \widetilde{\Psi} = \sum_{\ell=1}^{p} \lambda_\ell \gamma_\ell \gamma_\ell^\top,$$

with eigenvalues $\lambda_1 \geq \cdots \geq \lambda_p$. Assume that the first k eigenvalues $\lambda_1, \dots, \lambda_k$ are positive and large compared to the others. Then we can set

$$\widehat{q}_\ell = \sqrt{\lambda_\ell}\,\gamma_\ell\,, \quad \ell = 1, \dots, k$$

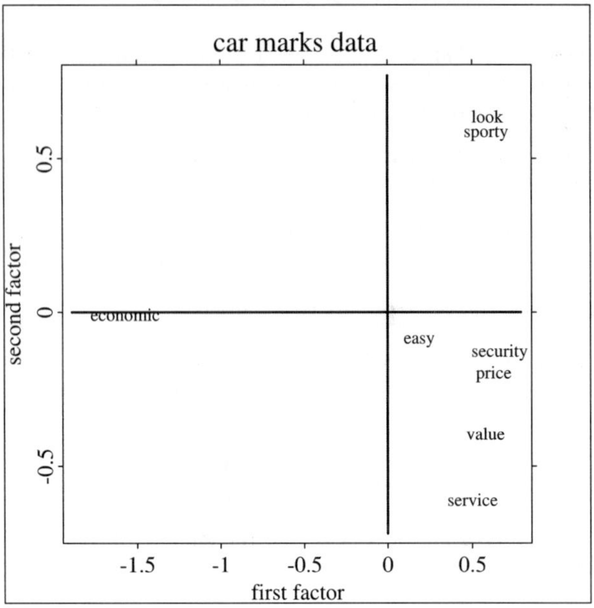

Figure 10.1. Loadings of the evaluated car qualities, factor analysis with $k = 2$. MVAfactcarm

or

$$\widehat{\mathcal{Q}} = \Gamma_1 \Lambda_1^{1/2}$$

with

$$\Gamma_1 = (\gamma_1, \ldots, \gamma_k) \quad \text{and} \quad \Lambda_1 = \operatorname{diag}(\lambda_1, \ldots, \lambda_k).$$

In the next step set

$$\widehat{\psi}_{jj} = 1 - \sum_{\ell=1}^{k} \widehat{q}_{j\ell}^{2}, \quad j = 1, \ldots, p.$$

Note that the procedure can be iterated: from $\widehat{\psi}_{jj}$ we can compute a new reduced correlation matrix $\mathcal{R} - \widehat{\Psi}$ following the same procedure. The iteration usually stops when the $\widehat{\psi}_{jj}$ have converged to a stable value.

EXAMPLE 10.4 *Consider once again the car data given in Table B.7. From Exercise 9.4 we know that the first PC is mainly influenced by X_2–X_7. Moreover, we know that most of the variance is already captured by the first PC. Thus we can conclude that the data are mainly determined by one factor ($k = 1$).*

The eigenvalues of $\mathcal{R} - \widehat{\Psi}$ for $\widehat{\Psi} = (\max_{j \neq i} |r_{X_i X_j}|)$ are

$$(5.448, 0.003, -.246, -0.646, -0.901, -0.911, -0.948, -0.964)^{\top}.$$

It would suffice to choose only one factor. Nevertheless, we have computed two factors. The result (the factor loadings for two factors) is shown in Figure 10.1.

We can clearly see a cluster of points to the right, which contain the factor loadings for the variables X_2–X_7. This shows, as did the PCA, that these variables are highly dependent and are thus more or less equivalent. The factor loadings for X_1 (economy) and X_8 (easy handling) are separate, but note the different scales on the horizontal and vertical axes! Although there are two or three sets of variables in the plot, the variance is already explained by the first factor, the "price+security" factor.

The Principal Component Method

The *principal factor method* involves finding an approximation $\tilde{\Psi}$ of Ψ, the matrix of specific variances, and then correcting $\mathcal{R}$, the correlation matrix of X, by $\tilde{\Psi}$. The *principal component method* starts with an approximation $\hat{\mathcal{Q}}$ of $\mathcal{Q}$, the factor loadings matrix. The sample covariance matrix is diagonalized, $\mathcal{S} = \Gamma\Lambda\Gamma^\top$. Then the first k eigenvectors are retained to build

$$\hat{\mathcal{Q}} = [\sqrt{\lambda_1}\gamma_1, \ldots, \sqrt{\lambda_k}\gamma_k]. \tag{10.16}$$

The estimated specific variances are provided by the diagonal elements of the matrix $\mathcal{S} - \hat{\mathcal{Q}}\hat{\mathcal{Q}}^\top$,

$$\hat{\Psi} = \begin{pmatrix} \hat{\psi}_{11} & & & 0 \\ & \hat{\psi}_{22} & & \\ & & \ddots & \\ 0 & & & \hat{\psi}_{pp} \end{pmatrix} \quad \text{with } \hat{\psi}_{jj} = s_{X_jX_j} - \sum_{\ell=1}^{k} \hat{q}_{j\ell}^2. \tag{10.17}$$

By definition, the diagonal elements of $\mathcal{S}$ are equal to the diagonal elements of $\hat{\mathcal{Q}}\hat{\mathcal{Q}}^\top + \hat{\Psi}$. The off-diagonal elements are not necessarily estimated. How good then is this approximation? Consider the residual matrix

$$\mathcal{S} - (\hat{\mathcal{Q}}\hat{\mathcal{Q}}^\top + \hat{\Psi})$$

resulting from the principal component solution. Analytically we have that

$$\sum_{i,j} (\mathcal{S} - \hat{\mathcal{Q}}\hat{\mathcal{Q}}^\top - \hat{\Psi})_{ij}^2 \leq \lambda_{k+1}^2 + \ldots + \lambda_p^2.$$

This implies that a small value of the neglected eigenvalues can result in a small approximation error. A heuristic device for selecting the number of factors is to consider the proportion of the total sample variance due to the j-th factor. This quantity is in general equal to

(A) $\lambda_j / \sum_{j=1}^{p} s_{jj}$ for a factor analysis of $\mathcal{S}$,

(B) λ_j / p for a factor analysis of $\mathcal{R}$.

EXAMPLE 10.5 *This example uses a consumer-preference study from Johnson and Wichern (1998). Customers were asked to rate several attributes of a new product. The responses were tabulated and the following correlation matrix $\mathcal{R}$ was constructed:*

Attribute (Variable)						
Taste	*1*	1.00	0.02	**0.96**	0.42	0.01
Good buy for money	*2*	0.02	1.00	0.13	0.71	**0.85**
Flavor	*3*	0.96	0.13	1.00	0.50	0.11
Suitable for snack	*4*	0.42	0.71	0.50	1.00	**0.79**
Provides lots of energy	*5*	0.01	0.85	0.11	0.79	1.00

Variable	Estimated factor loadings $\hat{q}_1$	$\hat{q}_2$	Communalities $\hat{h}_j^2$	Specific variances $\hat{\psi}_{jj} = 1 - \hat{h}_j^2$
1. Taste	0.56	0.82	0.98	0.02
2. Good buy for money	0.78	-0.53	0.88	0.12
3. Flavor	0.65	0.75	0.98	0.02
4. Suitable for snack	0.94	-0.11	0.89	0.11
5. Provides lots of energy	0.80	-0.54	0.93	0.07
Eigenvalues	2.85	1.81		
Cumulative proportion of total (standardized) sample variance	0.571	0.932		

Table 10.1. Estimated factor loadings, communalities, and specific variances

The bold entries of $\mathcal{R}$ show that variables 1 and 3 and variables 2 and 5 are highly correlated. Variable 4 is more correlated with variables 2 and 5 than with variables 1 and 3. Hence, a model with 2 (or 3) factors seems to be reasonable.

The first two eigenvalues $\lambda_1 = 2.85$ and $\lambda_2 = 1.81$ of $\mathcal{R}$ are the only eigenvalues greater than one. Moreover, $k = 2$ common factors account for a cumulative proportion

$$\frac{\lambda_1 + \lambda_2}{p} = \frac{2.85 + 1.81}{5} = 0.93$$

of the total (standardized) sample variance. Using the principal component method, the estimated factor loadings, communalities, and specific variances, are calculated from formulas (10.16) and (10.17), and the results are given in Table 10.1.

Take a look at:

$$\hat{Q}\hat{Q}^\top + \hat{\Psi} = \begin{pmatrix} 0.56 & 0.82 \\ 0.78 & -0.53 \\ 0.65 & 0.75 \\ 0.94 & -0.11 \\ 0.80 & -0.54 \end{pmatrix} \begin{pmatrix} 0.56 & 0.78 & 0.65 & 0.94 & 0.80 \\ 0.82 & -0.53 & 0.75 & -0.11 & -0.54 \end{pmatrix} +$$

$$+ \begin{pmatrix} 0.02 & 0 & 0 & 0 & 0 \\ 0 & 0.12 & 0 & 0 & 0 \\ 0 & 0 & 0.02 & 0 & 0 \\ 0 & 0 & 0 & 0.11 & 0 \\ 0 & 0 & 0 & 0 & 0.07 \end{pmatrix} = \begin{pmatrix} 1.00 & 0.01 & 0.97 & 0.44 & 0.00 \\ 0.01 & 1.00 & 0.11 & 0.79 & 0.91 \\ 0.97 & 0.11 & 1.00 & 0.53 & 0.11 \\ 0.44 & 0.79 & 0.53 & 1.00 & 0.81 \\ 0.00 & 0.91 & 0.11 & 0.81 & 1.00 \end{pmatrix}.$$

This nearly reproduces the correlation matrix $\mathcal{R}$. We conclude that the two-factor model provides a good fit of the data. The communalities $(0.98, 0.88, 0.98, 0.89, 0.93)$ indicate that the two factors account for a large percentage of the sample variance of each variable. Due to the nonuniqueness of factor loadings, the interpretation might be enhanced by rotation. This is the topic of the next subsection.

Rotation

The constraints (10.11) and (10.12) are given as a matter of mathematical convenience (to create unique solutions) and can therefore complicate the problem of interpretation. The interpretation of the loadings would be very simple if the variables could be split into disjoint sets, each being associated with one factor. A well known analytical algorithm to rotate the loadings is given by the *varimax rotation method* proposed by Kaiser (1985). In the simplest case of $k = 2$ factors, a rotation matrix $\mathcal{G}$ is given by

$$\mathcal{G}(\theta) = \begin{pmatrix} \cos\theta & \sin\theta \\ -\sin\theta & \cos\theta \end{pmatrix},$$

representing a clockwise rotation of the coordinate axes by the angle θ. The corresponding rotation of loadings is calculated via $\hat{\mathcal{Q}}^* = \hat{\mathcal{Q}}\mathcal{G}(\theta)$. The idea of the *varimax method* is to find the angle θ that maximizes the sum of the variances of the squared loadings $\hat{q}^*_{ij}$ within each column of $\hat{\mathcal{Q}}^*$. More precisely, defining $\tilde{q}_{jl} = \hat{q}^*_{jl}/\hat{h}^*_j$, the *varimax criterion* chooses θ so that

$$\mathcal{V} = \frac{1}{p}\sum_{\ell=1}^{k}\left[\sum_{j=1}^{p}\left(\tilde{q}^*_{jl}\right)^4 - \left\{\frac{1}{p}\sum_{j=1}^{p}\left(\tilde{q}^*_{jl}\right)^2\right\}^2\right]$$

is maximized.

EXAMPLE 10.6 *Let us return to the marketing example of Johnson and Wichern (1998) (Example 10.5). The basic factor loadings given in Table 10.1 of the first factor and a second factor are almost identical making it difficult to interpret the factors. Applying the varimax rotation we obtain the loadings* $\tilde{q}_1 = (0.02, \mathbf{0.94}, 0.13, \mathbf{0.84}, \mathbf{0.97})^\top$ *and* $\tilde{q}_2 = (\mathbf{0.99}, -0.01, \mathbf{0.98}, 0.43, -0.02)^\top$. *The high loadings, indicated as bold entries, show that variables 2, 4, 5 define factor 1, a nutricional factor. Variable 1 and 3 define factor 2 which might be referred to as a taste factor.*

Summary

- ↪ In practice, $\mathcal{Q}$ and Ψ have to be estimated from $\mathcal{S} = \widehat{\mathcal{Q}}\widehat{\mathcal{Q}}^\top + \widehat{\Psi}$. The number of parameters is $d = \frac{1}{2}(p-k)^2 - \frac{1}{2}(p+k)$.
- ↪ If $d = 0$, then there exists an exact solution. In practice, d is usually greater than 0, thus approximations must be considered.
- ↪ The maximum-likelihood method assumes a normal distribution for the data. A solution can be found using numerical algorithms.
- ↪ The method of principal factors is a two-stage method which calculates $\widehat{\mathcal{Q}}$ from the reduced correlation matrix $\mathcal{R} - \widetilde{\Psi}$, where $\widetilde{\Psi}$ is a pre-estimate of Ψ. The final estimate of Ψ is found by $\widehat{\psi}_{ii} = 1 - \sum_{j=1}^{k}\widehat{q}_{ij}^2$.
- ↪ The principal component method is based on an approximation, $\widehat{\mathcal{Q}}$, of $\mathcal{Q}$.
- ↪ Often a more informative interpretation of the factors can be found by rotating the factors.
- ↪ The varimax rotation chooses a rotation θ that maximizes $\mathcal{V} = \frac{1}{p}\sum_{\ell=1}^{k}\left[\sum_{j=1}^{p}\left(\tilde{q}^*_{jl}\right)^4 - \left\{\frac{1}{p}\sum_{j=1}^{p}\left(\tilde{q}^*_{jl}\right)^2\right\}^2\right]$.

10.3 Factor Scores and Strategies

Up to now strategies have been presented for factor analysis that have concentrated on the estimation of loadings and communalities and on their interpretations. This was a logical step since the factors F were considered to be normalized random sources of information and were explicitely addressed as nonspecific (common factors). The estimated values of the factors, called the *factor scores*, may also be useful in the interpretation as well as in the diagnostic analysis. To be more precise, the factor scores are estimates of the unobserved random vectors F_l, $l = 1, \dots, k$, for each individual x_i, $i = 1, \dots, n$. Johnson and Wichern (1998) describe three methods which in practice yield very similar results. Here, we present the regression method which has the advantage of being the simplest technique and is easy to implement.

The idea is to consider the joint distribution of $(X - \mu)$ and F, and then to proceed with the regression analysis presented in Chapter 5. Under the factor model (10.4), the joint covariance matrix of $(X - \mu)$ and F is:

$$Var\begin{pmatrix} X-\mu \\ F \end{pmatrix} = \begin{pmatrix} \mathcal{Q}\mathcal{Q}^\top + \Psi & \mathcal{Q} \\ \mathcal{Q}^\top & \mathcal{I}_k \end{pmatrix}. \tag{10.18}$$

Note that the upper left entry of this matrix equals Σ and that the matrix has size $(p+k) \times (p+k)$.

Assuming joint normality, the conditional distribution of $F|X$ is multinormal, see Theorem 5.1, with

$$E(F|X = x) = \mathcal{Q}^\top \Sigma^{-1}(X - \mu) \tag{10.19}$$

and using (5.7) the covariance matrix can be calculated:

$$Var(F|X = x) = \mathcal{I}_k - \mathcal{Q}^\top \Sigma^{-1} \mathcal{Q}. \tag{10.20}$$

In practice, we replace the unknown $\mathcal{Q}$, Σ and μ by corresponding estimators, leading to the estimated individual factor scores:

$$\widehat{f}_i = \widehat{\mathcal{Q}}^\top \mathcal{S}^{-1}(x_i - \overline{x}). \tag{10.21}$$

We prefer to use the original sample covariance matrix $\mathcal{S}$ as an estimator of Σ, instead of the factor analysis approximation $\widehat{\mathcal{Q}}\widehat{\mathcal{Q}}^\top + \widehat{\Psi}$, in order to be more robust against incorrect determination of the number of factors.

The same rule can be followed when using $\mathcal{R}$ instead of $\mathcal{S}$. Then (10.18) remains valid when standardized variables, i.e., $Z = \mathcal{D}_\Sigma^{-1/2}(X - \mu)$, are considered if $\mathcal{D}_\Sigma = \operatorname{diag}(\sigma_{11}, \dots, \sigma_{pp})$. In this case the factors are given by

$$\widehat{f}_i = \widehat{\mathcal{Q}}^\top \mathcal{R}^{-1}(z_i), \tag{10.22}$$

where $z_i = \mathcal{D}_S^{-1/2}(x_i - \overline{x})$, $\widehat{\mathcal{Q}}$ is the loading obtained with the matrix $\mathcal{R}$, and $\mathcal{D}_S = \operatorname{diag}(s_{11}, \dots, s_{pp})$.

If the factors are rotated by the orthogonal matrix $\mathcal{G}$, the factor scores have to be rotated accordingly, that is

$$\widehat{f}_i^* = \mathcal{G}^\top \widehat{f}_i. \tag{10.23}$$

A practical example is presented in Section 10.4 using the Boston Housing data.

Practical Suggestions

No one method outperforms another in the practical implementation of factor analysis. However, by applying the *tâtonnement* process, the factor analysis view of the data can be stabilized. This motivates the following procedure.

1. Fix a reasonable number of factors, say $k = 2$ or 3, based on the correlation structure of the data and/or screeplot of eigenvalues.
2. Perform several of the presented methods, including rotation. Compare the loadings, communalities, and factor scores from the respective results.
3. If the results show significant deviations, check for outliers (based on factor scores), and consider changing the number of factors k.

For larger data sets, cross-validation methods are recommended. Such methods involve splitting the sample into a training set and a validation data set. On the training sample one estimates the factor model with the desired methodology and uses the obtained parameters to predict the factor scores for the validation data set. The predicted factor scores should be comparable to the factor scores obtained using only the validation data set. This stability criterion may also involve the loadings and communalities.

Factor Analysis versus PCA

Factor analysis and principal component analysis use the same set of mathematical tools (spectral decomposition, projections, ...). One could conclude, on first sight, that they share the same view and strategy and therefore yield very similar results. This is not true. There are substantial differences between these two data analysis techniques that we would like to describe here.

The biggest difference between PCA and factor analysis comes from the model philosophy. Factor analysis imposes a strict structure of a fixed number of common (latent) factors whereas the PCA determines p factors in decreasing order of importance. The most important factor in PCA is the one that maximizes the projected variance. The most important factor in factor analysis is the one that (after rotation) gives the maximal interpretation. Often this is different from the direction of the first principal component.

From an implementation point of view, the PCA is based on a well-defined, unique algorithm (spectral decomposition), whereas fitting a factor analysis model involves a variety of numerical procedures. The non-uniqueness of the factor analysis procedure opens the door for subjective interpretation and yields therefore a spectrum of results. This data analysis philosophy makes factor analysis difficult especially if the model specification involves cross-validation and a data-driven selection of the number of factors.

10.4 Boston Housing

To illustrate how to implement factor analysis we will use the Boston housing data set and the by now well known set of transformations. Once again, the variable X_4 (Charles River indicator) will be excluded. As before, standardized variables are used and the analysis is based on the correlation matrix.

In Section 10.3, we described a practical implementation of factor analysis. Based on principal components, three factors were chosen and factor analysis was applied using the max-

		Estimated factor loadings			Communalities	Specific variances
		$\hat{q}_1$	$\hat{q}_2$	$\hat{q}_3$	$\hat{h}_j^2$	$\hat{\psi}_{jj} = 1 - \hat{h}_j^2$
1	crime	0.9295	−0.1653	0.1107	0.9036	0.0964
2	large lots	−0.5823	−0.0379	0.2902	0.4248	0.5752
3	nonretail acres	0.8192	0.0296	−0.1378	0.6909	0.3091
5	nitric oxides	0.8789	−0.0987	−0.2719	0.8561	0.1439
6	rooms	−0.4447	−0.5311	−0.0380	0.4812	0.5188
7	prior 1940	0.7837	0.0149	−0.3554	0.7406	0.2594
8	empl. centers	−0.8294	0.1570	0.4110	0.8816	0.1184
9	accessibility	0.7955	−0.3062	0.4053	0.8908	0.1092
10	tax-rate	0.8262	−0.1401	0.2906	0.7867	0.2133
11	pupil/teacher	0.5051	0.1850	0.1553	0.3135	0.6865
12	African American	−0.4701	−0.0227	−0.1627	0.2480	0.7520
13	lower status	0.7601	0.5059	−0.0070	0.8337	0.1663
14	value	−0.6942	−0.5904	−0.1798	0.8628	0.1371

Table 10.2. Estimated factor loadings, communalities, and specific variances, MLM. MVAfacthous

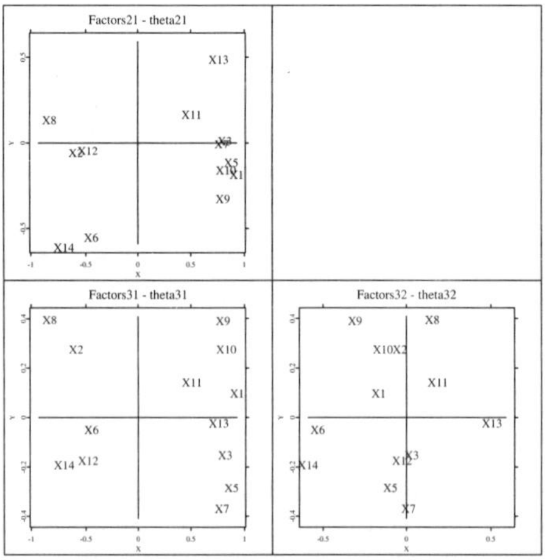

Figure 10.2. Factor analysis for Boston housing data, MLM. MVAfacthous

imum likelihood method (MLM), the principal factor method (PFM), and the principal component method (PCM). For illustration, the MLM will be presented with and without varimax rotation.

Table 10.2 gives the MLM factor loadings without rotation and Table 10.3 gives the varimax version of this analysis. The corresponding graphical representations of the loadings are displayed in Figures 10.2 and 10.3. We can see that the varimax does not significantly change the interpretation of the factors obtained by the MLM. Factor 1 can be roughly interpreted as a "quality of life factor" because it is positively correlated with variables like X_{11} and negatively correlated with X_8, both having low specific variances. The second factor

		Estimated factor loadings			Communalities	Specific variances
		$\hat{q}_1$	$\hat{q}_2$	$\hat{q}_3$	$\hat{h}_j^2$	$\hat{\psi}_{jj} = 1 - \hat{h}_j^2$
1	crime	0.8413	−0.0940	−0.4324	0.9036	0.0964
2	large lots	−0.3326	−0.1323	0.5447	0.4248	0.5752
3	nonretail acres	0.6142	0.1238	−0.5462	0.6909	0.3091
5	nitric oxides	0.5917	0.0221	−0.7110	0.8561	0.1439
6	rooms	−0.3950	−0.5585	0.1153	0.4812	0.5188
7	prior 1940	0.4665	0.1374	−0.7100	0.7406	0.2594
8	empl. centers	−0.4747	0.0198	0.8098	0.8816	0.1184
9	accessibility	0.8879	−0.2874	−0.1409	0.8908	0.1092
10	tax-rate	0.8518	−0.1044	−0.2240	0.7867	0.2133
11	pupil/teacher	0.5090	0.2061	−0.1093	0.3135	0.6865
12	African American	−0.4834	−0.0418	0.1122	0.2480	0.7520
13	lower status	0.6358	0.5690	−0.3252	0.8337	0.1663
14	value	−0.6817	−0.6193	0.1208	0.8628	0.1371

Table 10.3. Estimated factor loadings, communalities, and specific variances, MLM, varimax rotation. MVAfacthous

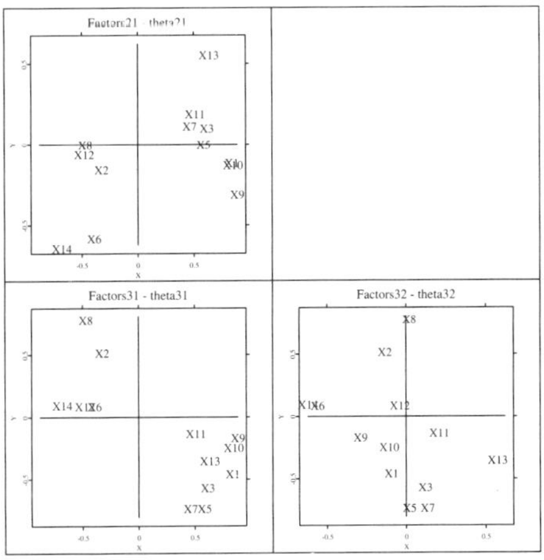

Figure 10.3. Factor analysis for Boston housing data, MLM after varimax rotation. MVAfacthous

may be interpreted as a "residential factor", since it is highly correlated with variables X_6, and X_{13}. The most striking difference between the results with and without varimax rotation can be seen by comparing the lower left corners of Figures 10.2 and 10.3. There is a clear separation of the variables in the varimax version of the MLM. Given this arrangement of the variables in Figure 10.3, we can interpret factor 3 as an employment factor, since we observe high correlations with X_8 and X_5.

We now turn to the PCM and PFM analyses. The results are presented in Tables 10.4 and 10.5 and in Figures 10.4 and 10.5. We would like to focus on the PCM, because this 3-factor model yields only one specific variance (unexplained variation) above 0.5. Looking

		Estimated factor loadings			Communalities	Specific variances
		$\hat{q}_1$	$\hat{q}_2$	$\hat{q}_3$	$\hat{h}_j^2$	$\hat{\psi}_{jj} = 1 - \hat{h}_j^2$
1	crime	0.9164	0.0152	0.2357	0.8955	0.1045
2	large lots	−0.6772	0.0762	0.4490	0.6661	0.3339
3	nonretail acres	0.8614	−0.1321	−0.1115	0.7719	0.2281
5	nitric oxides	0.9172	0.0573	−0.0874	0.8521	0.1479
6	rooms	−0.3590	0.7896	0.1040	0.7632	0.2368
7	prior 1940	0.8392	−0.0008	−0.2163	0.7510	0.2490
8	empl. centers	−0.8928	−0.1253	0.2064	0.8554	0.1446
9	accessibility	0.7562	0.0927	0.4616	0.7935	0.2065
10	tax-rate	0.7891	−0.0370	0.4430	0.8203	0.1797
11	pupil/teacher	0.4827	−0.3911	0.1719	0.4155	0.5845
12	African American	−0.4499	0.0368	−0.5612	0.5188	0.4812
13	lower status	0.6925	−0.5843	0.0035	0.8209	0.1791
14	value	−0.5933	0.6720	−0.1895	0.8394	0.1606

Table 10.4. Estimated factor loadings, communalities, and specific variances, PCM, varimax rotation. MVAfacthous

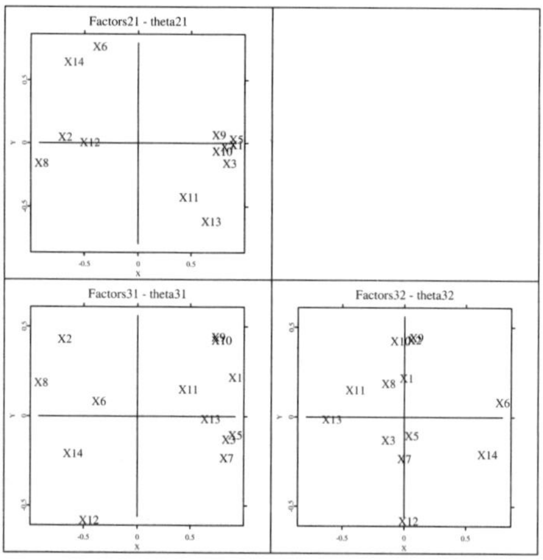

Figure 10.4. Factor analysis for Boston housing data, PCM after varimax rotation. MVAfacthous

at Figure 10.4, it turns out that factor 1 remains a "quality of life factor" which is clearly visible from the clustering of X_5, X_3, X_{10} and X_1 on the right-hand side of the graph, while the variables X_8, X_2, X_{14}, X_{12} and X_6 are on the left-hand side. Again, the second factor is a "residential factor", clearly demonstrated by the location of variables X_6, X_{14}, X_{11}, and X_{13}. The interpretation of the third factor is more difficult because all of the loadings (except for X_{12}) are very small.

		Estimated factor loadings			Communalities	Specific variances
		$\hat{q}_1$	$\hat{q}_2$	$\hat{q}_3$	$\hat{h}_j^2$	$\hat{\psi}_{jj} = 1 - \hat{h}_j^2$
1	crime	0.8579	−0.0270	−0.4175	0.9111	0.0889
2	large lots	−0.2953	0.2168	0.5756	0.4655	0.5345
3	nonretail acres	0.5893	−0.2415	−0.5666	0.7266	0.2734
5	nitric oxides	0.6050	−0.0892	−0.6855	0.8439	0.1561
6	rooms	−0.2902	0.6280	0.1296	0.4954	0.5046
7	prior 1940	0.4702	−0.1741	−0.6733	0.7049	0.2951
8	empl. centers	−0.4988	0.0414	0.7876	0.8708	0.1292
9	accessibility	0.8830	0.1187	−0.1479	0.8156	0.1844
10	tax-rate	0.8969	−0.0136	−0.1666	0.8325	0.1675
11	pupil/teacher	0.4590	−0.2798	−0.1412	0.3090	0.6910
12	African American	−0.4812	0.0666	0.0856	0.2433	0.7567
13	lower status	0.5433	−0.6604	−0.3193	0.8333	0.1667
14	value	−0.6012	0.7004	0.0956	0.8611	0.1389

Table 10.5. Estimated factor loadings, communalities, and specific variances, PFM, varimax rotation. MVAfacthous

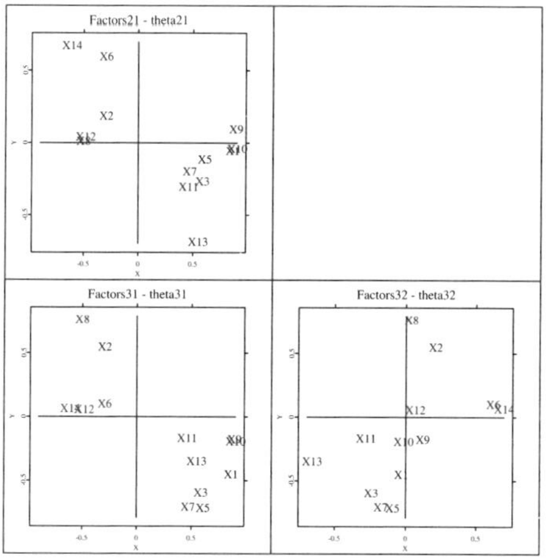

Figure 10.5. Factor analysis for Boston housing data, PFM after varimax rotation. MVAfacthous

10.5 Exercises

EXERCISE 10.1 *In Example 10.4 we have computed $\widehat{\mathcal{Q}}$ and $\widehat{\Psi}$ using the method of principal factors. We used a two-step iteration for $\widehat{\Psi}$. Perform the third iteration step and compare the results (i.e., use the given $\widehat{\mathcal{Q}}$ as a pre-estimate to find the final Ψ).*

EXERCISE 10.2 *Using the bank data set, how many factors can you find with the Method of Principal Factors?*

EXERCISE 10.3 *Repeat Exercise 10.2 with the U.S. company data set!*

EXERCISE 10.4 *Generalize the two-dimensional rotation matrix in Section 10.2 to n-dimensional space.*

EXERCISE 10.5 *Compute the orthogonal factor model for*

$$\Sigma = \begin{pmatrix} 1 & 0.9 & 0.7 \\ 0.9 & 1 & 0.4 \\ 0.7 & 0.4 & 1 \end{pmatrix}.$$

[Solution: $\psi_{11} = -0.575, q_{11} = 1.255$]

EXERCISE 10.6 *Perform a factor analysis on the type of families in the French food data set. Rotate the resulting factors in a way which provides the most reasonable interpretation. Compare your result with the varimax method.*

EXERCISE 10.7 *Perform a factor analysis on the variables X_3 to X_9 in the U.S. crime data set (Table B.10). Would it make sense to use all of the variables for the analysis?*

EXERCISE 10.8 *Analyze the athletic records data set (Table B.18). Can you recognize any patterns if you sort the countries according to the estimates of the factor scores?*

EXERCISE 10.9 *Perform a factor analysis on the U.S. health data set (Table B.16) and estimate the factor scores.*

EXERCISE 10.10 *Redo Exercise 10.9 using the U.S. crime data in Table B.10. Compare the estimated factor scores of the two data sets.*

EXERCISE 10.11 *Analyze the vocabulary data given in Table B.17.*

11 Cluster Analysis

The next two chapters address classification issues from two varying perspectives. When considering groups of objects in a multivariate data set, two situations can arise. Given a data set containing measurements on individuals, in some cases we want to see if some natural groups or classes of individuals exist, and in other cases, we want to classify the individuals according to a set of existing groups. Cluster analysis develops tools and methods concerning the former case, that is, given a data matrix containing multivariate measurements on a large number of individuals (or objects), the objective is to build some natural subgroups or clusters of individuals. This is done by grouping individuals that are "similar" according to some appropriate criterion. Once the clusters are obtained, it is generally useful to describe each group using some descriptive tool from Chapters 1, 8 or 9 to create a better understanding of the differences that exist among the formulated groups.

Cluster analysis is applied in many fields such as the natural sciences, the medical sciences, economics, marketing, etc. In marketing, for instance, it is useful to build and describe the different segments of a market from a survey on potential consumers. An insurance company, on the other hand, might be interested in the distinction among classes of potential customers so that it can derive optimal prices for its services. Other examples are provided below.

Discriminant analysis presented in Chapter 12 addresses the other issue of classification. It focuses on situations where the different groups are known *a priori*. Decision rules are provided in classifying a multivariate observation into one of the known groups.

Section 11.1 states the problem of cluster analysis where the criterion chosen to measure the similarity among objects clearly plays an important role. Section 11.2 shows how to precisely measure the proximity between objects. Finally, Section 11.3 provides some algorithms. We will concentrate on hierarchical algorithms only where the number of clusters is not known in advance.

11.1 The Problem

Cluster analysis is a set of tools for building groups (clusters) from multivariate data objects. The aim is to construct groups with homogeneous properties out of heterogeneous large samples. The groups or clusters should be as homogeneous as possible and the differences among the various groups as large as possible. Cluster analysis can be divided into two fundamental steps.

1. *Choice of a proximity measure:*
   One checks each pair of observations (objects) for the similarity of their values. A similarity (proximity) measure is defined to measure the "closeness" of the objects. The "closer" they are, the more homogeneous they are.
2. *Choice of group-building algorithm:*
   On the basis of the proximity measures the objects assigned to groups so that differences between groups become large and observations in a group become as close as possible.

In marketing, for exmaple, cluster analysis is used to select test markets. Other applications include the classification of companies according to their organizational structures, technologies and types. In psychology, cluster analysis is used to find types of personalities on the basis of questionnaires. In archaeology, it is applied to classify art objects in different time periods. Other scientific branches that use cluster analysis are medicine, sociology, linguistics and biology. In each case a heterogeneous sample of objects are analyzed with the aim to identify homogeneous subgroups.

Summary

↪ Cluster analysis is a set of tools for building groups (clusters) from multivariate data objects.

↪ The methods used are usually divided into two fundamental steps: The choice of a proximity measure and the choice of a group-building algorithm.

11.2 The Proximity between Objects

The starting point of a cluster analysis is a data matrix $\mathcal{X}(n \times p)$ with n measurements (objects) of p variables. The proximity (similarity) among objects is described by a matrix $\mathcal{D}(n \times n)$

$$\mathcal{D} = \begin{pmatrix} d_{11} & d_{12} & \dots & \dots & \dots & d_{1n} \\ \vdots & d_{22} & & & & \vdots \\ \vdots & \vdots & \ddots & & & \vdots \\ \vdots & \vdots & & \ddots & & \vdots \\ \vdots & \vdots & & & \ddots & \vdots \\ d_{n1} & d_{n2} & \dots & \dots & \dots & d_{nn} \end{pmatrix}. \tag{11.1}$$

The matrix $\mathcal{D}$ contains measures of similarity or dissimilarity among the n objects. If the values d_{ij} are distances, then they measure dissimilarity. The greater the distance, the less similar are the objects. If the values d_{ij} are proximity measures, then the opposite is true, i.e., the greater the proximity value, the more similar are the objects. A distance matrix, for example, could be defined by the L_2-norm: $d_{ij} = \|x_i - x_j\|_2$, where x_i and x_j denote the rows of the data matrix $\mathcal{X}$. Distance and similarity are of course dual. If d_{ij} is a distance, then $d'_{ij} = \max_{i,j}\{d_{ij}\} - d_{ij}$ is a proximity measure.

The nature of the observations plays an important role in the choice of proximity measure. Nominal values (like binary variables) lead in general to proximity values, whereas metric values lead (in general) to distance matrices. We first present possibilities for $\mathcal{D}$ in the binary case and then consider the continuous case.

Similarity of Objects with Binary Structure

In order to measure the similarity between objects we always compare pairs of observations (x_i, x_j) where $x_i^\top = (x_{i1}, \dots, x_{ip})$, $x_j^\top = (x_{j1}, \dots, x_{jp})$, and $x_{ik}, x_{jk} \in \{0, 1\}$. Obviously there

Name	δ	λ	Definition
Jaccard	0	1	$\frac{a_1}{a_1+a_2+a_3}$
Tanimoto	1	2	$\frac{a_1+a_4}{a_1+2(a_2+a_3)+a_4}$
Simple Matching (M)	1	1	$\frac{a_1+a_4}{p}$
Russel and Rao (RR)	–	–	$\frac{a_1}{p}$
Dice	0	0.5	$\frac{2a_1}{2a_1+(a_2+a_3)}$
Kulczynski	–	–	$\frac{a_1}{a_2+a_3}$

Table 11.1. The common similarity coefficients.

are four cases:

$$\begin{aligned} x_{ik} &= x_{jk} = 1, \\ x_{ik} &= 0, x_{jk} = 1, \\ x_{ik} &= 1, x_{jk} = 0, \\ x_{ik} &= x_{jk} = 0. \end{aligned}$$

Define

$$\begin{aligned} a_1 &= \sum_{k=1}^{p} \boldsymbol{I}(x_{ik} = x_{jk} = 1), \\ a_2 &= \sum_{k=1}^{p} \boldsymbol{I}(x_{ik} = 0, x_{jk} = 1), \\ a_3 &= \sum_{k=1}^{p} \boldsymbol{I}(x_{ik} = 1, x_{jk} = 0), \\ a_4 &= \sum_{k=1}^{p} \boldsymbol{I}(x_{ik} = x_{jk} = 0). \end{aligned}$$

Note that each a_ℓ, $\ell = 1, \ldots, 4$, depends on the pair (x_i, x_j).

The following proximity measures are used in practice:

$$d_{ij} = \frac{a_1 + \delta a_4}{a_1 + \delta a_4 + \lambda(a_2 + a_3)} \tag{11.2}$$

where δ and λ are weighting factors. Table 11.1 shows some similarity measures for given weighting factors.

These measures provide alternative ways of weighting mismatchings and positive (presence of a common character) or negative (absence of a common character) matchings. In principle, we could also consider the Euclidian distance. However, the disadvantage of this distance is that it treats the observations 0 and 1 in the same way. If $x_{ik} = 1$ denotes, say, knowledge of a certain language, then the contrary, $x_{ik} = 0$ (not knowing the language) should eventually be treated differently.

EXAMPLE 11.1 *Let us consider binary variables computed from the car data set (Table B.7). We define the new binary data by*

$$y_{ik} = \begin{cases} 1 & \text{if } x_{ik} > \overline{x}_k, \\ 0 & \text{otherwise,} \end{cases}$$

for $i = 1, \ldots, n$ and $k = 1, \ldots, p$. This means that we transform the observations of the k-th variable to 1 if it is larger than the mean value of all observations of the k-th variable. Let us only consider the data points 17 to 19 (Renault 19, Rover and Toyota Corolla) which lead to (3×3) distance matrices. The Jaccard measure gives the similarity matrix

$$\mathcal{D} = \begin{pmatrix} 1.000 & 0.000 & 0.333 \\ & 1.000 & 0.250 \\ & & 1.000 \end{pmatrix},$$

the Tanimoto measure yields

$$\mathcal{D} = \begin{pmatrix} 1.000 & 0.231 & 0.600 \\ & 1.000 & 0.455 \\ & & 1.000 \end{pmatrix},$$

whereas the Single Matching measure gives

$$\mathcal{D} = \begin{pmatrix} 1.000 & 0.375 & 0.750 \\ & 1.000 & 0.625 \\ & & 1.000 \end{pmatrix}.$$

Distance Measures for Continuous Variables

A wide variety of distance measures can be generated by the L_r-norms, $r \geq 1$,

$$d_{ij} = ||x_i - x_j||_r = \left\{ \sum_{k=1}^{p} |x_{ik} - x_{jk}|^r \right\}^{1/r}. \tag{11.3}$$

Here x_{ik} denotes the value of the k-th variable on object i. It is clear that $d_{ii} = 0$ for $i = 1, \ldots, n$. The class of distances (11.3) for varying r measures the dissimilarity of different weights. The L_1-metric, for example, gives less weight to outliers than the L_2-norm (Euclidean norm). It is common to consider the squared L_2-norm.

EXAMPLE 11.2 *Suppose we have $x_1 = (0, 0)$, $x_2 = (1, 0)$ and $x_3 = (5, 5)$. Then the distance matrix for the L_1-norm is*

$$\mathcal{D}_1 = \begin{pmatrix} 0 & 1 & 10 \\ 1 & 0 & 9 \\ 10 & 9 & 0 \end{pmatrix},$$

and for the squared L_2- or Euclidean norm

$$\mathcal{D}_2 = \begin{pmatrix} 0 & 1 & 50 \\ 1 & 0 & 41 \\ 50 & 41 & 0 \end{pmatrix}.$$

One can see that the third observation x_3 receives much more weight in the squared L_2-norm than in the L_1-norm.

An underlying assumption in applying distances based on L_r-norms is that the variables are measured on the same scale. If this is not the case, a standardization should first be applied. This corresponds to using a more general L_2- or Euclidean norm with a metric $\mathcal{A}$, where $\mathcal{A} > 0$ (see Section 2.6):

$$d_{ij}^2 = \|x_i - x_j\|_{\mathcal{A}} = (x_i - x_j)^\top \mathcal{A}(x_i - x_j). \tag{11.4}$$

L_2-norms are given by $\mathcal{A} = \mathcal{I}_p$, but if a standardization is desired, then the weight matrix $\mathcal{A} = \text{diag}(s_{X_1X_1}^{-1}, \ldots, s_{X_pX_p}^{-1})$ may be suitable. Recall that $s_{X_kX_k}$ is the variance of the k-th component. Hence we have

$$d_{ij}^2 = \sum_{k=1}^{p} \frac{(x_{ik} - x_{jk})^2}{s_{X_kX_k}}. \tag{11.5}$$

Here each component has the same weight in the computation of the distances and the distances do not depend on a particular choice of the units of measure.

EXAMPLE 11.3 *Consider the French Food expenditures (Table B.6). The Euclidean distance matrix (squared L_2-norm) is*

$$\mathcal{D} = 10^4 \cdot \begin{pmatrix} 0.00 & 5.82 & 58.19 & 3.54 & 5.15 & 151.44 & 16.91 & 36.15 & 147.99 & 51.84 & 102.56 & 271.83 \\ & 0.00 & 41.73 & 4.53 & 2.93 & 120.59 & 13.52 & 25.39 & 116.31 & 43.68 & 76.81 & 226.87 \\ & & 0.00 & 44.14 & 40.10 & 24.12 & 29.95 & 8.17 & 25.57 & 20.81 & 20.30 & 88.62 \\ & & & 0.00 & 0.76 & 127.85 & 5.62 & 21.70 & 124.98 & 31.21 & 72.97 & 231.57 \\ & & & & 0.00 & 121.05 & 5.70 & 19.85 & 118.77 & 30.82 & 67.39 & 220.72 \\ & & & & & 0.00 & 96.57 & 48.16 & 1.80 & 60.52 & 28.90 & 29.56 \\ & & & & & & 0.00 & 9.20 & 94.87 & 11.07 & 42.12 & 179.84 \\ & & & & & & & 0.00 & 46.95 & 6.17 & 18.76 & 113.03 \\ & & & & & & & & 0.00 & 61.08 & 29.62 & 31.86 \\ & & & & & & & & & 0.00 & 15.83 & 116.11 \\ & & & & & & & & & & 0.00 & 53.77 \\ & & & & & & & & & & & 0.00 \end{pmatrix}.$$

Taking the weight matrix $\mathcal{A} = \text{diag}(s_{X_1X_1}^{-1}, \ldots, s_{X_7X_7}^{-1})$, we obtain the distance matrix (squared L_2-norm)

$$\mathcal{D} = \begin{pmatrix} 0.00 & 6.85 & 10.04 & 1.68 & 2.66 & 24.90 & 8.28 & 8.56 & 24.61 & 21.55 & 30.68 & 57.48 \\ & 0.00 & 13.11 & 6.59 & 3.75 & 20.12 & 13.13 & 12.38 & 15.88 & 31.52 & 25.65 & 46.64 \\ & & 0.00 & 8.03 & 7.27 & 4.99 & 9.27 & 3.88 & 7.46 & 14.92 & 15.08 & 26.89 \\ & & & 0.00 & 0.64 & 20.06 & 2.76 & 3.82 & 19.63 & 12.81 & 19.28 & 45.01 \\ & & & & 0.00 & 17.00 & 3.54 & 3.81 & 15.76 & 14.98 & 16.89 & 39.87 \\ & & & & & 0.00 & 17.51 & 9.79 & 1.58 & 21.32 & 11.36 & 13.40 \\ & & & & & & 0.00 & 1.80 & 17.92 & 4.39 & 9.93 & 33.61 \\ & & & & & & & 0.00 & 10.50 & 5.70 & 7.97 & 24.41 \\ & & & & & & & & 0.00 & 24.75 & 11.02 & 13.07 \\ & & & & & & & & & 0.00 & 9.13 & 29.78 \\ & & & & & & & & & & 0.00 & 9.39 \\ & & & & & & & & & & & 0.00 \end{pmatrix}. \tag{11.6}$$

When applied to contingency tables, a χ^2-metric is suitable to compare (and cluster) rows and columns of a contingency table.

If $\mathcal{X}$ is a contingency table, row i is characterized by the conditional frequency distribution $\frac{x_{ij}}{x_{i\bullet}}$, where $x_{i\bullet} = \sum_{j=1}^{p} x_{ij}$ indicates the marginal distributions over the rows: $\frac{x_{i\bullet}}{x_{\bullet\bullet}}$, $x_{\bullet\bullet} =$

$\sum_{i=1}^{n} x_{i\bullet}$. Similarly, column j of $\mathcal{X}$ is characterized by the conditional frequencies $\frac{x_{ij}}{x_{\bullet j}}$, where $x_{\bullet j} = \sum_{i=1}^{n} x_{ij}$. The marginal frequencies of the columns are $\frac{x_{\bullet j}}{x_{\bullet\bullet}}$.

The distance between two rows, i_1 and i_2, corresponds to the distance between their respective frequency distributions. It is common to define this distance using the χ^2-metric:

$$d^2(i_1, i_2) = \sum_{j=1}^{p} \frac{1}{\left(\frac{x_{\bullet j}}{x_{\bullet\bullet}}\right)} \left(\frac{x_{i_1 j}}{x_{i_1\bullet}} - \frac{x_{i_2 j}}{x_{i_2\bullet}}\right)^2. \tag{11.7}$$

Note that this can be expressed as a distance between the vectors $x_1 = \left(\frac{x_{i_1 j}}{x_{\bullet\bullet}}\right)$ and $x_2 = \left(\frac{x_{i_2 j}}{x_{\bullet\bullet}}\right)$ as in (11.4) with weighting matrix $\mathcal{A} = \left\{diag\left(\frac{x_{\bullet j}}{x_{\bullet\bullet}}\right)\right\}^{-1}$. Similarly, if we are interested in clusters among the columns, we can define:

$$d^2(j_1, j_2) = \sum_{i=1}^{n} \frac{1}{\left(\frac{x_{i\bullet}}{x_{\bullet\bullet}}\right)} \left(\frac{x_{ij_1}}{x_{\bullet j_1}} - \frac{x_{ij_2}}{x_{\bullet j_2}}\right)^2.$$

Apart from the Euclidean and the L_r-norm measures one can use a proximity measure such as the Q-correlation coefficient

$$d_{ij} = \frac{\sum_{k=1}^{p}(x_{ik} - \overline{x}_i)(x_{jk} - \overline{x}_j)}{\{\sum_{k=1}^{p}(x_{ik} - \overline{x}_i)^2 \sum_{k=1}^{p}(x_{jk} - \overline{x}_j)^2\}^{1/2}}. \tag{11.8}$$

Here $\overline{x}_i$ denotes the mean over the variables $(x_{i1}, \ldots, x_{ip})$.

Summary

- ↪ The proximity between data points is measured by a distance or similarity matrix $\mathcal{D}$ whose components d_{ij} give the similarity coefficient or the distance between two points x_i and x_j.
- ↪ A variety of similarity (distance) measures exist for binary data (e.g., Jaccard, Tanimoto, Simple Matching coefficients) and for continuous data (e.g., L_r-norms).
- ↪ The nature of the data could impose the choice of a particular metric $\mathcal{A}$ in defining the distances (standardization, χ^2-metric etc.).

11.3 Cluster Algorithms

There are essentially two types of clustering methods: hierarchical algorithms and partioning algorithms. The hierarchical algorithms can be divided into agglomerative and splitting procedures. The first type of hierarchical clustering starts from the finest partition possible (each observation forms a cluster) and groups them. The second type starts with the coarsest partition possible: one cluster contains all of the observations. It proceeds by splitting the single cluster up into smaller sized clusters.

The partioning algorithms start from a given group definition and proceed by exchanging elements between groups until a certain score is optimized. The main difference between the two clustering techniques is that in hierarchical clustering once groups are found and elements are assigned to the groups, this assignment cannot be changed. In partitioning techniques, on the other hand, the assignment of objects into groups may change during the algorithm application.

Hierarchical Algorithms, Agglomerative Techniques

Agglomerative algorithms are used quite frequently in practice. The algorithm consists of the following steps:

Agglomerative Algorithm

1. Construct the finest partition.
2. Compute the distance matrix $\mathcal{D}$.

DO

3. Find the two clusters with the closest distance.
4. Put those two clusters into one cluster.
5. Compute the distance between the new groups and obtain a reduced distance matrix $\mathcal{D}$.

UNTIL all clusters are agglomerated into $\mathcal{X}$.

If two objects or groups say, P and Q, are united, one computes the distance between this new group (object) $P+Q$ and group R using the following distance function:

$$d(R, P+Q) = \delta_1 d(R,P) + \delta_2 d(R,Q) + \delta_3 d(P,Q) + \delta_4 |d(R,P) - d(R,Q)|. \tag{11.9}$$

The δ_j's are weighting factors that lead to different agglomerative algorithms as described in Table 11.2. Here $n_P = \sum_{i=1}^{n} \boldsymbol{I}(x_i \in P)$ is the number of objects in group P. The values of n_Q and n_R are defined analogously.

Name	δ_1	δ_2	δ_3	δ_4
Single linkage	1/2	1/2	0	-1/2
Complete linkage	1/2	1/2	0	1/2
Average linkage (unweighted)	1/2	1/2	0	0
Average linkage (weighted)	$\frac{n_P}{n_P+n_Q}$	$\frac{n_Q}{n_P+n_Q}$	0	0
Centroid	$\frac{n_P}{n_P+n_Q}$	$\frac{n_Q}{n_P+n_Q}$	$-\frac{n_P n_Q}{(n_P+n_Q)^2}$	0
Median	1/2	1/2	-1/4	0
Ward	$\frac{n_R+n_P}{n_R+n_P+n_Q}$	$\frac{n_R+n_Q}{n_R+n_P+n_Q}$	$-\frac{n_R}{n_R+n_P+n_Q}$	0

Table 11.2. Computations of group distances.

For the most common used Single and Complete linkages, below are the modified agglomerative algorithm steps:

Modified Agglomerative Algorithm

1. Construct the finest partition.
2. Compute the distance matrix $\mathcal{D}$.

DO

3. Find the smallest (Single linkage)/ largest (Complete linkage) value (between objects m and n) in $\mathcal{D}$.
4. If m and n are not in the same cluster, combine the clusters m and n belonging to together, and delete the smallest value.

UNTIL all clusters are agglomerated into $\mathcal{X}$ or the value in Step 3 exceeds the preset level.

As instead of computing new distance matrixes every step, a linear search in the original distance matrix is enough for clustering in the modified algorithm, it is more efficient in practice.

EXAMPLE 11.4 *Let us examine the agglomerative algorithm for the three points in Example 11.2, $x_1 = (0,0)$, $x_2 = (1,0)$ and $x_3 = (5,5)$, and the squared Euclidean distance matrix with single linkage weighting. The algorithm starts with $N = 3$ clusters: $P = \{x_1\}$, $Q = \{x_2\}$ and $R = \{x_3\}$. The distance matrix $\mathcal{D}_2$ is given in Example 11.2. The smallest distance in $\mathcal{D}_2$ is the one between the clusters P and Q. Therefore, applying step 4 in the above algorithm we combine these clusters to form $P + Q = \{x_1, x_2\}$. The single linkage distance between the remaining two clusters is from Table 11.2 and (11.9) equal to*

$$
\begin{aligned}
d(R, P+Q) &= \frac{1}{2}d(R,P) + \frac{1}{2}d(R,Q) - \frac{1}{2}|d(R,P) - d(R,Q)| \qquad (11.10)\\
&= \frac{1}{2}d_{13} + \frac{1}{2}d_{23} - \frac{1}{2}\cdot|d_{13} - d_{23}|\\
&= \frac{50}{2} + \frac{41}{2} - \frac{1}{2}\cdot|50 - 41|\\
&= 41.
\end{aligned}
$$

The reduced distance matrix is then $\begin{pmatrix} 0 & 41 \\ 41 & 0 \end{pmatrix}$. The next and last step is to unite the clusters R and $P + Q$ into a single cluster $\mathcal{X}$, the original data matrix.

When there are more data points than in the example above, a visualization of the implication of clusters is desirable. A graphical representation of the sequence of clustering is called a *dendrogram*. It displays the observations, the sequence of clusters and the distances between the clusters. The vertical axis displays the indices of the points, whereas the horizontal axis gives the distance between the clusters. Large distances indicate the clustering of heterogeneous groups. Thus, if we choose to "cut the tree" at a desired level, the branches describe the corresponding clusters.

EXAMPLE 11.5 *Here we describe the single linkage algorithm for the eight data points*

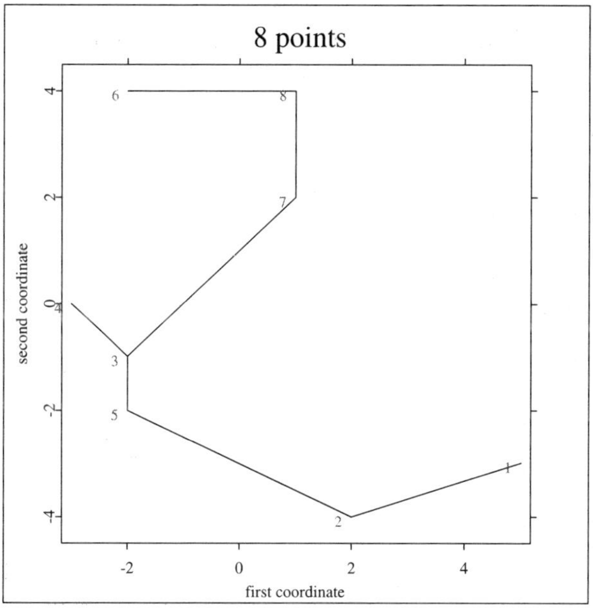

Figure 11.1. The 8-point example. MVAclus8p

displayed in Figure 11.1. The distance matrix (L_2-norms) is

$$\mathcal{D} = \begin{pmatrix} 0 & 10 & 53 & 73 & 50 & 98 & 41 & 65 \\ & 0 & 25 & 41 & 20 & 80 & 37 & 65 \\ & & 0 & 2 & 1 & 25 & 18 & 34 \\ & & & 0 & 5 & 17 & 20 & 32 \\ & & & & 0 & 36 & 25 & 45 \\ & & & & & 0 & 13 & 9 \\ & & & & & & 0 & 4 \\ & & & & & & & 0 \end{pmatrix}$$

and the dendrogram is shown in Figure 11.2.

If we decide to cut the tree at the level 10, *three clusters are defined:* $\{1, 2\}$, $\{3, 4, 5\}$ *and* $\{6, 7, 8\}$.

The single linkage algorithm defines the distance between two groups as the smallest value of the individual distances. Table 11.2 shows that in this case

$$d(R, P + Q) = \min\{d(R, P), d(R, Q)\}. \tag{11.11}$$

This algorithm is also called the *Nearest Neighbor* algorithm. As a consequence of its construction, single linkage tends to build large groups. Groups that differ but are not well separated may thus be classified into one group as long as they have two approximate points. The *complete linkage* algorithm tries to correct this kind of grouping by considering the largest (individual) distances. Indeed, the complete linkage distance can be written as

$$d(R, P + Q) = \max\{d(R, P),\ d(R, Q)\}. \tag{11.12}$$

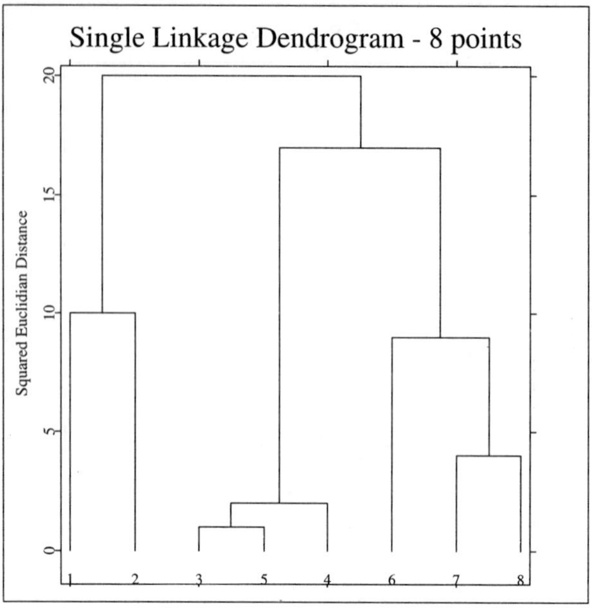

Figure 11.2. The dendrogram for the 8-point example, Single linkage algorithm. MVAclus8p

It is also called the *Farthest Neighbor* algorithm. This algorithm will cluster groups where all the points are proximate, since it compares the largest distances. The *average linkage* algorithm (weighted or unweighted) proposes a compromise between the two preceding algorithms, in that it computes an average distance:

$$d(R, P+Q) = \frac{n_P}{n_P+n_Q} d(R,P) + \frac{n_Q}{n_P+n_Q} d(R,Q). \tag{11.13}$$

The *centroid* algorithm is quite similar to the average linkage algorithm and uses the natural geometrical distance between R and the weighted center of gravity of P and Q (see Figure 11.3):

$$d(R, P+Q) = \frac{n_P}{n_P+n_Q} d(R,P) + \frac{n_Q}{n_P+n_Q} d(R,Q) - \frac{n_P n_Q}{(n_P+n_Q)^2} d(P,Q). \tag{11.14}$$

The *Ward clustering* algorithm computes the distance between groups according to the formula in Table 11.2. The main difference between this algorithm and the linkage procedures is in the unification procedure. The Ward algorithm does not put together groups with smallest distance. Instead, it joins groups that do not increase a given measure of heterogeneity "too much". The aim of the Ward procedure is to unify groups such that the variation inside these groups does not increase too drastically: the resulting groups are as homogeneous as possible.

The heterogeneity of group R is measured by the inertia inside the group. This inertia is defined as follows:

$$I_R = \frac{1}{n_R} \sum_{i=1}^{n_R} d^2(x_i, \overline{x}_R) \tag{11.15}$$

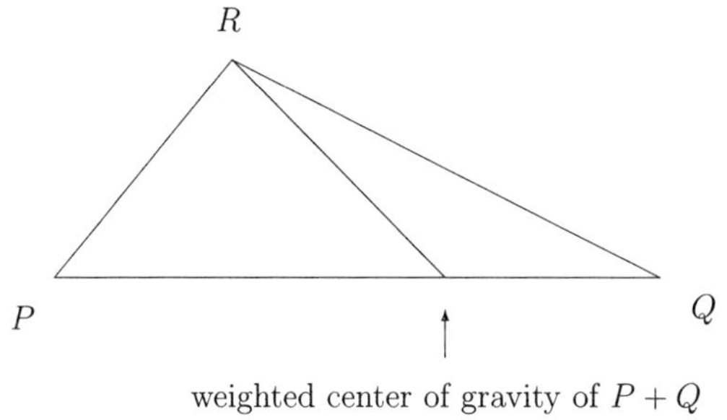

Figure 11.3. The centroid algorithm.

where $\overline{x}_R$ is the center of gravity (mean) over the groups. I_R clearly provides a scalar measure of the dispersion of the group around its center of gravity. If the usual Euclidean distance is used, then I_R represents the sum of the variances of the p components of x_i inside group R.

When two objects or groups P and Q are joined, the new group $P + Q$ has a larger inertia I_{P+Q}. It can be shown that the corresponding increase of inertia is given by

$$\Delta(P, Q) = \frac{n_P n_Q}{n_P + n_Q} d^2(P, Q). \tag{11.16}$$

In this case, the Ward algorithm is defined as an algorithm that "joins the groups that give the smallest increase in $\Delta(P, Q)$". It is easy to prove that when P and Q are joined, the new criterion values are given by (11.9) along with the values of δ_i given in Table 11.2, when the centroid formula is used to modify $d^2(R, P + Q)$. So, the Ward algorithm is related to the centroid algorithm, but with an "inertial" distance Δ rather than the "geometric" distance d^2.

As pointed out in Section 11.2, all the algorithms above can be adjusted by the choice of the metric $\mathcal{A}$ defining the geometric distance d^2. If the results of a clustering algorithm are illustrated as graphical representations of individuals in spaces of low dimension (using principal components (normalized or not) or using a correspondence analysis for contingency tables), it is important to be coherent in the choice of the metric used.

EXAMPLE 11.6 *As an example we randomly select 20 observations from the bank notes data and apply the Ward technique using Euclidean distances. Figure 11.4 shows the first two PCs of these data, Figure 11.5 displays the dendrogram.*

EXAMPLE 11.7 *Consider the French food expenditures. As in Chapter 9 we use standardized data which is equivalent to using* $\mathcal{A} = \mathrm{diag}(s_{X_1X_1}^{-1}, \ldots, s_{X_7X_7}^{-1})$ *as the weight matrix in the* L_2*-norm. The NPCA plot of the individuals was given in Figure 9.7. The Euclidean distance matrix is of course given by (11.6). The dendrogram obtained by using the Ward algorithm is shown in Figure 11.6.*

If the aim was to have only two groups, as can be seen in Figure 11.6 , they would be {CA2, CA3, CA4, CA5, EM5} and {MA2, MA3, MA4, MA5, EM2, EM3, EM4}. Clustering three groups is somewhat arbitrary (the levels of the distances are too similar). If we were interested in four groups, we would obtain {CA2, CA3, CA4}, {EM2, MA2, EM3, MA3}, {EM4, MA4, MA5} and {EM5, CA5}. This grouping shows a balance between socio-professional levels and size of the families in determining the clusters. The four groups are clearly well represented in the NPCA plot in Figure 9.7.

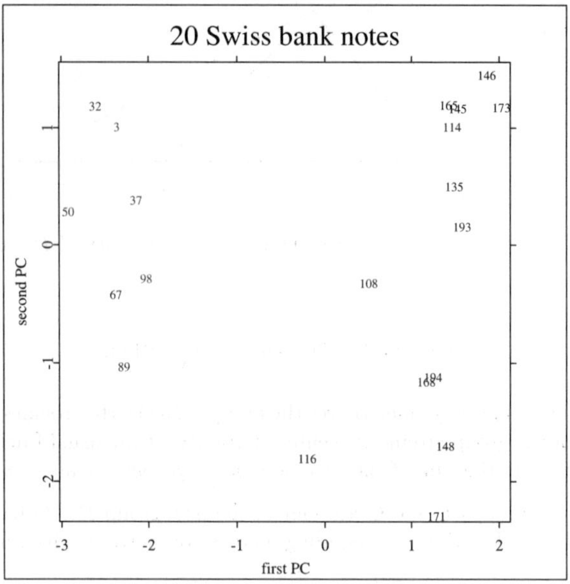

Figure 11.4. PCA for 20 randomly chosen bank notes. MVAclusbank

Summary

- ↪ The class of clustering algorithms can be divided into two types: hierarchical and partitioning algorithms. Hierarchical algorithms start with the finest (coarsest) possible partition and put groups together (split groups apart) step by step. Partitioning algorithms start from a preliminary clustering and exchange group elements until a certain score is reached.
- ↪ Hierarchical agglomerative techniques are frequently used in practice. They start from the finest possible structure (each data point forms a cluster), compute the distance matrix for the clusters and join the clusters that have the smallest distance. This step is repeated until all points are united in one cluster.
- ↪ The agglomerative procedure depends on the definition of the distance between two clusters. Single linkage, complete linkage, and Ward distance are frequently used distances.
- ↪ The process of the unification of clusters can be graphically represented by a dendrogram.

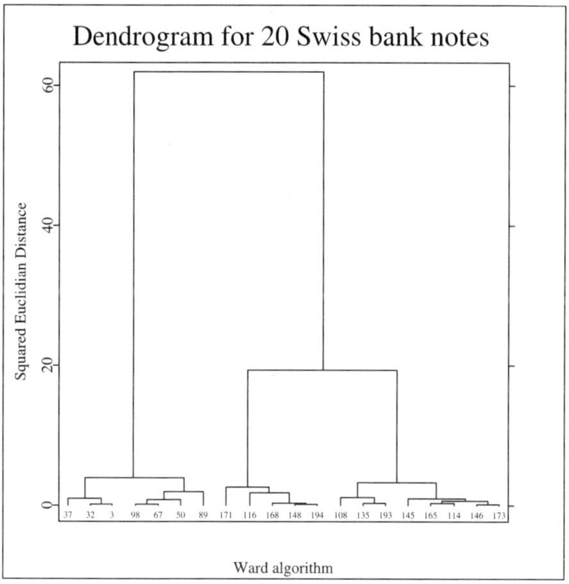

Figure 11.5. The dendrogram for the 20 bank notes, Ward algorithm. `MVAclusbank`

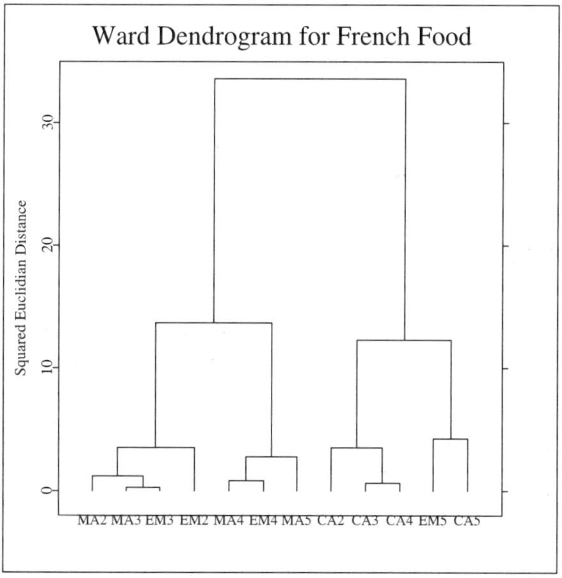

Figure 11.6. The dendrogram for the French food expenditures, Ward algorithm. `MVAclusfood`

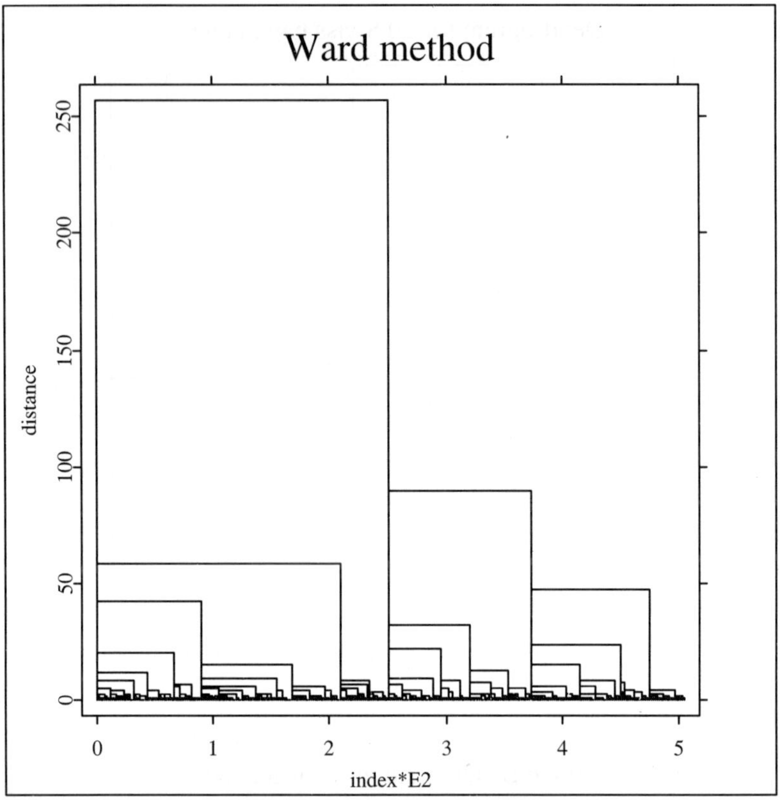

Figure 11.7. Dendrograms of the Boston housing data using the Ward algorithm. MVAclusbh

11.4 Boston Housing

We have motivated the transformation of the variables of the Boston housing data many times before. Now we illustrate the cluster algorithm with the transformed data $\widetilde{\mathcal{X}}$ excluding $\widetilde{X}_4$ (Charles River indicator). Among the various algorithms, the results from the Ward algorithm are presented since this algorithm gave the most sensible results. In order to be coherent with our previous analysis, we standardize each variable. The dendrogram of the Ward method is displayed in Figure 11.7. Two dominant clusters are visible. A further refinement of say, 4 clusters, could be considered at a lower level of distance.

To interprete the two clusters, we present the mean values and their respective standard errors of the thirteen $\widetilde{\mathcal{X}}$ variables by group in Table 11.3. Comparing the mean values for both groups shows that all the differences in the means are individually significant and that cluster one corresponds to housing districts with better living quality and higher house prices, whereas cluster two corresponds to less favored districts in Boston. This can be confirmed, for instance, by a lower crime rate, a higher proportion of residential land, lower proportion

Variable	Mean C1	*SE* C1	Mean C2	*SE* C2
1	−0.7105	0.0332	0.6994	0.0535
2	0.4848	0.0786	−0.4772	0.0047
3	−0.7665	0.0510	0.7545	0.0279
5	−0.7672	0.0365	0.7552	0.0447
6	0.4162	0.0571	−0.4097	0.0576
7	−0.7730	0.0429	0.7609	0.0378
8	0.7140	0.0472	−0.7028	0.0417
9	−0.5429	0.0358	0.5344	0.0656
10	−0.6932	0.0301	0.6823	0.0569
11	−0.5464	0.0469	0.5378	0.0582
12	0.3547	0.0080	−0.3491	0.0824
13	−0.6899	0.0401	0.6791	0.0509
14	0.5996	0.0431	−0.5902	0.0570

Table 11.3. Means and standard errors of the 13 standardized variables for Cluster 1 (251 observations) and Cluster 2 (255 observations). `MVAclusbh`

of African American, etc. for cluster one. Cluster two is identified by a higher proportion of older houses, a higher pupil/teacher ratio and a higher percentage of the lower status population.

This interpretation is underlined by visual inspection of all the variables presented on scatterplot matrices in Figures 11.8 and 11.9. For example, the lower right boxplot of Figure 11.9 and the correspondingly colored clusters in the last row confirm the role of each variable in determining the clusters. This interpretation perfectly coincides with the previous PC analysis (Figure 9.11). The quality of life factor is clearly visible in Figure 11.10, where cluster membership is distinguished by the shape and color of the points graphed according to the first two principal components. Clearly, the first PC completely separates the two clusters and corresponds, as we have discussed in Chapter 9, to a quality of life and house indicator.

11.5 Exercises

EXERCISE 11.1 *Prove formula (11.16).*

EXERCISE 11.2 *Prove that* $I_R = \operatorname{tr}(\mathcal{S}_R)$, *where* $\mathcal{S}_R$ *denotes the empirical covariance matrix of the observations contained in* R.

EXERCISE 11.3 *Prove that*

$$\Delta(R, P+Q) = \frac{n_R + n_P}{n_R + n_P + n_Q}\,\Delta(R,P) + \frac{n_R + n_Q}{n_R + n_P + n_Q}\,\Delta(R,Q) - \frac{n_R}{n_R + n_P + n_Q}\,\Delta(P,Q),$$

when the centroid formula is used to define $d^2(R, P+Q)$.

EXERCISE 11.4 *Repeat the 8-point example (Example 11.5) using the complete linkage and the Ward algorithm. Explain the difference to single linkage.*

EXERCISE 11.5 *Explain the differences between various proximity measures by means of an example.*

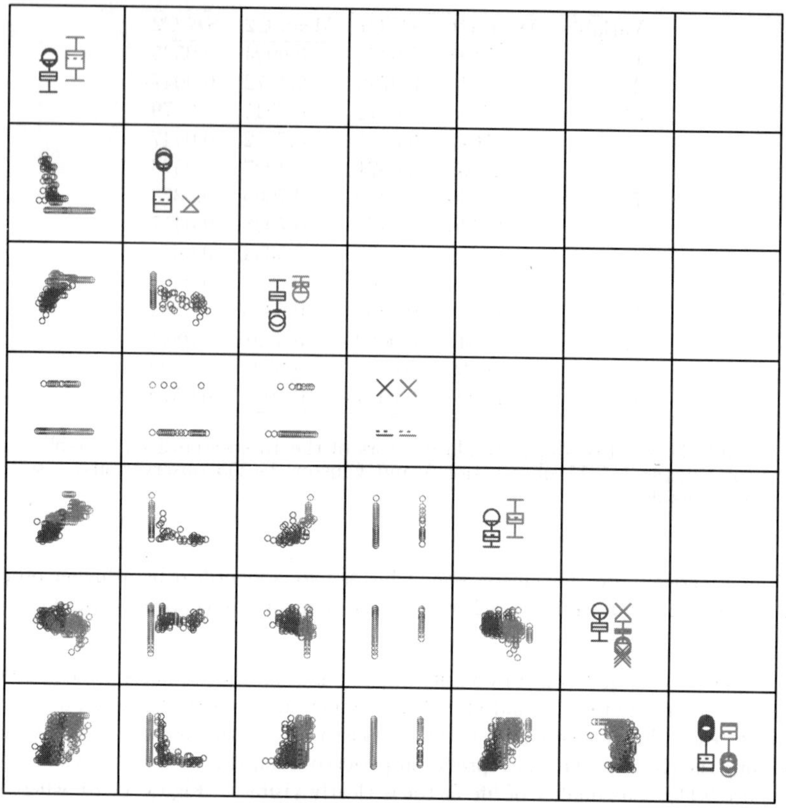

Figure 11.8. Scatterplot matrix for variables $\widetilde{X}_1$ to $\widetilde{X}_7$ of the Boston housing data. MVAclusbh

EXERCISE 11.6 *Repeat the bank notes example (Example 11.6) with another random sample of 20 notes.*

EXERCISE 11.7 *Repeat the bank notes example (Example 11.6) with another clustering algorithm.*

EXERCISE 11.8 *Repeat the bank notes example (Example 11.6) or the 8-point example (Example 11.5) with the L_1-norm.*

EXERCISE 11.9 *Analyze the U.S. companies example (Table B.5) using the Ward algorithm and the L_2-norm.*

EXERCISE 11.10 *Analyze the U.S. crime data set (Table B.10) with the Ward algorithm and the L_2-norm on standardized variables (use only the crime variables).*

EXERCISE 11.11 *Repeat Exercise 11.10 with the U.S. health data set (use only the number of deaths variables).*

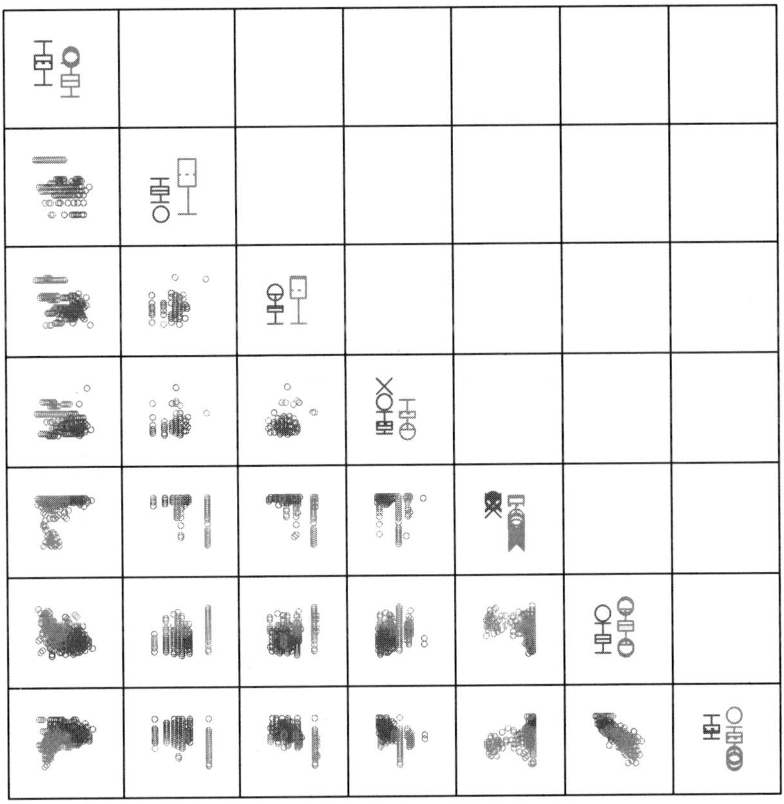

Figure 11.9. Scatterplot matrix for variables $\widetilde{X}_8$ to $\widetilde{X}_{14}$ of the Boston housing data. MVAclusbh

EXERCISE 11.12 *Redo Exercise 11.10 with the χ^2-metric. Compare the results.*

EXERCISE 11.13 *Redo Exercise 11.11 with the χ^2-metric and compare the results.*

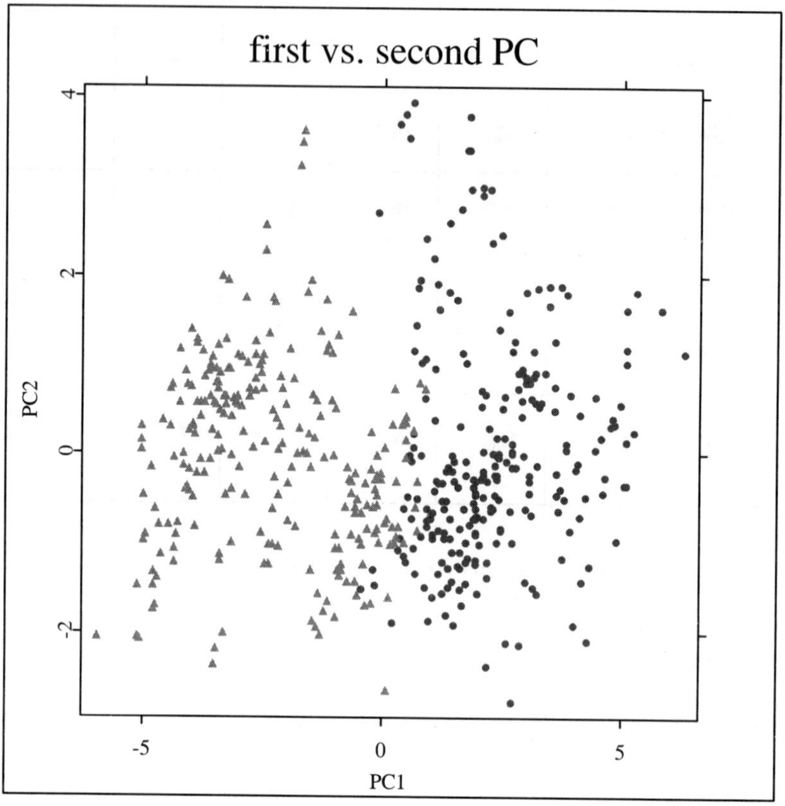

Figure 11.10. Scatterplot of the first two PCs displaying the two clusters.
MVAclusbh

12 Discriminant Analysis

Discriminant analysis is used in situations where the clusters are known *a priori*. The aim of discriminant analysis is to classify an observation, or several observations, into these known groups. For instance, in credit scoring, a bank knows from past experience that there are good customers (who repay their loan without any problems) and bad customers (who showed difficulties in repaying their loan). When a new customer asks for a loan, the bank has to decide whether or not to give the loan. The past records of the bank provides two data sets: multivariate observations x_i on the two categories of customers (including for example age, salary, marital status, the amount of the loan, etc.). The new customer is a new observation x with the same variables. The discrimination rule has to classify the customer into one of the two existing groups and the discriminant analysis should evaluate the risk of a possible "bad decision".

Many other examples are described below, and in most applications, the groups correspond to natural classifications or to groups known from history (like in the credit scoring example). These groups could have been formed by a cluster analysis performed on past data.

Section 12.1 presents the allocation rules when the populations are known, i.e., when we know the distribution of each population. As described in Section 12.2 in practice the population characteristics have to be estimated from history. The methods are illustrated in several examples.

12.1 Allocation Rules for Known Distributions

Discriminant analysis is a set of methods and tools used to distinguish between groups of populations Π_j and to determine how to allocate new observations into groups. In one of our running examples we are interested in discriminating between counterfeit and true bank notes on the basis of measurements of these bank notes, see Table B.2. In this case we have two groups (counterfeit and genuine bank notes) and we would like to establish an algorithm (rule) that can allocate a new observation (a new bank note) into one of the groups.

Another example is the detection of "fast" and "slow" consumers of a newly introduced product. Using a consumer's characteristics like education, income, family size, amount of previous brand switching, we want to classify each consumer into the two groups just identified.

In poetry and literary studies the frequencies of spoken or written words and lengths of sentences indicate profiles of different artists and writers. It can be of interest to attribute unknown literary or artistic works to certain writers with a specific profile. Anthropological measures on ancient sculls help in discriminating between male and female bodies. Good and poor credit risk ratings constitute a discrimination problem that might be tackled using observations on income, age, number of credit cards, family size etc.

In general we have populations $\Pi_j, j = 1, 2, ..., J$ and we have to allocate an observation x to one of these groups. A *discriminant rule* is a separation of the sample space (in general

$\mathbb{R}^p$) into sets R_j such that if $x \in R_j$, it is identified as a member of population Π_j.

The main task of discriminant analysis is to find "good" regions R_j such that the error of misclassification is small. In the following we describe such rules when the population distributions are known.

Maximum Likelihood Discriminant Rule

Denote the densities of each population Π_j by $f_j(x)$. The *maximum likelihood discriminant rule* (ML rule) is given by allocating x to Π_j maximizing the likelihood $L_j(x) = f_j(x) = \max_i f_i(x)$.

If several f_i give the same maximum then any of them may be selected. Mathematically, the sets R_j given by the ML discriminant rule are defined as

$$R_j = \{x : L_j(x) > L_i(x) \text{ for } i = 1, \ldots, J, i \neq j\}. \tag{12.1}$$

By classifying the observation into a certain group we may encounter a misclassification error. For $J = 2$ groups the probability of putting x into group 2 although it is from population 1 can be calculated as

$$p_{21} = P(X \in R_2|\Pi_1) = \int_{R_2} f_1(x)dx. \tag{12.2}$$

Similarly the conditional probability of classifying an object as belonging to the first population Π_1 although it actually comes from Π_2 is

$$p_{12} = P(X \in R_1|\Pi_2) = \int_{R_1} f_2(x)dx. \tag{12.3}$$

The misclassified observations create a cost $C(i|j)$ when a Π_j observation is assigned to R_i. In the credit risk example, this might be the cost of a "sour" credit. The cost structure can be pinned down in a cost matrix:

		Classified population	
		Π_1	Π_2
True population	Π_1	0	$C(2\|1)$
	Π_2	$C(1\|2)$	0

Let π_j be the prior probability of population Π_j, where "prior" means the *a priori* probability that an individual selected at random belongs to Π_j (i.e., before looking to the value x). Prior probabilities should be considered if it is clear ahead of time that an observation is more likely to stem from a certain population Π_j. An example is the classification of musical tunes. If it is known that during a certain period of time a majority of tunes were written by a certain composer, then there is a higher probability that a certain tune was composed by this composer. Therefore, he should receive a higher prior probability when tunes are assigned to a specific group.

The *expected cost of misclassification* (ECM) is given by

$$ECM = C(2|1)p_{21}\pi_1 + C(1|2)p_{12}\pi_2. \tag{12.4}$$

We will be interested in classification rules that keep the ECM small or minimize it over a class of rules. The discriminant rule minimizing the ECM (12.4) for two populations is given below.

THEOREM 12.1 *For two given populations, the rule minimizing the ECM is given by*

$$R_1 = \left\{ x : \frac{f_1(x)}{f_2(x)} \geq \left(\frac{C(1|2)}{C(2|1)}\right)\left(\frac{\pi_2}{\pi_1}\right)\right\}$$

$$R_2 = \left(x : \frac{f_1(x)}{f_2(x)} < \left\{\frac{C(1|2)}{C(2|1)}\right)\left(\frac{\pi_2}{\pi_1}\right)\right\}$$

The ML discriminant rule is thus a special case of the ECM rule for equal misclassification costs and equal prior probabilities. For simplicity the unity cost case, $C(1|2) = C(2|1) = 1$, and equal prior probabilities, $\pi_2 = \pi_1$, are assumed in the following.

Theorem 12.1 will be proven by an example from credit scoring.

EXAMPLE 12.1 *Suppose that Π_1 represents the population of bad clients who create the cost $C(2|1)$ if they are classified as good clients. Analogously, define $C(1|2)$ as the cost of loosing a good client classified as a bad one. Let γ denote the gain of the bank for the correct classification of a good client. The total gain of the bank is then*

$$\begin{aligned} G(R_2) &= -C(2|1)\pi_1 \int \boldsymbol{I}(x \in R_2) f_1(x)dx - C(1|2)\pi_2 \int \{1 - \boldsymbol{I}(x \in R_2)\} f_2(x)dx \\ &\quad + \gamma\pi_2 \int \boldsymbol{I}(x \in R_2) f_2(x)dx \\ &= -C(1|2)\pi_2 + \int \boldsymbol{I}(x \in R_2)\{-C(2|1)\pi_1 f_1(x) + (C(1|2) + \gamma)\pi_2 f_2(x)\}dx \end{aligned}$$

Since the first term in this equation is constant, the maximum is obviously obtained for

$$R_2 = \{\, x : -C(2|1)\pi_1 f_1(x) + \{C(1|2) + \gamma\}\pi_2 f_2(x) \geq 0\,\}.$$

This is equivalent to

$$R_2 = \left\{ x : \frac{f_2(x)}{f_1(x)} \geq \frac{C(2|1)\pi_1}{\{C(1|2) + \gamma\}\pi_2}\right\},$$

which corresponds to the set R_2 in Theorem 12.1 for a gain of $\gamma = 0$.

EXAMPLE 12.2 *Suppose $x \in \{0, 1\}$ and*

$$\begin{aligned} \Pi_1 &: P(X = 0) = P(X = 1) = \frac{1}{2} \\ \Pi_2 &: P(X = 0) = \frac{1}{4} = 1 - P(X = 1). \end{aligned}$$

The sample space is the set $\{0, 1\}$. The ML discriminant rule is to allocate $x = 0$ to Π_1 and $x = 1$ to Π_2, defining the sets $R_1 = \{0\}$, $R_2 = \{1\}$ and $R_1 \cup R_2 = \{0, 1\}$.

EXAMPLE 12.3 *Consider two normal populations*

$$\begin{aligned} \Pi_1 &: N(\mu_1, \sigma_1^2), \\ \Pi_2 &: N(\mu_2, \sigma_2^2). \end{aligned}$$

Then

$$L_i(x) = (2\pi\sigma_i^2)^{-1/2} \exp\left\{-\frac{1}{2}\left(\frac{x - \mu_i}{\sigma_i}\right)^2\right\}.$$

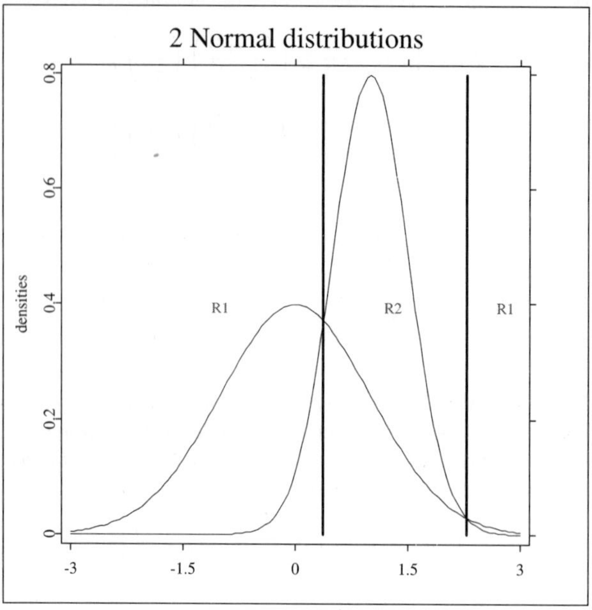

Figure 12.1. Maximum likelihood rule for normal distributions. MVAdisnorm

Hence x is allocated to Π_1 ($x \in R_1$) if $L_1(x) \geq L_2(x)$. Note that $L_1(x) \geq L_2(x)$ is equivalent to

$$\frac{\sigma_2}{\sigma_1} \exp\left\{-\frac{1}{2}\left[\left(\frac{x-\mu_1}{\sigma_1}\right)^2 - \left(\frac{x-\mu_2}{\sigma_2}\right)^2\right]\right\} \geq 1$$

or

$$x^2\left(\frac{1}{\sigma_1^2} - \frac{1}{\sigma_2^2}\right) - 2x\left(\frac{\mu_1}{\sigma_1^2} - \frac{\mu_2}{\sigma_2^2}\right) + \left(\frac{\mu_1^2}{\sigma_1^2} - \frac{\mu_2^2}{\sigma_2^2}\right) \leq 2\log\frac{\sigma_2}{\sigma_1}. \tag{12.5}$$

Suppose that $\mu_1 = 0$, $\sigma_1 = 1$ and $\mu_2 = 1$, $\sigma_2 = \frac{1}{2}$. Formula (12.5) leads to

$$\begin{aligned} R_1 &= \left\{x : x \leq \frac{1}{3}\left(4 - \sqrt{4 + 6\log(2)}\right) \text{ or } x \geq \frac{1}{3}\left(4 + \sqrt{4 + 6\log(2)}\right)\right\}, \\ R_2 &= \mathbb{R} \setminus R_1. \end{aligned}$$

This situation is shown in Figure 12.1.

The situation simplifies in the case of equal variances $\sigma_1 = \sigma_2$. The discriminant rule (12.5) is then (for $\mu_1 < \mu_2$)

$$\begin{array}{lll} x \to \Pi_1, & \text{if} & x \in R_1 = \{x : x \leq \frac{1}{2}(\mu_1 + \mu_2)\}, \\ x \to \Pi_2, & \text{if} & x \in R_2 = \{x : x > \frac{1}{2}(\mu_1 + \mu_2)\}. \end{array} \tag{12.6}$$

Theorem 12.2 shows that the ML discriminant rule for multinormal observations is intimately connected with the Mahalanobis distance. The discriminant rule is based on linear combinations and belongs to the family of Linear Discriminant Analysis (LDA) methods.

THEOREM 12.2 *Suppose* $\Pi_i = N_p(\mu_i, \Sigma)$.

(a) *The ML rule allocates* x *to* Π_j*, where* $j \in \{1, \ldots, J\}$ *is the value minimizing the square Mahalanobis distance between* x *and* μ_i*:*

$$\delta^2(x, \mu_i) = (x - \mu_i)^\top \Sigma^{-1} (x - \mu_i)\ ,\ i = 1, \ldots, J\ .$$

(b) *In the case of* $J = 2$,

$$x \in R_1 \iff \alpha^\top (x - \mu) \geq 0\ ,$$

where $\alpha = \Sigma^{-1}(\mu_1 - \mu_2)$ *and* $\mu = \frac{1}{2}(\mu_1 + \mu_2)$.

Proof:
Part (a) of the Theorem follows directly from comparison of the likelihoods.

For $J = 2$, part (a) says that x is allocated to Π_1 if

$$(x - \mu_1)^\top \Sigma^{-1}(x - \mu_1) \leq (x - \mu_2)^\top \Sigma^{-1}(x - \mu_2)$$

Rearranging terms leads to

$$-2\mu_1^\top \Sigma^{-1} x + 2\mu_2^\top \Sigma^{-1} x + \mu_1^\top \Sigma^{-1} \mu_1 - \mu_2^\top \Sigma^{-1} \mu_2 \leq 0,$$

which is equivalent to

$$2(\mu_2 - \mu_1)^\top \Sigma^{-1} x + (\mu_1 - \mu_2)^\top \Sigma^{-1}(\mu_1 + \mu_2) \leq 0,$$

$$(\mu_1 - \mu_2)^\top \Sigma^{-1}\{x - \frac{1}{2}(\mu_1 + \mu_2)\} \geq 0,$$

$$\alpha^\top (x - \mu) \geq 0.$$

□

Bayes Discriminant Rule

We have seen an example where prior knowledge on the probability of classification into Π_j was assumed. Denote the prior probabilities by π_j and note that $\sum_{j=1}^J \pi_j = 1$. The Bayes rule of discrimination allocates x to the Π_j that gives the largest value of $\pi_i f_i(x)$, $\pi_j f_j(x) = \max_i \pi_i f_i(x)$. Hence, the discriminant rule is defined by $R_j = \{x : \pi_j f_j(x) \geq \pi_i f_i(x) \text{ for } i = 1, \ldots, J\}$. Obviously the Bayes rule is identical to the ML discriminant rule for $\pi_j = 1/J$.

A further modification is to allocate x to Π_j with a certain probability $\phi_j(x)$, such that $\sum_{j=1}^J \phi_j(x) = 1$ for all x. This is called a *randomized discriminant rule*. A randomized discriminant rule is a generalization of deterministic discriminant rules since

$$\phi_j(x) = \begin{cases} 1 & \text{if } \pi_j f_j(x) = \max_i \pi_i f_i(x), \\ 0 & \text{otherwise} \end{cases}$$

reflects the deterministic rules.

Which discriminant rules are good? We need a measure of comparison. Denote

$$p_{ij} = \int \phi_i(x) f_j(x) dx \tag{12.7}$$

as the probability of allocating x to Π_i if it in fact belongs to Π_j. A discriminant rule with probabilities p_{ij} is as good as any other discriminant rule with probabilities p'_{ij} if

$$p_{ii} \geq p'_{ii} \quad \text{for all} \quad i = 1, \ldots, J. \tag{12.8}$$

We call the first rule better if the strict inequality in (12.8) holds for at least one i. A discriminant rule is called *admissible* if there is no better discriminant rule.

THEOREM 12.3 *All Bayes discriminant rules (including the ML rule) are admissible.*

Probability of Misclassification for the ML rule ($J = 2$)

Suppose that $\Pi_i = N_p(\mu_i, \Sigma)$. In the case of two groups, it is not difficult to derive the probabilities of misclassification for the ML discriminant rule. Consider for instance $p_{12} = P(x \in R_1 \mid \Pi_2)$. By part (b) in Theorem 12.2 we have

$$p_{12} = P\{\alpha^\top (x - \mu) > 0 \mid \Pi_2\}.$$

If $X \in R_2$, $\alpha^\top (X - \mu) \sim N\left(-\frac{1}{2}\delta^2, \delta^2\right)$ where $\delta^2 = (\mu_1 - \mu_2)^\top \Sigma^{-1} (\mu_1 - \mu_2)$ is the squared Mahalanobis distance between the two populations, we obtain

$$p_{12} = \Phi\left(-\frac{1}{2}\delta\right).$$

Similarly, the probability of being classified into population 2 although x stems from Π_1 is equal to $p_{21} = \Phi\left(-\frac{1}{2}\delta\right)$.

Classification with Different Covariance Matrices

The minimum ECM depends on the ratio of the densities $\frac{f_1(x)}{f_2(x)}$ or equivalently on the difference $\ln\{f_1(x)\} - \ln\{f_2(x)\}$. When the covariance for both density functions differ, the allocation rule becomes more complicated:

$$\begin{aligned}
R_1 &= \left\{ x : -\frac{1}{2} x^\top (\Sigma_1^{-1} - \Sigma_2^{-1}) x + (\mu_1^\top \Sigma_1^{-1} - \mu_2^\top \Sigma_2^{-1}) x - k \geq \ln\left[\left(\frac{C(1|2)}{C(2|1)} \right) \left(\frac{\pi_2}{\pi_1} \right) \right] \right\}, \\
R_2 &= \left\{ x : -\frac{1}{2} x^\top (\Sigma_1^{-1} - \Sigma_2^{-1}) x + (\mu_1^\top \Sigma_1^{-1} - \mu_2^\top \Sigma_2^{-1}) x - k < \ln\left[\left(\frac{C(1|2)}{C(2|1)} \right) \left(\frac{\pi_2}{\pi_1} \right) \right] \right\},
\end{aligned}$$

where $k = \frac{1}{2} \ln\left(\frac{|\Sigma_1|}{|\Sigma_2|} \right) + \frac{1}{2}(\mu_1^T \Sigma_1^{-1} \mu_1 - \mu_2^T \Sigma_2^{-1} \mu_2)$. The classification regions are defined by *quadratic* functions. Therefore they belong to the family of Quadratic Discriminant Analysis (QDA) methods. This *quadratic* classification rule coincides with the rules used when $\Sigma_1 = \Sigma_2$, since the term $\frac{1}{2} x^T (\Sigma_1^{-1} - \Sigma_2^{-1}) x$ disappears.

Summary

↪ Discriminant analysis is a set of methods used to distinguish among groups in data and to allocate new observations into the existing groups.

Summary (continued)
↪ Given that data are from populations Π_j with densities f_j, $j = 1, \dots, J$, the maximum likelihood discriminant rule (ML rule) allocates an observation x to that population Π_j which has the maximum likelihood $L_j(x) = f_j(x) = \max_i f_i(x)$.
↪ Given prior probabilities π_j for populations Π_j, Bayes discriminant rule allocates an observation x to the population Π_j that maximizes $\pi_i f_i(x)$ with respect to i. All Bayes discriminant rules (incl. the ML rule) are admissible.
↪ For the ML rule and $J = 2$ normal populations, the probabilities of misclassification are given by $p_{12} = p_{21} = \Phi\left(-\frac{1}{2}\delta\right)$ where δ is the Mahalanobis distance between the two populations.
↪ Classification of two normal populations with different covariance matrices (ML rule) leads to regions defined by a quadratic function.
↪ Desirable discriminant rules have a low expected cost of misclassification (*ECM*).

12.2 Discrimination Rules in Practice

The ML rule is used if the distribution of the data is known up to parameters. Suppose for example that the data come from multivariate normal distributions $N_p(\mu_j, \Sigma)$. If we have J groups with n_j observations in each group, we use $\overline{x}_j$ to estimate μ_j, and $\mathcal{S}_j$ to estimate Σ. The common covariance may be estimated by

$$\mathcal{S}_u = \sum_{j=1}^{J} n_j \left(\frac{\mathcal{S}_j}{n-J}\right), \tag{12.9}$$

with $n = \sum_{j=1}^{J} n_j$. Thus the empirical version of the ML rule of Theorem 12.2 is to allocate a new observation x to Π_j such that j minimizes

$$(x - \overline{x}_i)^{\top}\mathcal{S}_u^{-1}(x - \overline{x}_i) \quad \text{for} \quad i \in \{1, \dots, J\}.$$

EXAMPLE 12.4 *Let us apply this rule to the Swiss bank notes. The 20 randomly chosen bank notes which we had clustered into two groups in Example 11.6 are used. First the covariance Σ is estimated by the average of the covariances of Π_1 (cluster 1) and Π_2 (cluster 2). The hyperplane $\widehat{\alpha}^{\top}(x - \overline{x}) = 0$ which separates the two populations is given by*

$$\begin{aligned}\widehat{\alpha} &= \mathcal{S}_u^{-1}(\overline{x}_1 - \overline{x}_2) = (-12.18, 20.54, -19.22, -15.55, -13.06, 21.43)^{\top},\\ \overline{x} &= \frac{1}{2}(\overline{x}_1 + \overline{x}_2) = (214.79, 130.05, 129.92, 9.23, 10.48, 140.46)^{\top}.\end{aligned}$$

Now let us apply the discriminant rule to the entire bank notes data set. Counting the number of misclassifications by

$$\sum_{i=1}^{100} \boldsymbol{I}\{\widehat{\alpha}^{\top}(x_i - \overline{x}) < 0\}, \sum_{i=101}^{200} \boldsymbol{I}\{\widehat{\alpha}^{\top}(x_i - \overline{x}) > 0\},$$

we obtain 1 misclassified observation for the conterfeit bank notes and 0 misclassification for the genuine bank notes.

When $J = 3$ groups, the allocation regions can be calculated using

$$\begin{aligned} h_{12}(x) &= (\bar{x}_1 - \bar{x}_2)^\top \mathcal{S}_u^{-1} \left\{ x - \frac{1}{2}(\bar{x}_1 + \bar{x}_2) \right\} \\ h_{13}(x) &= (\bar{x}_1 - \bar{x}_3)^\top \mathcal{S}_u^{-1} \left\{ x - \frac{1}{2}(\bar{x}_1 + \bar{x}_3) \right\} \\ h_{23}(x) &= (\bar{x}_2 - \bar{x}_3)^\top \mathcal{S}_u^{-1} \left\{ x - \frac{1}{2}(\bar{x}_2 + \bar{x}_3) \right\}. \end{aligned}$$

The rule is to allocate x to

$$\begin{cases} \Pi_1 & \text{if } h_{12}(x) \geq 0 \text{ and } h_{13}(x) \geq 0 \\ \Pi_2 & \text{if } h_{12}(x) < 0 \text{ and } h_{23}(x) \geq 0 \\ \Pi_3 & \text{if } h_{13}(x) < 0 \text{ and } h_{23}(x) < 0. \end{cases}$$

Estimation of the Probabilities of Misclassifications

Misclassification probabilities are given by (12.7) and can be estimated by replacing the unknown parameters by their corresponding estimators.

For the ML rule for two normal populations we obtain

$$\hat{p}_{12} = \hat{p}_{21} = \Phi\left(-\frac{1}{2}\hat{\delta}\right)$$

where $\hat{\delta}^2 = (\bar{x}_1 - \bar{x}_2)^\top \mathcal{S}_u^{-1}(\bar{x}_1 - \bar{x}_2)$ is the estimator for δ^2.

The probabilities of misclassification may also be estimated by the *re-substitution method.* We reclassify each original observation x_i, $i = 1, \cdots, n$ into $\Pi_1, \cdots, \Pi_J$ according to the chosen rule. Then denoting the number of individuals coming from Π_j which have been classified into Π_i by n_{ij}, we have $\hat{p}_{ij} = \frac{n_{ij}}{n_j}$, an estimator of p_{ij}. Clearly, this method leads to too optimistic estimators of p_{ij}, but it provides a rough measure of the quality of the discriminant rule. The matrix $(\hat{p}_{ij})$ is called the *confussion matrix* in Johnson and Wichern (1998).

EXAMPLE 12.5 *In the above classification problem for the Swiss bank notes (Table B.2), we have the following confussion matrix:*

		true membership	
		genuine (Π_1)	*counterfeit (Π_2)*
predicted	Π_1	*100*	*1*
	Π_2	*0*	*99*

Q MVAaper

The *apparent error rate* (APER) is defined as the fraction of observations that are misclassified. The APER, expressed as a percentage, is

$$\text{APER} = \left(\frac{1}{200}\right) 100\% = 0.5\%.$$

For the calculation of the APER we use the observations twice: the first time to construct the classification rule and the second time to evaluate this rule. An APER of 0.5% might therefore be too optimistic. An approach that corrects for this bias is based on the holdout procedure of Lachenbruch and Mickey (1968). For two populations this procedure is as follows:

1. Start with the first population Π_1. Omit one observation and develop the classification rule based on the remaining $n_1 - 1$, n_2 observations.
2. Classify the "holdout" observation using the discrimination rule in Step 1.
3. Repeat steps 1 and 2 until all of the Π_1 observations are classified. Count the number n'_{21} of misclassified observations.
4. Repeat steps 1 through 3 for population Π_2. Count the number n'_{12} of misclassified observations.

Estimates of the misclassification probabilities are given by

$$\hat{p}'_{12} = \frac{n'_{12}}{n_2}$$

and

$$\hat{p}'_{21} = \frac{n'_{21}}{n_1}.$$

A more realistic estimator of the actual error rate (AER) is given by

$$\frac{n'_{12} + n'_{21}}{n_2 + n_1}. \tag{12.10}$$

Statisticians favor the AER (for its unbiasedness) over the APER. In large samples, however, the computational costs might counterbalance the statistical advantage. This is not a real problem since the two misclassification measures are asymptotically equivalent.

MVAaer

Fisher's Linear Discrimination Function

Another approach stems from R. A. Fisher. His idea was to base the discriminant rule on a projection $a^\top x$ such that a good separation was achieved. This LDA projection method is called *Fisher's linear discrimination function.* If

$$\mathcal{Y} = \mathcal{X}a$$

denotes a linear combination of observations, then the total sum of squares of y, $\sum_{i=1}^{n}(y_i - \bar{y})^2$, is equal to

$$\mathcal{Y}^\top \mathcal{H} \mathcal{Y} = a^\top \mathcal{X}^\top \mathcal{H} \mathcal{X} a = a^\top \mathcal{T} a \tag{12.11}$$

with the centering matrix $\mathcal{H} = \mathcal{I} - n^{-1} 1_n 1_n^\top$ and $\mathcal{T} = \mathcal{X}^\top \mathcal{H} \mathcal{X}$.

Suppose we have samples $\mathcal{X}_j$, $j = 1, \ldots, J$, from J populations. Fisher's suggestion was to find the linear combination $a^\top x$ which maximizes the ratio of the *between-group-sum of squares* to the *within-group-sum of squares.*

The within-group-sum of squares is given by

$$\sum_{j=1}^{J} \mathcal{Y}_j^\top \mathcal{H}_j \mathcal{Y}_j = \sum_{j=1}^{J} a^\top \mathcal{X}_j^\top \mathcal{H}_j \mathcal{X}_j a = a^\top \mathcal{W} a, \tag{12.12}$$

where $\mathcal{Y}_j$ denotes the j-th sub-matrix of $\mathcal{Y}$ corresponding to observations of group j and $\mathcal{H}_j$ denotes the $(n_j \times n_j)$ centering matrix. The within-group-sum of squares measures the sum of variations within each group.

The between-group-sum of squares is

$$\sum_{j=1}^{J} n_j(\overline{y}_j - \overline{y})^2 = \sum_{j=1}^{J} n_j\{a^\top(\overline{x}_j - \overline{x})\}^2 = a^\top \mathcal{B} a, \tag{12.13}$$

where $\overline{y}_j$ and $\overline{x}_j$ denote the means of $\mathcal{Y}_j$ and $\mathcal{X}_j$ and $\overline{y}$ and $\overline{x}$ denote the sample means of $\mathcal{Y}$ and $\mathcal{X}$. The between-group-sum of squares measures the variation of the means across groups.

The total sum of squares (12.11) is the sum of the within-group-sum of squares and the between-group-sum of squares, i.e.,

$$a^\top \mathcal{T} a = a^\top \mathcal{W} a + a^\top \mathcal{B} a.$$

Fisher's idea was to select a projection vector a that maximizes the ratio

$$\frac{a^\top \mathcal{B} a}{a^\top \mathcal{W} a}. \tag{12.14}$$

The solution is found by applying Theorem 2.5.

THEOREM 12.4 *The vector a that maximizes (12.14) is the eigenvector of $\mathcal{W}^{-1}\mathcal{B}$ that corresponds to the largest eigenvalue.*

Now a discrimination rule is easy to obtain:
classify x into group j where $a^\top \overline{x}_j$ is closest to $a^\top x$, i.e.,

$$x \rightarrow \Pi_j \text{ where } j = \arg\min_i |a^\top(x - \overline{x}_i)|.$$

When $J = 2$ groups, the discriminant rule is easy to compute. Suppose that group 1 has n_1 elements and group 2 has n_2 elements. In this case

$$\mathcal{B} = \left(\frac{n_1 n_2}{n}\right) d d^\top,$$

where $d = (\overline{x}_1 - \overline{x}_2)$. $\mathcal{W}^{-1}\mathcal{B}$ has only one eigenvalue which equals

$$\operatorname{tr}(\mathcal{W}^{-1}\mathcal{B}) = \left(\frac{n_1 n_2}{n}\right) d^\top \mathcal{W}^{-1} d,$$

and the corresponding eigenvector is $a = \mathcal{W}^{-1}d$. The corresponding discriminant rule is

$$\begin{array}{ll} x \rightarrow \Pi_1 & \text{if } a^\top\{x - \frac{1}{2}(\overline{x}_1 + \overline{x}_2)\} > 0, \\ x \rightarrow \Pi_2 & \text{if } a^\top\{x - \frac{1}{2}(\overline{x}_1 + \overline{x}_2)\} \le 0. \end{array} \tag{12.15}$$

The Fisher LDA is closely related to projection pursuit (Chapter 18) since the statistical technique is based on a *one dimensional* index $a^\top x$.

EXAMPLE 12.6 *Consider the bank notes data again. Let us use the subscript "g" for the genuine and "f" for the conterfeit bank notes, e.g., $\mathcal{X}_g$ denotes the first hundred observations*

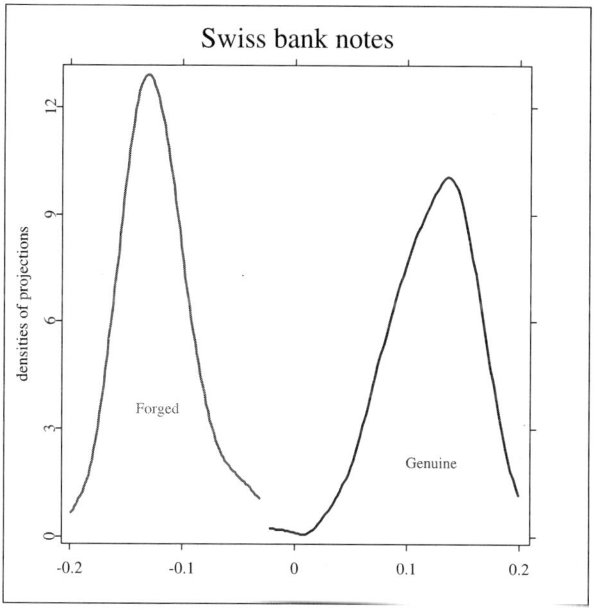

Figure 12.2. Densities of projections of genuine and counterfeit bank notes by Fisher's discrimination rule. MVAdisfbank

of $\mathcal{X}$ and $\mathcal{X}_f$ the second hundred. In the context of the bank data set the "between-group-sum of squares" is defined as

$$100\left\{(\overline{y}_g - \overline{y})^2 + (\overline{y}_f - \overline{y})^2\right\} = a^\top \mathcal{B} a \qquad (12.16)$$

for some matrix $\mathcal{B}$. Here, $\overline{y}_g$ and $\overline{y}_f$ denote the means for the genuine and counterfeit bank notes and $\overline{y} = \frac{1}{2}(\overline{y}_g + \overline{y}_f)$. The "within-group-sum of squares" is

$$\sum_{i=1}^{100}\{(y_g)_i - \overline{y}_g\}^2 + \sum_{i=1}^{100}\{(y_f)_i - \overline{y}_f\}^2 = a^\top \mathcal{W} a, \qquad (12.17)$$

with $(y_g)_i = a^\top x_i$ and $(y_f)_i = a^\top x_{i+100}$ for $i = 1, \ldots, 100$.

The resulting discriminant rule consists of allocating an observation x_0 to the genuine sample space if

$$a^\top(x_0 - x) > 0,$$

with $a = \mathcal{W}^{-1}(\overline{x}_g - \overline{x}_f)$ (see Exercise 12.8) and of allocating x_0 to the counterfeit sample space when the opposite is true. In our case

$$a = (0.000, 0.029, -0.029, -0.039, -0.041, 0.054)^\top \cdot$$

One genuine and no counterfeit bank notes are misclassified. Figure 12.2 shows the estimated densities for $y_g = a^\top \mathcal{X}_g$ and $y_f = a^\top \mathcal{X}_f$. They are separated better than those of the diagonals in Figure 1.9.

Note that the allocation rule (12.15) is exactly the same as the ML rule for $J = 2$ groups and for normal distributions with the same covariance. For $J = 3$ groups this rule will be different, except for the special case of collinear sample means.

Summary

↪ A discriminant rule is a separation of the sample space into sets R_j. An observation x is classified as coming from population Π_j if it lies in R_j.

↪ The expected cost of misclassification (ECM) for two populations is given by ECM $= C(2|1)p_{21}\pi_1 + C(1|2)p_{12}\pi_2$.

↪ The ML rule is applied if the distributions in the populations are known up to parameters, e.g., for normal distributions $N_p(\mu_j, \Sigma)$.

↪ The ML rule allocates x to the population that exhibits the smallest Mahalanobis distance

$$\delta^2(x; \mu_i) = (x - \mu_i)^\top \Sigma^{-1} (x - \mu_i).$$

↪ The probability of misclassification is given by

$$p_{12} = p_{21} = \Phi\left(-\frac{1}{2}\delta\right),$$

where δ is the Mahalanobis distance between μ_1 and μ_2.

↪ Classification for different covariance structures in the two populations leads to quadratic discrimination rules.

↪ A different approach is Fisher's linear discrimination rule which finds a linear combination $a^\top x$ that maximizes the ratio of the "between-group-sum of squares" and the "within-group-sum of squares". This rule turns out to be identical to the ML rule when $J = 2$ for normal populations.

12.3 Boston Housing

One interesting application of discriminant analysis with respect to the Boston housing data is the classification of the districts according to the house values. The rationale behind this is that certain observables must determine the value of a district, as in Section 3.7 where the house value was regressed on the other variables. Two groups are defined according to the median value of houses $\widetilde{X}_{14}$: in group Π_1 the value of $\widetilde{X}_{14}$ is greater than or equal to the median of $\widetilde{X}_{14}$ and in group Π_2 the value of $\widetilde{X}_{14}$ is less than the median of $\widetilde{X}_{14}$.

The linear discriminant rule, defined on the remaining 12 variables (excluding $\widetilde{X}_4$ and $\widetilde{X}_{14}$) is applied. After reclassifying the 506 observations, we obtain an apparent error rate of 0.146. The details are given in Table 12.3. The more appropriate error rate, given by the AER, is 0.160 (see Table 12.4).

Let us now turn to a group definition suggested by the Cluster Analysis in Section 11.4. Group Π_1 was defined by higher quality of life and house. We define the linear discriminant rule using the 13 variables from $\widetilde{\mathcal{X}}$ excluding $\widetilde{X}_4$. Then we reclassify the 506 observations and we obtain an APER of 0.0395. Details are summarized in Table 12.5. The AER turns

		True	
		Π_1	Π_2
Predicted	Π_1	216	40
	Π_2	34	216

Table 12.3. APER for price of Boston houses. `MVAdiscbh`

		True	
		Π_1	Π_2
Predicted	Π_1	211	42
	Π_2	39	214

Table 12.4. AER for price of Boston houses. `MVAaerbh`

		True	
		Π_1	Π_2
Predicted	Π_1	244	13
	Π_2	7	242

Table 12.5. APER for clusters of Boston houses. `MVAdiscbh`

		True	
		Π_1	Π_2
Predicted	Π_1	244	14
	Π_2	7	241

Table 12.6. AER for clusters of Boston houses. `MVAaerbh`

out to be 0.0415 (see Table 12.6).

Figure 12.3 displays the values of the linear discriminant scores (see Theorem 12.2) for all of the 506 observations, colored by groups. One can clearly see the APER is derived from the 7 observations from group Π_1 with a negative score and the 13 observations from group Π_2 with positive score.

12.4 Exercises

EXERCISE 12.1 *Prove Theorem 12.2 (a) and 12.2 (b).*

EXERCISE 12.2 *Apply the rule from Theorem 12.2 (b) for $p = 1$ and compare the result with that of Example 12.3.*

EXERCISE 12.3 *Calculate the ML discrimination rule based on observations of a one-dimensional variable with an exponential distribution.*

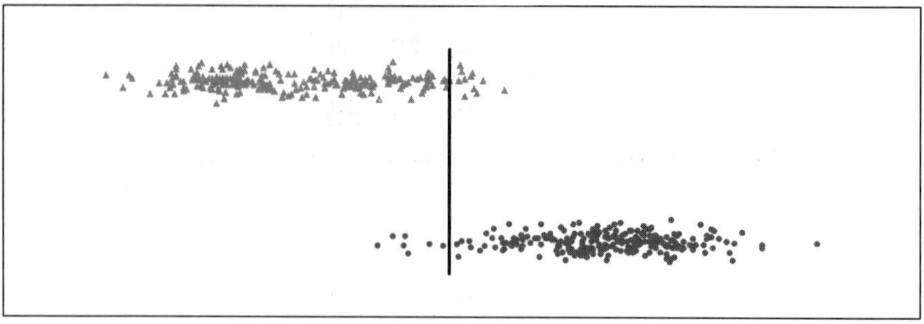

Figure 12.3. Discrimination scores for the two clusters created from the Boston housing data. MVAdiscbh

EXERCISE 12.4 *Calculate the ML discrimination rule based on observations of a two-dimensional random variable, where the first component has an exponential distribution and the other has an alternative distribution. What is the difference between the discrimination rule obtained in this exercise and the Bayes discrimination rule?*

EXERCISE 12.5 *Apply the Bayes rule to the car data (Table B.3) in order to discriminate between Japanese, European and U.S. cars, i.e., $J = 3$. Consider only the "miles per gallon" variable and take the relative frequencies as prior probabilities.*

EXERCISE 12.6 *Compute Fisher's linear discrimination function for the 20 bank notes from Example 11.6. Apply it to the entire bank data set. How many observations are misclassified?*

EXERCISE 12.7 *Use the Fisher's linear discrimination function on the WAIS data set (Table B.12) and evaluate the results by re-substitution the probabilities of misclassification.*

EXERCISE 12.8 *Show that in Example 12.6*

(a) $\mathcal{W} = 100\,(\mathcal{S}_g + \mathcal{S}_f)$, *where $\mathcal{S}_g$ and $\mathcal{S}_f$ denote the empirical covariances (3.6) and (3.5) w.r.t. the genuine and counterfeit bank notes,*

(b) $\mathcal{B} = 100\left\{(\overline{x}_g - \overline{x})(\overline{x}_g - \overline{x})^\top + (\overline{x}_f - \overline{x})(\overline{x}_f - \overline{x})^\top\right\}$, *where* $\overline{x} = \frac{1}{2}(\overline{x}_g + \overline{x}_f)$,

(c) $a = \mathcal{W}^{-1}(\overline{x}_g - \overline{x}_f)$.

EXERCISE 12.9 *Recalculate Example 12.3 with the prior probability $\pi_1 = \frac{1}{3}$ and $C(2|1) = 2C(1|2)$.*

EXERCISE 12.10 *Explain the effect of changing π_1 or $C(1|2)$ on the relative location of the region $R_j, j = 1, 2$.*

EXERCISE 12.11 *Prove that Fisher's linear discrimination function is identical to the ML rule when the covariance matrices are identical ($J = 2$).*

EXERCISE 12.12 *Suppose that* $x \in \{0, 1, 2, 3, 4, 5, 6, 7, 8, 9, 10\}$ *and*

$$\begin{aligned} \Pi_1 &: X \sim \text{Bi}(10, 0.2) \quad \textit{with the prior probability } \pi_1 = 0.5; \\ \Pi_2 &: X \sim \text{Bi}(10, 0.3) \quad \textit{with the prior probability } \pi_2 = 0.3; \\ \Pi_3 &: X \sim \text{Bi}(10, 0.5) \quad \textit{with the prior probability } \pi_3 = 0.2. \end{aligned}$$

Determine the sets R_1, R_2 *and* R_3. *(Use the Bayes discriminant rule.)*

13 Correspondence Analysis

Correspondence analysis provides tools for analyzing the associations between rows and columns of contingency tables. A contingency table is a two-entry frequency table where the joint frequencies of two qualitative variables are reported. For instance a (2×2) table could be formed by observing from a sample of n individuals two qualitative variables: the individual's sex and whether the individual smokes. The table reports the observed joint frequencies. In general $(n \times p)$ tables may be considered.

The main idea of correspondence analysis is to develop simple indices that will show the relations between the row and the columns categories. These indices will tell us simultaneously which column categories have more weight in a row category and vice-versa. Correspondence analysis is also related to the issue of reducing the dimension of the table, similar to principal component analysis in Chapter 9, and to the issue of decomposing the table into its factors as discussed in Chapter 8. The idea is to extract the indices in decreasing order of importance so that the main information of the table can be summarized in spaces with smaller dimensions. For instance, if only two factors (indices) are used, the results can be shown in two-dimensional graphs, showing the relationship between the rows and the columns of the table.

Section 13.1 defines the basic notation and motivates the approach and Section 13.2 gives the basic theory. The indices will be used to describe the χ^2 statistic measuring the associations in the table. Several examples in Section 13.3 show how to provide and interpret, in practice, the two-dimensional graphs displaying the relationship between the rows and the columns of a contingency table.

13.1 Motivation

The aim of correspondence analysis is to develop simple indices that show relations between the row and columns of a contingency tables. Contingency tables are very useful to describe the association between two variables in very general situations. The two variables can be qualitative (nominal), in which case they are also referred to as categorical variables. Each row and each column in the table represents one category of the corresponding variable. The entry x_{ij} in the table $\mathcal{X}$ (with dimension $(n \times p)$) is the number of observations in a sample which simultaneously fall in the i-th row category and the j-th column category, for $i = 1, \ldots, n$ and $j = 1, \ldots, p$. Sometimes a "category" of a nominal variable is also called a "modality" of the variable.

The variables of interest can also be discrete quantitative variables, such as the number of family members or the number of accidents an insurance company had to cover during one year, etc. Here, each possible value that the variable can have defines a row or a column category. Continuous variables may be taken into account by defining the categories in terms of intervals or classes of values which the variable can take on. Thus contingency tables can be used in many situations, implying that correspondence analysis is a very useful tool in many applications.

The graphical relationships between the rows and the columns of the table $\mathcal{X}$ that result from correspondence analysis are based on the idea of representing all the row and column categories and interpreting the relative positions of the points in terms of the weights corresponding to the column and the row. This is achieved by deriving a system of simple indices providing the coordinates of each row and each column. These row and column coordinates are simultaneously represented in the same graph. It is then clear to see which column categories are more important in the row categories of the table (and the other way around).

As was already eluded to, the construction of the indices is based on an idea similar to that of PCA. Using PCA the total variance was partitioned into independent contributions stemming from the principal components. Correspondence analysis, on the other hand, decomposes a measure of association, typically the total χ^2 value used in testing independence, rather than decomposing the total variance.

EXAMPLE 13.1 *The French "baccalauréat" frequencies have been classified into regions and different baccalauréat categories, see Appendix, Table B.8. Altogether* $n = 202100$ *baccalauréats were observed. The joint frequency of the region* Ile-de-France *and the modality* Philosophy, *for example, is 9724. That is, 9724 baccalauréats were in Ile-de-France and the category Philosophy.*

The question is whether certain regions prefer certain baccalauréat types. If we consider, for instance, the region Lorraine, *we have the following percentages:*

A	B	C	D	E	F	G	H
20.5	7.6	15.3	19.6	3.4	14.5	18.9	0.2

The total percentages of the different modalities of the variable baccalauréat are as follows:

A	B	C	D	E	F	G	H
22.6	10.7	16.2	22.8	2.6	9.7	15.2	0.2

One might argue that the region Lorraine seems to prefer the modalities E, F, G and dislike the specializations A, B, C, D relative to the overall frequency of baccalauréat type.

In correspondence analysis we try to develop an index for the regions so that this over- or underrepresentation can be measured in just one single number. Simultaneously we try to weight the regions so that we can see in which region certain baccalauréat types are preferred.

EXAMPLE 13.2 *Consider* n *types of companies and* p *locations of these companies. Is there a certain type of company that prefers a certain location? Or is there a location index that corresponds to a certain type of company?*

Assume that $n = 3$, $p = 3$, *and that the frequencies are as follows:*

$$\mathcal{X} = \begin{pmatrix} 4 & 0 & 2 \\ 0 & 1 & 1 \\ 1 & 1 & 4 \end{pmatrix} \begin{matrix} \leftarrow \textit{Finance} \\ \leftarrow \textit{Energy} \\ \leftarrow \textit{HiTech} \end{matrix}$$

$\uparrow$ *Frankfurt*
$\uparrow$ *Berlin*
$\uparrow$ *Munich*

The frequencies imply that four type 3 companies (HiTech) are in location 3 (Munich), and so on. Suppose there is a (company) weight vector $r = (r_1, \ldots, r_n)^\top$ such that a location index s_j could be defined as

$$s_j = c \sum_{i=1}^{n} r_i \frac{x_{ij}}{x_{\bullet j}}, \tag{13.1}$$

where $x_{\bullet j} = \sum_{i=1}^{n} x_{ij}$ is the number of companies in location j and c is a constant. s_1, for example, would give the average weighted frequency (by r) of companies in location 1 (Frankfurt).

Given a location weight vector $s^* = \left(s_1^*, \ldots, s_p^*\right)^\top$, we can define a company index in the same way as

$$r_i^* = c^* \sum_{j=1}^{p} s_j^* \frac{x_{ij}}{x_{i\bullet}}, \tag{13.2}$$

where c^* is a constant and $x_{i\bullet} = \sum_{j=1}^{p} x_{ij}$ is the sum of the i-th row of $\mathcal{X}$, i.e., the number of type i companies. Thus r_2^*, for example, would give the average weighted frequency (by s^*) of energy companies.

If (13.1) and (13.2) can be solved simultaneously for a "row weight" vector $r = (r_1, \ldots, r_n)^\top$ and a "column weight" vector $s = (s_1, \ldots, s_p)^\top$, we may represent each row category by r_i, $i = 1, \ldots, n$ and each column category by s_j, $j = 1, \ldots, p$ in a one-dimensional graph. If in this graph r_i and s_j are in close proximity (far from the origin), this would indicate that the i-th row category has an important conditional frequency $x_{ij}/x_{\bullet j}$ in (13.1) and that the j-th column category has an important conditional frequency $x_{ij}/x_{i\bullet}$ in (13.2). This would indicate a positive association between the i-th row and the j-th column. A similar line of argument could be used if r_i was very far away from s_j (and far from the origin). This would indicate a small conditional frequency contribution, or a negative association between the i-th row and the j-th column.

Summary

- ↪ The aim of correspondence analysis is to develop simple indices that show relations among qualitative variables in a contingency table.
- ↪ The joint representation of the indices reveals relations among the variables.

13.2 Chi-square Decomposition

An alternative way of measuring the association between the row and column categories is a decomposition of the value of the χ^2-test statistic. The well known χ^2-test for independence in a two-dimensional contingency table consists of two steps. First the expected value of each cell of the table is estimated under the hypothesis of independence. Second, the corresponding observed values are compared to the expected values using the statistic

$$t = \sum_{i=1}^{n} \sum_{j=1}^{p} (x_{ij} - E_{ij})^2 / E_{ij}, \tag{13.3}$$

where x_{ij} is the observed frequency in cell (i, j) and E_{ij} is the corresponding estimated expected value under the assumption of independence, i.e.,

$$E_{ij} = \frac{x_{i\bullet}\, x_{\bullet j}}{x_{\bullet\bullet}}. \tag{13.4}$$

Here $x_{\bullet\bullet} = \sum_{i=1}^{n} x_{i\bullet}$. Under the hypothesis of independence, t has a $\chi^2_{(n-1)(p-1)}$ distribution. In the industrial location example introduced above the value of $t = 6.26$ is almost significant at the 5% level. It is therefore worth investigating the special reasons for departure from independence.

The method of χ^2 decomposition consists of finding the SVD of the matrix $\mathcal{C}$ $(n \times p)$ with elements

$$c_{ij} = (x_{ij} - E_{ij})/E_{ij}^{1/2}. \tag{13.5}$$

The elements c_{ij} may be viewed as measuring the (weighted) departure between the observed x_{ij} and the theoretical values E_{ij} under independence. This leads to the factorial tools of Chapter 8 which describe the rows and the columns of $\mathcal{C}$.

For simplification define the matrics $\mathcal{A}\,(n \times n)$ and $\mathcal{B}\,(p \times p)$ as

$$\mathcal{A} = \mathrm{diag}(x_{i\bullet}) \text{ and } \mathcal{B} = \mathrm{diag}(x_{\bullet j}). \tag{13.6}$$

These matrices provide the marginal row frequencies $a(n \times 1)$ and the marginal column frequencies $b(p \times 1)$:

$$a = \mathcal{A}1_n \text{ and } b = \mathcal{B}1_p. \tag{13.7}$$

It is easy to verify that

$$\mathcal{C}\sqrt{b} = 0 \text{ and } \mathcal{C}^\top\sqrt{a} = 0, \tag{13.8}$$

where the square root of the vector is taken element by element and $R = \mathrm{rank}(\mathcal{C}) \leq \min\{(n-1), (p-1)\}$. From (8.14) of Chapter 8, the SVD of $\mathcal{C}$ yields

$$\mathcal{C} = \Gamma\Lambda\Delta^\top, \tag{13.9}$$

where Γ contains the eigenvectors of $\mathcal{C}\mathcal{C}^\top$, Δ the eigenvectors of $\mathcal{C}^\top\mathcal{C}$ and $\Lambda = \mathrm{diag}(\lambda_1^{1/2}, \ldots, \lambda_R^{1/2})$ with $\lambda_1 \geq \lambda_2 \geq \ldots \geq \lambda_R$ (the eigenvalues of $\mathcal{C}\mathcal{C}^\top$). Equation (13.9) implies that

$$c_{ij} = \sum_{k=1}^{R} \lambda_k^{1/2}\gamma_{ik}\delta_{jk}. \tag{13.10}$$

Note that (13.3) can be rewritten as

$$\mathrm{tr}(\mathcal{C}\mathcal{C}^\top) = \sum_{k=1}^{R}\lambda_k = \sum_{i=1}^{n}\sum_{j=1}^{p} c_{ij}^2 = t. \tag{13.11}$$

This relation shows that the SVD of $\mathcal{C}$ decomposes the total χ^2 value rather than, as in Chapter 8, the total variance.

The duality relations between the row and the column space (8.11) are now for $k = 1, \ldots, R$ given by

$$\begin{aligned} \delta_k &= \frac{1}{\sqrt{\lambda_k}}\mathcal{C}^\top\gamma_k, \\ \gamma_k &= \frac{1}{\sqrt{\lambda_k}}\mathcal{C}\delta_k. \end{aligned} \tag{13.12}$$

The projections of the rows and the columns of $\mathcal{C}$ are given by

$$\begin{aligned} \mathcal{C}\delta_k &= \sqrt{\lambda_k}\gamma_k, \\ \mathcal{C}^\top\gamma_k &= \sqrt{\lambda_k}\delta_k. \end{aligned} \tag{13.13}$$

Note that the eigenvectors satisfy

$$\delta_k^\top \sqrt{b} = 0, \quad \gamma_k^\top \sqrt{a} = 0. \tag{13.14}$$

From (13.10) we see that the eigenvectors δ_k and γ_k are the objects of interest when analyzing the correspondence between the rows and the columns. Suppose that the first eigenvalue in (13.10) is dominant so that

$$c_{ij} \approx \lambda_1^{1/2} \gamma_{i1} \delta_{j1}. \tag{13.15}$$

In this case when the coordinates γ_{i1} and δ_{j1} are both large (with the same sign) relative to the other coordinates, then c_{ij} will be large as well, indicating a positive association between the i-th row and the j-th column category of the contingency table. If γ_{i1} and δ_{j1} were both large with opposite signs, then there would be a negative association between the i-th row and j-th column.

In many applications, the first two eigenvalues, λ_1 and λ_2, dominate and the percentage of the total χ^2 explained by the eigenvectors γ_1 and γ_2 and δ_1 and δ_2 is large. In this case (13.13) and (γ_1, γ_2) can be used to obtain a graphical display of the n rows of the table $((\delta_1, \delta_2)$ play a similar role for the p columns of the table). The interpretation of the proximity between row and column points will be interpreted as above with respect to (13.10).

In correspondence analysis, we use the projections of weighted rows of $\mathcal{C}$ and the projections of weighted columns of $\mathcal{C}$ for graphical displays. Let $r_k(n \times 1)$ be the projections of $\mathcal{A}^{-1/2}\mathcal{C}$ on δ_k and $s_k(p \times 1)$ be the projections of $\mathcal{B}^{-1/2}\mathcal{C}^\top$ on γ_k $(k = 1, \ldots, R)$:

$$\begin{aligned} r_k &= \mathcal{A}^{-1/2}\mathcal{C}\delta_k = \sqrt{\lambda_k}\mathcal{A}^{-1/2}\gamma_k, \\ s_k &= \mathcal{B}^{-1/2}\mathcal{C}^\top\gamma_k = \sqrt{\lambda_k}\mathcal{B}^{-1/2}\delta_k. \end{aligned} \tag{13.16}$$

These vectors have the property that

$$\begin{aligned} r_k^\top a &= 0, \\ s_k^\top b &= 0. \end{aligned} \tag{13.17}$$

The obtained projections on each axis $k = 1, \ldots, R$ are centered at zero with the natural weights given by a (the marginal frequencies of the rows of $\mathcal{X}$) for the row coordinates r_k and by b (the marginal frequencies of the columns of $\mathcal{X}$) for the column coordinates s_k (compare this to expression (13.14)). As a result, the origin is the center of gravity for all of the representations. We also know from (13.16) and the SVD of $\mathcal{C}$ that

$$\begin{aligned} r_k^\top \mathcal{A} r_k &= \lambda_k, \\ s_k^\top \mathcal{B} s_k &= \lambda_k. \end{aligned} \tag{13.18}$$

From the duality relation between δ_k and γ_k (see (13.12)) we obtain

$$\begin{aligned} r_k &= \frac{1}{\sqrt{\lambda_k}}\mathcal{A}^{-1/2}\mathcal{C}\mathcal{B}^{1/2}s_k, \\ s_k &= \frac{1}{\sqrt{\lambda_k}}\mathcal{B}^{-1/2}\mathcal{C}^\top\mathcal{A}^{1/2}r_k, \end{aligned} \tag{13.19}$$

which can be simplified to

$$\begin{aligned} r_k &= \sqrt{\frac{x_{\bullet\bullet}}{\lambda_k}}\mathcal{A}^{-1}\mathcal{X}s_k, \\ s_k &= \sqrt{\frac{x_{\bullet\bullet}}{\lambda_k}}\mathcal{B}^{-1}\mathcal{X}^\top r_k. \end{aligned} \tag{13.20}$$

These vectors satisfy the relations (13.1) and (13.2) for each $k = 1, \ldots, R$ simultaneously.

As in Chapter 8, the vectors r_k and s_k are referred to as factors (row factor and column factor respectively). They have the following means and variances:

$$\begin{aligned} \overline{r}_k &= \frac{1}{x_{\bullet\bullet}} r_k^\top a = 0, \\ \overline{s}_k &= \frac{1}{x_{\bullet\bullet}} s_k^\top b = 0, \end{aligned} \tag{13.21}$$

and

$$
\begin{aligned}
Var(r_k) &= \frac{1}{x_{\bullet\bullet}} \sum_{i=1}^{n} x_{i\bullet} r_{ki}^2 = \frac{r_k^\top \mathcal{A} r_k}{x_{\bullet\bullet}} = \frac{\lambda_k}{x_{\bullet\bullet}}, \\
Var(s_k) &= \frac{1}{x_{\bullet\bullet}} \sum_{j=1}^{p} x_{\bullet j} s_{kj}^2 = \frac{s_k^\top \mathcal{B} s_k}{x_{\bullet\bullet}} = \frac{\lambda_k}{x_{\bullet\bullet}}.
\end{aligned}
\tag{13.22}
$$

Hence, $\lambda_k / \sum_{k=1}^{j} \lambda_j$, which is the part of the k-th factor in the decomposition of the χ^2 statistic t, may also be interpreted as the proportion of the variance explained by the factor k. The proportions

$$
C_a(i, r_k) = \frac{x_{i\bullet} r_{ki}^2}{\lambda_k}, \text{ for } i = 1, \dots, n,\ k = 1, \dots, R \tag{13.23}
$$

are called the absolute contributions of row i to the variance of the factor r_k. They show which row categories are most important in the dispersion of the k-th row factors. Similarly, the proportions

$$
C_a(j, s_k) = \frac{x_{\bullet j} s_{kj}^2}{\lambda_k}, \text{ for } j = 1, \dots, p,\ k = 1, \dots, R \tag{13.24}
$$

are called the absolute contributions of column j to the variance of the column factor s_k. These absolute contributions may help to interpret the graph obtained by correspondence analysis.

13.3 Correspondence Analysis in Practice

The graphical representations on the axes $k = 1, 2, \dots, R$ of the n rows and of the p columns of $\mathcal{X}$ are provided by the elements of r_k and s_k. Typically, two-dimensional displays are often satisfactory if the cumulated percentage of variance explained by the first two factors, $\Psi_2 = \frac{\lambda_1 + \lambda_2}{\sum_{k=1}^{R} \lambda_k}$, is sufficiently large.

The interpretation of the graphs may be summarized as follows:

- The proximity of two rows (two columns) indicates a similar profile in these two rows (two columns), where "profile" referrs to the conditional frequency distribution of a row (column); those two rows (columns) are almost proportional. The opposite interpretation applies when the two rows (two columns) are far apart.
- The proximity of a particular row to a particular column indicates that this row (column) has a particularly important weight in this column (row). In contrast to this, a row that is quite distant from a particular column indicates that there are almost no observations in this column for this row (and vice versa). Of course, as mentioned above, these conclusions are particularly true when the points are far away from 0.
- The origin is the average of the factors r_k and s_k. Hence, a particular point (row or column) projected close to the origin indicates an average profile.
- The absolute contributions are used to evaluate the weight of each row (column) in the variances of the factors.
- All the interpretations outlined above must be carried out in view of the quality of the graphical representation which is evaluated, as in PCA, using the cumulated percentage of variance.

REMARK 13.1 Note that correspondence analysis can also be applied to more general $(n \times p)$ tables $\mathcal{X}$ which in a "strict sense" are not contingency tables.

λ_j	percentage of variance	cumulated percentage
183.40	0.653	0.653
43.75	0.156	0.809
25.21	0.090	0.898
11.74	0.042	0.940
8.04	0.029	0.969
4.68	0.017	0.985
2.13	0.008	0.993
1.20	0.004	0.997
0.82	0.003	1.000
0.00	0.000	1.000

Table 13.1. Eigenvalues and percentages of the variance (Example 13.3) .

As long as statistical (or natural) meaning can be given to sums over rows and columns, Remark 13.1 holds. This implies, in particular, that all of the variables are measured in the same units. In that case, $x_{\bullet\bullet}$ constitutes the total frequency of the observed phenomenon, and is shared between individuals (n rows) and between variables (p columns). Representations of the rows and columns of $\mathcal{X}$, r_k and s_k, have the basic property (13.19) and show which variables have important weights for each individual and vice versa. This type of analysis is used as an alternative to PCA. PCA is mainly concerned with covariances and correlations, whereas correspondence analysis analyzes a more general kind of association. (See Exercises 13.3 and 13.11.)

EXAMPLE 13.3 *A survey of Belgium citizens who regularly read a newspaper was conducted in the 1980's. They were asked where they lived. The possible answers were 10 regions: 7 provinces (Antwerp, Western Flanders, Eastern Flanders, Hainant, Liège, Limbourg, Luxembourg) and 3 regions around Brussels (Flemish-Brabant, Wallon-Brabant and the city of Brussels). They were also asked what kind of newspapers they read on a regular basis. There were 15 possible answers split up into 3 classes: Flemish newspapers (label begins with the letter v), French newspapers (label begins with f) and both languages together (label begins with b). The data set is given in Table B.9. The eigenvalues of the factorial correspondence analysis are given in Table 13.1.*

Two-dimensional representations will be quite satisfactory since the first two eigenvalues account for 81% of the variance. Figure 13.1 shows the projections of the rows (the 15 newspapers) and of the columns (the 10 regions).

As expected, there is a high association between the regions and the type of newspapers which is read. In particular, v_b (Gazet van Antwerp) is almost exclusively read in the province of Antwerp (this is an extreme point in the graph). The points on the left all belong to Flanders, whereas those on the right all belong to Wallonia. Notice that the Wallon-Brabant and the Flemish-Brabant are not far from Brussels. Brussels is close to the center (average) and also close to the bilingual newspapers. It is shifted a little to the right of the origin due to the majority of French speaking people in the area.

The absolute contributions of the first 3 factors are listed in Tables 13.2 and 13.3. The row factors r_k are in Table 13.2 and the column factors s_k are in Table 13.3.

They show, for instance, the important role of Antwerp and the newspaper v_b in determining the variance of both factors. Clearly, the first axis expresses linguistic differences between the 3 parts of Belgium. The second axis shows a larger dispersion between the Flemish

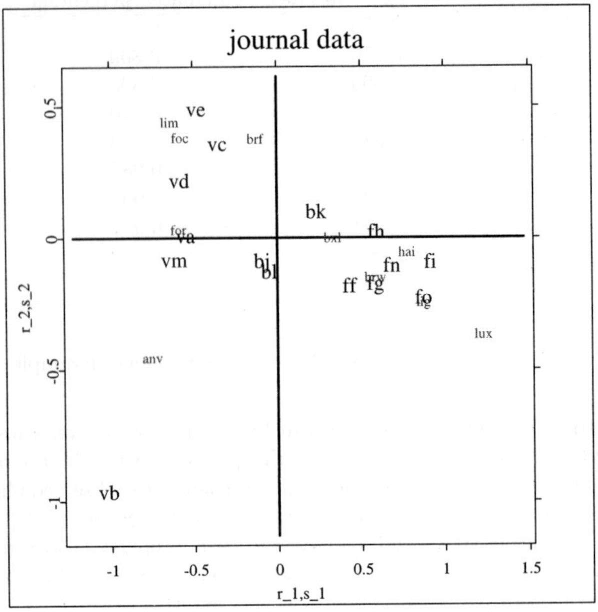

Figure 13.1. Projection of rows (the 15 newspapers) and columns (the 10 regions) MVAcorrjourn

	$C_a(i,r_1)$	$C_a(i,r_2)$	$C_a(i,r_3)$
v_a	0.0563	0.0008	0.0036
v_b	0.1555	0.5567	0.0067
v_c	0.0244	0.1179	0.0266
v_d	0.1352	0.0952	0.0164
v_e	0.0253	0.1193	0.0013
f_f	0.0314	0.0183	0.0597
f_g	0.0585	0.0162	0.0122
f_h	0.1086	0.0024	0.0656
f_i	0.1001	0.0024	0.6376
b_j	0.0029	0.0055	0.0187
b_k	0.0236	0.0278	0.0237
b_l	0.0006	0.0090	0.0064
v_m	0.1000	0.0038	0.0047
f_n	0.0966	0.0059	0.0269
f_o	0.0810	0.0188	0.0899
Total	1.0000	1.0000	1.0000

Table 13.2. Absolute contributions of row factors r_k.

region than the French speaking regions. Note also that the 3-rd axis shows an important role of the category "f_i" (other French newspapers) with the Wallon-Brabant "brw" and the Hainant "hai" showing the most important contributions. The coordinate of "f_i" on this axis

	$C_a(j, s_1)$	$C_a(j, s_2)$	$C_a(j, s_3)$
brw	0.0887	0.0210	0.2860
bxl	0.1259	0.0010	0.0960
anv	0.2999	0.4349	0.0029
brf	0.0064	0.2370	0.0090
foc	0.0729	0.1409	0.0033
for	0.0998	0.0023	0.0079
hai	0.1046	0.0012	0.3141
lig	0.1168	0.0355	0.1025
lim	0.0562	0.1162	0.0027
lux	0.0288	0.0101	0.1761
Total	1.0000	1.0000	1.0000

Table 13.3. Absolute contributions of column factors s_k.

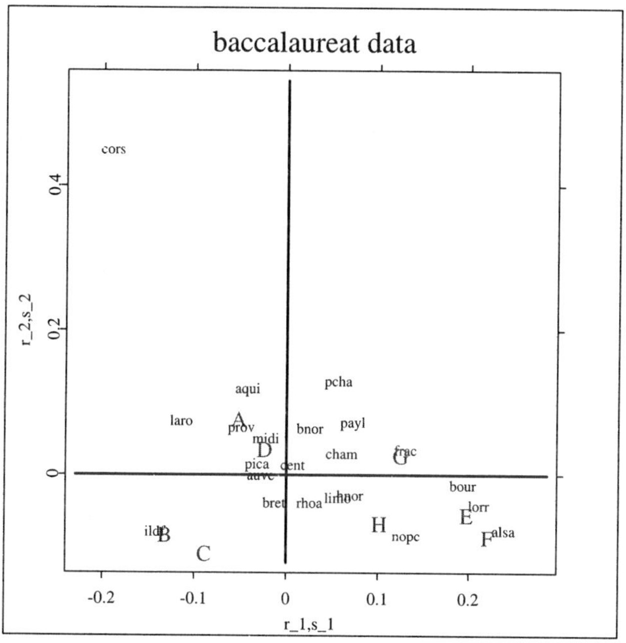

Figure 13.2. Correspondence analysis including Corsica MVAcorrbac

is negative (not shown here) so are the coordinates of "brw" and "hai". Apparently, these two regions also seem to feature a greater proportion of readers of more local newspapers.

EXAMPLE 13.4 *Applying correspondence analysis to the French baccalauréat data (Table B.8) leads to Figure 13.2. Excluding Corsica we obtain Figure 13.3. The different modalities are labeled* `A`, ..., `H` *and the regions are labeled* `ILDF`, ..., `CORS`. *The results of the correspondence analysis are given in Table 13.4 and Figure 13.2.*

eigenvalues λ	percentage of variances	cumulated percentage
2436.2	0.5605	0.561
1052.4	0.2421	0.803
341.8	0.0786	0.881
229.5	0.0528	0.934
152.2	0.0350	0.969
109.1	0.0251	0.994
25.0	0.0058	1.000
0.0	0.0000	1.000

Table 13.4. Eigenvalues and percentages of explained variance (including Corsica).

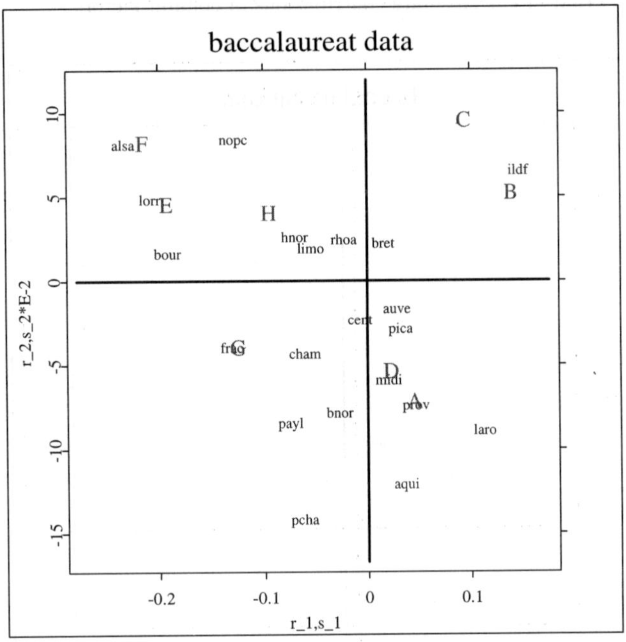

Figure 13.3. Correspondence analysis excluding Corsica. MVAcorrbac

The first two factors explain 80 % of the total variance. It is clear from Figure 13.2 that Corsica (in the upper left) is an outlier. The analysis is therefore redone without Corsica and the results are given in Table 13.5 and Figure 13.3. Since Corsica has such a small weight in the analysis, the results have not changed much.

The projections on the first three axes, along with their absolute contribution to the variance of the axis, are summarized in Table 13.6 for the regions and in Table 13.7 for baccalauréats.

The interpretation of the results may be summarized as follows. Table 13.7 shows that the baccalauréats B on one side and F on the other side are most strongly responsible for the variation on the first axis. The second axis mostly characterizes an opposition between

eigenvalues λ	percentage of variances	cumulated percentage
2408.6	0.5874	0.587
909.5	0.2218	0.809
318.5	0.0766	0.887
195.9	0.0478	0.935
149.3	0.0304	0.971
96.1	0.0234	0.994
22.8	0.0056	1.000
0.0	0.0000	1.000

Table 13.5. Eigenvalues and percentages of explained variance (excluding Corsica).

Region	r_1	r_2	r_3	$C_a(i, r_1)$	$C_a(i, r_2)$	$C_a(i, r_3)$
ILDF	0.1464	0.0677	0.0157	0.3839	0.2175	0.0333
CHAM	-0.0603	-0.0410	-0.0187	0.0064	0.0078	0.0047
PICA	0.0323	-0.0258	-0.0318	0.0021	0.0036	0.0155
HNOR	-0.0692	0.0287	0.1156	0.0096	0.0044	0.2035
CENT	-0.0068	-0.0205	-0.0145	0.0001	0.0030	0.0043
BNOR	-0.0271	-0.0762	0.0061	0.0014	0.0284	0.0005
BOUR	-0.1921	0.0188	0.0578	0.0920	0.0023	0.0630
NOPC	-0.1278	0.0863	-0.0570	0.0871	0.1052	0.1311
LORR	-0.2084	0.0511	0.0467	0.1606	0.0256	0.0608
ALSA	-0.2331	0.0838	0.0655	0.1283	0.0439	0.0767
FRAC	-0.1304	-0.0368	-0.0444	0.0265	0.0056	0.0232
PAYL	-0.0743	-0.0816	-0.0341	0.0232	0.0743	0.0370
BRET	0.0158	0.0249	-0.0469	0.0011	0.0070	0.0708
PCHA	-0.0610	-0.1391	-0.0178	0.0085	0.1171	0.0054
AQUI	0.0368	-0.1183	0.0455	0.0055	0.1519	0.0643
MIDI	0.0208	-0.0567	0.0138	0.0018	0.0359	0.0061
LIMO	-0.0540	0.0221	-0.0427	0.0033	0.0014	0.0154
RHOA	-0.0225	0.0273	-0.0385	0.0042	0.0161	0.0918
AUVE	0.0290	-0.0139	-0.0554	0.0017	0.0010	0.0469
LARO	0.0290	-0.0862	-0.0177	0.0383	0.0595	0.0072
PROV	0.0469	-0.0717	0.0279	0.0142	0.0884	0.0383

Table 13.6. Coefficients and absolute contributions for regions, Example 13.4.

baccalauréats A and C. Regarding the regions, Ile de France plays an important role on each axis. On the first axis, it is opposed to Lorraine and Alsace, whereas on the second axis, it is opposed to Poitou-Charentes and Aquitaine. All of this is confirmed in Figure 13.3.

On the right side are the more classical baccalauréats and on the left, more technical ones. The regions on the left side have thus larger weights in the technical baccalauréats. Note also that most of the southern regions of France are concentrated in the lower part of the graph near the baccalauréat A.

Finally, looking at the 3-rd axis, we see that it is dominated by the baccalauréat E (negative sign) and to a lesser degree by H (negative) (as opposed to A (positive sign)). The dominating regions are HNOR (positive sign), opposed to NOPC and AUVE (negative sign). For

Baccal	s_1	s_2	s_3	$C_a(j, s_1)$	$C_a(j, s_2)$	$C_a(j, s_3)$
A	0.0447	-0.0679	0.0367	0.0376	0.2292	0.1916
B	0.1389	0.0557	0.0011	0.1724	0.0735	0.0001
C	0.0940	0.0995	0.0079	0.1198	0.3556	0.0064
D	0.0227	-0.0495	-0.0530	0.0098	0.1237	0.4040
E	-0.1932	0.0492	-0.1317	0.0825	0.0141	0.2900
F	-0.2156	0.0862	0.0188	0.3793	0.1608	0.0219
G	-0.1244	-0.0353	0.0279	0.1969	0.0421	0.0749
H	-0.0945	0.0438	-0.0888	0.0017	0.0010	0.0112

Table 13.7. Coefficients and absolute contributions for baccalauréats, Example 13.4.

λ_j	percentage of variance	cumulated percentage
4399.0	0.4914	0.4914
2213.6	0.2473	0.7387
1382.4	0.1544	0.8932
870.7	0.0973	0.9904
51.0	0.0057	0.9961
34.8	0.0039	1.0000
0.0	0.0000	0.0000

Table 13.8. Eigenvalues and explained proportion of variance, Example 13.5.

instance, HNOR is particularly poor in baccalauréat D.

EXAMPLE 13.5 *The U.S. crime data set (Table B.10) gives the number of crimes in the 50 states of the U.S. classified in 1985 for each of the following seven categories: murder, rape, robbery, assault, burglary, larceny and auto-theft. The analysis of the contingency table, limited to the first two factors, provides the following results (see Table 13.8).*

Looking at the absolute contributions (not reproduced here, see Exercise 13.6), it appears that the first axis is robbery (+) versus larceny (-) and auto-theft (-) axis and that the second factor contrasts assault (-) to auto-theft (+). The dominating states for the first axis are the North-Eastern States MA (+) and NY (+) constrasting the Western States WY (-)and ID (-). For the second axis, the differences are seen between the Northern States (MA (+) and RI (+)) and the Southern States AL (-), MS (-) and AR (-). These results can be clearly seen in Figure 13.4 where all the states and crimes are reported. The figure also shows in which states the proportion of a particular crime category is higher or lower than the national average (the origin).

Biplots

The biplot is a low-dimensional display of a data matrix $\mathcal{X}$ where the rows and columns are represented by points. The interpretation of a biplot is specifically directed towards the scalar products of lower dimensional factorial variables and is designed to approximately recover the individual elements of the data matrix in these scalar products. Suppose that we have a (10×5) data matrix with elements x_{ij}. The idea of the biplot is to find 10 row

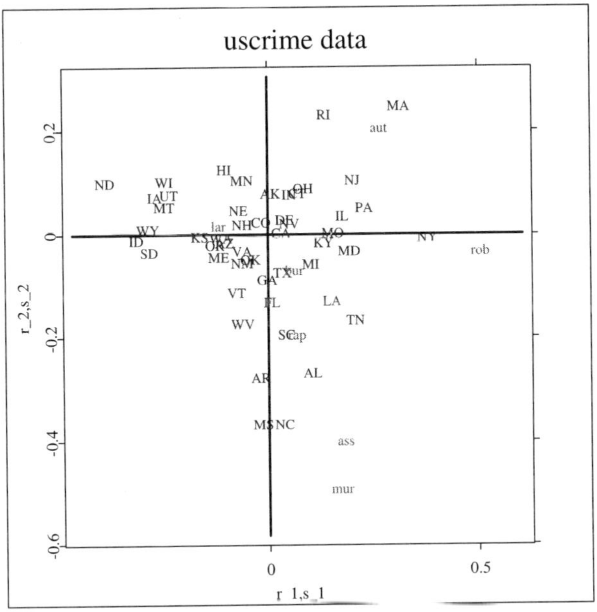

Figure 13.4. Projection of rows (the 50 states) and columns (the 7 crime categories). MVAcorrcrime

points $q_i \in \mathbb{R}^k$ $(k < p,\ i = 1, \ldots, 10)$ and 5 column points $t_j \in \mathbb{R}^k$ $(j = 1, \ldots, 5)$ such that the 50 scalar products between the row and the column vectors closely approximate the 50 corresponding elements of the data matrix $\mathcal{X}$. Usually we choose $k = 2$. For example, the scalar product between q_7 and t_4 should approximate the data value x_{74} in the seventh row and the fourth column. In general, the biplot models the data x_{ij} as the sum of a scalar product in some low-dimensional subspace and a residual "error" term:

$$\begin{aligned} x_{ij} &= q_i^\top t_j + e_{ij} \\ &= \sum_k q_{ik} t_{jk} + e_{ij}. \end{aligned} \tag{13.25}$$

To understand the link between correspondence analysis and the biplot, we need to introduce a formula which expresses x_{ij} from the original data matrix (see (13.3)) in terms of row and column frequencies. One such formula, known as the "reconstitution formula", is (13.10):

$$x_{ij} = E_{ij} \left(1 + \frac{\sum_{k=1}^{R} \lambda_k^{\frac{1}{2}} \gamma_{ik} \delta_{jk}}{\sqrt{\frac{x_{i\bullet} x_{\bullet j}}{x_{\bullet\bullet}}}} \right) \tag{13.26}$$

Consider now the row profiles $x_{ij}/x_{i\bullet}$ (the conditional frequencies) and the average row profile $x_{i\bullet}/x_{\bullet\bullet}$. From (13.26) we obtain the difference between each row profile and this average:

$$\left(\frac{x_{ij}}{x_{i\bullet}} - \frac{x_{i\bullet}}{x_{\bullet\bullet}} \right) = \sum_{k=1}^{R} \lambda_k^{\frac{1}{2}} \gamma_{ik} \left(\sqrt{\frac{x_{\bullet j}}{x_{i\bullet} x_{\bullet\bullet}}} \right) \delta_{jk}. \tag{13.27}$$

By the same argument we can also obtain the difference between each column profile and the average column profile:

$$\left(\frac{x_{ij}}{x_{\bullet j}} - \frac{x_{\bullet j}}{x_{\bullet\bullet}}\right) = \sum_{k=1}^{R} \lambda_k^{\frac{1}{2}} \gamma_{ik} \left(\sqrt{\frac{x_{i\bullet}}{x_{\bullet j} x_{\bullet\bullet}}}\right) \delta_{jk}. \tag{13.28}$$

Now, if $\lambda_1 \gg \lambda_2 \gg \lambda_3 \ldots$, we can approximate these sums by a finite number of K terms (usually $K = 2$) using (13.16) to obtain

$$\left(\frac{x_{ij}}{x_{\bullet j}} - \frac{x_{i\bullet}}{x_{\bullet\bullet}}\right) = \sum_{k=1}^{K} \left(\frac{x_{\bullet i}}{\sqrt{\lambda_k x_{\bullet\bullet}}} r_{ki}\right) s_{kj} + e_{ij}, \tag{13.29}$$

$$\left(\frac{x_{ij}}{x_{i\bullet}} - \frac{x_{\bullet j}}{x_{\bullet\bullet}}\right) = \sum_{k=1}^{K} \left(\frac{x_{\bullet j}}{\sqrt{\lambda_k x_{\bullet\bullet}}} s_{kj}\right) r_{ki} + e'_{ij}, \tag{13.30}$$

where e_{ij} and e'_{ij} are error terms. (13.30) shows that if we consider displaying the differences between the row profiles and the average profile, then the projection of the row profile r_k and a rescaled version of the projections of the column profile s_k constitute a biplot of these differences. (13.29) implies the same for the differences between the column profiles and this average.

Summary

- ↪ Correspondence analysis is a factorial decomposition of contingency tables. The p-dimensional individuals and the n-dimensional variables can be graphically represented by projecting onto spaces of smaller dimension.
- ↪ The practical computation consists of first computing a spectral decomposition of $\mathcal{A}^{-1}\mathcal{X}\mathcal{B}^{-1}\mathcal{X}^{\top}$ and $\mathcal{B}^{-1}\mathcal{X}^{\top}\mathcal{A}^{-1}\mathcal{X}$ which have the same first p eigenvalues. The graphical representation is obtained by plotting $\sqrt{\lambda_1} r_1$ vs. $\sqrt{\lambda_2} r_2$ and $\sqrt{\lambda_1} s_1$ vs. $\sqrt{\lambda_2} s_2$. Both plots maybe displayed in the same graph taking into account the appropriate orientation of the eigenvectors r_i, s_j.
- ↪ Correspondence analysis provides a graphical display of the association measure $c_{ij} = (x_{ij} - E_{ij})^2/E_{ij}$.
- ↪ Biplot is a low-dimensional display of a data matrix where the rows and columns are represented by points

13.4 Exercises

EXERCISE 13.1 *Show that the matrices $\mathcal{A}^{-1}\mathcal{X}\mathcal{B}^{-1}\mathcal{X}^{\top}$ and $\mathcal{B}^{-1}\mathcal{X}^{\top}\mathcal{A}^{-1}\mathcal{X}$ have an eigenvalue equal to 1 and that the corresponding eigenvectors are proportional to $(1, \ldots, 1)^{\top}$.*

EXERCISE 13.2 *Verify the relations in (13.8), (13.14) and (13.17).*

EXERCISE 13.3 *Do a correspondence analysis for the car marks data (Table B.7)! Explain how this table can be considered as a contingency table.*

EXERCISE 13.4 *Compute the χ^2-statistic of independence for the French baccalauréat data.*

EXERCISE 13.5 *Prove that $\mathcal{C} = \mathcal{A}^{-1/2}(\mathcal{X} - E)\mathcal{B}^{-1/2}\sqrt{x_{\bullet\bullet}}$ and $E = \frac{ab^\top}{x_{\bullet\bullet}}$ and verify (13.20).*

EXERCISE 13.6 *Do the full correspondence analysis of the U.S. crime data (Table B.10), and determine the absolute contributions for the first three axes. How can you interpret the third axis? Try to identify the states with one of the four regions to which it belongs. Do you think the four regions have a different behavior with respect to crime?*

EXERCISE 13.7 *Repeat Exercise 13.6 with the U.S. health data (Table B.16). Only analyze the columns indicating the number of deaths per state.*

EXERCISE 13.8 *Consider a $(n \times n)$ contingency table being a diagonal matrix $\mathcal{X}$. What do you expect the factors r_k, s_k to be like?*

EXERCISE 13.9 *Assume that after some reordering of the rows and the columns, the contingency table has the following structure:*

$\mathcal{X} =$

	J_1	J_2
I_1	$*$	0
I_2	0	$*$

That is, the rows I_i only have weights in the columns J_i, for $i = 1, 2$. What do you expect the graph of the first two factors to look like?

EXERCISE 13.10 *Redo Exercise 13.9 using the following contingency table:*

$\mathcal{X} =$

	J_1	J_2	J_3
I_1	$*$	0	0
I_2	0	$*$	0
I_3	0	0	$*$

EXERCISE 13.11 *Consider the French food data (Table B.6). Given that all of the variables are measured in the same units (Francs), explain how this table can be considered as a contingency table. Perform a correspondence analysis and compare the results to those obtained in the NPCA analysis in Chapter 9.*

14 Canonical Correlation Analysis

Complex multivariate data structures are better understood by studying low-dimensional projections. For a joint study of two data sets, we may ask what type of low-dimensional projection helps in finding possible joint structures for the two samples. The canonical correlation analysis is a standard tool of multivariate statistical analysis for discovery and quantification of associations between two sets of variables.

The basic technique is based on projections. One defines an index (projected multivariate variable) that maximally correlates with the index of the other variable for each sample separately. The aim of canonical correlation analysis is to maximize the association (measured by correlation) between the low-dimensional projections of the two data sets. The canonical correlation vectors are found by a joint covariance analysis of the two variables. The technique is applied to a marketing example where the association of a price factor and other variables (like design, sportiness etc.) is analysed. Tests are given on how to evaluate the significance of the discovered association.

14.1 Most Interesting Linear Combination

The associations between two sets of variables may be identified and quantified by canonical correlation analysis. The technique was originally developed by Hotelling (1935) who analyzed how arithmetic speed and arithmetic power are related to reading speed and reading power. Other examples are the relation between governmental policy variables and economic performance variables and the relation between job and company characteristics.

Suppose we are given two random variables $X \in \mathbb{R}^q$ and $Y \in \mathbb{R}^p$. The idea is to find an index describing a (possible) link between X and Y. Canonical correlation analysis (CCA) is based on linear indices, i.e., linear combinations

$$a^\top X \quad \text{and} \quad b^\top Y$$

of the random variables. Canonical correlation analysis searches for vectors a and b such that the relation of the two indices $a^\top x$ and $b^\top y$ is quantified in some interpretable way. More precisely, one is looking for the "most interesting" projections a and b in the sense that they maximize the correlation

$$\rho(a, b) = \rho_{a^\top X\, b^\top Y} \tag{14.1}$$

between the two indices.

Let us consider the correlation $\rho(a, b)$ between the two projections in more detail. Suppose that

$$\begin{pmatrix} X \\ Y \end{pmatrix} \sim \left(\begin{pmatrix} \mu \\ \nu \end{pmatrix}, \begin{pmatrix} \Sigma_{XX} & \Sigma_{XY} \\ \Sigma_{YX} & \Sigma_{YY} \end{pmatrix} \right)$$

where the sub-matrices of this covariance structure are given by

$$Var(X) = \Sigma_{XX} \ (q \times q)$$

$$\begin{aligned} Var(Y) &= \Sigma_{YY}\ (p \times p) \\ Cov(X,Y) &= E(X-\mu)(Y-\nu)^\top = \Sigma_{XY} = \Sigma_{YX}^\top \quad (q \times p). \end{aligned}$$

Using (3.7) and (4.26),

$$\rho(a,b) = \frac{a^\top \Sigma_{XY} b}{(a^\top \Sigma_{XX} a)^{1/2}\,(b^\top \Sigma_{YY} b)^{1/2}}. \tag{14.2}$$

Therefore, $\rho(ca,b) = \rho(a,b)$ for any $c \in \mathbb{R}^+$. Given the invariance of scale we may rescale projections a and b and thus we can equally solve

$$\max_{a,b} = a^\top \Sigma_{XY} b$$

under the constraints

$$\begin{aligned} a^\top \Sigma_{XX} a &= 1 \\ b^\top \Sigma_{YY} b &= 1. \end{aligned}$$

For this problem, define

$$\mathcal{K} = \Sigma_{XX}^{-1/2} \Sigma_{XY} \Sigma_{YY}^{-1/2}. \tag{14.3}$$

Recall the singular value decomposition of $\mathcal{K}(q \times p)$ from Theorem 2.2. The matrix $\mathcal{K}$ may be decomposed as

$$\mathcal{K} = \Gamma \Lambda \Delta^\top$$

with

$$\begin{aligned} \Gamma &= (\gamma_1, \ldots, \gamma_k) \\ \Delta &= (\delta_1, \ldots, \delta_k) \\ \Lambda &= \operatorname{diag}(\lambda_1^{1/2}, \ldots, \lambda_k^{1/2}) \end{aligned} \tag{14.4}$$

where by (14.3) and (2.15),

$$k = \operatorname{rank}(\mathcal{K}) = \operatorname{rank}(\Sigma_{XY}) = \operatorname{rank}(\Sigma_{YX}),$$

and $\lambda_1 \geq \lambda_2 \geq \ldots \lambda_k$ are the nonzero eigenvalues of $\mathcal{N}_1 = \mathcal{K}\mathcal{K}^\top$ and $\mathcal{N}_2 = \mathcal{K}^\top\mathcal{K}$ and γ_i and δ_j are the standardized eigenvectors of $\mathcal{N}_1$ and $\mathcal{N}_2$ respectively.

Define now for $i = 1, \ldots, k$ the vectors

$$a_i = \Sigma_{XX}^{-1/2} \gamma_i, \tag{14.5}$$

$$b_i = \Sigma_{YY}^{-1/2} \delta_i, \tag{14.6}$$

which are called the *canonical correlation vectors.* Using these canonical correlation vectors we define the *canonical correlation variables*

$$\eta_i = a_i^\top X \tag{14.7}$$

$$\varphi_i = b_i^\top Y. \tag{14.8}$$

The quantities $\rho_i = \lambda_i^{1/2}$ for $i = 1, \ldots, k$ are called the *canonical correlation coefficients.*

From the properties of the singular value decomposition given in (14.4) we have

$$Cov(\eta_i, \eta_j) = a_i^\top \Sigma_{XX} a_j = \gamma_i^\top \gamma_j = \begin{cases} 1 & i = j, \\ 0 & i \neq j. \end{cases} \tag{14.9}$$

The same is true for $Cov(\varphi_i, \varphi_j)$. The following theorem tells us that the canonical correlation vectors are the solution to the maximization problem of (14.1).

THEOREM 14.1 *For any given r, $1 \le r \le k$, the maximum*

$$C(r) = \max_{a,b} a^\top \Sigma_{XY} b \tag{14.10}$$

subject to

$$a^\top \Sigma_{XX} a = 1, \quad b^\top \Sigma_{YY} b = 1$$

and

$$a_i^\top \Sigma_{XX} a = 0 \textit{ for } i = 1, \ldots, r-1$$

is given by

$$C(r) = \rho_r = \lambda_r^{1/2}$$

and is attained when $a = a_r$ and $b = b_r$.

Proof:
The proof is given in three steps.

(i) Fix a and maximize over b, i.e., solve:

$$\max_b \left(a^\top \Sigma_{XY} b\right)^2 = \max_b \left(b^\top \Sigma_{YX} a\right)\left(a^\top \Sigma_{XY} b\right)$$

subject to $b^\top \Sigma_{YY} b = 1$. By Theorem 2.5 the maximum is given by the largest eigenvalue of the matrix

$$\Sigma_{YY}^{-1} \Sigma_{YX} a a^\top \Sigma_{XY}.$$

By Corollary 2.2, the only nonzero eigenvalue equals

$$a^\top \Sigma_{XY} \Sigma_{YY}^{-1} \Sigma_{YX} a. \tag{14.11}$$

(ii) Maximize (14.11) over a subject to the constraints of the Theorem. Put $\gamma = \Sigma_{XX}^{1/2} a$ and observe that (14.11) equals

$$\gamma^\top \Sigma_{XX}^{-1/2} \Sigma_{XY} \Sigma_{YY}^{-1} \Sigma_{YX} \Sigma_{XX}^{-1/2} \gamma = \gamma^\top \mathcal{K}^\top \mathcal{K} \gamma.$$

Thus, solve the equivalent problem

$$\max_\gamma \gamma^\top \mathcal{N}_1 \gamma \tag{14.12}$$

subject to $\gamma^\top \gamma = 1$, $\gamma_i^\top \gamma = 0$ for $i = 1, \ldots, r-1$.

Note that the γ_i's are the eigenvectors of $\mathcal{N}_1$ corresponding to its first $r-1$ largest eigenvalues. Thus, as in Theorem 9.3, the maximum in (14.12) is obtained by setting γ equal to the eigenvector corresponding to the r-th largest eigenvalue, i.e., $\gamma = \gamma_r$ or equivalently $a = a_r$. This yields

$$C^2(r) = \gamma_r^\top \mathcal{N}_1 \gamma_r = \lambda_r \gamma_r^\top \gamma = \lambda_r.$$

(iii) Show that the maximum is attained for $a = a_r$ and $b = b_r$. From the SVD of $\mathcal{K}$ we conclude that $\mathcal{K}\delta_r = \rho_r \gamma_r$ and hence

$$a_r^\top \Sigma_{XY} b_r = \gamma_r^\top \mathcal{K} \delta_r = \rho_r \gamma_r^\top \gamma_r = \rho_r.$$

□

Let

$$\begin{pmatrix} X \\ Y \end{pmatrix} \sim \left(\begin{pmatrix} \mu \\ \nu \end{pmatrix}, \begin{pmatrix} \Sigma_{XX} & \Sigma_{XY} \\ \Sigma_{YX} & \Sigma_{YY} \end{pmatrix} \right).$$

The canonical correlation vectors

$$a_1 = \Sigma_{XX}^{-1/2} \gamma_1,$$

$$b_1 = \Sigma_{YY}^{-1/2} \delta_1$$

maximize the correlation between the canonical variables

$$\eta_1 = a_1^\top X,$$

$$\varphi_1 = b_1^\top Y.$$

The covariance of the canonical variables η and φ is given in the next theorem.

THEOREM 14.2 *Let η_i and φ_i be the i-th canonical correlation variables ($i = 1, \dots, k$). Define $\eta = (\eta_1, \dots, \eta_k)$ and $\varphi = (\varphi_1, \dots, \varphi_k)$. Then*

$$Var \begin{pmatrix} \eta \\ \varphi \end{pmatrix} = \begin{pmatrix} \mathcal{I}_k & \Lambda \\ \Lambda & \mathcal{I}_k \end{pmatrix}$$

with Λ given in (14.4).

This theorem shows that the canonical correlation coefficients, $\rho_i = \lambda_i^{1/2}$, are the covariances between the canonical variables η_i and φ_i and that the indices $\eta_1 = a_1^\top X$ and $\varphi_1 = b_1^\top Y$ have the maximum covariance $\sqrt{\lambda_1} = \rho_1$.

The following theorem shows that canonical correlations are invariant w.r.t. linear transformations of the original variables.

THEOREM 14.3 *Let $X^* = \mathcal{U}^\top X + u$ and $Y^* = \mathcal{V}^\top Y + v$ where $\mathcal{U}$ and $\mathcal{V}$ are nonsingular matrices. Then the canonical correlations between X^* and Y^* are the same as those between X and Y. The canonical correlation vectors of X^* and Y^* are given by*

$$\begin{aligned} a_i^* &= \mathcal{U}^{-1} a_i, \\ b_i^* &= \mathcal{V}^{-1} b_i. \end{aligned} \tag{14.13}$$

Summary

↪ Canonical correlation analysis aims to identify possible links between two (sub-)sets of variables $X \in \mathbb{R}^q$ and $Y \in \mathbb{R}^p$. The idea is to find indices $a^\top X$ and $b^\top Y$ such that the correlation $\rho(a, b) = \rho_{a^\top X b^\top Y}$ is maximal.

↪ The maximum correlation (under constraints) is attained by setting $a_i = \Sigma_{XX}^{-1/2} \gamma_i$ and $b_i = \Sigma_{YY}^{-1/2} \delta_i$, where γ_i and δ_i denote the eigenvectors of $\mathcal{K}\mathcal{K}^\top$ and $\mathcal{K}^\top\mathcal{K}$, $\mathcal{K} = \Sigma_{XX}^{-1/2} \Sigma_{XY} \Sigma_{YY}^{-1/2}$ respectively.

↪ The vectors a_i and b_i are called canonical correlation vectors.

Summary (continued)
↪ The indices $\eta_i = a_i^\top X$ and $\varphi_i = b_i^\top Y$ are called canonical correlation variables.
↪ The values $\rho_1 = \sqrt{\lambda_1}, \ldots, \rho_k = \sqrt{\lambda_k}$, which are the square roots of the nonzero eigenvalues of $\mathcal{K}\mathcal{K}^\top$ and $\mathcal{K}^\top\mathcal{K}$, are called the canonical correlation coefficients. The covariance between the canonical correlation variables is $Cov(\eta_i, \varphi_i) = \sqrt{\lambda_i}$, $i = 1, \ldots, k$.
↪ The first canonical variables, $\eta_1 = a_1^\top X$ and $\varphi_1 = b_1^\top Y$, have the maximum covariance $\sqrt{\lambda_1}$.
↪ Canonical correlations are invariant w.r.t. linear transformations of the original variables X and Y.

14.2 Canonical Correlation in Practice

In practice we have to estimate the covariance matrices Σ_{XX}, Σ_{XY} and Σ_{YY}. Let us apply the canonical correlation analysis to the car marks data (see Table B.7). In the context of this data set one is interested in relating price variables with variables such as sportiness, safety, etc. In particular, we would like to investigate the relation between the two variables *non-depreciation of value* and *price of the car* and all other variables.

EXAMPLE 14.1 *We perform the canonical correlation analysis on the data matrices $\mathcal{X}$ and $\mathcal{Y}$ that correspond to the set of values {Price, Value Stability} and {Economy, Service, Design, Sporty car, Safety, Easy handling}, respectively. The estimated covariance matrix $\mathcal{S}$ is given by*

$$\mathcal{S} = \begin{array}{c} \begin{array}{cccccccc} \text{Price} & \text{Value} & \text{Econ.} & \text{Serv.} & \text{Design} & \text{Sport.} & \text{Safety} & \text{Easy h.} \end{array} \\ \left(\begin{array}{rr|rrrrrr} 1.41 & -1.11 & 0.78 & -0.71 & -0.90 & -1.04 & -0.95 & 0.18 \\ -1.11 & 1.19 & -0.42 & 0.82 & 0.77 & 0.90 & 1.12 & 0.11 \\ \hline 0.78 & -0.42 & 0.75 & -0.23 & -0.45 & -0.42 & -0.28 & 0.28 \\ -0.71 & 0.82 & -0.23 & 0.66 & 0.52 & 0.57 & 0.85 & 0.14 \\ -0.90 & 0.77 & -0.45 & 0.52 & 0.72 & 0.77 & 0.68 & -0.10 \\ -1.04 & 0.90 & -0.42 & 0.57 & 0.77 & 1.05 & 0.76 & -0.15 \\ -0.95 & 1.12 & -0.28 & 0.85 & 0.68 & 0.76 & 1.26 & 0.22 \\ 0.18 & 0.11 & 0.28 & 0.14 & -0.10 & -0.15 & 0.22 & 0.32 \end{array}\right) \end{array}.$$

Hence,

$$\mathcal{S}_{XX} = \begin{pmatrix} 1.41 & -1.11 \\ -1.11 & 1.19 \end{pmatrix}, \quad \mathcal{S}_{XY} = \begin{pmatrix} 0.78 & -0.71 & -0.90 & -1.04 & -0.95 & 0.18 \\ -0.42 & 0.82 & 0.77 & 0.90 & 1.12 & 0.11 \end{pmatrix},$$

$$\mathcal{S}_{YY} = \begin{pmatrix} 0.75 & -0.23 & -0.45 & -0.42 & -0.28 & 0.28 \\ -0.23 & 0.66 & 0.52 & 0.57 & 0.85 & 0.14 \\ -0.45 & 0.52 & 0.72 & 0.77 & 0.68 & -0.10 \\ -0.42 & 0.57 & 0.77 & 1.05 & 0.76 & -0.15 \\ -0.28 & 0.85 & 0.68 & 0.76 & 1.26 & 0.22 \\ 0.28 & 0.14 & -0.10 & -0.15 & 0.22 & 0.32 \end{pmatrix}.$$

It is interesting to see that value stability and price have a negative covariance. This makes sense since highly priced vehicles tend to loose their market value at a faster pace than medium priced vehicles.

Now we estimate $\mathcal{K} = \Sigma_{XX}^{-1/2}\,\Sigma_{XY}\,\Sigma_{YY}^{-1/2}$ *by*

$$\widehat{\mathcal{K}} = \mathcal{S}_{XX}^{-1/2}\,\mathcal{S}_{XY}\,\mathcal{S}_{YY}^{-1/2}$$

and perform a singular value decomposition of $\widehat{\mathcal{K}}$*:*

$$\widehat{\mathcal{K}} = \mathcal{G}\mathcal{L}\mathcal{D}^{\top} = (g_1, g_2)\,\mathrm{diag}(\ell_1^{1/2}, \ell_2^{1/2})\,(d_1, d_2)^{\top}$$

where the ℓ_i*'s are the eigenvalues of* $\widehat{\mathcal{K}}\widehat{\mathcal{K}}^{\top}$ *and* $\widehat{\mathcal{K}}^{\top}\widehat{\mathcal{K}}$ *with* $\mathrm{rank}(\widehat{\mathcal{K}}) = 2$*, and* g_i *and* d_i *are the eigenvectors of* $\widehat{\mathcal{K}}\widehat{\mathcal{K}}^{\top}$ *and* $\widehat{\mathcal{K}}^{\top}\widehat{\mathcal{K}}$*, respectively. The canonical correlation coefficients are*

$$r_1 = \ell_1^{1/2} = 0.98, \quad r_2 = \ell_2^{1/2} = 0.89.$$

The high correlation of the first two canonical variables can be seen in Figure 14.1. The first canonical variables are

$$\begin{aligned} \widehat{\eta}_1 &= \widehat{a}_1^{\top} x = 1.602\, x_1 + 1.686\, x_2 \\ \widehat{\varphi}_1 &= \widehat{b}_1^{\top} y = 0.568\, y_1 + 0.544\, y_2 - 0.012\, y_3 - 0.096\, y_4 - 0.014\, y_5 + 0.915\, y_6. \end{aligned}$$

Note that the variables y_1 *(economy),* y_2 *(service) and* y_6 *(easy handling) have positive coefficients on* $\widehat{\varphi}_1$*. The variables* y_3 *(design),* y_4 *(sporty car) and* y_5 *(safety) have a negative influence on* $\widehat{\varphi}_1$*.*

The canonical variable η_1 *may be interpreted as a price and value index. The canonical variable* φ_1 *is mainly formed from the qualitative variables economy, service and handling with negative weights on design, safety and sportiness. These variables may therefore be interpreted as an appreciation of the value of the car. The sportiness has a negative effect on the price and value index, as do the design and the safety features.*

Testing the Canonical Correlation Coefficients

The hypothesis that the two sets of variables $\mathcal{X}$ and $\mathcal{Y}$ are uncorrelated may be tested (under normality assumptions) with Wilk's likelihood ratio statistic (Gibbins, 1985):

$$T^{2/n} = \left|\mathcal{I} - \mathcal{S}_{YY}^{-1}\mathcal{S}_{YX}\mathcal{S}_{XX}^{-1}\mathcal{S}_{XY}\right| = \prod_{i=1}^{k}(1 - l_i).$$

This statistic unfortunately has a rather complicated distribution. Bartlett (1939) provides an approximation for large n:

$$-\{n - (p+q+3)/2\}\log\prod_{i=1}^{k}(1 - l_i) \sim \chi^2_{pq}. \tag{14.14}$$

A test of the hypothesis that only s of the canonical correlation coefficients are non-zero may be based (asymptotically) on the statistic

$$-\{n - (p+q+3)/2\}\log\prod_{i=s+1}^{k}(1 - l_i) \sim \chi^2_{(p-s)(q-s)}. \tag{14.15}$$

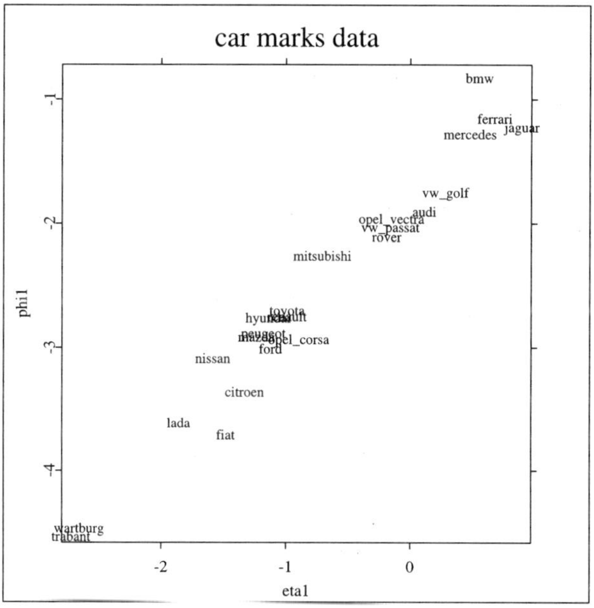

Figure 14.1. The first canonical variables for the car marks data. MVAcancarm

EXAMPLE 14.2 *Consider Example 14.1 again. There are $n = 40$ persons that have rated the cars according to different categories with $p = 2$ and $q = 6$. The canonical correlation coefficients were found to be $r_1 = 0.98$ and $r_2 = 0.89$. Bartlett's statistic (14.14) is therefore*

$$-\{40 - (2 + 6 + 3)/2\} \log\{(1 - 0.98^2)(1 - 0.89^2)\} = 165.59 \sim \chi^2_{12}$$

which is highly significant (the 99% quantile of the χ^2_{12} is 26.23). The hypothesis of no correlation between the variables $\mathcal{X}$ and $\mathcal{Y}$ is therefore rejected.

Let us now test whether the second canonical correlation coefficient is different from zero. We use Bartlett's statistic (14.15) with $s = 1$ and obtain

$$-\{40 - (2 + 6 + 3)/2\} \log\{(1 - 0.89^2)\} = 54.19 \sim \chi^2_5$$

which is again highly significant with the χ^2_5 distribution.

Canonical Correlation Analysis with Qualitative Data

The canonical correlation technique may also be applied to qualitative data. Consider for example the contingency table $\mathcal{N}$ of the French baccalauréat data. The dataset is given in Table B.8 in Appendix B.8. The CCA cannot be applied directly to this contingency table since the table does not correspond to the usual data matrix structure. We may wish, however, to explain the relationship between the row r and column c categories. It is possible to represent the data in a $(n \times (r + c))$ data matrix $\mathcal{Z} = (\mathcal{X}, \mathcal{Y})$ where n is the total number of frequencies in the contingency table $\mathcal{N}$ and $\mathcal{X}$ and $\mathcal{Y}$ are matrices of zero-one dummy variables. More precisely, let

$$x_{ki} = \begin{cases} 1 & \text{if the } k\text{-th individual belongs to the } i\text{-th row category} \\ 0 & \text{otherwise} \end{cases}$$

and

$$y_{kj} = \begin{cases} 1 & \text{if the } k\text{-th individual belongs to the } j\text{-th column category} \\ 0 & \text{otherwise} \end{cases}$$

where the indices range from $k = 1, \ldots, n$, $i = 1, \ldots, r$ and $j = 1, \ldots, c$. Denote the cell frequencies by n_{ij} so that $\mathcal{N} = (n_{ij})$ and note that

$$x_{(i)}^{\top} y_{(j)} = n_{ij},$$

where $x_{(i)}$ ($y_{(j)}$) denotes the i-th (j-th) column of $\mathcal{X}$ ($\mathcal{Y}$).

EXAMPLE 14.3 *Consider the following example where*

$$\mathcal{N} = \begin{pmatrix} 3 & 2 \\ 1 & 4 \end{pmatrix}.$$

The matrix $\mathcal{X}$ is therefore

$$\mathcal{X} = \begin{pmatrix} 1 & 0 \\ 1 & 0 \\ 1 & 0 \\ 1 & 0 \\ 1 & 0 \\ 0 & 1 \\ 0 & 1 \\ 0 & 1 \\ 0 & 1 \\ 0 & 1 \end{pmatrix},$$

the matrix $\mathcal{Y}$ is

$$\mathcal{Y} = \begin{pmatrix} 1 & 0 \\ 1 & 0 \\ 1 & 0 \\ 0 & 1 \\ 0 & 1 \\ 1 & 0 \\ 0 & 1 \\ 0 & 1 \\ 0 & 1 \\ 0 & 1 \end{pmatrix}$$

and the data matrix $\mathcal{Z}$ is

$$\mathcal{Z} = (\mathcal{X}, \mathcal{Y}) = \begin{pmatrix} 1 & 0 & 1 & 0 \\ 1 & 0 & 1 & 0 \\ 1 & 0 & 1 & 0 \\ 1 & 0 & 0 & 1 \\ 1 & 0 & 0 & 1 \\ 0 & 1 & 1 & 0 \\ 0 & 1 & 0 & 1 \\ 0 & 1 & 0 & 1 \\ 0 & 1 & 0 & 1 \\ 0 & 1 & 0 & 1 \end{pmatrix}.$$

The element n_{12} of $\mathcal{N}$ may be obtained by multiplying the first column of $\mathcal{X}$ with the second column of $\mathcal{Y}$ to yield

$$x_{(1)}^{\top} y_{(2)} = 2.$$

The purpose is to find the canonical variables $\eta = a^{\top}x$ and $\varphi = b^{\top}y$ that are maximally correlated. Note, however, that x has only one non-zero component and therefore an "individual" may be directly associated with its canonical variables or score (a_i, b_j). There will be n_{ij} points at each (a_i, b_j) and the correlation represented by these points may serve as a measure of dependence between the rows and columns of $\mathcal{N}$.

Let $\mathcal{Z} = (\mathcal{X}, \mathcal{Y})$ denote a data matrix constructed from a contingency table $\mathcal{N}$. Similar to Chapter 12 define

$$c = x_{i\bullet} = \sum_{j=1}^{c} n_{ij},$$

$$d = x_{\bullet j} = \sum_{i=1}^{r} n_{ij},$$

and define $\mathcal{C} = \mathrm{diag}(c)$ and $\mathcal{D} = \mathrm{diag}(d)$. Suppose that $x_{i\bullet} > 0$ and $x_{\bullet j} > 0$ for all i and j. It is not hard to see that

$$nS = \mathcal{Z}^{\top}\mathcal{H}\mathcal{Z} = \mathcal{Z}^{\top}\mathcal{Z} - n\bar{z}\bar{z}^{\top} = \begin{pmatrix} nS_{XX} & nS_{XY} \\ nS_{YX} & nS_{YY} \end{pmatrix}$$

$$= \left(\frac{n}{n-1}\right) \begin{pmatrix} \mathcal{C} - n^{-1}cc^{\top} & \mathcal{N} - \widehat{\mathcal{N}} \\ \mathcal{N}^{\top}\widehat{\mathcal{N}}^{\top} & \mathcal{D} - n^{-1}dd^{\top} \end{pmatrix}$$

where $\widehat{\mathcal{N}} = cd^{\top}/n$ is the estimated value of $\mathcal{N}$ under the assumption of independence of the row and column categories.

Note that

$$(n-1)S_{XX}1_r - \mathcal{C}1_r - n^{-1}cc^{\top}1_r = c - c(n^{-1}c^{\top}1_r) = c - c(n^{-1}n) = 0$$

and therefore S_{XX}^{-1} does not exist. The same is true for S_{YY}^{-1}. One way out of this difficulty is to drop one column from both $\mathcal{X}$ and $\mathcal{Y}$, say the first column. Let $\bar{c}$ and $\bar{d}$ denote the vectors obtained by deleting the first component of c and d.

Define $\bar{\mathcal{C}}$, $\bar{\mathcal{D}}$ and $\bar{S}_{XX}$, $\bar{S}_{YY}$, $\bar{S}_{XY}$ accordingly and obtain

$$(n\bar{S}_{XX})^{-1} = \bar{\mathcal{C}}^{-1} + n_{i\bullet}^{-1}1_r1_r^{\top}$$

$$(n\bar{S}_{YY})^{-1} = \bar{\mathcal{D}}^{-1} + n_{\bullet j}^{-1}1_c1_c^{\top}$$

so that (14.3) exists. The score associated with an individual contained in the first row (column) category of $\mathcal{N}$ is 0.

The technique described here for purely qualitative data may also be used when the data is a mixture of qualitative and quantitative characteristics. One has to "blow up" the data matrix by dummy zero-one values for the qualitative data variables.

Summary

↪ In practice we estimate Σ_{XX}, Σ_{XY}, Σ_{YY} by the empirical covariances and use them to compute estimates ℓ_i, g_i, d_i for λ_i, γ_i, δ_i from the SVD of $\widehat{\mathcal{K}} = \mathcal{S}_{XX}^{-1/2}\mathcal{S}_{XY}\mathcal{S}_{YY}^{-1/2}$.

↪ The signs of the coefficients of the canonical variables tell us the direction of the influence of these variables.

14.3 Exercises

EXERCISE 14.1 *Show that the eigenvalues of $\mathcal{K}\mathcal{K}^\top$ and $\mathcal{K}^\top\mathcal{K}$ are identical. (Hint: Use Theorem 2.6)*

EXERCISE 14.2 *Perform the canonical correlation analysis for the following subsets of variables: $\mathcal{X}$ corresponding to {price} and $\mathcal{Y}$ corresponding to {economy, easy handling} from the car marks data (Table B.7).*

EXERCISE 14.3 *Calculate the second canonical variables for Example 14.1. Interpret the coefficients.*

EXERCISE 14.4 *Use the SVD of matrix $\mathcal{K}$ to show that the canonical variables η_1 and η_2 are not correlated.*

EXERCISE 14.5 *Verify that the number of nonzero eigenvalues of matrix $\mathcal{K}$ is equal to* $\operatorname{rank}(\Sigma_{XY})$.

EXERCISE 14.6 *Express the singular value decomposition of matrices $\mathcal{K}$ and $\mathcal{K}^\top$ using eigenvalues and eigenvectors of matrices $\mathcal{K}^\top\mathcal{K}$ and $\mathcal{K}\mathcal{K}^\top$.*

EXERCISE 14.7 *What will be the result of CCA for $Y = X$?*

EXERCISE 14.8 *What will be the results of CCA for $Y = 2X$ and for $Y = -X$?*

EXERCISE 14.9 *What results do you expect if you perform CCA for X and Y such that $\Sigma_{XY} = 0$? What if $\Sigma_{XY} = \mathcal{I}_p$?*

15 Multidimensional Scaling

One major aim of multivariate data analysis is dimension reduction. For data measured in Euclidean coordinates, Factor Analysis and Principal Component Analysis are dominantly used tools. In many applied sciences data is recorded as ranked information. For example, in marketing, one may record "product A is better than product B". High-dimensional observations therefore often have mixed data characteristics and contain relative information (w.r.t. a defined standard) rather than absolute coordinates that would enable us to employ one of the multivariate techniques presented so far.

Multidimensional scaling (MDS) is a method based on proximities between objects, subjects, or stimuli used to produce a spatial representation of these items. Proximities express the similarity or dissimilarity between data objects. It is a dimension reduction technique since the aim is to find a set of points in low dimension (typically 2 dimensions) that reflect the relative configuration of the high-dimensional data objects. The metric MDS is concerned with such a representation in Euclidean coordinates. The desired projections are found via an appropriate spectral decomposition of a distance matrix.

The metric MDS solution may result in projections of data objects that conflict with the ranking of the original observations. The nonmetric MDS solves this problem by iterating between a monotizing algorithmic step and a least squares projection step. The examples presented in this chapter are based on reconstructing a map from a distance matrix and on marketing concerns such as ranking of the outfit of cars.

15.1 The Problem

Multidimensional scaling (MDS) is a mathematical tool that uses proximities between objects, subjects or stimuli to produce a spatial representation of these items. The proximities are defined as any set of numbers that express the amount of similarity or dissimilarity between pairs of objects, subjects or stimuli. In contrast to the techniques considered so far, MDS does not start from the raw multivariate data matrix $\mathcal{X}$, but from a $(n \times n)$ dissimilarity or distance matrix, $\mathcal{D}$, with the elements δ_{ij} and d_{ij} respectively. Hence, the underlying dimensionality of the data under investigation is in general not known.

MDS is a data reduction technique because it is concerned with the problem of finding a set of points in low dimension that represents the "configuration" of data in high dimension. The "configuration" in high dimension is represented by the distance or dissimilarity matrix $\mathcal{D}$.

MDS-techniques are often used to understand how people perceive and evaluate certain signals and information. For instance, political scientists use MDS techniques to understand why political candidates are perceived by voters as being similar or dissimilar. Psychologists use MDS to understand the perceptions and evaluations of speech, colors and personality traits, among other things. Last but not least, in marketing researchers use MDS techniques to shed light on the way consumers evaluate brands and to assess the relationship between product attributes.

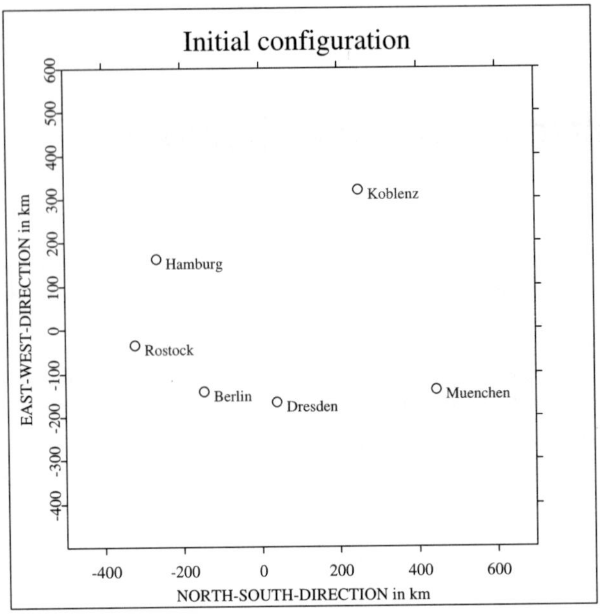

Figure 15.1. Metric MDS solution for the inter-city road distances. MVAMDScity1

In short, the primary purpose of all MDS-techniques is to uncover structural relations or patterns in the data and to represent it in a simple geometrical model or picture. One of the aims is to determine the dimension of the model (the goal is a low-dimensional, easily interpretable model) by finding the d-dimensional space in which there is maximum correspondence between the observed proximities and the distances between points measured on a metric scale.

Multidimensional scaling based on proximities is usually referred to as metric MDS, whereas the more popular nonmetric MDS is used when the proximities are measured on an ordinal scale.

EXAMPLE 15.1 *A good example of how MDS works is given by Dillon and Goldstein (1984) (Page 108). Suppose one is confronted with a map of Germany and asked to measure, with the use of a ruler and the scale of the map, some inter-city distances. Admittedly this is quite an easy exercise. However, let us now reverse the problem: One is given a set of distances, as in Table 15.1, and is asked to recreate the map itself. This is a far more difficult exercise, though it can be solved with a ruler and a compass in two dimensions. MDS is a method for solving this reverse problem in arbitrary dimensions. In Figure 15.2 you can see the graphical representation of the metric MDS solution to Table 15.1 after rotating and reflecting the points representing the cities. Note that the distances given in Table 15.1 are road distances that in general do not correspond to Euclidean distances. In real-life applications, the problems are exceedingly more complex: there are usually errors in the data and the dimensionality is rarely known in advance.*

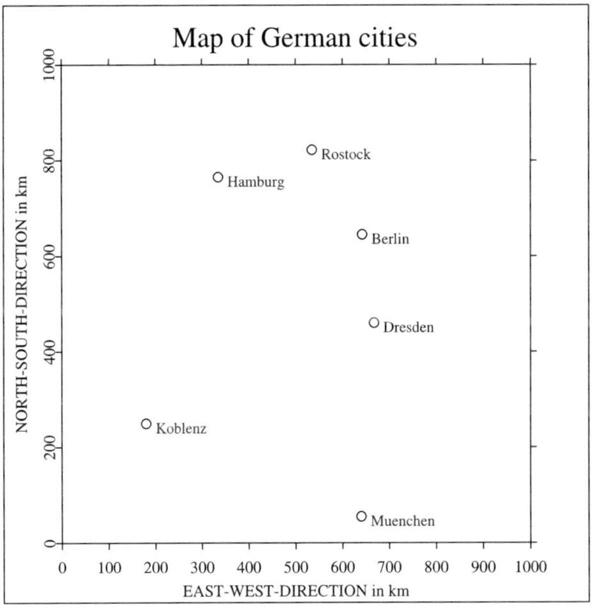

Figure 15.2. Metric MDS solution for the inter-city road distances after reflection and 90° rotation. MVAMDScity2

	Berlin	Dresden	Hamburg	Koblenz	Munich	Rostock
Berlin	0	214	279	610	596	237
Dresden		0	492	533	496	444
Hamburg			0	520	772	140
Koblenz				0	521	687
Munich					0	771
Rostock						0

Table 15.1. Inter-city distances.

	Audi 100	BMW 5	Citroen AX	Ferrari	...
Audi 100	0	2.232	3.451	3.689	...
BMW 5	2.232	0	5.513	3.167	...
Citroen AX	3.451	5.513	0	6.202	...
Ferrari	3.689	3.167	6.202	0	...
⋮	⋮	⋮	⋮	⋮	⋱

Table 15.2. Dissimilarities for cars.

EXAMPLE 15.2 *A further example is given in Table 15.2 where consumers noted their impressions of the dissimilarity of certain cars. The dissimilarities in this table were in*

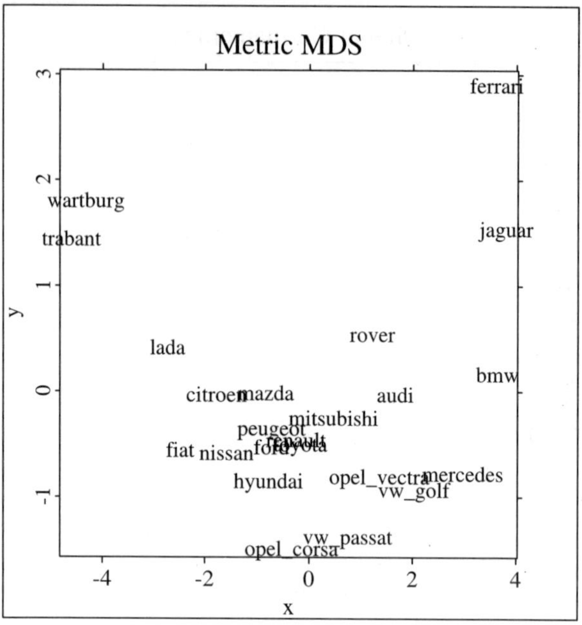

Figure 15.3. MDS solution on the car data. MVAmdscarm

fact computed from Table B.7 as Euclidean distances

$$d_{ij} = \sqrt{\sum_{l=1}^{8} (x_{il} - x_{jl})^2}.$$

MDS produces Figure 15.3 which shows a nonlinear relationship for all the cars in the projection. This enables us to build a nonlinear (quadratic) index with the Wartburg and the Trabant on the left and the Ferrari and the Jaguar on the right. We can construct an order or ranking of the cars based on the subjective impression of the consumers.

What does the ranking describe? The answer is given by Figure 15.4 which shows the correlation between the MDS projection and the variables. Apparently, the first MDS direction is highly correlated with service(-), value(-), design(-), sportiness(-), safety(-) and price(+). We can interpret the first direction as the price direction since a bad mark in price ("high price") obviously corresponds with a good mark, say, in sportiness ("very sportive"). The second MDS direction is highly positively correlated with practicability. We observe from this data an almost orthogonal relationship between price and practicability.

In MDS a map is constructed in Euclidean space that corresponds to given distances. Which solution can we expect? The solution is determined only up to rotation, reflection and shifts. In general, if $P_1, ..., P_n$ with coordinates $x_i = (x_{i1}, ..., x_{ip})^\top$ for $i = 1, ..., n$ represents a MDS solution in p dimensions, then $y_i = \mathcal{A}x_i + b$ with an orthogonal matrix $\mathcal{A}$ and a shift vector b also represents a MDS solution. A comparison of Figure 15.1 and Figure 15.2 illustrates this fact.

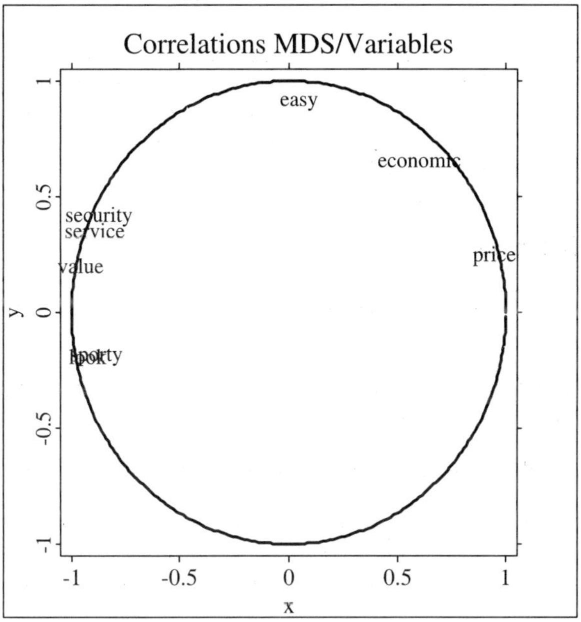

Figure 15.4. Correlation between the MDS direction and the variables. MVAmdscarm

Solution methods that use only the rank order of the distances are termed *nonmetric methods* of MDS. Methods aimed at finding the points P_i directly from a distance matrix like the one in the Table 15.2 are called *metric methods*.

Summary

↪ MDS is a set of techniques which use distances or dissimilarities to project high-dimensional data into a low-dimensional space essential in understanding respondents perceptions and evaluations for all sorts of items.

↪ MDS starts with a $(n \times n)$ proximity matrix $\mathcal{D}$ consisting of dissimilarities $\delta_{i,j}$ or distances d_{ij}.

↪ MDS is an explorative technique and focuses on data reduction.

↪ The MDS-solution is indeterminate with respect to rotation, reflection and shifts.

↪ The MDS-techniques are divided into metric MDS and nonmetric MDS.

15.2 Metric Multidimensional Scaling

Metric MDS begins with a $(n \times n)$ distance matrix $\mathcal{D}$ with elements d_{ij} where $i, j = 1, \dots, n$. The objective of metric MDS is to find a configuration of points in p-dimensional space from the distances between the points such that the coordinates of the n points along the p dimensions yield a Euclidean distance matrix whose elements are as close as possible to the elements of the given distance matrix $\mathcal{D}$.

The Classical Solution

The classical solution is based on a distance matrix that is computed from a *Euclidean geometry.*

DEFINITION 15.1 *A $(n \times n)$ distance matrix $\mathcal{D} = (d_{ij})$ is Euclidean if for some points $x_1, \dots, x_n \in \mathbb{R}^p$; $d_{ij}^2 = (x_i - x_j)^\top (x_i - x_j)$.*

The following result tells us whether a distance matrix is Euclidean or not.

THEOREM 15.1 *Define $\mathcal{A} = (a_{ij}), a_{ij} = -\frac{1}{2} d_{ij}^2$ and $\mathcal{B} = \mathcal{HAH}$ where $\mathcal{H}$ is the centering matrix. $\mathcal{D}$ is Euclidean if and only if $\mathcal{B}$ is positive semidefinite. If $\mathcal{D}$ is the distance matrix of a data matrix $\mathcal{X}$, then $\mathcal{B} = \mathcal{HXX}^T\mathcal{H}$. $\mathcal{B}$ is called the inner product matrix.*

Recovery of coordinates

The task of MDS is to find the original Euclidean coordinates from a given distance matrix. Let the coordinates of n points in a p dimensional Euclidean space be given by x_i $(i = 1, \dots, n)$ where $x_i = (x_{i1}, \dots, x_{ip})^\top$. Call $\mathcal{X} = (x_1, \dots, x_n)^\top$ the coordinate matrix and assume $\overline{x} = 0$. The Euclidean distance between the i-th and j-th points is given by:

$$d_{ij}^2 = \sum_{k=1}^{p} (x_{ik} - x_{jk})^2. \tag{15.1}$$

The general b_{ij} term of $\mathcal{B}$ is given by:

$$b_{ij} = \sum_{k=1}^{p} x_{ik} x_{jk} = x_i^\top x_j. \tag{15.2}$$

It is possible to derive $\mathcal{B}$ from the known squared distances d_{ij}, and then from $\mathcal{B}$ the unknown coordinates.

$$\begin{aligned} d_{ij}^2 &= x_i^\top x_i + x_j^\top x_j - 2x_i^\top x_j \\ &= b_{ii} + b_{jj} - 2b_{ij}. \end{aligned} \tag{15.3}$$

Centering of the coordinate matrix $\mathcal{X}$ implies that $\sum_{i=1}^{n} b_{ij} = 0$. Summing (15.3) over i, over j, and over i and j, we find:

$$\begin{aligned} \frac{1}{n} \sum_{i=1}^{n} d_{ij}^2 &= \frac{1}{n} \sum_{i=1}^{n} b_{ii} + b_{jj} \\ \frac{1}{n} \sum_{j=1}^{n} d_{ij}^2 &= b_{ii} + \frac{1}{n} \sum_{j=1}^{n} b_{jj} \\ \frac{1}{n^2} \sum_{i=1}^{n} \sum_{j=1}^{n} d_{ij}^2 &= \frac{2}{n} \sum_{i=1}^{n} b_{ii}. \end{aligned} \tag{15.4}$$

Solving (15.3) and (15.4) gives:

$$b_{ij} = -\frac{1}{2}(d_{ij}^2 - d_{i\bullet}^2 - d_{\bullet j}^2 + d_{\bullet\bullet}^2). \tag{15.5}$$

With $a_{ij} = -\frac{1}{2}d_{ij}^2$, and

$$\begin{aligned} a_{i\bullet} &= \frac{1}{n}\sum_{j=1}^{n} a_{ij} \\ a_{\bullet j} &= \frac{1}{n}\sum_{i=1}^{n} a_{ij} \\ a_{\bullet\bullet} &= \frac{1}{n^2}\sum_{i=1}^{n}\sum_{j=1}^{n} a_{ij} \end{aligned} \tag{15.6}$$

we get:

$$b_{ij} = a_{ij} - a_{i\bullet} - a_{\bullet j} + a_{\bullet\bullet}. \tag{15.7}$$

Define the matrix $\mathcal{A}$ as (a_{ij}), and observe that:

$$\mathcal{B} = \mathcal{H}\mathcal{A}\mathcal{H}. \tag{15.8}$$

The inner product matrix $\mathcal{B}$ can be expressed as:

$$\mathcal{B} = \mathcal{X}\mathcal{X}^\top, \tag{15.9}$$

where $\mathcal{X} = (x_1, \ldots, x_n)^\top$ is the $(n \times p)$ matrix of coordinates. The rank of $\mathcal{B}$ is then

$$rank(\mathcal{B}) = rank(\mathcal{X}\mathcal{X}^\top) = rank(\mathcal{X}) = p. \tag{15.10}$$

As required in Theorem 15.1 the matrix $\mathcal{B}$ is symmetric, positive semidefinite and of rank p, and hence it has p non-negative eigenvalues and $n-p$ zero eigenvalues. $\mathcal{B}$ can now be written as:

$$\mathcal{B} = \Gamma\Lambda\Gamma^\top \tag{15.11}$$

where $\Lambda = \text{diag}(\lambda_1, \ldots, \lambda_p)$, the diagonal matrix of the eigenvalues of $\mathcal{B}$, and $\Gamma = (\gamma_1, \ldots, \gamma_p)$, the matrix of corresponding eigenvectors. Hence the coordinate matrix $\mathcal{X}$ containing the point configuration in $\mathbb{R}^p$ is given by:

$$\mathcal{X} = \Gamma\Lambda^{\frac{1}{2}}. \tag{15.12}$$

How many dimensions?

The number of desired dimensions is small in order to provide practical interpretations, and is given by the rank of $\mathcal{B}$ or the number of nonzero eigenvalues λ_i. If $\mathcal{B}$ is positive semidefinite, then the number of nonzero eigenvalues gives the number of eigenvalues required for representing the distances d_{ij}.

The proportion of variation explained by p dimensions is given by

$$\frac{\sum_{i=1}^{p}\lambda_i}{\sum_{i=1}^{n-1}\lambda_i}. \tag{15.13}$$

It can be used for the choice of p. If $\mathcal{B}$ is not positive semidefinite we can modify (15.13) to

$$\frac{\sum_{i=1}^{p}\lambda_i}{\sum(\text{"positive eigenvalues"})}. \tag{15.14}$$

In practice the eigenvalues λ_i are almost always unequal to zero. To be able to represent the objects in a space with dimensions as small as possible we may modify the distance matrix to:

$$\mathcal{D}^* = d^*_{ij} \tag{15.15}$$

with

$$d^*_{ij} = \begin{cases} 0 & ; i = j \\ d_{ij} + e \geq 0 & ; i \neq j \end{cases} \tag{15.16}$$

where e is determined such that the inner product matrix $\mathcal{B}$ becomes positive semidefinite with a small rank.

Similarities

In some situations we do not start with distances but with similarities. The standard transformation (see Chapter 11) from a similarity matrix $\mathcal{C}$ to a distance matrix $\mathcal{D}$ is:

$$d_{ij} = (c_{ii} - 2c_{ij} + c_{jj})^{\frac{1}{2}}. \tag{15.17}$$

THEOREM 15.2 *If $\mathcal{C} \leq 0$, then the distance matrix $\mathcal{D}$ defined by (15.17) is Euclidean with centered inner product matrix $\mathcal{B} = \mathcal{HCH}$.*

Relation to Factorial Analysis

Suppose that the $(n \times p)$ data matrix $\mathcal{X}$ is centered so that $\mathcal{X}^\top\mathcal{X}$ equals a multiple of the covariance matrix $n\mathcal{S}$. Suppose that the p eigenvalues $\lambda_1, \ldots, \lambda_p$ of $n\mathcal{S}$ are distinct and non zero. Using the duality Theorem 8.4 of factorial analysis we see that $\lambda_1, \ldots, \lambda_p$ are also eigenvalues of $\mathcal{X}\mathcal{X}^\top = \mathcal{B}$ when $\mathcal{D}$ is the Euclidean distance matrix between the rows of $\mathcal{X}$. The k-dimensional solution to the metric MDS problem is thus given by the k first principal components of $\mathcal{X}$.

Optimality properties of the classical MDS solution

Let $\mathcal{X}$ be a $(n \times p)$ data matrix with some inter-point distance matrix $\mathcal{D}$. The objective of MDS is thus to find $\mathcal{X}_1$, a representation of $\mathcal{X}$ in a lower dimensional Euclidean space $\mathbb{R}^k$ whose inter-point distance matrix $\mathcal{D}_1$ is not far from $\mathcal{D}$. Let $\mathcal{L} = (\mathcal{L}_1, \mathcal{L}_2)$ be a $(p \times p)$ orthogonal matrix where $\mathcal{L}_1$ is $(p \times k)$. $\mathcal{X}_1 = \mathcal{X}\mathcal{L}_1$ represents a projection of $\mathcal{X}$ on the column space of $\mathcal{L}_1$; in other words, $\mathcal{X}_1$ may be viewed as a fitted configuration of $\mathcal{X}$ in $\mathbb{R}^k$. A measure of discrepancy between $\mathcal{D}$ and $\mathcal{D}_1 = (d^{(1)}_{ij})$ is given by

$$\phi = \sum_{i,j=1}^{n} (d_{ij} - d^{(1)}_{ij})^2. \tag{15.18}$$

THEOREM 15.3 *Among all projections $\mathcal{X}\mathcal{L}_1$ of $\mathcal{X}$ onto k-dimensional subspaces of $\mathbb{R}^p$ the quantity ϕ in (15.18) is minimized when $\mathcal{X}$ is projected onto its first k principal factors.*

We see therefore that the metric MDS is identical to principal factor analysis as we have defined it in Chapter 8.

Summary

↪ Metric MDS starts with a distance matrix $\mathcal{D}$.

↪ The aim of metric MDS is to construct a map in Euclidean space that corresponds to the given distances.

↪ A practical algorithm is given as:

1. start with distances d_{ij}
2. define $\mathcal{A} = -\frac{1}{2}d_{ij}^2$
3. put $\mathcal{B} = (a_{ij} - a_{i\bullet} - a_{\bullet j} + a_{\bullet\bullet})$
4. find the eigenvalues $\lambda_1, \dots, \lambda_p$ and the associated eigenvectors $\gamma_1, \dots, \gamma_p$ where the eigenvectors are normalized so that $\gamma_i^\top \gamma_i = 1$.
5. Choose an appropriate number of dimensions p (ideally $p = 2$)
6. The coordinates of the n points in the Euclidean space are given by $x_{ij} = \gamma_{ij}\lambda_j^{1/2}$ for $i = 1, \dots, n$ and $j = 1, \dots, p$.

↪ Metric MDS is identical to principal components analysis.

15.3 Nonmetric Multidimensional Scaling

The object of nonmetric MDS, as well as of metric MDS, is to find the coordinates of the points in p-dimensional space, so that there is a good agreement between the observed proximities and the inter-point distances. The development of nonmetric MDS was motivated by two main weaknesses in the metric MDS (Fahrmeir and Hamerle, 1984, Page 679):

1. the definition of an explicit functional connection between dissimilarities and distances in order to derive distances out of given dissimilarities, and
2. the restriction to Euclidean geometry in order to determine the object configurations.

The idea of a nonmetric MDS is to demand a less rigid relationship between the dissimilarities and the distances. Suppose that an unknown monotonic increasing function f,

$$d_{ij} = f(\delta_{ij}), \tag{15.19}$$

is used to generate a set of distances d_{ij} as a function of given dissimilarities δ_{ij}. Here f has the property that if $\delta_{ij} < \delta_{rs}$, then $f(\delta_{ij}) < f(\delta_{rs})$. The scaling is based on the rank order of the dissimilarities. Nonmetric MDS is therefore ordinal in character.

The most common approach used to determine the elements d_{ij} and to obtain the coordinates of the objects $x_1, x_2, \dots, x_n$ given only rank order information is an iterative process commonly referred to as the Shepard-Kruskal algorithm.

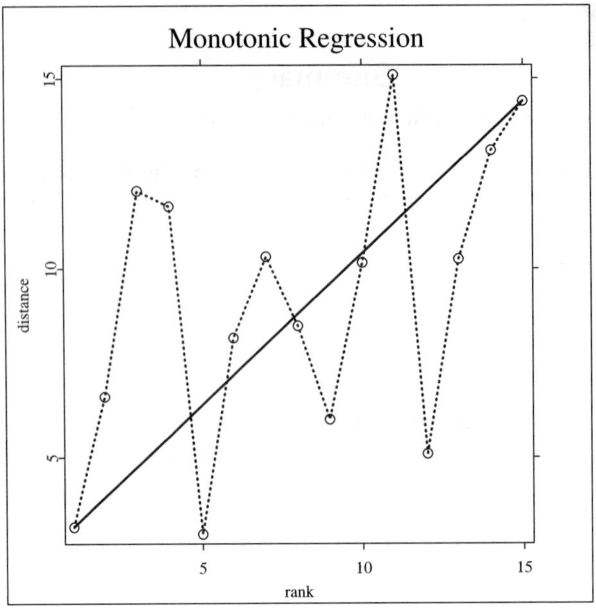

Figure 15.5. Ranks and distances. MVAMDSnonmstart

Shepard-Kruskal Algorithm

In a first step, called the initial phase, we calculate Euclidean distances $d_{ij}^{(0)}$ from an arbitrarily chosen initial configuration $\mathcal{X}_0$ in dimension p^*, provided that all objects have different coordinates. One might use metric MDS to obtain these initial coordinates. The second step or nonmetric phase determines disparities $\widehat{d}_{ij}^{(0)}$ from the distances $d_{ij}^{(0)}$ by constructing a monotone regression relationship between the $d_{ij}^{(0)}$'s and δ_{ij}'s, under the requirement that if $\delta_{ij} < \delta_{rs}$, then $\widehat{d}_{ij}^{(0)} \leq \widehat{d}_{rs}^{(0)}$. This is called the weak monotonicity requirement. To obtain the disparities $\widehat{d}_{ij}^{(0)}$, a useful approximation method is the *pool-adjacent violators* (PAV) algorithm (see Figure 15.6). Let

$$(i_1, j_1) > (i_2, j_2) > ... > (i_k, j_k) \tag{15.20}$$

be the rank order of dissimilarities of the $k = n(n-1)/2$ pairs of objects. This corresponds to the points in Figure 15.5. The PAV algorithm is described as follows: "beginning with the lowest ranked value of δ_{ij}, the adjacent $d_{ij}^{(0)}$ values are compared for each δ_{ij} to determine if they are monotonically related to the δ_{ij}'s. Whenever a block of consecutive values of $d_{ij}^{(0)}$ are encountered that violate the required monotonicity property the $d_{ij}^{(0)}$ values are averaged together with the most recent non-violator $d_{ij}^{(0)}$ value to obtain an estimator. Eventually this value is assigned to all points in the particular block".

In a third step, called the metric phase, the spatial configuration of $\mathcal{X}_0$ is altered to obtain $\mathcal{X}_1$. From $\mathcal{X}_1$ the new distances $d_{ij}^{(1)}$ can be obtained which are more closely related to the disparities $\widehat{d}_{ij}^{(0)}$ from step two.

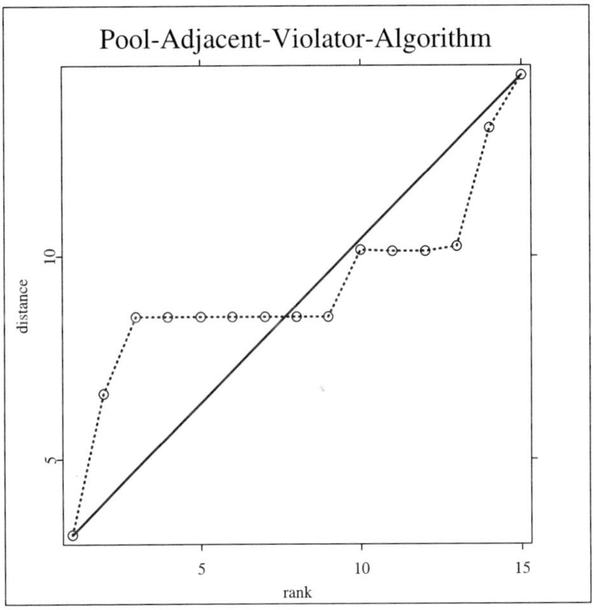

Figure 15.6. Pool-adjacent violators algorithm. MVAMDSpooladj

	j	1	2	3	4
i		Mercedes	Jaguar	Ferrari	VW
1	Mercedes	-			
2	Jaguar	3	-		
3	Ferrari	2	1	-	
4	VW	5	4	6	-

Table 15.3. Dissimilarities δ_{ij} for car marks.

i		x_{i1}	x_{i2}
1	Mercedes	3	2
2	Jaguar	2	7
3	Ferrari	1	3
4	VW	10	4

Table 15.4. Initial coordinates for MDS.

EXAMPLE 15.3 *Consider a small example with 4 objects based on the car marks data set. Our aim is to find a representation with* $p^* = 2$ *via MDS. Suppose that we choose as an initial configuration of* $\mathcal{X}_0$ *the coordinates given in Table 15.4. The corresponding distances* $d_{ij} = \sqrt{(x_i - x_j)^\top (x_i - x_j)}$ *are calculated in Table 15.5*

A plot of the dissimilarities of Table 15.5 against the distance yields Figure 15.8. This relation is not satisfactory since the ranking of the δ_{ij} *did not result in a monotone relation*

i,j	d_{ij}	$rank(d_{ij})$	δ_{ij}
1,2	5.1	3	3
1,3	2.2	1	2
1,4	7.3	4	5
2,3	4.1	2	1
2,4	8.5	5	4
3,4	9.1	6	6

Table 15.5. Ranks and distances.

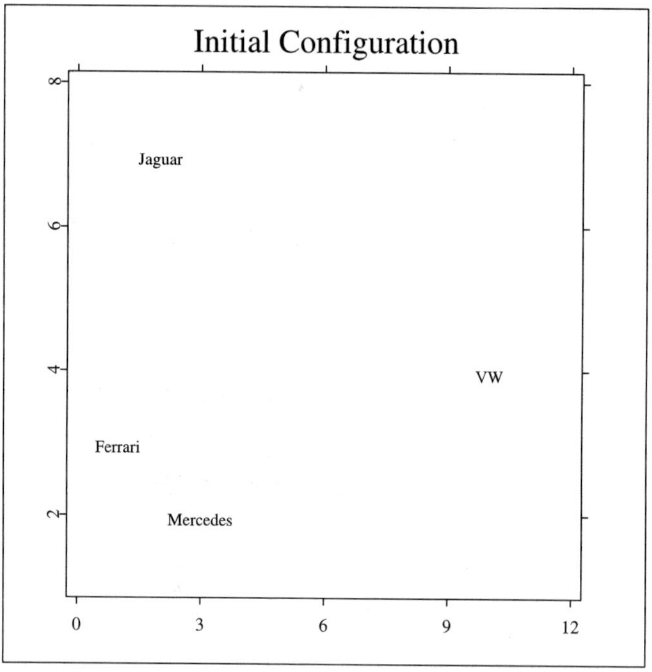

Figure 15.7. Initial configuration of the MDS of the car data. MVAnmdscar1

of the corresponding distances d_{ij}. We apply therefore the PAV algorithm.

The first violator of monotonicity is the second point $(1,3)$. Therefore we average the distances d_{13} and d_{23} to obtain the disparities

$$\hat{d}_{13} = \hat{d}_{23} = \frac{d_{13} + d_{23}}{2} = \frac{2.2 + 4.1}{2} = 3.17.$$

Applying the same procedure to $(2,4)$ and $(1,4)$ we obtain $\hat{d}_{24} = \hat{d}_{14} = 7.9$. The plot of δ_{ij} versus the disparities $\hat{d}_{ij}$ represents a monotone regression relationship.

In the initial configuration (Figure 15.7), the third point (Ferrari) could be moved so that the distance to object 2 (Jaguar) is reduced. This procedure however also alters the distance

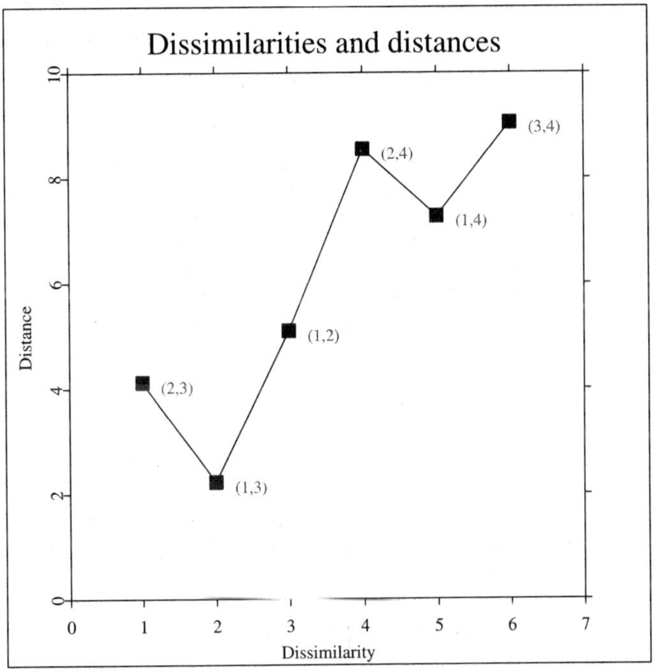

Figure 15.8. Scatterplot of dissimilarities against distances. MVAnmdscar2

between objects 3 and 4. Care should be given when establishing a monotone relation between δ_{ij} *and* d_{ij}.

In order to assess how well the derived configuration fits the given dissimilarities Kruskal suggests a measure called STRESS1 that is given by

$$STRESS1 = \left(\frac{\sum_{i<j}(d_{ij} - \hat{d}_{ij})^2}{\sum_{i<j} d_{ij}^2} \right)^{\frac{1}{2}}. \tag{15.21}$$

An alternative stress measure is given by

$$STRESS2 = \left(\frac{\sum_{i<j}(d_{ij} - \hat{d}_{ij})^2}{\sum_{i<j}(d_{ij} - d)^2} \right)^{\frac{1}{2}}, \tag{15.22}$$

where $\overline{d}$ denotes the average distance.

EXAMPLE 15.4 *The Table 15.6 presents the STRESS calculations for the car example. The average distance is* $\overline{d} = 36.4/6 = 6.1$. *The corresponding STRESS measures are:*

$$STRESS1 = \sqrt{2.6/256} = 0.1$$

$$STRESS2 = \sqrt{2.6/36.4} = 0.27.$$

(i,j)	δ_{ij}	d_{ij}	$\hat{d}_{ij}$	$(d_{ij}-\hat{d}_{ij})^2$	d_{ij}^2	$(d_{ij}-\overline{d})^2$
(2,3)	1	4.1	3.15	0.9	16.8	3.8
(1,3)	2	2.2	3.15	0.9	4.8	14.8
(1,2)	3	5.1	5.1	0	26.0	0.9
(2,4)	4	8.5	7.9	0.4	72.3	6.0
(1,4)	5	7.3	7.9	0.4	53.3	1.6
(3,4)	6	9.1	9.1	0	82.8	9.3
Σ		36.3		2.6	256.0	36.4

Table 15.6. STRESS calculations for car marks example.

The goal is to find a point configuration that balances the effects STRESS and non monotonicity. This is achieved by an iterative procedure. More precisely, one defines a new position of object i relative to object j by

$$x_{il}^{NEW} = x_{il} + \alpha\left(1 - \frac{\hat{d}_{ij}}{d_{ij}}\right)(x_{jl} - x_{il}), \qquad l = 1, \ldots, p^*. \tag{15.23}$$

Here α denotes the step width of the iteration.

By (15.23) the configuration of object i is improved relative to object j. In order to obtain an overall improvement relative to all remaining points one uses:

$$x_{il}^{NEW} = x_{il} + \frac{\alpha}{n-1}\sum_{j=1,j\neq i}^{n}\left(1 - \frac{\hat{d}_{ij}}{d_{ij}}\right)(x_{jl} - x_{il}), \qquad l = 1, \ldots, p^*. \tag{15.24}$$

The choice of step width α is crucial. Kruskal proposes a starting value of $\alpha = 0.2$. The iteration is continued by a numerical approximation procedure, such as steepest descent or the Newton-Raphson procedure.

In a fourth step, the evaluation phase, the STRESS measure is used to evaluate whether or not its change as a result of the last iteration is sufficiently small that the procedure is terminated. At this stage the optimal fit has been obtained for a given dimension. Hence, the whole procedure needs to be carried out for several dimensions.

EXAMPLE 15.5 *Let us compute the new point configuration for* $i = 3$ *(Ferrari). The initial coordinates from Table 15.4 are*

$$x_{31} = 1 \text{ and } x_{32} = 3.$$

Applying (15.24) yields (for $\alpha = 3$*):*

$$\begin{aligned} x_{31}^{NEW} &= 1 + \frac{3}{4-1}\sum_{j=1,j\neq 3}^{4}\left(1 - \frac{\hat{d}_{3j}}{d_{3j}}\right)(x_{j1} - 1) \\ &= 1 + \left(1 - \frac{3.15}{2.2}\right)(3-1) + \left(1 - \frac{3.15}{4.1}\right)(2-1) + \left(1 - \frac{9.1}{9.1}\right)(10-1) \\ &= 1 - 0.86 + 0.23 + 0 \\ &= 0.37. \end{aligned}$$

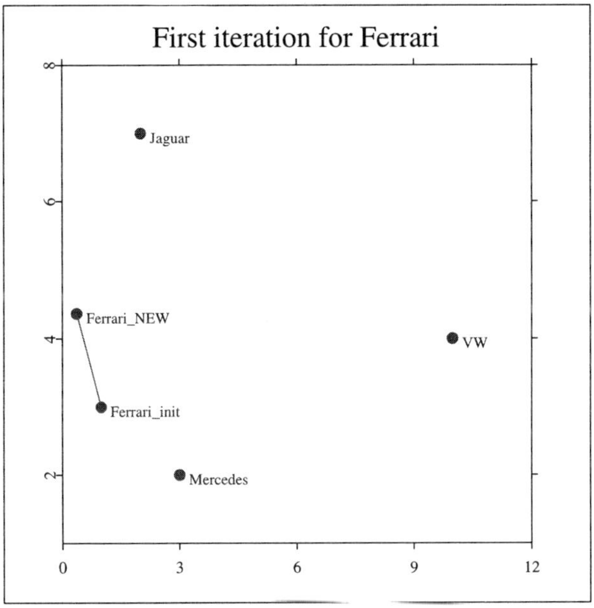

Figure 15.9. First iteration for Ferrari. MVAnmdscar3

Similarly we obtain $x_{32}^{NEW} = 4.36$.

To find the appropriate number of dimensions, p^*, a plot of the minimum STRESS value as a function of the dimensionality is made. One possible criterion in selecting the appropriate dimensionality is to look for an elbow in the plot. A rule of thumb that can be used to decide if a STRESS value is sufficiently small or not is provided by Kruskal:

$$S > 20\%, \text{ poor}; \ S = 10\%, \text{ fair}; \ S < 5\%, \text{ good}; \ S = 0, \text{ perfect}. \tag{15.25}$$

Summary

↪ Nonmetric MDS is only based on the rank order of dissimilarities.

↪ The object of nonmetric MDS is to create a spatial representation of the objects with low dimensionality.

Summary (continued)

↪ A practical algorithm is given as:

1. Choose an initial configuration.
2. Find d_{ij} from the configuration.
3. Fit $\hat{d}_{ij}$, the disparities, by the PAV algorithm.
4. Find a new configuration $\mathcal{X}_{n+1}$ by using the steepest descent.
5. Go to 2.

15.4 Exercises

EXERCISE 15.1 *Apply the MDS method to the Swiss bank note data. What do you expect to see ?*

EXERCISE 15.2 *Using (15.6), show that (15.7) can be written in the form (15.2).*

EXERCISE 15.3 *Show that*

1. $b_{ii} = a_{\bullet\bullet} - 2a_{i\bullet};\ b_{ij} = a_{ij} - a_{i\bullet} - a_{\bullet j} + a_{\bullet\bullet};\ i \neq j$
2. $\mathcal{B} = \sum_{i=1}^{p} x_i x_i^\top$
3. $\sum_{i=1}^{n} \lambda_i = \sum_{i=1}^{n} b_{ii} = \frac{1}{2n \sum_{i,j=1}^{n} d_{ij}^2}$.

EXERCISE 15.4 *Redo a careful analysis of the car marks data based on the following dissimilarity matrix:*

	j	*1*	*2*	*3*	*4*
i		*Nissan*	*Kia*	*BMW*	*Audi*
1	*Nissan*	-			
2	*Kia*	*2*	-		
3	*BMW*	*4*	*6*	-	
4	*Audi*	*3*	*5*	*1*	-

EXERCISE 15.5 *Apply the MDS method to the U.S. health data. Is the result in accordance with the geographic location of the U.S. states?*

EXERCISE 15.6 *Redo Exercise 15.5 with the U.S. crime data.*

EXERCISE 15.7 *Perform the MDS analysis on the Athletic Records data in Appendix B.18. Can you see which countries are "close to each other"?*

16 Conjoint Measurement Analysis

Conjoint Measurement Analysis plays an important role in marketing. In the design of new products it is valuable to know which components carry what kind of utility for the customer. Marketing and advertisement strategies are based on the perception of the new product's overall utility. It can be valuable information for a car producer to know whether a change in sportiness or a change in safety or comfort equipment is perceived as a higher increase in overall utility. The Conjoint Measurement Analysis is a method for attributing utilities to the components (part worths) on the basis of ranks given to different outcomes (stimuli) of the product. An important assumption is that the overall utility is decomposed as a sum of the utilities of the components.

In Section 16.1 we introduce the idea of Conjoint Measurement Analysis. We give two examples from the food and car industries. In Section 16.2 we shed light on the problem of designing questionnaires for ranking different product outcomes. In Section 16.3 we see that the metric solution of estimating the part-worths is given by solving a least squares problem. The estimated preference ordering may be nonmonotone. The nonmetric solution strategy takes care of this inconsistency by iterating between a least squares solution and the pool adjacent violators algorithm.

16.1 Introduction

In the design and perception of new products it is important to specify the contributions made by different facets or elements. The overall utility and acceptance of such a new product can then be estimated and understood as a possibly additive function of the elementary utilities. Examples are the design of cars, a food article or the program of a political party. For a new type of margarine one may ask whether a change in taste or presentation will enhance the overall perception of the product. The elementary utilities are here the presentation style and the taste (e.g., calory content). For a party program one may want to investigate whether a stronger ecological or a stronger social orientation gives a better overall profile of the party. For the marketing of a new car one may be interested in whether this new car should have a stronger active safety or comfort equipment or a more sporty note or combinations of both.

In Conjoint Measurement Analysis one assumes that the overall utility can be explained as an additive decomposition of the utilities of different elements. In a sample of questionnaires people ranked the product types and thus revealed their preference orderings. The aim is to find the decomposition of the overall utility on the basis of observed data and to interpret the elementary or marginal utilities.

EXAMPLE 16.1 *A car producer plans to introduce a new car with features that appeal to the customer and that may help in promoting future sales. The new elements that are considered are comfort/safety components (e.g. active steering or GPS) and a sporty look (leather steering wheel and additional KW of the engine). The car producer has thus 4 lines of cars.*

car 1:	*basic safety equipment*	*and*	*low sportiness*
car 2:	*basic safety equipment*	*and*	*high sportiness*
car 3:	*high safety equipment*	*and*	*low sportiness*
car 4:	*high safety equipment*	*and*	*high sportiness*

For the car producer it is important to rank these cars and to find out customers' attitudes toward a certain product line in order to develop a suitable marketing scheme. A tester may rank the cars as follows:

car	1	2	3	4
ranking	1	2	4	3

Table 16.1. Tester's ranking of cars.

The elementary utilities here are the comfort equipment and the level of sportiness. Conjoint Measurement Analysis aims at explaining the rank order given by the test person as a function of these elementary utilities.

EXAMPLE 16.2 *A food producer plans to create a new margarine and varies the product characteristics "calories" (low vs. high) and "presentation" (a plastic pot vs. paper package) (Backhaus, Erichson, Plinke and Weiber, 1996). We can view this in fact as ranking four products.*

product 1:	*low calories*	*and*	*plastic pot*
product 2:	*low calories*	*and*	*paper package*
product 3:	*high calories*	*and*	*plastic pot*
product 4:	*high calories*	*and*	*paper package*

These four fictive products may now be ordered by a set of sample testers as described in Table 16.2.

Product	1	2	3	4
ranking	3	4	1	2

Table 16.2. Tester's ranking of margarine.

The Conjoint Measurement Analysis aims to explain such a preference ranking by attributing *part-worths* to the different elements of the product. The part-worths are the utilities of the elementary components of the product.

In interpreting the part-worths one may find that for a test person one of the elements has a higher value or utility. This may lead to a new design or to the decision that this utility should be emphasized in advertisement schemes.

Summary

↪ Conjoint Measurement Analysis is used in the design of new products.

↪ Conjoint Measurement Analysis tries to identify part-worth utilities that contribute to an overall utility.

↪ The part-worths enter additively into an overall utility.

↪ The interpretation of the part-worths gives insight into the perception and acceptance of the product.

16.2 Design of Data Generation

The product is defined through the properties of the components. A *stimulus* is defined as a combination of the different components. Examples 16.1 and 16.2 had four stimuli each. In the margarine example they were the possible combinations of the factors X_1 (calories) and X_2 (presentation). If a product property such as

$$X_3(\text{usage}) = \begin{cases} 1 & \text{bread} \\ 2 & \text{cooking} \\ 3 & \text{universal} \end{cases}$$

is added, then there are $3 \cdot 2 \cdot 2 = 12$ stimuli.

For the automobile Example 16.1 additional characteristics may be engine power and the number of doors. Suppose that the engines offered for the new car have $50, 70, 90$ kW and that the car may be produced in 2-, 4-, or 5-door versions. These categories may be coded as

$$X_3(\text{power of engine}) = \begin{cases} 1 & 50 \text{ kW} \\ 2 & 70 \text{ kW} \\ 3 & 90 \text{ kW} \end{cases}$$

and

$$X_4(\text{doors}) = \begin{cases} 1 & 2 \text{ doors} \\ 2 & 4 \text{ doors} \\ 3 & 5 \text{ doors} \end{cases} .$$

Both X_3 and X_4 have three factor levels each, whereas the first two factors X_1 (safety) and X_2 (sportiness) have only two levels. Altogether $2 \cdot 2 \cdot 3 \cdot 3 = 36$ stimuli are possible. In a questionnaire a tester would have to rank all 36 different products.

The *profile method* asks for the utility of each stimulus. This may be time consuming and tiring for a test person if there are too many factors and factor levels. Suppose that there are 6 properties of components with 3 levels each. This results in $3^6 = 729$ stimuli (i.e., 729 different products) that a tester would have to rank.

The *two factor method* is a simplification and considers only two factors simultaneously. It is also called trade-off analysis. The idea is to present just two stimuli at a time and then to recombine the information. Trade-off analysis is performed by defining the trade-off matrices corresponding to stimuli of two factors only.

The trade-off matrices for the levels X_1, X_2 and X_3 from the margarine Example 16.2 are given below.

X_3	X_1	
1	1	2
2	1	2
3	1	2

X_3	X_2	
1	1	2
2	1	2
3	1	2

X_1	X_2	
1	1	2
2	1	2

Table 16.3. Trade-off matrices for margarine.

The trade-off matrices for the new car outfit are as follows:

X_4	X_3		
1	1	2	3
2	1	2	3
3	1	2	3

X_4	X_2	
1	1	2
2	1	2
3	1	2

X_4	X_1	
1	1	2
2	1	2
3	1	2

X_3	X_2	
1	1	2
2	1	2
3	1	2

X_3	X_1	
1	1	2
2	1	2
3	1	2

X_2	X_1	
1	1	2
2	1	2

Table 16.4. Trade-off matrices for car design.

The choice between the profile method and the trade-off analysis should be guided by consideration of the following aspects:

1. requirements on the test person,
2. time consumption,
3. product perception.

The first aspect relates to the ability of the test person to judge the different stimuli. It is certainly an advantage of the trade-off analysis that one only has to consider two factors simultaneously. The two factor method can be carried out more easily in a questionnaire without an interview.

The profile method incorporates the possibility of a complete product perception since the test person is not confronted with an isolated aspect (2 factors) of the product. The stimuli may be presented visually in its final form (e.g., as a picture). With the number of levels and properties the number of stimuli rise exponentially with the profile method. The time to complete a questionnaire is therefore a factor in the choice of method.

In general the product perception is the most important aspect and is therefore the profile method that is used the most. The time consumption aspect speaks for the trade-off analysis. There exist, however, clever strategies on selecting representation subsets of all profiles that bound the time investment. We therefore concentrate on the profile method in the following.

Summary

↪ A stimulus is a combination of different properties of a product.

↪ Conjoint measurement analysis is based either on a list of all factors (profile method) or on trade-off matrices (two factor method).

↪ Trade-off matrices are used if there are too many factor levels.

↪ Presentation of trade-off matrices makes it easier for testers since only two stimuli have to be ranked at a time.

16.3 Estimation of Preference Orderings

On the basis of the reported preference values for each stimulus conjoint analysis determines the part-worths. Conjoint analysis uses an additive model of the form

$$Y_k = \sum_{j=1}^{J} \sum_{l=1}^{L_j} \beta_{jl} I(X_j = x_{jl}) + \mu, \text{ for } k = 1, \dots, K \text{ and } \forall\, j \sum_{l=1}^{L_j} \beta_{jl} = 0. \tag{16.1}$$

X_j $(j = 1, \dots, J)$ denote the factors, x_{jl} $(l = 1, \dots, L_j)$ are the levels of each factor X_j and the coefficients β_{jl} are the part-worths. The constant μ denotes an overall level and Y_k is the observed preference for each stimulus and the total number of stimuli are:

$$K = \prod_{j=1}^{J} L_j.$$

Equation (16.1) is without an error term for the moment. In order to explain how (16.1) may be written in the standard linear model form we first concentrate on $J = 2$ factors. Suppose that the factors engine power and airbag safety equipment have been ranked as follows:

			airbag 1	airbag 2
engine	50 kW	1	1	3
	70 kW	2	2	6
	90 kW	3	4	5

There are $K = 6$ preferences altogether. Suppose that the stimuli have been sorted so that Y_1 corresponds to engine level 1 and airbag level 1, Y_2 corresponds to engine level 1 and airbag level 2, and so on. Then model (16.1) reads:

$$\begin{aligned}
Y_1 &= \beta_{11} + \beta_{21} + \mu \\
Y_2 &= \beta_{11} + \beta_{22} + \mu \\
Y_3 &= \beta_{12} + \beta_{21} + \mu \\
Y_4 &= \beta_{12} + \beta_{22} + \mu \\
Y_5 &= \beta_{13} + \beta_{21} + \mu \\
Y_6 &= \beta_{13} + \beta_{22} + \mu.
\end{aligned}$$

Now we would like to estimate the part-worths β_{jl}.

EXAMPLE 16.3 *In the margarine example let us consider the part-worths of* X_1 = *usage and* X_2 = *calories. We have* $x_{11} = 1$, $x_{12} = 2$, $x_{13} = 3$, $x_{21} = 1$ *and* $x_{22} = 2$. *(We momentarily re-labeled the factors:* X_3 *became* X_1*). Hence* $L_1 = 3$ *and* $L_2 = 2$. *Suppose that a person has ranked the six different products as in Table 16.5.*

			X_2 (calories)	
			low	high
			1	2
	bread	1	2	1
X_1 (usage)	cooking	2	3	4
	universal	3	6	5

Table 16.5. Ranked products.

If we order the stimuli as follows:

$$\begin{aligned}
Y_1 &= \textit{Utility}(X_1 = 1 \wedge X_2 = 1)\\
Y_2 &= \textit{Utility}(X_1 = 1 \wedge X_2 = 2)\\
Y_3 &= \textit{Utility}(X_1 = 2 \wedge X_2 = 1)\\
Y_4 &= \textit{Utility}(X_1 = 2 \wedge X_2 = 2)\\
Y_5 &= \textit{Utility}(X_1 = 3 \wedge X_2 = 1)\\
Y_6 &= \textit{Utility}(X_1 = 3 \wedge X_2 = 2),
\end{aligned}$$

we obtain from equation (16.1) the same decomposition as above:

$$\begin{aligned}
Y_1 &= \beta_{11} + \beta_{21} + \mu\\
Y_2 &= \beta_{11} + \beta_{22} + \mu\\
Y_3 &= \beta_{12} + \beta_{21} + \mu\\
Y_4 &= \beta_{12} + \beta_{22} + \mu\\
Y_5 &= \beta_{13} + \beta_{21} + \mu\\
Y_6 &= \beta_{13} + \beta_{22} + \mu.
\end{aligned}$$

Our aim is to estimate the part-worths β_{jl} as well as possible from a collection of tables like Table 16.5 that have been generated by a sample of test persons. First, the so-called metric solution to this problem is discussed and then a non-metric solution.

Metric Solution

The problem of conjoint measurement analysis can be solved by the technique of Analysis of Variance (ANOVA). An important assumption underlying this technique is that the "distance" between any two adjacent preference orderings corresponds to the same difference in utility. That is, the difference in utility between the products ranked 1st and 2nd is the same as the difference in utility between the products ranked 4th and 5th. Put differently, we treat the ranking of the products—which is a cardinal variable—as if it were a metric variable.

Introducing a mean utility μ equation (16.1) can be rewritten. The mean utility in the above Example 16.3 is $\mu = (1+2+3+4+5+6)/6 = 21/6 = 3.5$. In order to check the deviations of the utilities from this mean, we enlarge Table 16.5 by the mean utility $\bar{p}_{x_j \bullet}$, given a certain level of the other factor. The metric solution for the car example is given in Table 16.6:

			X_2 (airbags) 1	2	$\bar{p}_{x_1\bullet}$	β_{1l}
X_1 (engine)	50 kW	1	1	3	2	−1.5
	70 kW	2	2	6	4	0.5
	90 kW	3	4	5	4.5	1
$\bar{p}_{x_2\bullet}$			2.33	4.66	3.5	
β_{2l}			−1.16	1.16		

Table 16.6. Metric solution for car example.

			X_2 (calories) low 1	high 2	$\bar{p}_{x_1\bullet}$	β_{1l}
X_1 (usage)	bread	1	2	1	1.5	−2
	cooking	2	3	4	3.5	0
	universal	3	6	5	5.5	2
$\bar{p}_{x_2\bullet}$			3.66	3.33	3.5	
β_{2l}			0.16	−0.16		

Table 16.7. Metric solution for Table 16.5.

EXAMPLE 16.4 *In the margarine example the resulting part-worths for $\mu = 3.5$ are*

$$\begin{array}{rclrcl} \beta_{11} &=& -2 & \beta_{21} &=& 0.16 \\ \beta_{12} &=& 0 & \beta_{22} &=& -0.16\ . \\ \beta_{13} &=& 2 & & & \end{array}$$

Note that $\sum_{l=1}^{L_j} \beta_{jl} = 0$ $(j = 1, \dots, J)$. The estimated utility $\hat{Y}_1$ for the product with low calories and usage of bread, for example, is:

$$\hat{Y}_1 = \beta_{11} + \beta_{21} + \mu = -2 + 0.16 + 3.5 = 1.66.$$

The estimated utility $\hat{Y}_4$ for product 4 (cooking ($X_1 = 2$) and high calories ($X_2 = 2$)) is:

$$\hat{Y}_4 = \beta_{12} + \beta_{22} + \mu = 0 - 0.16 + 3.5 = 3.33.$$

The coefficients β_{jl} are computed as $\bar{p}_{x_{jl}} - \mu$, where $\bar{p}_{x_{jl}}$ is the average preference ordering for each factor level. For instance, $\bar{p}_{x_{11}} = 1/2 * (2 + 1) = 1.5$.

The fit can be evaluated by calculating the deviations of the fitted values to the observed preference orderings. In the rightmost column of Table 16.8 the quadratic deviations between the observed rankings (utilities) Y_k and the estimated utilities $\hat{Y}_k$ are listed.

The technique described that generated Table 16.7 is in fact the solution to a least squares problem. The conjoint measurement problem (16.1) may be rewritten as a linear regression model (with error $\varepsilon = 0$):

$$Y = \mathcal{X}\beta + \varepsilon \tag{16.2}$$

with $\mathcal{X}$ being a design matrix with dummy variables. $\mathcal{X}$ has the row dimension $K = \prod_{j=1}^{J} L_j$ (the number of stimuli) and the column dimension $D = \sum_{j=1}^{J} L_j - J$. The reason for the

Stimulus	Y_k	$\hat{Y}_k$	$Y_k - \hat{Y}_k$	$(Y_k - \hat{Y}_k)^2$
1	2	1.66	0.33	0.11
2	1	1.33	−0.33	0.11
3	3	3.66	−0.66	0.44
4	4	3.33	0.66	0.44
5	6	5.66	0.33	0.11
6	5	5.33	−0.33	0.11
$\sum$	21	21	0	1.33

Table 16.8. Deviations between model and data.

reduced column number is that per factor only $(L_j - 1)$ vectors are linearly independent. Without loss of generality we may standardize the problem so that the last coefficient of each factor is omitted. The error term ε is introduced since even for one person the preference orderings may not fit the model (16.1).

EXAMPLE 16.5 *If we rewrite the β coefficients in the form*

$$\begin{pmatrix} \beta_1 \\ \beta_2 \\ \beta_3 \\ \beta_4 \end{pmatrix} = \begin{pmatrix} \mu + \beta_{13} + \beta_{22} \\ \beta_{11} - \beta_{13} \\ \beta_{12} - \beta_{13} \\ \beta_{21} - \beta_{22} \end{pmatrix} \tag{16.3}$$

and define the design matrix $\mathcal{X}$ as

$$\mathcal{X} = \left(\begin{array}{c|cc|c} 1 & 1 & 0 & 1 \\ 1 & 1 & 0 & 0 \\ \hline 1 & 0 & 1 & 1 \\ 1 & 0 & 1 & 0 \\ \hline 1 & 0 & 0 & 1 \\ 1 & 0 & 0 & 0 \end{array}\right), \tag{16.4}$$

then equation (16.1) leads to the linear model (with error $\varepsilon = 0$):

$$Y = \mathcal{X}\beta + \varepsilon. \tag{16.5}$$

The least squares solution to this problem is the technique used for Table 16.7.

In practice we have more than one person to answer the utility rank question for the different factor levels. The design matrix is then obtained by stacking the above design matrix n times. Hence, for n persons we have as a final design matrix:

$$\mathcal{X}^* = 1_n \otimes \mathcal{X} = \left.\begin{pmatrix} \mathcal{X} \\ \vdots \\ \vdots \\ \mathcal{X} \end{pmatrix}\right\} n - \text{times}$$

which has dimension $(nK)(L - J)$ (where $L = \sum_{j=1}^{J} L_j$) and $Y^* = (Y_1^\top, ..., Y_n^\top)^\top$. The linear model (16.5) can now be written as:

$$Y^* = \mathcal{X}^*\beta + \varepsilon^*. \tag{16.6}$$

Given that the test people assign different rankings, the error term ε^* is a necessary part of the model.

EXAMPLE 16.6 *If we take the β vector as defined in (16.3) and the design matrix $\mathcal{X}$ from (16.4), we obtain the coefficients:*

$$\begin{array}{rcrcl}
\hat{\beta}_1 & = & 5.33 & = & \hat{\mu}+\hat{\beta}_{13}+\hat{\beta}_{22} \\
\hat{\beta}_2 & = & -4 & = & \hat{\beta}_{11}-\hat{\beta}_{13} \\
\hat{\beta}_3 & = & -2 & = & \hat{\beta}_{12}-\hat{\beta}_{13} \\
\hat{\beta}_4 & = & 0.33 & = & \hat{\beta}_{21}-\hat{\beta}_{22} \\
\sum_{l=1}^{L_j}\hat{\beta}_{jl} & = & 0. & &
\end{array} \tag{16.7}$$

Solving (16.7) we have:

$$\begin{array}{rcl}
\hat{\beta}_{11} & = & \hat{\beta}_2-\frac{1}{3}\left(\hat{\beta}_2+\hat{\beta}_3\right) & = & -2 \\
\hat{\beta}_{12} & = & \hat{\beta}_3-\frac{1}{3}\left(\hat{\beta}_2+\hat{\beta}_3\right) & = & 0 \\
\hat{\beta}_{13} & = & -\frac{1}{3}\left(\hat{\beta}_2+\hat{\beta}_3\right) & = & 2 \\
\hat{\beta}_{21} & = & \hat{\beta}_4-\frac{1}{2}\hat{\beta}_4=\frac{1}{2}\hat{\beta}_4 & = & 0.16 \\
\hat{\beta}_{31} & = & -\frac{1}{2}\hat{\beta}_4 & = & -0.16 \\
\hat{\mu} & = & \hat{\beta}_1+\frac{1}{3}\left(\hat{\beta}_2+\hat{\beta}_3\right)+\frac{1}{2}(\hat{\beta}_4) & = & 3.5.
\end{array} \tag{16.8}$$

In fact, we obtain the same estimated part-worths as in Table 16.7. The stimulus $k=2$ corresponds to adding up β_{11}, β_{22}, and μ (see (16.3)). Adding $\hat{\beta}_1$ and $\hat{\beta}_2$ gives:

$$\hat{Y}_2 = 5.33 - 4 = 1.33.$$

Nonmetric solution

If we drop the assumption that utilities are measured on a metric scale, we have to use (16.1) to estimate the coefficients from an adjusted set of estimated utilities. More precisely, we may use the monotone ANOVA as developed by Kruskal (1965). The procedure works as follows. First, one estimates model (16.1) with the ANOVA technique described above. Then one applies a monotone transformation $\hat{Z}=f(\hat{Y})$ to the estimated stimulus utilities. The monotone transformation f is used because the fitted values $\hat{Y}_k$ from (16.2) of the reported preference orderings Y_k may not be monotone. The transformation $\hat{Z}_k = f(\hat{Y}_k)$ is introduced to guarantee monotonicity of preference orderings. For the car example the reported Y_k values were $Y=(1,\ 3,\ 2,\ 6,\ 4,\ 5)^\top$. The estimated values are computed as:

$$\begin{array}{rcrl}
\hat{Y}_1 & = & -1.5-1.16+3.5 & =0.84 \\
\hat{Y}_2 & = & -1.5+1.16+3.5 & =3.16 \\
\hat{Y}_3 & = & 0.5-1.16+3.5 & =2.84 \\
\hat{Y}_4 & = & -0.5+1.16+3.5 & =5.16 \\
\hat{Y}_5 & = & 1.5-1.16+3.5 & =3.34 \\
\hat{Y}_6 & = & 1.5+1.16+3.5 & =5.66.
\end{array}$$

If we make a plot of the estimated preference orderings against the revealed ones, we obtain Figure 16.1.

We see that the estimated $\hat{Y}_6 = 5.16$ is below the estimated $\hat{Y}_5 = 5.66$ and thus an inconsistency in ranking the utilities occurrs. The monotone transformation $\hat{Z}_k = f(\hat{Y}_k)$ is

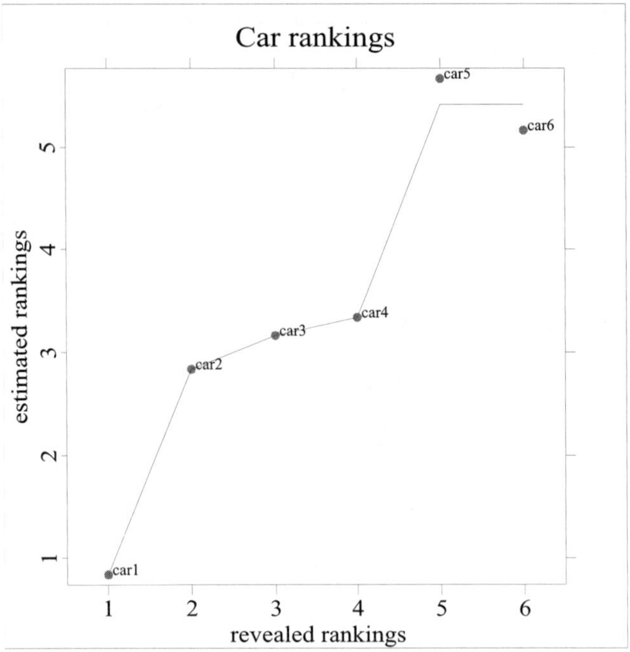

Figure 16.1. Plot of estimated preference orderings vs. revealed rankings and PAV fit. MVAcarrankings

introduced to make the relationship in Figure 16.1 monotone. A very simple procedure consists of averaging the "violators" $\hat{Y}_6$ and $\hat{Y}_5$ to obtain 5.41. The relationship is then monotone but the model (16.1) may now be violated. The idea is therefore to iterate these two steps. This procedure is iterated until the stress measure (see Chapter 15)

$$\text{STRESS} = \frac{\sum_{k=1}^{K} (\hat{Z}_k - \hat{Y}_k)^2}{\sum_{k=1}^{K} (\hat{Y}_k - \bar{\hat{Y}})^2} \tag{16.9}$$

is minimized over β and the monotone transformation f. The monotone transformation can be computed by the so called pool-adjacent-violators (PAV) algorithm.

Summary

↪ The part-worths are estimated via the least squares method.

↪ The metric solution corresponds to analysis of variance in a linear model.

↪ The non-metric solution iterates between a monotone regression curve fitting and determining the part-worths by ANOVA methodology.

↪ The fitting of data to a monotone function is done via the PAV algorithm.

16.4 Exercises

EXERCISE 16.1 *Compute the part-worths for the following table of rankings*

		X_2	
		1	2
	1	1	2
X_1	2	4	3
	3	6	5

EXERCISE 16.2 *Consider again Example 16.5. Rewrite the design matrix $\mathcal{X}$ and the parameter vector β so that the overall mean effect μ is part of $\mathcal{X}$ and β, i.e., find the matrix $\mathcal{X}'$ and β' such that $Y = \mathcal{X}'\beta'$.*

EXERCISE 16.3 *Compute the design matrix for Example 16.5 for $n = 3$ persons ranking the margarine with X_1 and X_2.*

EXERCISE 16.4 *Construct an analog for Table 16.8 for the car example.*

EXERCISE 16.5 *Compute the part-worths on the basis of the following tables of rankings observed on $n = 3$ persons.*

		X_2	
	1	1	2
X_1	2	4	3
	3	6	5

	X_2	
	1	3
X_1	4	2
	5	6

	X_2	
	3	1
X_1	5	2
	6	4

EXERCISE 16.6 *Suppose that in the car example a person has ranked cars by the profile method on the following characteristics:*

$$
\begin{aligned}
X_1 &= \textit{motor} \\
X_2 &= \textit{safety} \\
X_3 &= \textit{doors}
\end{aligned}
$$

There are $k = 18$ stimuli.

X_1	X_2	X_3	*preference*
1	1	1	1
1	1	2	3
1	1	3	2
1	2	1	5
1	2	2	4
1	2	3	6

X_1	X_2	X_3	*preference*
2	1	1	7
2	1	2	8
2	1	3	9
2	2	1	10
2	2	2	12
2	2	3	11

X_1	X_2	X_3	*preference*
3	*1*	*1*	*13*
3	*1*	*2*	*15*
3	*1*	*3*	*14*
3	*2*	*1*	*16*
3	*2*	*2*	*17*
3	*2*	*3*	*18*

Estimate and analyze the part-worths.

17 Applications in Finance

A portfolio is a linear combination of assets. Each asset contributes with a weight c_j to the portfolio. The performance of such a portfolio is a function of the various returns of the assets and of the weights $c = (c_1, \ldots, c_p)^\top$. In this chapter we investigate the "optimal choice" of the portfolio weights c. The optimality criterion is the mean-variance efficiency of the portfolio. Usually investors are risk-averse, therefore, we can define a mean-variance efficient portfolio to be a portfolio that has a minimal variance for a given desired mean return. Equivalently, we could try to optimize the weights for the portfolios with maximal mean return for a given variance (risk structure). We develop this methodology in the situations of (non)existence of riskless assets and discuss relations with the Capital Assets Pricing Model (CAPM).

17.1 Portfolio Choice

Suppose that one has a portfolio of p assets. The price of asset j at time i is denoted as p_{ij}. The return from asset j in a single time period (day, month, year etc.) is:

$$x_{ij} = \frac{p_{ij} - p_{i-1,j}}{p_{ij}}.$$

We observe the vectors $x_i = (x_{i1}, \ldots, x_{ip})^\top$ (i.e., the returns of the assets which are contained in the portfolio) over several time periods. We stack these observations into a data matrix $\mathcal{X} = (x_{ij})$ consisting of observations of a random variable

$$X \sim (\mu, \Sigma).$$

The return of the portfolio is the weighted sum of the returns of the p assets:

$$Q = c^\top X, \tag{17.1}$$

where $c = (c_1, \ldots, c_p)^\top$ (with $\sum_{j=1}^{p} c_j = 1$) denotes the proportions of the assets in the portfolio. The mean return of the portfolio is given by the expected value of Q, which is $c^\top \mu$. The *risk* or *volatility* of the portfolio is given by the variance of Q (Theorem 4.6), which is equal to two times

$$\frac{1}{2}\, c^\top \Sigma c. \tag{17.2}$$

The reason for taking *half of* the variance of Q is merely technical. The optimization of (17.2) with respect to c is of course equivalent to minimizing $c^\top \Sigma c$. Our aim is to maximize the portfolio returns (17.1) given a bound on the volatility (17.2) or vice versa to minimize risk given a (desired) mean return of the portfolio.

Summary

↪ Given a matrix of returns $\mathcal{X}$ from p assets in n time periods, and that the underlying distribution is stationary, i.e., $X \sim (\mu, \Sigma)$, then the (theoretical) return of the portfolio is a weighted sum of the returns of the p assets, namely $Q = c^\top X$.

↪ The expected value of Q is $c^\top \mu$. For technical reasons one considers optimizing $\frac{1}{2}\, c^\top \Sigma c$. The risk or volatility is $c^\top \Sigma c = Var(c^\top X)$.

↪ The portfolio choice, i.e., the selection of c, is such that the return is maximized for a given risk bound.

17.2 Efficient Portfolio

A variance efficient portfolio is one that keeps the risk (17.2) minimal under the constraint that the weights sum to 1, i.e., $c^\top 1_p = 1$. For a variance efficient portfolio, we therefore try to find the value of c that minimizes the Lagrangian

$$\mathcal{L} = \frac{1}{2}\, c^\top \Sigma c - \lambda(c^\top 1_p - 1). \tag{17.3}$$

A mean-variance efficient portfolio is defined as one that has minimal variance among all portfolios with the same mean. More formally, we have to find a vector of weights c such that the variance of the portfolio is minimal subject to two constraints:

1. a certain, pre-specified mean return $\overline{\mu}$ has to be achieved,
2. the weights have to sum to one.

Mathematically speaking, we are dealing with an optimization problem under two constraints.

The Lagrangian function for this problem is given by

$$\mathcal{L} = c^\top \Sigma c + \lambda_1(\overline{\mu} - c^\top \mu) + \lambda_2(1 - c^\top 1_p).$$

With tools presented in Section 2.4 we can calculate the first order condition for a minimum:

$$\frac{\partial \mathcal{L}}{\partial c} = 2\Sigma c - \lambda_1 \mu - \lambda_2 1_p = 0. \tag{17.4}$$

EXAMPLE 17.1 *Figure 17.1 shows the returns from January 1978 to December 1987 of six stocks traded on the New York stock exchange (Berndt, 1990). For each stock we have chosen the same scale on the vertical axis (which gives the return of the stock). Note how the return of some stocks, such as Pan American Airways and Delta Airlines, are much more volatile than the returns of other stocks, such as IBM or Consolidated Edison (Electric utilities).*

As a very simple example consider two differently weighted portfolios containing only two assets, IBM and PanAm. Figure 17.2 displays the monthly returns of the two portfolios. The portfolio in the upper panel consists of approximately 10% PanAm assets and 90% IBM assets. The portfolio in the lower panel contains an equal proportion of each of the assets. The text windows on the right of Figure 17.2 show the exact weights which were used. We

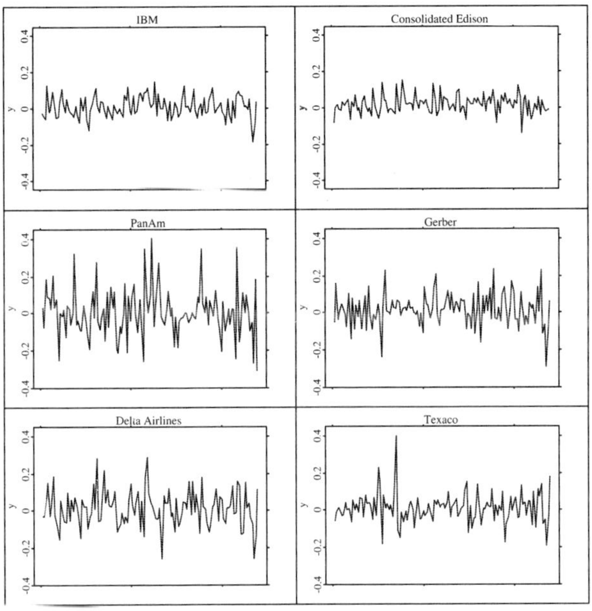

Figure 17.1. Returns of six firms from January 1978 to December 1987.
MVAreturns

can clearly see that the returns of the portfolio with a higher share of the IBM assets (which have a low variance) are much less volatile.

For an exact analysis of the optimization problem (17.4) we distinguish between two cases: the existence and nonexistence of a riskless asset. A riskless asset is an asset such as a zero bond, i.e., a financial instrument with a fixed nonrandom return (Franke, Härdle and Hafner, 2004).

Nonexistence of a Riskless Asset

Assume that the covariance matrix Σ is invertible (which implies positive definiteness). This is equivalent to the nonexistence of a portfolio c with variance $c^\top \Sigma c = 0$. If all assets are uncorrelated, Σ is invertible if all of the asset returns have positive variances. A riskless asset (uncorrelated with all other assets) would have zero variance since it has fixed, nonrandom returns. In this case Σ would not be positive definite.

The optimal weights can be derived from the first order condition (17.4) as

$$c = \frac{1}{2}\Sigma^{-1}(\lambda_1 \mu + \lambda_2 1_p). \tag{17.5}$$

Multiplying this by a $(p \times 1)$ vector 1_p of ones, we obtain

$$1 = 1_p^\top c = \frac{1}{2} 1_p^\top \Sigma^{-1}(\lambda_1 \mu + \lambda_2 1_p^\top),$$

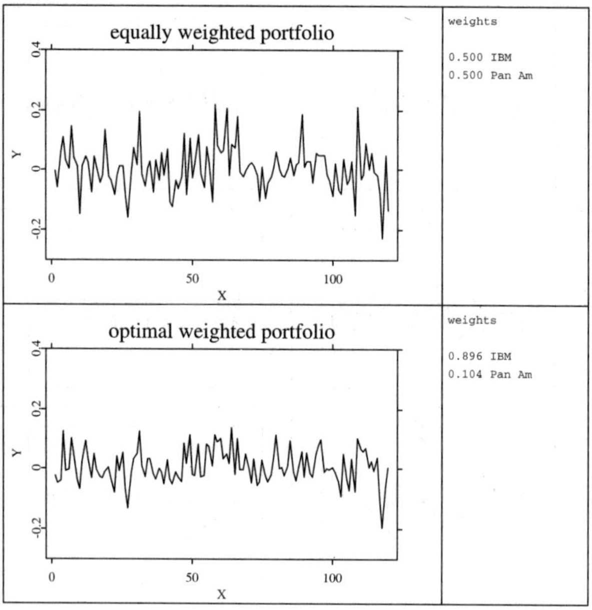

Figure 17.2. Portfolio of IBM and PanAm assets, equal and efficient weights. `MVAportfol`

which can be solved for λ_2 to get:

$$\lambda_2 = \frac{2 - \lambda_1 1_p^\top \Sigma^{-1}\mu}{1_p^\top \Sigma^{-1} 1_p}.$$

Plugging this expression into (17.5) yields

$$c = \frac{1}{2}\lambda_1 \left(\Sigma^{-1}\mu - \frac{1_p^\top \Sigma^{-1}\mu}{1_p^\top \Sigma^{-1} 1_p} \Sigma^{-1} 1_p \right) + \frac{\Sigma^{-1} 1_p}{1_p^\top \Sigma^{-1} 1_p}. \tag{17.6}$$

For the case of a variance efficient portfolio there is no restriction on the mean of the portfolio ($\lambda_1 = 0$). The optimal weights are therefore:

$$c = \frac{\Sigma^{-1} 1_p}{1_p^\top \Sigma^{-1} 1_p}. \tag{17.7}$$

This formula is identical to the solution of (17.3). Indeed, differentiation with respect to c gives

$$\begin{aligned} \Sigma c &= \lambda 1_p \\ c &= \lambda \Sigma^{-1} 1_p. \end{aligned}$$

If we plug this into (17.3), we obtain

$$\begin{aligned} \mathcal{L} &= \frac{1}{2}\lambda^2 1_p \Sigma^{-1} 1_p - \lambda(\lambda 1_p \Sigma^{-1} 1_p - 1) \\ &= \lambda - \frac{1}{2}\lambda^2 1_p \Sigma^{-1} 1_p. \end{aligned}$$

This quantity is a function of λ and is minimal for

$$\lambda = (1_p \Sigma^{-1} 1_p)^{-1}$$

since

$$\frac{\partial^2 \mathcal{L}}{\partial c^\top \partial c} = \Sigma > 0.$$

THEOREM 17.1 *The variance efficient portfolio weights for returns $X \sim (\mu, \Sigma)$ are*

$$c_{opt} = \frac{\Sigma^{-1} 1_p}{1_p^\top \Sigma^{-1} 1_p}. \tag{17.8}$$

Existence of a Riskless Asset

If an asset exists with variance equal to zero, then the covariance matrix Σ is not invertible. The notation can be adjusted for this case as follows: denote the return of the riskless asset by r (under the absence of arbitrage this is the interest rate), and partition the vector and the covariance matrix of returns such that the last component is the riskless asset. Thus, the last equation of the system (17.4) becomes

$$2\, Cov(r, X) - \lambda_1 r - \lambda_2 = 0,$$

and, because the covariance of the riskless asset with any portfolio is zero, we have

$$\lambda_2 = -r\lambda_1. \tag{17.9}$$

Let us for a moment modify the notation in such a way that in each vector and matrix the components corresponding to the riskless asset are excluded. For example, c is the weight vector of the *risky* assets (i.e., assets with positive variance), and c_0 denotes the proportion invested in the riskless asset. Obviously, $c_0 = 1 - 1_p^\top c$, and Σ the covariance matrix of the *risky* assets, is assumed to be invertible. Solving (17.4) using (17.9) gives

$$c = \frac{\lambda_1}{2} \Sigma^{-1} (\mu - r 1_p). \tag{17.10}$$

This equation may be solved for λ_1 by plugging it into the condition $\mu^\top c = \overline{\mu}$. This is the mean-variance efficient weight vector of the risky assets if a riskless asset exists. The final solution is:

$$c = \frac{\overline{\mu} \Sigma^{-1} (\mu - r 1_p)}{\mu^\top \Sigma^{-1} (\mu - r 1_p)}. \tag{17.11}$$

The variance optimal weighting of the assets in the portfolio depends on the structure of the covariance matrix as the following corollaries show.

COROLLARY 17.1 *A portfolio of uncorrelated assets whose returns have equal variances ($\Sigma = \sigma^2 \mathcal{I}_p$) needs to be weighted equally:*

$$c_{opt} = \frac{1}{p} 1_p.$$

Proof:
Here we obtain $1_p^\top \Sigma^{-1} 1_p = \sigma^{-2} 1_p^\top 1_p = \sigma^{-2} p$ and therefore $c = \frac{\sigma^{-2} 1_p}{\sigma^{-2} p} = \frac{1}{p} 1_p$. □

COROLLARY 17.2 *A portfolio of correlated assets whose returns have equal variances, i.e.,*

$$\Sigma = \sigma^2 \begin{pmatrix} 1 & \rho & \cdots & \rho \\ \rho & 1 & \cdots & \rho \\ \vdots & \vdots & \ddots & \vdots \\ \rho & \rho & \cdots & 1 \end{pmatrix}, \qquad -\frac{1}{p-1} < \rho < 1$$

needs to be weighted equally:

$$c_{opt} = \frac{1}{p} 1_p.$$

Proof:
Σ can be rewritten as $\Sigma = \sigma^2 \left\{(1-\rho)\mathcal{I}_p + \rho 1_p 1_p^\top\right\}$. The inverse is

$$\Sigma^{-1} = \frac{\mathcal{I}_p}{\sigma^2(1-\rho)} - \frac{\rho 1_p 1_p^\top}{\sigma^2(1-\rho)\{1+(p-1)\rho\}}$$

since for a $(p \times p)$ matrix $\mathcal{A}$ of the form $\mathcal{A} = (a-b)\mathcal{I}_p + b 1_p 1_p^\top$ the inverse is generally given by

$$\mathcal{A}^{-1} = \frac{\mathcal{I}_p}{(a-b)} - \frac{b\, 1_p 1_p^\top}{(a-b)\{a+(p-1)b\}}.$$

Hence

$$\begin{aligned} \Sigma^{-1} 1_p &= \frac{1_p}{\sigma^2(1-\rho)} - \frac{\rho 1_p 1_p^\top 1_p}{\sigma^2(1-\rho)\{1+(p-1)\rho\}} \\ &= \frac{[\{1+(p-1)\rho\} - \rho p] 1_p}{\sigma^2(1-\rho)\{1+(p-1)\rho\}} = \frac{\{1-\rho\} 1_p}{\sigma^2(1-\rho)\{1+(p-1)\rho\}} \\ &= \frac{1_p}{\sigma^2\{1+(p-1)\rho\}} \end{aligned}$$

which yields

$$1_p^\top \Sigma^{-1} 1_p^\top = \frac{p}{\sigma^2\{1+(p-1)\rho\}}$$

and thus $c = \frac{1}{p} 1_p$. □

Let us now consider assets with different variances. We will see that in this case the weights are adjusted to the risk.

COROLLARY 17.3 *A portfolio of uncorrelated assets with returns of different variances, i.e.,* $\Sigma = \operatorname{diag}(\sigma_1^2, \ldots, \sigma_p^2)$*, has the following optimal weights*

$$c_{j,opt} = \frac{\sigma_j^{-2}}{\sum\limits_{l=1}^{p} \sigma_l^{-2}}, \quad j = 1, \ldots, p.$$

Proof:
From $\Sigma^{-1} = \operatorname{diag}(\sigma_1^{-2}, \ldots, \sigma_p^{-2})$ we have $1_p^\top \Sigma^{-1} 1_p^\top = \sum_{l=1}^{p} \sigma_l^{-2}$ and therefore the optimal weights are $c_j = \sigma_j^{-2} / \sum\limits_{l=1}^{p} \sigma_l^{-2}$. □

This result can be generalized for covariance matrices with block structures.

COROLLARY 17.4 *A portfolio of assets with returns $X \sim (\mu, \Sigma)$, where the covariance matrix has the form:*

$$\Sigma = \begin{pmatrix} \Sigma_1 & 0 & \dots & 0 \\ 0 & \Sigma_2 & \ddots & \vdots \\ \vdots & \ddots & \ddots & \vdots \\ 0 & \dots & 0 & \Sigma_r \end{pmatrix}$$

has optimal weights $c = (c_1, \dots, c_r)^\top$ given by

$$c_{j,opt} = \frac{\Sigma_j^{-1} 1}{1^T \Sigma_j^{-1} 1}, \quad j = 1, \dots, r.$$

Summary

- ↪ An efficient portfolio is one that keeps the risk minimal under the constraint that a given mean return is achieved and that the weights sum to 1, i.e., that minimizes $\mathcal{L} = c^\top \Sigma c + \lambda_1(\overline{\mu} - c^\top \mu) + \lambda_2(1 - c^\top 1_p)$.
- ↪ If a riskless asset does not exist, the variance efficient portfolio weights are given by
$$c = \frac{\Sigma^{-1} 1_p}{1_p^\top \Sigma^{-1} 1_p}.$$
- ↪ If a riskless asset exists, the mean-variance efficient portfolio weights are given by
$$c = \frac{\overline{\mu} \Sigma^{-1}(\mu - r 1_p)}{\mu^\top \Sigma^{-1}(\mu - r 1_p)}.$$
- ↪ The efficient weighting depends on the structure of the covariance matrix Σ. Equal variances of the assets in the portfolio lead to equal weights, different variances lead to weightings proportional to these variances:
$$c_{j,opt} = \frac{\sigma_j^{-2}}{\sum_{l=1}^{p} \sigma_l^{-2}}, \quad j = 1, \dots, p.$$

17.3 Efficient Portfolios in Practice

We can now demonstrate the usefulness of this technique by applying our method to the monthly market returns computed on the basis of transactions at the New York stock market between January 1978 to December 1987 (Berndt, 1990).

EXAMPLE 17.2 *Recall that we had shown the portfolio returns with uniform and optimal*

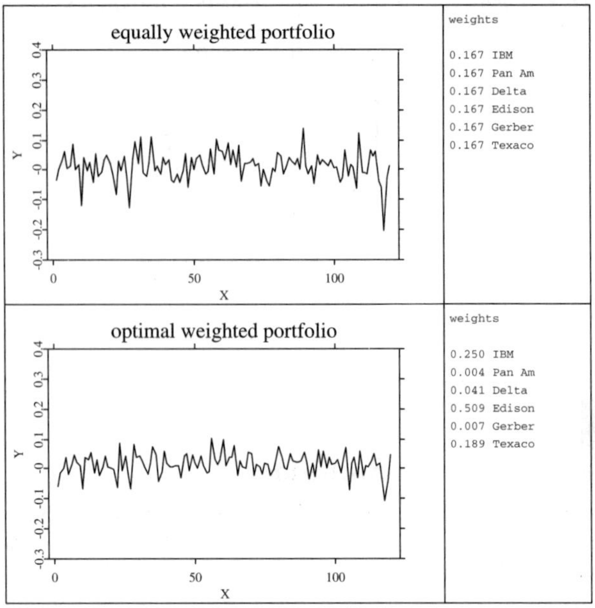

Figure 17.3. Portfolio of all six assets, equal and efficient weights. MVAportfol

weights in Figure 17.2. The covariance matrix of the returns of IBM and PanAm is

$$\mathcal{S} = \begin{pmatrix} 0.0034 & 0.0016 \\ 0.0016 & 0.0172 \end{pmatrix}.$$

Hence by (17.7) the optimal weighting is

$$\widehat{c} = \frac{\mathcal{S}^{-1} 1_2}{1_2^\top \mathcal{S}^{-1} 1_2} = (0.8957, 0.1043)^\top.$$

The effect of efficient weighting becomes even clearer when we expand the portfolio to six assets. The covariance matrix for the returns of all six firms introduced in Example 17.1 is

$$\mathcal{S} = \begin{pmatrix} 0.0035 & 0.0016 & 0.0019 & 0.0003 & 0.0015 & 0.0010 \\ 0.0016 & 0.0172 & 0.0049 & 0.0011 & 0.0019 & 0.0003 \\ 0.0019 & 0.0049 & 0.0091 & 0.0004 & 0.0016 & 0.0010 \\ 0.0003 & 0.0011 & 0.0004 & 0.0025 & 0.0007 & -0.0004 \\ 0.0015 & 0.0019 & 0.0016 & 0.0007 & 0.0076 & 0.0021 \\ 0.0010 & 0.0003 & 0.0010 & -0.0004 & 0.0021 & 0.0063 \end{pmatrix}.$$

Hence the optimal weighting is

$$\widehat{c} = \frac{\mathcal{S}^{-1} 1_6}{1_6^\top \mathcal{S}^{-1} 1_6} = (0.2504, 0.0039, 0.0409, 0.5087, 0.0072, 0.1890)^\top.$$

As we can clearly see, the optimal weights are quite different from the equal weights ($c_j = 1/6$). The weights which were used are shown in text windows on the right hand side of Figure 17.3.

This efficient weighting assumes stable covariances between the assets over time. Changing covariance structure over time implies weights that depend on time as well. This is part of a large body of literature on multivariate volatility models. For a review refer to Franke et al. (2004).

Summary

- ↪ Efficient portfolio weighting in practice consists of estimating the covariances of the assets in the portfolio and then computing efficient weights from this empirical covariance matrix.
- ↪ Note that this efficient weighting assumes stable covariances between the assets over time.

17.4 The Capital Pricing Model (CAPM)

The CAPM considers the relation between a mean-variance efficient portfolio and an asset uncorrelated with this portfolio. Let us denote this specific asset return by y_0. The riskless asset with constant return $y_0 \equiv r$ may be such an asset. Recall from (17.4) the condition for a mean-variance efficient portfolio:

$$2\Sigma c - \lambda_1 \mu - \lambda_2 1_p = 0.$$

In order to eliminate λ_2, we can multiply (17.4) by $c^\top$ to get:

$$2c^\top \Sigma c - \lambda_1 \bar{\mu} = \lambda_2.$$

Plugging this into (17.4), we obtain:

$$\begin{aligned} 2\Sigma c - \lambda_1 \mu &= 2c^\top \Sigma c 1_p - \lambda_1 \bar{\mu} 1_p \\ \mu &= \bar{\mu} 1_p + \frac{2}{\lambda_1}(\Sigma c - c^\top \Sigma c 1_p). \end{aligned} \tag{17.12}$$

For the asset that is uncorrelated with the portfolio, equation (17.12) can be written as:

$$y_0 = \bar{\mu} - \frac{2}{\lambda_1} c^\top \Sigma c$$

since $y_0 = r$ is the mean return of this asset and is otherwise uncorrelated with the risky assets. This yields:

$$\lambda_1 = 2 \frac{c^\top \Sigma c}{\bar{\mu} - y_0} \tag{17.13}$$

and if (17.13) is plugged into (17.12):

$$\begin{aligned} \mu &= \bar{\mu} 1_p + \frac{\bar{\mu} - y_0}{c^\top \Sigma c}(\Sigma c - c^\top \Sigma c 1_p) \\ \mu &= y_0 1_p + \frac{\Sigma c}{c^\top \Sigma c}(\bar{\mu} - y_0) \\ \mu &= y_0 1_p + \beta(\bar{\mu} - y_0) \end{aligned} \tag{17.14}$$

with

$$\beta \stackrel{\text{def}}{=} \frac{\Sigma c}{c^\top \Sigma c}.$$

The relation (17.14) holds if there exists any asset that is uncorrelated with the mean-variance efficient portfolio c. The existence of a riskless asset is not a necessary condition for deriving (17.14). However, for this special case we arrive at the well-known expression

$$\mu = r1_p + \beta(\bar{\mu} - r), \tag{17.15}$$

which is known as the *Capital Asset Pricing Model* (CAPM), see Franke et al. (2004). The *beta factor* β measures the relative performance with respect to riskless assets or an index. It reflects the sensitivity of an asset with respect to the whole market. The beta factor is close to 1 for most assets. A factor of 1.16, for example, means that the asset reacts in relation to movements of the whole market (expressed through an index like DAX or DOW JONES) 16 percents stronger than the index. This is of course true for both positive and negative fluctuations of the whole market.

Summary

- ↪ The weights of the mean-variance efficient portfolio satisfy $2\Sigma c - \lambda_1 \mu - \lambda_2 1_p = 0$.
- ↪ In the CAPM the mean of X depends on the riskless asset and the pre-specified mean $\overline{\mu}$ as follows $\mu = r1_p + \beta(\overline{\mu} - r)$.
- ↪ The beta factor β measures the relative performance with respect to riskless assets or an index and reflects the sensitivity of an asset with respect to the whole market.

17.5 Exercises

EXERCISE 17.1 *Prove that the inverse of* $\mathcal{A} = (a-b)\mathcal{I}_p + b1_p1_p^\top$ *is given by*

$$\mathcal{A}^{-1} = \frac{\mathcal{I}_p}{(a-b)} - \frac{b\,1_p1_p^\top}{(a-b)\{a+(p-1)b\}}.$$

EXERCISE 17.2 *The empirical covariance between the 120 returns of IBM and PanAm is* 0.0016 *(see Example 17.2). Test if the true covariance is zero. Hint: Use Fisher's Z-transform.*

EXERCISE 17.3 *Explain why in both Figures 17.2 and 17.3 the portfolios have negative returns just before the end of the series, regardless of whether they are optimally weighted or not! (What happened in December 1987?)*

EXERCISE 17.4 *Apply the method used in Example 17.2 on the same data (Table B.5) including also the Digital Equipment company. Obviously one of the weights is negative. Is this an efficient weighting?*

EXERCISE 17.5 *In the CAPM the β value tells us about the performance of the portfolio relative to the riskless asset. Calculate the β value for each single stock price series relative to the "riskless" asset IBM.*

18 Computationally Intensive Techniques

It is generally accepted that training in statistics must include some exposure to the mechanics of computational statistics. This exposure to computational methods is of an essential nature when we consider extremely high dimensional data. Computer aided techniques can help us to discover dependencies in high dimensions without complicated mathematical tools. A draftman's plot (i.e. a matrix of pairwise scatterplots like in Figure 1.14) may lead us immediately to a theoretical hypothesis (on a lower dimensional space) on the relationship of the variables. Computer aided techniques are therefore at the heart of multivariate statistical analysis.

With the rapidly increasing amount of data statistics faces a new challenge. While in the 20th century the focus was on the mathematical precision of statistical modeling, the 21th century relies more and more on data analytic procedures that provide information (even for extremely large data bases) on the fingertip. This demand on fast availability of condensed statistical information has changed the statistical paradigm and has shifted energy from mathematical analysis to computational analysis of course without loosing sight of the statistical core questions.

In this chapter we first present the concept of Simplicial Depth—a multivariate extension of the data depth concept of Section 1.1. We then present Projection Pursuit—a semiparametric technique which is based on a one-dimensional, flexible regression or on the idea of density smoothing applied to PCA type projections. A similar model is underlying the Sliced Inverse Regression (SIR) technique which we discuss in Section 18.3.

The next technique is called support vector machines and is motivated by nonlinear classification (discrimination) problems. Support Vector Machines (SVM) are classification methods based on statistical learning theory. A quadratic optimization problem determines so-called support vectors with high margin that guarantee maximal separability. Nonlinear classification is achieved by mapping the data into a feature space and finding a linear separating hyperplane in this feature space. Another advanced technique is CART – Classification and Regression Trees, a decision tree procedure developed by (Breiman, Friedman, Olshen and Stone, 1984).

18.1 Simplicial Depth

Simplicial depth generalizes the notion of data depth as introduced in Section 1.1. This general definition allows us to define a multivariate median and to visually present high dimensional data in low dimension. For univariate data we have well known parameters of location which describe the center of a distribution of a random variable X. These parameters are for example the *mean*

$$\bar{x} = \frac{1}{n}\sum_{i=1}^{n} x_i, \tag{18.1}$$

or the *mode*

$$x_{mod} = \arg\max_x \hat{f}(x),$$

where $\hat{f}$ is the estimated density function of X (see Section 1.3). The *median*

$$x_{med} = \begin{cases} x_{((n+1)/2)} & \text{if } n \text{ odd} \\ \frac{x_{(n/2)}+x_{(n/2+1)}}{2} & \text{otherwise,} \end{cases}$$

where $x_{(i)}$ is the order statistics of the n observations x_i, is yet another measure of location.

The first two parameters can be easily extended to multivariate random variables. The mean in higher dimensions is defined as in (18.1) and the mode accordingly,

$$x_{mod} = \arg\max_x \hat{f}(x)$$

with $\hat{f}$ the estimated multidimensional density function of X (see Section 1.3). The median poses a problem though since in a multivariate sense we cannot interpret the element-wise median

$$x_{med,j} = \begin{cases} x_{((n+1)/2),j} & \text{if } n \text{ odd} \\ \frac{x_{(n/2),j}+x_{(n/2+1),j}}{2} & \text{otherwise} \end{cases} \tag{18.2}$$

as a point that is "most central". The same argument applies to other observations of a sample that have a certain "depth" as defined in Section 1.1. The "fourths" or the "extremes" are not defined in a straightforward way in higher (not even for two) dimensions.

An equivalent definition of the median in one dimension is given by the *simplicial depth.* It is defined as follows: For each pair of datapoints x_i and x_j we generate a closed interval, a one-dimensional simplex, which contains x_i and x_j as border points. Redefine the median as the datapoint x_{med}, which is enclosed in the maximum number of intervals:

$$x_{med} = \arg\max_i \#\{k, l; x_i \in [x_k, x_l]\}. \tag{18.3}$$

With this definition of the median, the median is the "deepest" and "most central" point in a data set as discussed in Section 1.1. This definition involves a computationally intensive operation since we generate $n(n-1)/2$ intervals for n observations.

In two dimensions, the computation is even more intensive since the interval $[x_k, x_l]$ is replaced by a triangle constructed from three different datapoints. The median as the deepest point is then defined by that datapoint that is covered by the maximum number of triangles. In three dimensions triangles become pyramids formed from 4 points and the median is that datapoint that lies in the maximum number of pyramids.

An example for the depth in 2 dimensions is given by the constellation of points given in Figure 18.1. If we build for example the traingle of the points 1, 3, 5 (denoted as $\triangle$ 135 in Table 18.1), it contains the point 4. From Table 18.1 we count the number of coverages to obtain the simplicial depth values of Table 18.2.

In arbitrary dimension p, we look for datapoints that lie inside a simplex (or convex *hull*) formed from $p+1$ points. We therefore extend the definition of the median to the multivariate case as follows

$$x_{med} = \arg\max_i \#\{k_0, \ldots, k_p; x_i \in hull(x_{k_0}, ..., x_{k_p})\}. \tag{18.4}$$

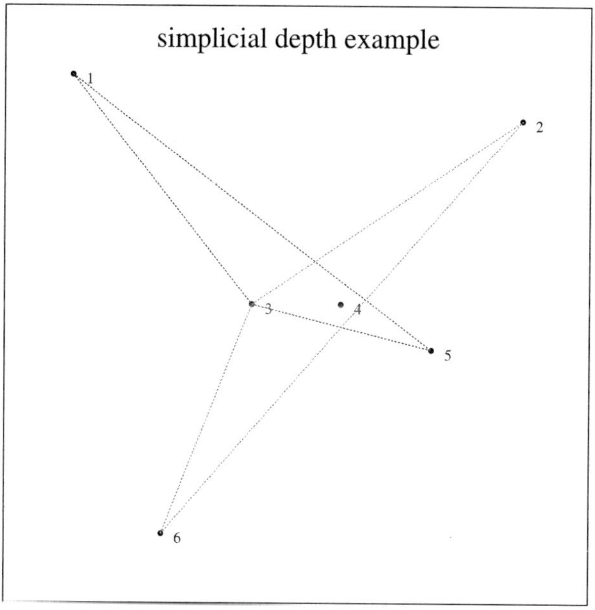

Figure 18.1. Construction of simplicial depth. MVAsimdep1

	Triangle	Coverages					
1	△ 123	1	2	3			
2	△ 124	1	2		4		
3	△ 125	1	2			5	
4	△ 126	1	2	3	4		6
5	△ 134	1		3	4		
6	△ 135	1		3	4	5	
7	△ 136	1		3			6
8	△ 145	1			4	5	
9	△ 146	1		3	4		6
10	△ 156	1		3	4	5	6
11	△ 234		2	3	4		
12	△ 235		2	3	4	5	
13	△ 236		2	3	4		6
14	△ 245		2		4	5	
15	△ 246		2		4		6
16	△ 256		2			5	6
17	△ 345			3	4	5	
18	△ 346			3	4		6
19	△ 356			3		5	6
20	△ 456				4	5	6

Table 18.1. Coverages for artificial configuration of points.

point	1	2	3	4	5	6
depth	10	10	12	14	8	8

Table 18.2. Simplicial depths for artificial configuration of points.

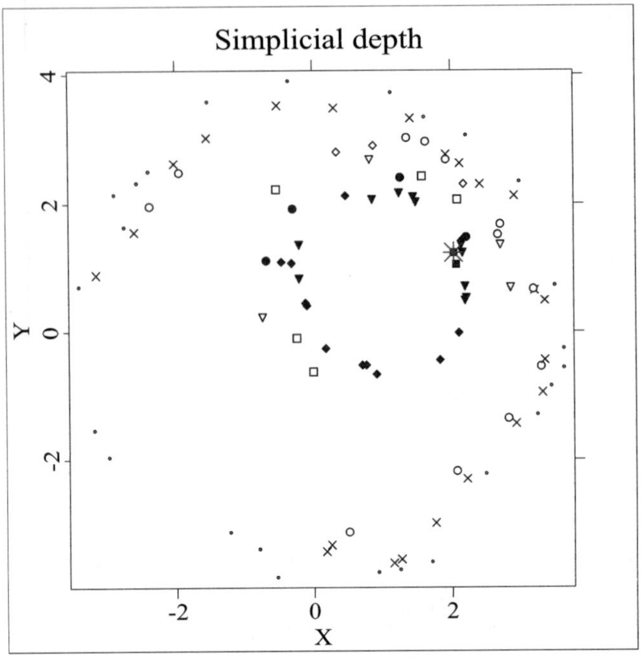

Figure 18.2. 10 point distribution according to depth with the median shown as a big star in the center. MVAsimdepex

Here $k_0, ..., k_p$ denote the indices of $p+1$ datapoints. Thus for each datapoint we have a multivariate data depth. If we compute all the necessary simplices $hull(x_{k_0}, \dots, x_{k_p})$, the computing time will unfortunately be exponential as the dimension increases.

In Figure 18.2 we calculate the simplicial depth for a two-dimensional, 10 point distribution according to depth. It contains 100 data points with corresponding parameters controlling its spread. The deepest point, the two-dimensional median, is indicated as a big star in the center. The points with less depth are indicated via grey shades.

Summary

↪ The "depth" of a datapoint in one dimension can be computed by counting all (closed) intervals of two datapoints which contain the datapoint.

↪ The "deepest" datapoint is the central point of the distribution, the median.

Summary (continued)
$\hookrightarrow$ The "depth" of a datapoint in arbitrary dimension p is defined as the number of simplices (constructed from $p+1$ points) covering this point. It is called simplicial depth.
$\hookrightarrow$ A multivariate extension of the median is to take the "deepest" datapoint of the distribution.
$\hookrightarrow$ In the bivariate case we count all triangles of datapoints which contain the datapoint to compute its depth.

18.2 Projection Pursuit

"Projection Pursuit" stands for a class of exploratory projection techniques. This class contains statistical methods designed for analyzing high-dimensional data using low-dimensional projections. The aim of projection pursuit is to reveal possible nonlinear and therefore interesting structures hidden in the high-dimensional data. To what extent these structures are "interesting" is measured by an index. Exploratory Projection Pursuit (EPP) goes back to Kruskal(1969; 1972). The approach was successfully implemented for exploratory purposes by various other authors. The idea has been applied to regression analysis, density estimation, classification and discriminant analysis.

Exploratory Projection Pursuit

In EPP, we try to find "interesting" low-dimensional projections of the data. For this purpose, a suitable index function $I(\alpha)$, depending on a normalized projection vector α, is used. This function will be defined such that "interesting" views correspond to local and global maxima of the function. This approach naturally accompanies the technique of principal component analysis (PCA) of the covariance structure of a random vector X. In PCA we are interested in finding the axes of the covariance ellipsoid. The index function $I(\alpha)$ is in this case the variance of a linear combination $\alpha^\top X$ subject to the normalizing constraint $\alpha^\top\alpha = 1$ (see Theorem 9.2). If we analyze a sample with a p-dimensional normal distribution, the "interesting" high-dimensional structure we find by maximizing this index is of course linear.

There are many possible projection indices, for simplicity the kernel based and polynomial based indices are reported. Assume that the p-dimensional random variable X is sphered and centered, that is, $E(X) = 0$ and $Var(X) = \mathcal{I}_p$. This will remove the effect of location, scale, and correlation structure. This covariance structure can be achieved easily by the Mahalanobis transformation (3.26).

Friedman and Tukey (1974) proposed to investigate the high-dimensional distribution of X by considering the index

$$I_{\mathrm{FT},h}(\alpha) = n^{-1}\sum_{i=1}^{n} \hat{f}_{h,\alpha}(\alpha^\top X_i) \tag{18.5}$$

where $\hat{f}_{h,\alpha}$ denotes the kernel estimator (see Section 1.3)

$$\hat{f}_{h,\alpha}(z) = n^{-1}\sum_{j=1}^{n} K_h(z-\alpha^\top X_j) \tag{18.6}$$

of the projected data. Note that (18.5) is an estimate of $\int f^2(z)dz$ where $z = \alpha^\top X$ is a one-dimensional random variable with mean zero and unit variance. If the high-dimensional distribution of X is normal, then each projection $z = \alpha^\top X$ is standard normal since $||\alpha|| = 1$ and since X has been centered and sphered by, e.g., the Mahalanobis transformation.

The index should therefore be stable as a function of α if the high-dimensional data is in fact normal. Changes in $I_{\text{FT},h}(\alpha)$ with respect to α therefore indicate deviations from normality. Hodges and Lehman (1956) showed that, given a mean of zero and unit variance, the (compact support) density which minimizes $\int f^2$ is uniquely given by

$$f(z) = \max\{0, c(b^2 - z^2)\},$$

where $c = 3/(20\sqrt{5})$ and $b = \sqrt{5}$. This is a parabolic density function, which is equal to zero outside the interval $(-\sqrt{5}, \sqrt{5})$. A high value of the Friedman-Tukey index indicates a larger departure from the parabolic form.

An alternative index is based on the negative of the entropy measure, i.e., $\int -f \log f$. The density for zero mean and unit variance which minimizes the index

$$\int f \log f$$

is the standard normal density, a far more plausible candidate than the parabolic density as a norm from which departure is to be regarded as "interesting". Thus in using $\int f \log f$ as a projection index we are really implementing the viewpoint of seeing "interesting" projections as departures from normality. Yet another index could be based on the Fisher information (see Section 6.2)

$$\int (f')^2/f.$$

To optimize the entropy index, it is necessary to recalculate it at each step of the numerical procedure. There is no method of obtaining the index via summary statistics of the multivariate data set, so the workload of the calculation at each iteration is determined by the number of observations. It is therefore interesting to look for approximations to the entropy index. Jones and Sibson (1987) suggested that deviations from the normal density should be considered as

$$f(x) = \varphi(x)\{1 + \varepsilon(x)\} \tag{18.7}$$

where the function ε satisfies

$$\int \varphi(u)\varepsilon(u)u^{-r}du = 0, \text{ for } r = 0, 1, 2. \tag{18.8}$$

In order to develop the Jones and Sibson index it is convenient to think in terms of cumulants $\kappa_3 = \mu_3 = E(X^3)$, $\kappa_4 = \mu_4 = E(X^4) - 3$ (see Section 1.3). The standard normal density satisfies $\kappa_3 = \kappa_4 = 0$, an index with any hope of tracking the entropy index must at least incorporate information up to the level of symmetric departures (κ_3 or κ_4 not zero) from normality. The simplest of such indices is a positive definite quadratic form in κ_3 and κ_4. It must be invariant under sign-reversal of the data since both $\alpha^\top X$ and $-\alpha^\top X$ should show the same kind of departure from normality. Note that κ_3 is odd under sign-reversal, i.e., $\kappa_3(\alpha^\top X) = -\kappa_3(-\alpha^\top X)$. The cumulant κ_4 is even under sign-reversal, i.e., $\kappa_4(\alpha^\top X) = \kappa_4(-\alpha^\top X)$. The quadratic form in κ_3 and κ_4 measuring departure from normality cannot include a mixed $\kappa_3\kappa_4$ term.

For the density (18.7) one may conclude with (18.8) that

$$\int f(u)\log(u)du \approx \frac{1}{2}\int \varphi(u)\varepsilon(u)du.$$

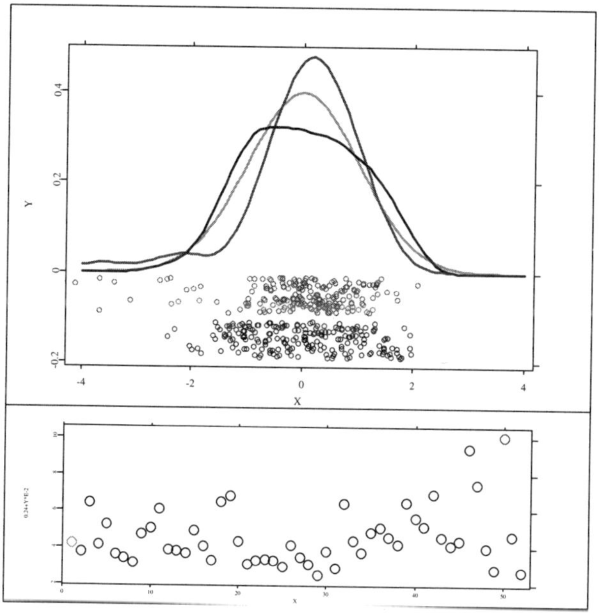

Figure 18.3. Exploratory Projection Pursuit for the Swiss bank notes data (green = standard normal, red = best, blue = worst). MVAppexample

Now if f is expressed as a Gram-Charliér expansion

$$f(x)\varphi(x) = \{1 + \kappa_3 H_3(x)/6 + \kappa_4 H_4(x)/24...\} \tag{18.9}$$

(Kendall and Stuart, 1977, p. 169) where H_r is the r-th Hermite polynomial, then the truncation of (18.9) and use of orthogonality and normalization properties of Hermite polynomials with respect to φ yields

$$\frac{1}{2}\int \varphi(x)\varepsilon^2(x)dx = \left(\kappa_3^2 + \kappa_4^2/4\right)/12.$$

The index proposed by Jones and Sibson (1987) is therefore

$$I_{JS}(\alpha) = \{\kappa_3^2(\alpha^\top X) + \kappa_4^2(\alpha^\top X)/4\}/12.$$

This index measures in fact the difference $\int f \log f - \int \varphi \log \varphi$.

EXAMPLE 18.1 *The exploratory Projection Pursuit is used on the Swiss bank note data. For 50 randomly chosen one-dimensional projections of this six-dimensional dataset we calculate the Friedman-Tukey index to evaluate how "interesting" their structures are.*

Figure 18.3 shows the density for the standard, normally distributed data (green) and the estimated densities for the best (red) and the worst (blue) projections found. A dotplot of the projections is also presented. In the lower part of the figure we see the estimated value of the Friedman-Tukey index for each computed projection. From this information we can judge the non normality of the bank note data set since there is a lot of variation across the 50 random projections.

Projection Pursuit Regression

The problem in projection pursuit regression is to estimate a response surface

$$f(x) = E(Y \mid x)$$

via approximating functions of the form:

$$\hat{f}(x) = \sum_{k=1}^{M} g_k(\Lambda_k^\top x)$$

with non-parametric regression functions g_k and projection indices Λ_k. Given observations $\{(x_1, y_1), ..., (x_n, y_n)\}$ with $x_i \in \mathbb{R}^p$ and $y_i \in \mathbb{R}$ the basic algorithm works as follows.

1. Set $r_i^{(0)} = y_i$ and $k = 1$.
2. Minimize
$$E_k = \sum_{i=1}^{n} \left\{ r_i^{(k-1)} - g_k(\Lambda_k^\top x_i) \right\}^2$$
where Λ_k is an orthogonal projection matrix and g_k is a non-parametric regression estimator.
3. Compute new residuals
$$r_i^{(k)} = r_i^{(k-1)} - g_k(\Lambda_k^\top x_i).$$
4. Increase k and repeat the last two steps until E_k becomes small.

Although this approach seems to be simple, we encounter some problems. One of the most serious is that the decomposition of a function into sums of functions of projections may not be unique. An example is

$$z_1 z_2 = \frac{1}{4ab}\{(az_1 + bz_2)^2 - (az_1 - bz_2)^2\}.$$

Numerical improvements of this algorithm were suggested by Friedman and Stuetzle (1981).

Summary

↪ Exploratory Projection Pursuit is a technique used to find interesting structures in high-dimensional data via low-dimensional projections. Since the Gaussian distribution represents a standard situation, we define the Gaussian distribution as the most uninteresting.

↪ The search for interesting structures is done via a projection score like the Friedman-Tukey index $I_{\mathrm{FT}}(\alpha) = \int f^2$. The parabolic distribution has the minimal score. We maximize this score over all projections.

↪ The Jones-Sibson index maximizes

$$I_{\mathrm{JS}}(\alpha) = \{\kappa_3(\alpha^\top X) + \kappa_4^2(\alpha^\top X)/4\}/12$$

as a function of α.

Summary (continued)

↪ The entropy index maximizes

$$I_{\mathrm{E}}(\alpha) = \int f(\alpha^\top X) \log f(\alpha^\top X)$$

where f is the density of $\alpha^\top X$.

↪ In Projection Pursuit Regression the idea is to represent the unknown function by a sum of non-parametric regression functions on projections. The key problem is in choosing the number of terms and often the interpretability.

18.3 Sliced Inverse Regression

Sliced inverse regression (SIR) is a dimension reduction method proposed by Duan and Li (1991). The idea is to find a smooth regression function that operates on a variable set of projections. Given a response variable Y and a (random) vector $X \in \mathbb{R}^p$ of explanatory variables, SIR is based on the model:

$$Y = m(\beta_1^\top X, \ldots, \beta_k^\top X, \varepsilon), \tag{18.10}$$

where $\beta_1, \ldots, \beta_k$ are unknown projection vectors, k is unknown and assumed to be less than p, $m : \mathbb{R}^{k+1} \to \mathbb{R}$ is an unknown function, and ε is the noise random variable with $\mathrm{E}(\varepsilon \mid X) = 0$.

Model (18.10) describes the situation where the response variable Y depends on the p-dimensional variable X only through a k-dimensional subspace. The unknown β_i's, which span this space, are called *effective dimension reduction directions* (EDR-directions). The span is denoted as *effective dimension reduction space* (EDR-space). The aim is to estimate the base vectors of this space, for which neither the length nor the direction can be identified. Only the space in which they lie is identifiable.

SIR tries to find this k-dimensional subspace of $\mathbb{R}^p$ which under the model (18.10) carries the essential information of the regression between X and Y. SIR also focuses on small k, so that nonparametric methods can be applied for the estimation of m. A direct application of nonparametric smoothing to X is for high dimension p generally not possible due to the sparseness of the observations. This fact is well known as the *curse of dimensionality*, see Huber (1985).

The name of SIR comes from computing the inverse regression (IR) curve. That means instead of looking for $\mathrm{E}(Y \mid X = x)$, we investigate $\mathrm{E}(X \mid Y = y)$, a curve in $\mathbb{R}^p$ consisting of p one-dimensional regressions. What is the connection between the IR and the SIR model (18.10)? The answer is given in the following theorem from Li (1991).

THEOREM 18.1 *Given the model (18.10) and the assumption*

$$\forall b \in \mathbb{R}^p : \; E\left(b^\top X \mid \beta_1^\top X = \beta_1^\top x, \ldots, \beta_k^\top X = \beta_k^\top x\right) = c_0 + \sum_{i=1}^{k} c_i \beta_i^\top x, \tag{18.11}$$

the centered IR curve $E(X \mid Y = y) - E(X)$ lies in the linear subspace spanned by the vectors $\Sigma\beta_i$, $i = 1, \ldots, k$, where $\Sigma = Cov(X)$.

Assumption (18.11) is equivalent to the fact that X has an elliptically symmetric distribution, see Cook and Weisberg (1991). Hall and Li (1993) have shown that assumption (18.11) only needs to hold for the EDR-directions.

It is easy to see that for the standardized variable $Z = \Sigma^{-1/2}\{X - \mathrm{E}(X)\}$ the IR curve $m_1(y) = \mathrm{E}(Z \,|\, Y = y)$ lies in $\mathrm{span}(\eta_1, \ldots, \eta_k)$, where $\eta_i = \Sigma^{1/2}\beta_i$. This means that the conditional expectation $m_1(y)$ is moving in $\mathrm{span}(\eta_1, \ldots, \eta_k)$ depending on y. With b orthogonal to $\mathrm{span}(\eta_1, \ldots, \eta_k)$, it follows that

$$b^\top m_1(y) \;=\; 0,$$

and further that

$$m_1(y)m_1(y)^\top b \;=\; Cov\{m_1(y)\}b \;=\; 0.$$

As a consequence $Cov\{\mathrm{E}(Z \mid y)\}$ is degenerated in each direction orthogonal to all EDR-directions η_i of Z. This suggests the following algorithm.

First, estimate $Cov\{m_1(y)\}$ and then calculate the orthogonal directions of this matrix (for example, with eigenvalue/eigenvector decomposition). In general, the estimated covariance matrix will have full rank because of random variability, estimation errors and numerical imprecision. Therefore, we investigate the eigenvalues of the estimate and ignore eigenvectors having small eigenvalues. These eigenvectors $\hat{\eta}_i$ are estimates for the EDR-direction η_i of Z. We can easily rescale them to estimates $\hat{\beta}_i$ for the EDR-directions of X by multiplying by $\hat{\Sigma}^{-1/2}$, but then they are not necessarily orthogonal. SIR is strongly related to PCA. If all of the data falls into a single interval, which means that $\widehat{Cov}\{m_1(y)\}$ is equal to $\widehat{Cov}(Z)$, SIR coincides with PCA. Obviously, in this case any information about y is ignored.

The SIR Algorithm

The algorithm to estimate the EDR-directions via SIR is as follows:

1. Standardize x:
$$z_i \;=\; \hat{\Sigma}^{-1/2}(x_i - \bar{x}).$$
2. Divide the range of y_i into S nonoverlapping intervals *(slices)* H_s, $s = 1, \ldots, S$. n_s denotes the number of observations within slice H_s, and $\boldsymbol{I}_{H_s}$ the indicator function for this slice:
$$n_s \;=\; \sum_{i=1}^{n} \boldsymbol{I}_{H_s}(y_i).$$
3. Compute the mean of z_i over all slices. This is a crude estimate $\widehat{m}_1$ for the *inverse regression curve* m_1:
$$\bar{z}_s \;=\; \frac{1}{n_s}\sum_{i=1}^{n} z_i \;\boldsymbol{I}_{H_s}(y_i).$$
4. Calculate the estimate for $Cov\{m_1(y)\}$:
$$\widehat{V} \;=\; n^{-1}\sum_{s=1}^{S} n_s \bar{z}_s \bar{z}_s^\top.$$
5. Identify the eigenvalues $\hat{\lambda}_i$ and eigenvectors $\hat{\eta}_i$ of $\widehat{V}$.

6. Transform the standardized EDR-directions $\hat{\eta}_i$ back to the original scale. Now the estimates for the EDR-directions are given by

$$\hat{\beta}_i = \hat{\Sigma}^{-1/2}\hat{\eta}_i.$$

REMARK 18.1 *The number of different eigenvalues unequal to zero depends on the number of slices. The rank of $\hat{V}$ cannot be greater than the number of slices-1 (the z_i sum up to zero). This is a problem for categorical response variables, especially for a binary response—where only one direction can be found.*

SIR II

In the previous section we learned that it is interesting to consider the IR curve, that is, $E(X \mid y)$. In some situations however SIR does not find the EDR-direction. We overcome this difficulty by considering the conditional covariance $Cov(X \mid y)$ instead of the IR curve. An example where the EDR directions are not found via the SIR curve is given below.

EXAMPLE 18.2 *Suppose that $(X_1, X_2)^\top \sim N(0, \mathcal{I}_2)$ and $Y = X_1^2$. Then $E(X_2 \mid y) = 0$ because of independence and $E(X_1 \mid y) = 0$ because of symmetry. Hence, the EDR-direction $\beta = (1, 0)^\top$ is not found when the IR curve $E(X \mid y) = 0$ is considered.*

The conditional variance

$$Var(X_1 \mid Y = y) = E(X_1^2 \mid Y = y) = y,$$

offers an alternative way to find β. It is a function of y while $Var(X_2 \mid y)$ is a constant.

The idea of SIR II is to consider the conditional covariances. The principle of SIR II is the same as before: investigation of the IR curve (here the conditional covariance instead of the conditional expectation). Unfortunately, the theory of SIR II is more complicated. The assumption of the elliptical symmetrical distribution of X has to be more restrictive, i.e., assuming the normality of X.

Given this assumption, one can show that the vectors with the largest distance to $Cov(Z \mid Y = y) - E\{Cov(Z \mid Y = y)\}$ for all y are the most interesting for the EDR-space. An appropriate measure for the overall mean distance is, according to Li (1992),

$$E\left(||\,[Cov(Z \mid Y = y) - E\{Cov(Z \mid Y = y)\}]\,b||^2\right) = \tag{18.12}$$

$$= b^\top E\left(||\,Cov(Z \mid y) - E\{Cov(Z \mid y)\}||^2\right) b. \tag{18.13}$$

Equipped with this distance, we conduct again an eigensystem decomposition, this time for the above expectation $E\left(||\,Cov(Z \mid y) - E\{Cov(Z \mid y)\}||^2\right)$. Then we take the rescaled eigenvectors with the largest eigenvalues as estimates for the unknown EDR-directions.

The SIR II Algorithm

The algorithm of SIR II is very similar to the one for SIR, it differs in only two steps. Instead of merely computing the mean, the covariance of each slice has to be computed. The estimate for the above expectation (18.12) is calculated after computing all slice covariances. Finally, decomposition and rescaling are conducted, as before.

$\hat{\beta}_1$	$\hat{\beta}_2$	$\hat{\beta}_3$
0.578	-0.723	-0.266
0.586	0.201	0.809
0.568	0.661	-0.524

Table 18.3. SIR: EDR-directions for simulated data.

1. Do steps 1 to 3 of the SIR algorithm.
2. Compute the slice covariance matrix $\widehat{V}_s$:

$$\widehat{V}_s = \frac{1}{n_s - 1} \sum_{i=1}^{n} I_{H_s}(y_i) z_i z_i^\top - n_s \bar{z}_s \bar{z}_s^\top .$$

3. Calculate the mean over all slice covariances:

$$\bar{V} = \frac{1}{n} \sum_{s=1}^{S} n_s \widehat{V}_s .$$

4. Compute an estimate for (18.12):

$$\widehat{V} = \frac{1}{n} \sum_{s=1}^{S} n_s \left(\widehat{V}_s - \bar{V} \right)^2 = \frac{1}{n} \sum_{s=1}^{S} n_s \widehat{V}_s^2 - \bar{V}^2 .$$

5. Identify the eigenvectors and eigenvalues of $\widehat{V}$ and scale back the eigenvectors. This gives estimates for the SIR II EDR-directions:

$$\hat{\beta}_i = \hat{\Sigma}^{-1/2} \hat{\eta}_i .$$

EXAMPLE 18.3 *The result of SIR is visualized in four plots in Figure 18.4: the left two show the response variable versus the first respectively second direction. The upper right plot consists of a three-dimensional plot of the first two directions and the response. The last picture shows $\hat{\Psi}_k$, the ratio of the sum of the first k eigenvalues and the sum of all eigenvalues, similar to principal component analysis.*

The data are generated according to the following model:

$$y_i = \beta_1^\top x_i + (\beta_1^\top x_i)^3 + 4 \left(\beta_2^\top x_i \right)^2 + \varepsilon_i ,$$

where the x_i's follow a three-dimensional normal distribution with zero mean, the covariance equal to the identity matrix, $\beta_2 = (1, -1, -1)^\top$, and $\beta_1 = (1, 1, 1)^\top$. ε_i is standard, normally distributed and $n = 300$. Corresponding to model (18.10), $m(u, v, \varepsilon) = u + u^3 + v^2 + \varepsilon$. The situation is depicted in Figure 18.5 and Figure 18.6.

Both algorithms were conducted using the slicing method with 20 elements in each slice. The goal was to find β_1 and β_2 with SIR. The data are designed such that SIR can detect β_1 because of the monotonic shape of $\{\beta_1^\top x_i + (\beta_1^\top x_i)^3\}$, while SIR II will search for β_2, as in this direction the conditional variance on y is varying.

If we normalize the eigenvalues for the EDR-directions in Table 18.3 such that they sum up to one, the resulting vector is $(0.852, 0.086, 0.062)$. As can be seen in the upper left

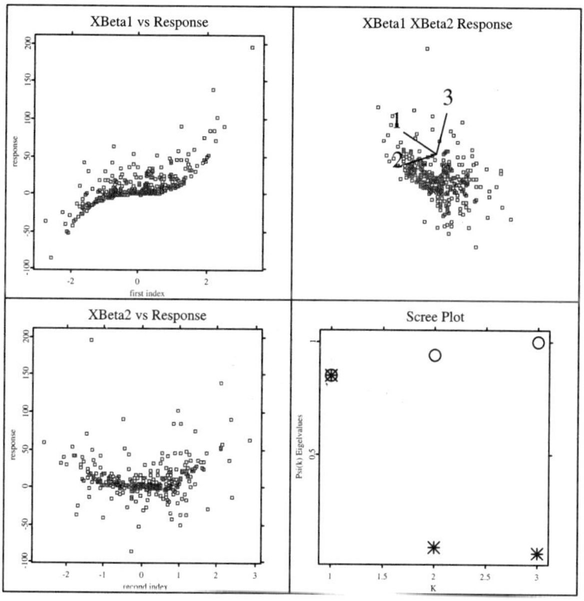

Figure 18.4. SIR: The left plots show the response versus the estimated EDR-directions. The upper right plot is a three-dimensional plot of the first two directions and the response. The lower right plot shows the eigenvalues $\hat{\lambda}_i$ $(*)$ and the cumulative sum $(\circ)$. MVAsirdata

$\hat{\beta}_1$	$\hat{\beta}_2$	$\hat{\beta}_3$
0.821	0.180	0.446
-0.442	-0.826	0.370
-0.361	-0.534	0.815

Table 18.4. SIR II: EDR-directions for simulated data.

plot of Figure 18.4, there is a functional relationship found between the first index $\hat{\beta}_1^\top x$ *and the response. Actually,* β_1 *and* $\hat{\beta}_1$ *are nearly parallel, that is, the normalized inner product* $\hat{\beta}_1^\top \beta_1/\{||\hat{\beta}_1|| ||\beta_1||\} = 0.9894$ *is very close to one.*

The second direction along β_2 *is probably found due to the good approximation, but SIR does not provide it clearly, because it is "blind" with respect to the change of variance, as the second eigenvalue indicates.*

For SIR II, the normalized eigenvalues are $(0.706, 0.185, 0.108)$, *that is, about 69% of the variance is explained by the first EDR-direction (Table 18.4). Here, the normalized inner product of* β_2 *and* $\hat{\beta}_1$ *is* 0.9992. *The estimator* $\hat{\beta}_1$ *estimates in fact* β_2 *of the simulated model. In this case, SIR II found the direction where the second moment varies with respect to* $\beta_2^\top x$.

In summary, SIR has found the direction which shows a strong relation regarding the conditional expectation between $\beta_1^\top x$ and y, and SIR II has found the direction where the

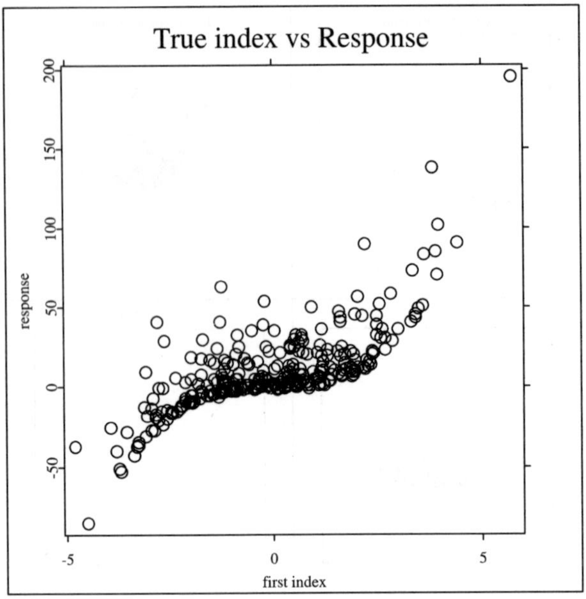

Figure 18.5. Plot of the true response versus the true indices. The monotonic and the convex shapes can be clearly seen. MVAsirdata

conditional variance is varying, namely, $\beta_2^\top x$.

The behavior of the two SIR algorithms is as expected. In addition, we have seen that it is worthwhile to apply both versions of SIR. It is possible to combine SIR and SIR II (Cook and Weisberg, 1991; Li, 1991; Schott, 1994) directly, or to investigate higher conditional moments. For the latter it seems to be difficult to obtain theoretical results.

Summary

↪ SIR serves as a dimension reduction tool for regression problems.

↪ Inverse regression avoids the *curse of dimensionality.*

↪ The dimension reduction can be conducted without estimation of the regression function $y = m(x)$.

↪ SIR searches for the effective dimension reduction (EDR) by computing the inverse regression IR.

↪ SIR II uses the EDR on computing the inverse conditional variance.

↪ SIR might miss EDR directions that are found by SIR II.

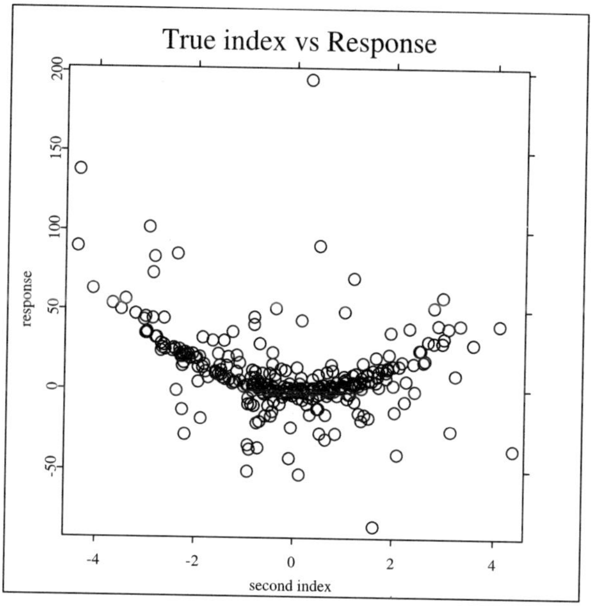

Figure 18.6. Plot of the true response versus the true indices. The monotonic and the convex shapes can be clearly seen. MVAsirdata

18.4 Support Vector Machines

The purpose of this section is to introduce one of the most promising among recently developed multivariate non-linear statistical techniques: the support vector machine (SVM). The SVM is a classification method that is based on statistical learning theory. It has been successfully applied to optical character recognition, early medical diagnostics, and text classification. One application where SVMs outperformed other methods is electric load prediction (EUNITE, 2001), another one is optical character recognition (Vapnik, 1995). In a variety of applications SVMs produce better classification results than parametric methods (e.g. logit analysis) and are outperforming widely used nonparametric techniques, such as neural networks. Here we apply SVMs to corporate bankruptcy analysis.

Classification Methodology

In order to illustrate the classification methodology we focus for the moment on a company rating example that we will treat further in more detail. Investment risks are evaluated via the default probability (PD) for a company. Each company is described by a set of variables (predictors) x, such as financial ratios, and its class y that can be either $y = -1$ ('successful') or $y = 1$ ('bankrupt'). Financial ratios are constructed from the variables like net income, total assets, interest payments, etc. A training set represents a sample of data for companies which are known to have survived or gone bankrupt. From the training set one estimates a classifier function f that is then applied to computing PDs. These PDs can be uniquely translated into a company rating.

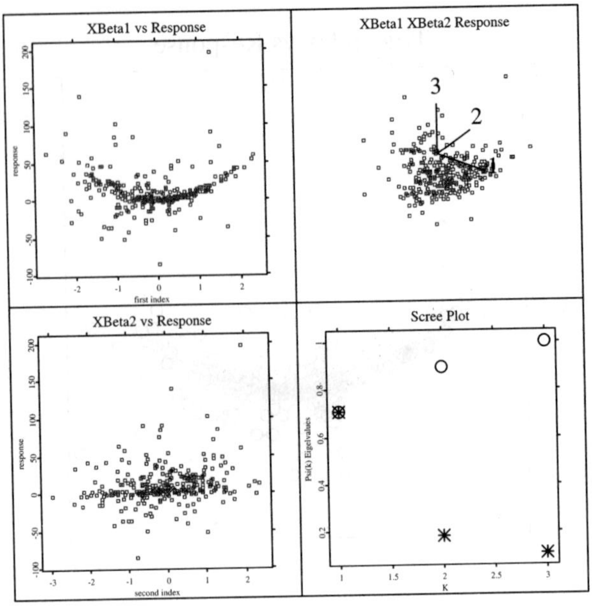

Figure 18.7. SIR II mainly sees the direction β_2. The left plots show the response versus the estimated EDR-directions. The upper right plot is a three-dimensional plot of the first two directions and the response. The lower right plot shows the eigenvalues $\hat{\lambda}_i$ $(*)$ and the cumulative sum $(\circ)$. MVAsir2data

Classical discriminant analysis is based on the assumption that each group of observations is normally distributed with the same variance-covariance matrix but different means. Under such a formulation the discriminating function will be linear, see Theorem 12.2. Figure 18.8 displays this situation: if some linear combination of predictors (called Z-score in the context of bankruptcy analysis) is greater than a particular threshold value z_0 the observation under consideration is regarded as belonging to $y = 1$; if $Z < z_0$ the observation would belong to $y = -1$ (successful). One can change the labels "-1,+1" to the more standard notation "0,1". The current labeling is done only for mathematical convenience.

The Z-score is:

$$Z_i = a_1 x_{i1} + a_2 x_{i2} + ... + a_p x_{ip} = a^\top x_i,$$

where $x_i = (x_{i1}, ..., x_{ip})^\top \in \mathbb{R}^p$ are predictors for the i-th company. The classification based on the Z-score are necessarily linear and, therefore, may not handle more complex situations as in Figure 18.9 when non-linear classifiers, such as those generated by SVMs, can produce better results.

Expected vs. Empirical Risk Minimization

A nonlinear classifier function f may be described by a function class $\mathcal{F}$. $\mathcal{F}$ is fixed *a priori*, e.g. it can be the class of linear classifiers (hyperplanes). A good classifier optimizes some criterion that tells us how well f separates the classes.

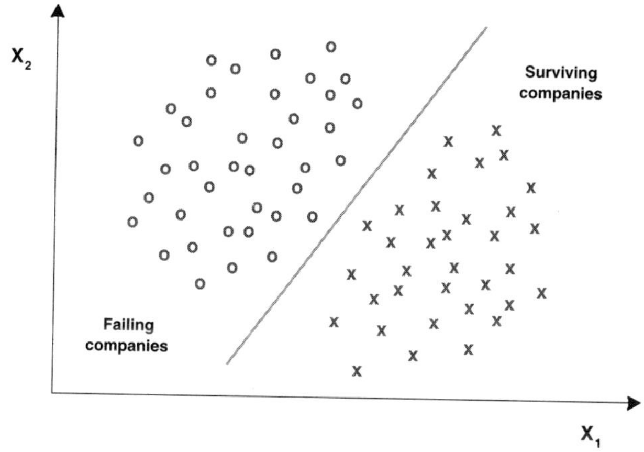

Figure 18.8. A linear classification function in the case of linearly separable data.

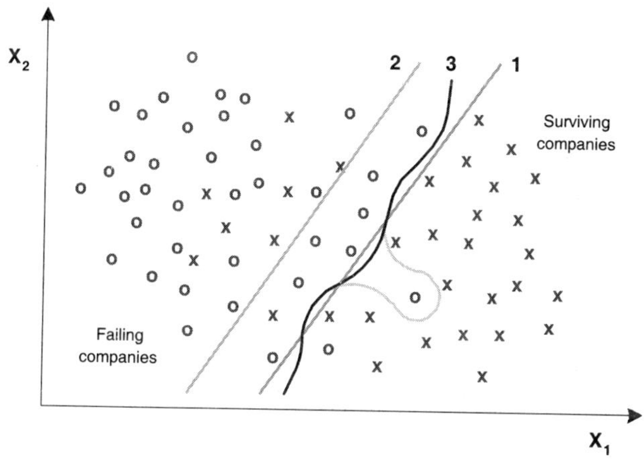

Figure 18.9. Different linear classification functions (1) and (2) and a non-linear one (3) in the linearly non-separable case.

As in (12.4) one considers the minimization of the expected risk:

$$R(f) = \int \frac{1}{2} |f(x) - y| \, dF(x,y). \tag{18.14}$$

The joint distribution $F(x, y)$, however, is never known in practical applications and must be estimated from the *training set* $\{x_i, y_i\}_{i=1}^{n}$. By replacing $F(x, y)$ with the empirical cdf $F_n(x, y)$ one obtains the empirical risk:

$$\hat{R}(f) = \frac{1}{n} \sum_{i=1}^{n} \frac{1}{2} |f(x_i) - y_i| \,. \tag{18.15}$$

The empirical risk is an average value of loss over the training set, while the expected risk is the expected value of loss under the true probability measure. The loss is given by:

$$L(x,y) = \frac{1}{2}\left|f(x)-y\right| = \begin{cases} 0, & \text{if classification is correct,} \\ 1, & \text{if classification is wrong.} \end{cases}$$

One sees here that it is convenient to work with the labels "-1,1" for y. The solutions to the problems of expected and empirical risk minimization:

$$f_{opt} = \arg\min_{f \in \mathcal{F}} R(f), \qquad (18.16)$$

$$\hat{f}_n = \arg\min_{f \in \mathcal{F}} \hat{R}(f), \qquad (18.17)$$

generally do not coincide (Figure 18.10), although converge as $n \to \infty$ if $\mathcal{F}$ is not too large. According to statistical learning theory (Vapnik, 1995), it is possible to get a uniform upper bound on the difference between $R(f)$ and $\hat{R}(f)$ via the Vapnik-Chervonenkis (VC) theory. The VC bound states that there is a function ϕ (monotone increasing in h) so that for all $f \in \mathcal{F}$ with a probability $1-\eta$:

$$R(f) \le \hat{R}(f) + \phi\left(\frac{h}{n}, \frac{\log(\eta)}{n}\right). \qquad (18.18)$$

Here h denotes the VC dimension, a measure of complexity of the involved function class $\mathcal{F}$. For a linear classification rule $g(x) = \operatorname{sign}(x^\top w + b)$:

$$\phi\left(\frac{h}{n}, \frac{\log(\eta)}{n}\right) = \sqrt{\frac{h\left(\log\frac{2n}{h}\right) - \log\frac{\eta}{4}}{n}}, \qquad (18.19)$$

where h is the VC dimension. By plotting the function $\phi(u,v) = \{-u \cdot \log 2u + \log 4 - v\}^{-1/2}$ for small u one sees the monotonicity of $\varphi(u,v)$. In fact one can show that

$$\frac{\partial \phi\left(\frac{h}{n}, \frac{\log(\eta)}{n}\right)}{\partial h} \geqslant 0$$

if and only if $2n \geqslant h$. For a linear classifier with $h = p+1$ this is an easy condition to meet.

The VC dimension of a set $\mathcal{F}$ of functions in a d-dimensional space is h if some function $f \in \mathcal{F}$ can shatter h objects $\left\{x_i \in \mathbb{R}^d, i = 1, ..., h\right\}$, in all 2^h possible configurations and no set $\left\{x_j \in \mathbb{R}^d, j = 1, ..., q\right\}$ with $q > h$, exists that satisfies this property. For example, three points on a plane ($d = 2$) can be shattered by linear indicator functions in $2^h = 2^3 = 8$ ways, whereas 4 points can not be shattered in $2^q = 2^4 = 16$ ways. Thus, the VC dimension of the set of linear indicator functions in a two-dimensional space is $h = 3$, see Figure 18.11. The expression for the VC bound (18.18) involves the VC dimension h, a parameter controlling complexity of $\mathcal{F}$. The term $\phi\left(\frac{h}{n}, \frac{\ln(\eta)}{n}\right)$ introduces a penalty for excessive complexity of a classifier function. The higher is the complexity of $f \in \mathcal{F}$ the higher are h and therefore ϕ. There is a trade-off between the number of classification errors on the training set and the complexity of the classifier function. If the complexity were not controlled for, it would be possible to construct a classifier function with no classification errors on the training set notwithstanding how low its generalization ability would be.

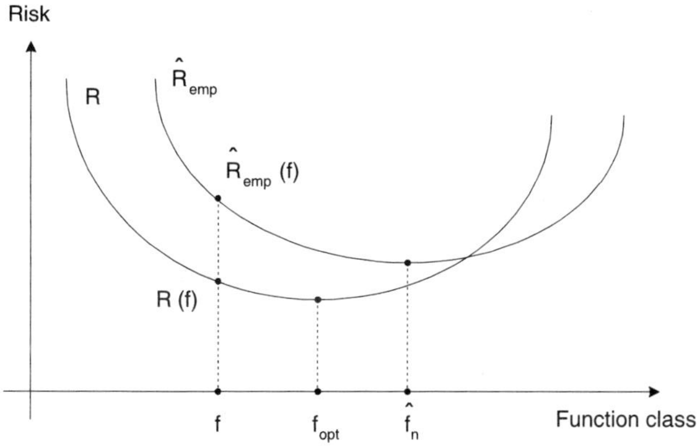

Figure 18.10. The minima f_{opt} and $\hat{f}_n$ of the expected (R) and empirical ($\hat{R}$) risk functions generally do not coincide.

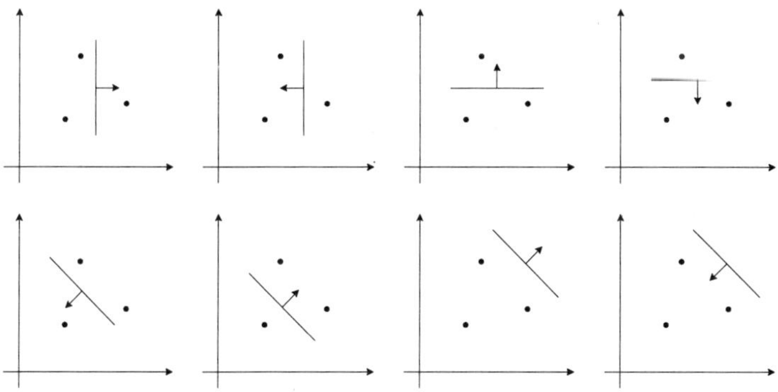

Figure 18.11. Eight possible ways of shattering 3 points on the plane with a linear indicator function.

The SVM in the Linearly Separable Case

First we will describe the SVM in the linearly separable case. The family $\mathcal{F}$ of classification functions in the data space is given by:

$$\mathcal{F} = \left\{x^\top w + b, w \in \mathbb{R}^p, b \in \mathbb{R}\right\} \tag{18.20}$$

In order to determine the support vectors we choose $f \in \mathcal{F}$ (or equivalently (w, b)) such that the so called margin – the corridor between the separating hyperplanes – is maximal. This situation is illustrated in Figure 18.12. The margin is equal to $d_- + d_+$. The classification function is a hyperplane plus the margin zone, where, in the separable case, no observations can lie. It separates the points from both classes with the highest 'safest' distance (margin) between them. It can be shown that margin maximization corresponds to the reduction of complexity as given by the VC-dimension of the SVM classifier. Apparently, the separating hyperplane is defined only by the *support vectors* that hold the hyperplanes parallel to the

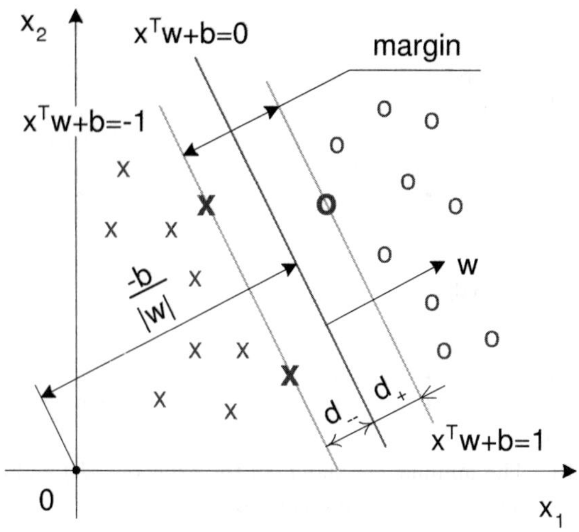

Figure 18.12. The separating hyperplane $x^\top w + b = 0$ and the margin in the linearly separable case.

separating one. In Figure 18.12 there are three support vectors that are marked with bold style: two crosses and one circle. We come now to the description of the SVM selection.

Let $x^\top w + b = 0$ be a separating hyperplane. Then d_+ (d_-) will be the shortest distance to the closest objects from the classes $+1$ (-1). Since the separation can be done without errors, all observations $i = 1, 2, ..., n$ must satisfy:

$$\begin{array}{rcrl} x_i^\top w + b & \geq & +1 & \text{for} \quad y_i = +1 \\ x_i^\top w + b & \leq & -1 & \text{for} \quad y_i = -1 \end{array}$$

We can combine both constraints into one:

$$y_i(x_i^\top w + b) - 1 \geq 0 \qquad i = 1, 2, ..., n \tag{18.21}$$

The *canonical hyperplanes* $x_i^\top w + b = \pm 1$ are parallel and the distance between each of them and the separating hyperplane is $d_+ = d_- = 1/\|w\|$. To maximize the margin $d_+ + d_- = 2/\|w\|$ one therefore minimizes the Euclidean norm $\|w\|$ or its square $\|w\|^2$.

The Lagrangian for the primal problem that corresponds to margin maximization subject to constraint (18.21) is:

$$L_P(w, b) = \frac{1}{2}\|w\|^2 - \sum_{i=1}^{n} \alpha_i \{y_i(x_i^\top w + b) - 1\} \tag{18.22}$$

The Karush-Kuhn-Tucker (KKT) (Gale, Kuhn and Tucker, 1951) first order optimality conditions are:

$$\frac{\partial L_P}{\partial w_k} = 0: \qquad w_k - \sum_{i=1}^{n} \alpha_i y_i x_{ik} = 0, \quad k = 1, ..., d$$

$$\frac{\partial L_P}{\partial b} = 0: \qquad \sum_{i=1}^{n} \alpha_i y_i = 0$$

$$y_i(x_i^\top w + b) - 1 \geq 0, \quad i = 1, ..., n$$

$$\alpha_i \geq 0$$

$$\alpha_i \{y_i(x_i^\top w + b) - 1\} = 0$$

From these first order condition, we can derive $w = \sum_{i=1}^{n} \alpha_i y_i x_i$ and therefore the summands in (18.22) read:

$$\begin{aligned} \frac{1}{2}\|w\|^2 &= \frac{1}{2} \sum_{i=1}^{n} \sum_{j=1}^{n} \alpha_i \alpha_j y_i y_j x_i^\top x_j \\ -\sum_{i=1}^{n} \alpha_i \{y_i(x_i^\top w + b) - 1\} &= -\sum_{i=1}^{n} \alpha_i y_i x_i^\top \sum_{j=1}^{n} \alpha_j y_j x_j + \sum_{i=1}^{n} \alpha_i \\ &= -\sum_{i=1}^{n} \sum_{j=1}^{n} \alpha_i \alpha_j y_i y_j x_i^\top x_j + \sum_{i=1}^{n} \alpha_i \end{aligned}$$

Substituting this into (18.22) we obtain the Lagrangian for the dual problem:

$$L_D(\alpha) = \sum_{i=1}^{n} \alpha_i - \frac{1}{2} \sum_{i=1}^{n} \sum_{j=1}^{n} \alpha_i \alpha_j y_i y_j x_i^\top x_j. \tag{18.23}$$

The primal and dual problems are:

$$\min_{w,b} L_P(w, b)$$

$$\max_{\alpha} L_D(\alpha) \quad \text{s.t.} \quad \alpha_i \geq 0, \quad \sum_{i=1}^{n} \alpha_i y_i = 0.$$

Since the optimization problem is convex the dual and primal formulations give the same solution.

Those points i for which the equation $y_i(x_i^\top w + b) = 1$ holds are called support vectors. After "training the support vector machine" i.e. solving the dual problem above and deriving Lagrange multipliers (they are equal to 0 for non-support vectors) one can classify a company. One uses the classification rule:

$$y(x) = \operatorname{sign}\left(x^\top w + b\right), \tag{18.24}$$

where $w = \sum_{i=1}^{n} \alpha_i y_i x_i$ and $b = \frac{1}{2}(x_{+1} + x_{-1})\, w$. x_{+1} and x_{-1} are two support vectors belonging to different classes for which $y(x^\top w + b) = 1$. The value of the classification function (the score of a company) can be computed as

$$f(x) = x^\top w + b. \tag{18.25}$$

Each score $f(x)$ uniquely corresponds to a default probability (PD). The higher $f(x)$ the higher the PD .

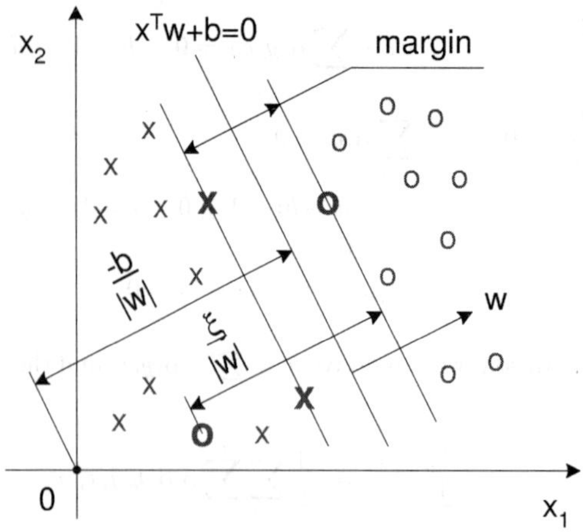

Figure 18.13. The separating hyperplane $x^\top w + b = 0$ and the margin in the linearly non-separable case.

SVMs in the Linearly Non-separable Case

In the linearly non-separable case the situation is like in Figure 18.13. The slack variables ξ_i represent the violation from strict separation. In this case the following inequalities can be induced from Figure 18.13:

$$\begin{aligned} x_i^\top w + b &\geq 1 - \xi_i \quad \text{for} \quad y_i = 1, \\ x_i^\top w + b &\leq -1 + \xi_i \quad \text{for} \quad y_i = -1, \\ \xi_i &\geq 0. \end{aligned}$$

They can be combined into two constraints:

$$y_i(x_i^\top w + b) \geq 1 - \xi_i \tag{18.26}$$

$$\xi_i \geq 0. \tag{18.27}$$

SVM classification again maximizes the margin given a family of classification functions $\mathcal{F}$.

The penalty for misclassification, the classification error $\xi_i \geq 0$, is related to the distance from a misclassified point x_i to the canonical hyperplane bounding its class. If $\xi_i > 0$, an error in separating the two sets occurs. The objective function corresponding to penalized margin maximization is then formulated as:

$$\frac{1}{2} \|w\|^2 + C \sum_{i=1}^{n} \xi_i, \tag{18.28}$$

where the parameter C characterizes the weight given to the classification errors. The minimization of the objective function with constraint (18.26) and (18.27) provides the highest possible margin in the case when classification errors are inevitable due to the linearity of the separating hyperplane. Under such a formulation the problem is convex.

The Lagrange function for the primal problem is:

$$L_P(w, b, \xi) = \frac{1}{2}\|w\|^2 + C\sum_{i=1}^{n}\xi_i - \sum_{i=1}^{n}\alpha_i\{y_i(x_i^\top w + b) - 1 + \xi_i\} - \sum_{i=1}^{n}\mu_i\xi_i, \tag{18.29}$$

where $\alpha_i \geq 0$ and $\mu_i \geq 0$ are Lagrange multipliers. The primal problem is formulated as:

$$\min_{w,b,\xi} L_P(w, b, \xi).$$

The first order conditions in this case are:

$$\begin{aligned}
\frac{\partial L_P}{\partial w_k} = 0: &\qquad w_k - \sum_{i=1}^{n}\alpha_i y_i x_{ik} = 0\\
\frac{\partial L_P}{\partial b} = 0: &\qquad \sum_{i=1}^{n}\alpha_i y_i = 0\\
\frac{\partial L_P}{\partial \xi_i} = 0: &\qquad C - \alpha_i - \mu_i = 0
\end{aligned}$$

With the conditions for the Lagrange multipliers:

$$\begin{aligned}
&\alpha_i \geq 0\\
&\mu_i \geq 0\\
&\alpha_i\{y_i(x_i^\top w + b) - 1 + \xi_i\} = 0\\
&\mu_i\xi_i = 0
\end{aligned}$$

Note that $\sum_{i=1}^{n}\alpha_i y_i b = 0$ therefore similar to the linear separable case the primal problem translates into :

$$\begin{aligned}
L_D(\alpha) &= \frac{1}{2}\sum_{i=1}^{n}\sum_{j=1}^{n}\alpha_i\alpha_j y_i y_j x_i^\top x_j - \sum_{i=1}^{n}\alpha_i y_i x_i^\top\sum_{j=1}^{n}\alpha_j y_j x_j\\
&\quad + C\sum_{i=1}^{n}\xi_i + \sum_{i=1}^{n}\alpha_i - \sum_{i=1}^{n}\alpha_i\xi_i - \sum_{i=1}^{n}\mu_i\xi_i\\
&= \sum_{i=1}^{n}\alpha_i - \frac{1}{2}\sum_{i=1}^{n}\sum_{j=1}^{n}\alpha_i\alpha_j y_i y_j x_i^\top x_j + \sum_{i=1}^{n}\xi_i(C - \alpha_i - \mu_i)
\end{aligned}$$

Since the last term is 0 we derive the dual problem as:

$$L_D(\alpha) = \sum_{i=1}^{n}\alpha_i - \frac{1}{2}\sum_{i=1}^{n}\sum_{j=1}^{n}\alpha_i\alpha_j y_i y_j x_i^\top x_j, \tag{18.30}$$

and the dual problem is posed as:

$$\max_{\alpha} L_D(\alpha),$$

subject to:

$$\begin{aligned}
&0 \leq \alpha_i \leq C,\\
&\sum_{i=1}^{n}\alpha_i y_i = 0.
\end{aligned}$$

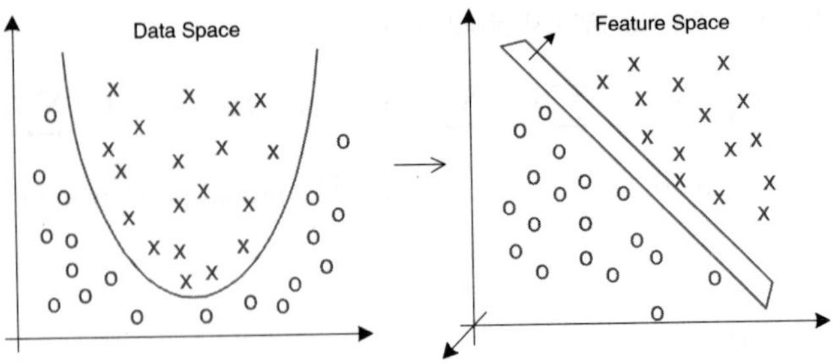

Figure 18.14. Mapping into a three dimensional feature space from a two dimensional data space $\mathbb{R}^2 \mapsto \mathbb{R}^3$. The transformation $\Psi(x_1, x_2) = (x_1^2, \sqrt{2}x_1x_2, x_2^2)^\top$ corresponds to the kernel function $K(x_i, x_j) = (x_i^\top x_j)^2$

Nonlinear Classification

The SVMs can also be generalized to the nonlinear case. In order to obtain non-linear classifiers as in Figure 18.14 one maps the data with a non-linear structure via a function $\Psi : \mathbb{R}^p \mapsto \mathbb{H}$ into a very large dimensional space $\mathbb{H}$ where the classification rule is (almost) linear. Note that all the training vectors x_i appear in L_D (18.30) only as scalar products of the form $x_i^\top x_j$. In the nonlinear SVM situations this transforms to $\psi(x_i)^\top \psi(x_j)$.

The so called *kernel trick* is to compute this scalar product via a kernel function. These kernel functions are actually related to those we presented in section 1.3. If a kernel function K exists such that $K(x_i, x_j) = \Psi(x_i)^\top \Psi(x_j)$, then it can be used without knowing the transformation Ψ explicitly. A necessary and sufficient condition for a symmetric function $K(x_i, x_j)$ to be a kernel is given by Mercer's theorem (Mercer, 1909). It requires positive definiteness, i.e. for any data set $x_1, ..., x_n$ and any real numbers $\lambda_1, ..., \lambda_n$ the function K must satisfy

$$\sum_{i=1}^{n}\sum_{j=1}^{n} \lambda_i \lambda_j K(x_i, x_j) \geq 0. \tag{18.31}$$

Some examples of kernel functions are:

- $K(x_i, x_j) = e^{-\|x_i - x_j\|/2\sigma^2}$ – the isotropic Gaussian kernel;
- $K(x_i, x_j) = e^{-(x_i - x_j)^\top r^{-2}\Sigma^{-1}(x_i - x_j)/2}$ – the stationary Gaussian kernel with an anisotropic radial basis; we will apply this kernel in our study taking Σ equal to the covariance matrix of the training set; r is a constant;
- $K(x_i, x_j) = (x_i^\top x_j + 1)^p$ – the polynomial kernel;
- $K(x_i, x_j) = \tanh(kx_i^\top x_j - \delta)$ – the hyperbolic tangent kernel.

SVMs for Simulated Data

The basic parameters of SVMs on the scaling r of the radial basis functions (in the kernel) and the capacity C. The parameter r controls the local resolution of the SVM in the sense that smaller r create smaller curvature of the margin. The capacity C controls the amount of slack to allow for unclassified observations. A large C would create a very rough and curved margin where C close to zero makes the margin more smooth.

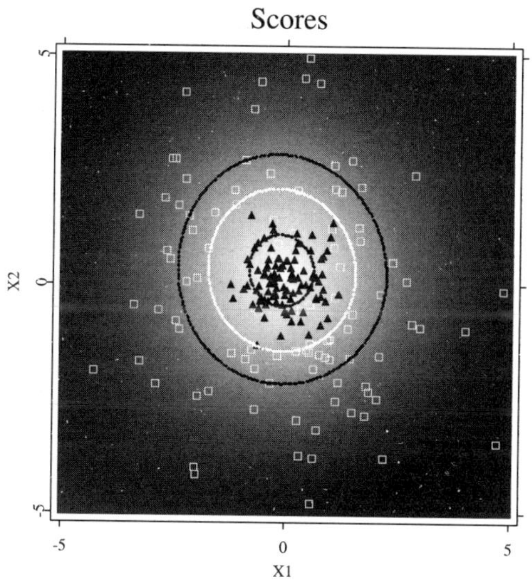

Figure 18.15. SVM classification results for the "orange peel" data ($RB = 2\Sigma^{1/2}$, $C = 1/n$). $d = 2$, $n_{-1} = n_{+1} = 100$, $n = 200$; $x_{+1,i} \sim \mathrm{N}((0,0)^\top, 2^2\mathcal{I})$, $x_{-1,i} \sim \mathrm{N}((0,0)^\top, 0.5^2\mathcal{I})$.

One of the guinea pig tests for a classification algorithm is the data described as "orange peel", i.e. when two groups of observations have similar means, their variance, however, being different. The classification results in this case are presented in Figure 18.15. An SVM with a radial basis kernel is highly suitable for such a kind of data.

Another popular non-linear test is the classification of "spiral data". We generated two spirals with the distance between them equal 1.0 that span over 3π radian. The SVM was chosen with $RB = 0.4\Sigma^{1/2}$ and $C = 1.2/n$. The RB is relatively small since the spiral data have a fine structure. The SVM was able to separate the classes without an error if noise with parameters $\varepsilon_i \sim \mathrm{N}(0, 0.1^2\mathcal{I})$ was injected into the pure spiral data (Figure 18.16). Obviously, both the "orange peel" and the "spiral data" are not linearly separable.

Solution of the SVM Classification Problem

The standard SVM optimization problem (18.30), which is a quadratic optimization problem, is usually solved by means of quadratic programming (QP). This technique, however, is notorious for (i) its bad scaling properties (the time required to solve the problem is proportional to n^3, where n is the number of observations), (ii) implementation difficulty and (iii) enormous memory requirements. With the QP technique the whole kernel matrix of the size $n \times n$ has to be fit in the memory, which, assuming that each variable takes up 10 bytes of memory, will require $10 \times n \times n$ bytes. This means that 1 million observation (which is not unusual for practical applications such as credit scoring) will require 12000 TBytes (terabytes) or 10000000 MBytes of operating memory to store. With a typical size of the computer memory of 512 MBytes no more than around 5000 observations can be processed. Thus, the main emphasis in designing new algorithms was made on using special properties

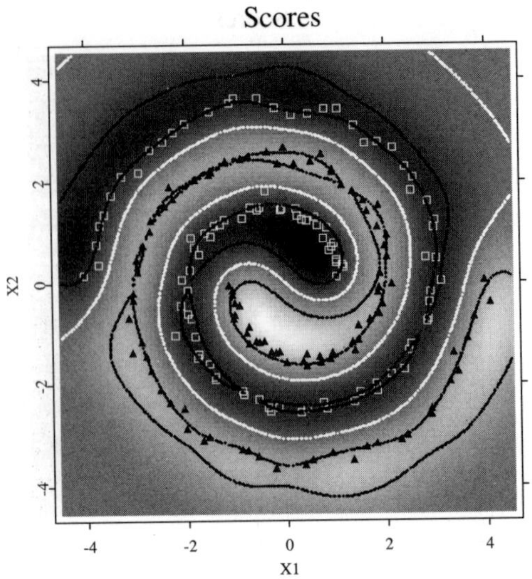

Figure 18.16. SVM classification results for the noisy spiral data ($RB = 0.4\Sigma^{1/2}$, $C = 1.2/n$). The spirals spread over 3π radian; the distance between the spirals equals 1.0. $d = 2$, $n_{-1} = n_{+1} = 100$, $n = 200$. The noise was injected with the parameters $\varepsilon_i \sim N(0, 0.1^2\mathcal{I})$. The separation is perfect.

of SVMs to speed up the solution and reduce memory requirements.

Scoring Companies

For our illustration we selected the largest bankrupt companies with the capitalization of no less than $1 billion that filed for protection against creditors under Chapter 11 of the US Bankruptcy Code in 2001–2002 after the stock marked crash of 2000. We excluded a few companies due to incomplete data, leaving us with 42 companies. They were matched with 42 surviving companies with the closest capitalizations and the same US industry classification codes available through the Division of Corporate Finance of the Securities and Exchange Commission (CF SEC, 2004).

From the selected 84 companies 28 belonged to various manufacturing industries, 20 to telecom and IT industries, 8 to energy industries, 4 to retail industries, 6 to air transportation industries, 6 to miscellaneous service industries, 6 to food production and processing industries and 6 to construction and construction material industries. From all information collected for each company from the annual reports for 1998–1999 (Securities and Exchange Commission, 2004), i.e. three years prior to defaults of bankrupt companies, the following four variables were kept for the illustration: (i) NI – net income (loss); (ii) TA – total assets; (iii) TL – total liabilities; (iv) Bankr – bankruptcy (1 if a company went bankrupt, −1 otherwise). Based on these financial indicators we constructed the profitability (NI/TA) and leverage (TL/TA) ratios.

The applied SVMs differed in two aspects: (i) their capacity that is controlled by the coeffi-

Variable	Min	Max	Mean	Std. Dev.
TA	0.367	91.072	8.122	13.602
TL	0.115	36.437	4.880	6.537
NI	-2.022	4.013	0.161	0.628
NI/TA	-0.599	0.186	-0.003	0.110
TL/TA	0.270	1.463	0.706	0.214

Table 18.5. Descriptive statistics for the companies. All data are given in billions of dollars.

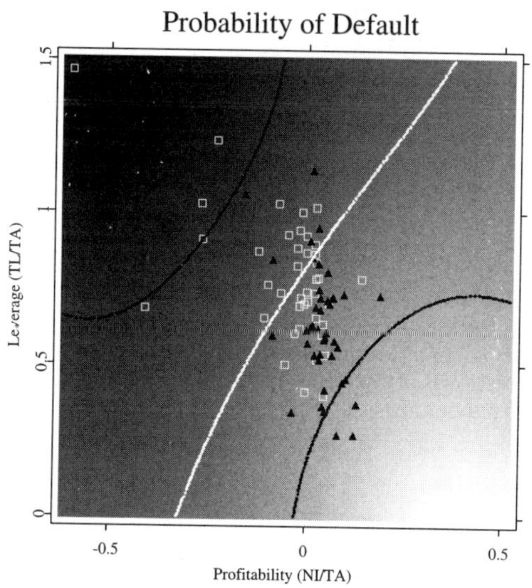

Figure 18.17. Ratings of companies in two dimensions. Low complexity of classifier functions (the radial basis is $5\Sigma^{1/2}$). The capacity is fixed at $C = 1/n$. Percentage of correctly cross-validated out-of-sample observations is 0.607.

cient C in (18.29) and (ii) the complexity of classifier functions controlled in our case by the anisotropic radial basis in the Gaussian kernel. Triangles and squares in Figures 18.17–18.21 represent successful and failing companies from the training set, respectively. The white line is the separating line corresponding to the score value $f = 0$. The area between the black lines is the margin zone. The black lines correspond to $f = 1$ and $f = -1$. The intensity of the gray background represents different score values f. The darker the area, the higher the score and the greater is the probability of default. Most successful companies lying in the bright area have positive profitability and a reasonable leverage TL/TA of around 0.4, which makes economic sense.

Figure 18.17 represents the classification results for an SVM using locally near linear classifier functions (the anisotropic radial basis is $5\Sigma^{1/2}$) with the capacity fixed at $C = 1/n$. The discriminating rule in this case can be approximated by a linear combination of predictors and is similar to that suggested by discriminant analysis, although the coefficients of the predictors may be different. If the complexity of classifying functions increases (the radial

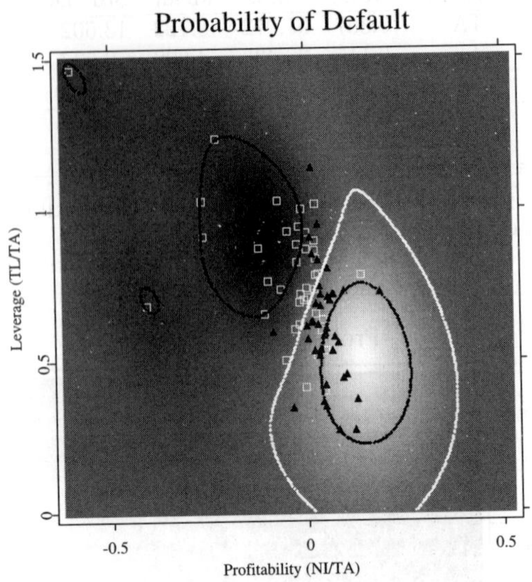

Figure 18.18. Ratings of companies in two dimensions. Optimal complexity of classifier functions (the radial basis is $1.2\Sigma^{1/2}$). The capacity is $C = 1/n$. Percentage of correctly cross-validated out-of-sample observations is 0.750 (maximum).

basis goes down to $1.2\Sigma^{1/2}$) as illustrated in Figure 18.18, we get a more detailed picture. Now the areas of successful and failing companies become localized. If the radial basis is decreased further down to $0.5\Sigma^{1/2}$ (Figure 18.19), the SVM will try to track each observation. The complexity in this case is too high for the given data set.

Figures 18.20 and 18.21 demonstrates the effects of low ($C = 0.3/n$) and high ($C = 10/n$) capacities on the classification results. As capacity is growing, the solution becomes more and more dependent on local distribution of the data. Since we have been applying a radial basis kernel, we get a clusterization of the solution. In our example we got a distinct cluster of successful companies. This persistense in the result may indicate that a company must possess certain characteristics in order to be successful and failing companies can be located elsewhere.

Figure 18.22 shows a power (Lorenz) curve (Lorenz, 1905) – the cumulative number of defaults as a function of the score – for the training set of companies. In most cases our model would be useless if we could not calibrate it, i.e. estimate the relationship between the score and PD. The calibration of the model or estimation of the mapping $f \to \text{PD}$ can be illustrated by the following example (the SVM with the radial basis $2\Sigma^{1/2}$ and capacity $C = 1/n$ will be applied). We can set three rating grades: safe, neutral and risky which correspond to the values of the score $f < -0.0115$, $-0.0115 < f < 0.0115$ and $f > 0.0115$, respectively, and calculate the total number of companies and the number of failing companies in each of the three groups. If the training set were representative of the whole population of companies, the ratio of failing to all companies in a group would give the estimated probability of default. For the abovementioned three rating grades we derive $\text{PD}_{\text{safe}} = 0.24$, $\text{PD}_{\text{neutral}} = 0.50$ and $\text{PD}_{\text{risky}} = 0.76$.

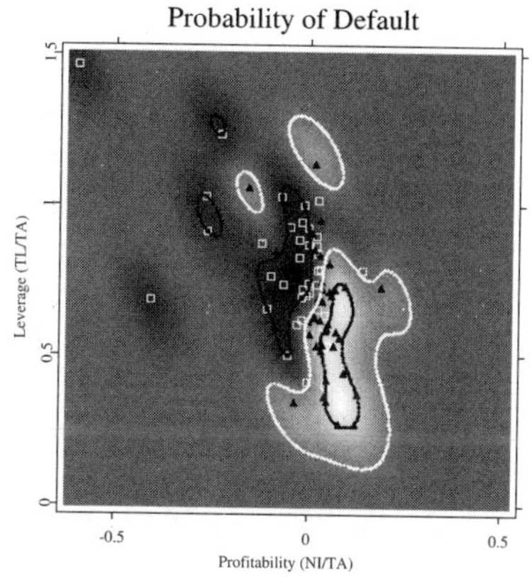

Figure 18.19. Ratings of companies in two dimensions. Excessively complex classification functions (the radial basis is $0.5\Sigma^{1/2}$). The capacity is fixed at $C = 1/n$. Percentage of correctly cross-validated out-of-sample observations is 0.690.

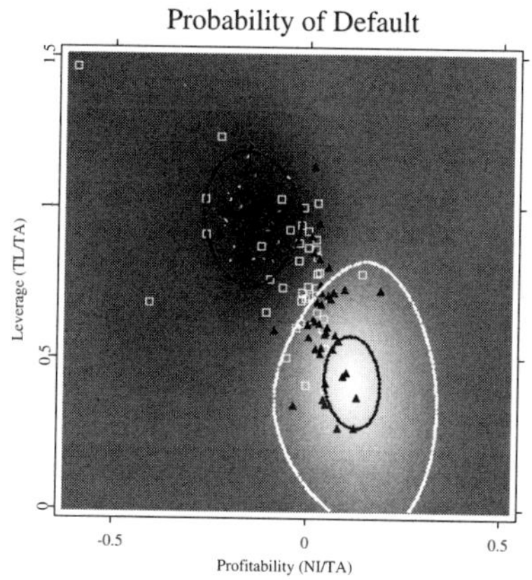

Figure 18.20. Ratings of companies in two dimensions. Low capacity ($C = 0.3/n$). The radial basis is fixed at $1.2\Sigma^{1/2}$. Percentage of correctly cross-validated out-of-sample observations is 0.714.

Cluster center locations, as they were estimated using cluster analysis, are presented in Table 18.6. The results of the cluster analysis indicate that two clusters are likely to correspond to successful and failing companies.

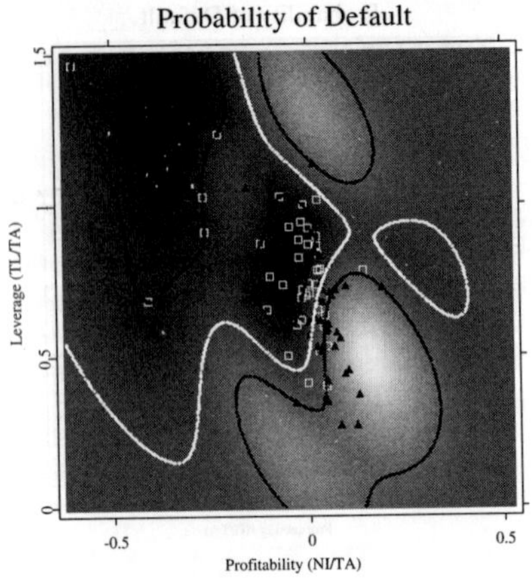

Figure 18.21. Ratings of companies in two dimensions. High capacity ($C = 10/n$). The radial basis is fixed at $1.2\Sigma^{1/2}$. Percentage of correctly cross-validated out-of-sample observations is 0.631.

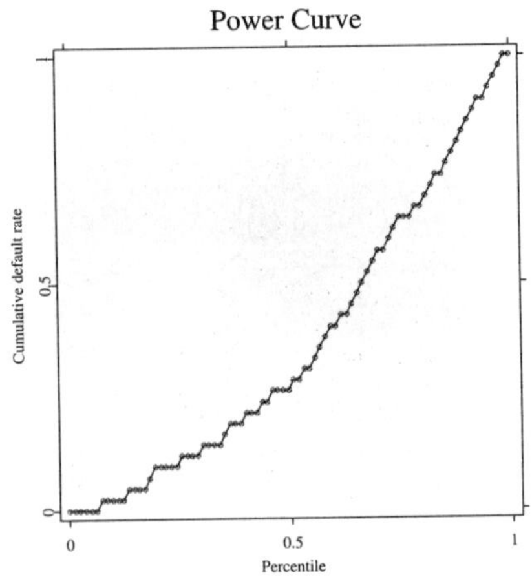

Figure 18.22. Power (Lorenz) curve (Lorenz, 1905) – the cumulative default rate as a function of the percentile of companies sorted according to their score – for the training set of companies. An SVM is applied with the radial basis $2\Sigma^{1/2}$ and capacity $C = 1/n$.

Cluster	{-1}	{1}
NI/TA	0.078	-0.027
TL/TA	0.549	0.752

Table 18.6. Cluster centre locations. There are 19 members in class {-1} – successful companies, and 65 members in class {1} – failing companies.

Summary

- ↪ SVM classification is done by mapping the data into feature space and finding a separating hyperplane there.
- ↪ The support vectors are determined via a quadratic optimization problem.
- ↪ SVM produces highly nonlinear classification boundaries.

18.5 Classification and Regression Trees

Classification and Regression Trees (CART) is a relatively new method of data analysis developed by a group of American statisticians (Breiman et al., 1984). The aim of CART is to classify observations into a subset of *known classes* or to predict levels of regression functions. CART is a non-parametric tool which is designed to represent decision rules in a form of so called *binary trees.* Binary trees split a learning sample parallel to the coordinate axis and represent the resulting data clusters hierarchically starting from a *root node* for the whole learning sample itself and ending with relatively homogenous buckets of observations.

Regression trees are constructed in a similar way but the final buckets do not represent classes but rather approximations to an unknown regression functions at a particular point of the independent variable. In this sense regression trees are estimates via a non-parametric regression model. Here we provide an outlook of how decision trees are created, what challenges arise during practical applications and, of course, a number of examples will illustrate the power of CART.

How Does CART Work ?

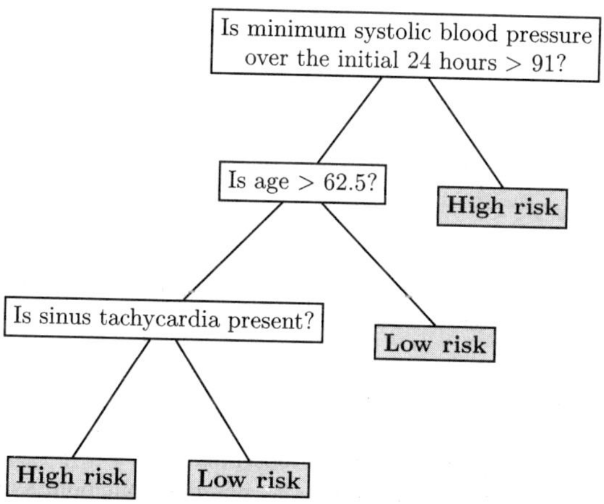

Figure 18.23. Decision tree for low/high patients

Consider the following real life example of how high risk patients (those who will not survive at least 30 days after a heart attack is admitted) were identified at San Diego Medical Center, University of California on the basis of initial 24-hour data. A classification rule using at most three decisions (questions) is presented in Figure 18.23. Left branches of the tree represent cases of positive answers, right branches – negative ones so that e.g. if minimum systolic blood pressure over the last 24 hours is less or equal 91, then the patient belongs to *the high risk* group. In this example the dependant variable is binary: low risk (0) and high risk (1).

A different situation occurs when we are interested in the expected *amount* of days the patient will be able to survive. The decision tree will probably change and the *terminal nodes* will now indicate a mean expected number of days the patient will survive. This situation describes a regression tree rather than a classification tree.

In a more formal setup let Y be a dependent variable – binary or continuous and $X \in \mathbb{R}^d$. We are interested in approximating

$$f(x) = E(Y|X = x)$$

For the definition of conditional expectations we refer to Section 4.2. CART estimates this function f by a step function that is constructed via splits along the coordinate axis. An illustration is given in Figure 18.24. The regression function $f(x)$ is approximated by the values of the step function. The splits along the coordinate axes are to be determined from the data.

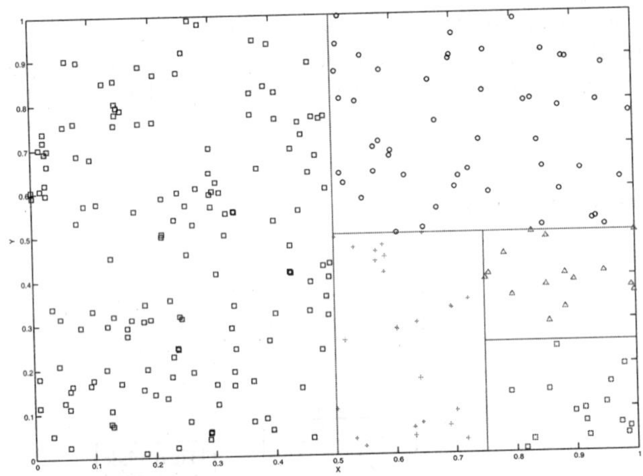

Figure 18.24. CART orthogonal splitting example where each color corresponds to one cluster

The following simple one dimensional example shows that the choice of splits points involves some decisions. Suppose that $f(x) = I(x \in [0,1]) + 2I(x \in [1,2])$ is a simple step function with a step at $x = 1$. Assume now that one observes $Y_i = f(x_i) + \varepsilon_i, X_i \sim U[0,2], \varepsilon_i \sim N(0,1)$. By going through the X data points as possible split points one sees that in the neighborhood of $x = 1$ one has two possibilities: one simply takes the X_i left to 1 or the observation right to 1. In order to make such splits unique one averages these neighboring points.

Impurity Measures

A more formal framework on how to split and where to split needs to be developed. Suppose there are n observations in the learning sample and n_j is the overall number of observations belonging to class j, $j = 1, \dots, J$. The *class probabilities* are:

$$\pi(j) = \frac{n_j}{n}, j = 1, \dots, J \tag{18.32}$$

$\pi(j)$ is the proportion of observations belonging to a particular class. Let $n(t)$ be the number of observations at node t and $n_j(t)$ – the number of observations belonging to the j-th class at t. The frequency of the event that an observation of the j-th class falls into node t is:

$$p(j,t) = \pi(j)\frac{n_j(t)}{n_j} \tag{18.33}$$

The proportion of observations at t are $p(t) = \sum_{j=1}^{J} p(j,t)$ the *conditional probability* of an observation to belong to class j given that it is at node t is:

$$p(j|\,t) = \frac{p(j,t)}{p(t)} = \frac{n_j(t)}{n(t)} \tag{18.34}$$

Define now a degree of class homogeneity in a given node. This characteristic – an *impurity measure* $i(t)$ – will represent a class homogeneity indicator for a given tree node and hence will help to find optimal splits. Define an *impurity function* $\iota(t)$ which is determined on $(p_1, \dots, p_J) \in [0,1]^J$ with $\sum_{j=1}^{J} p_j = 1$ so that:

1. ι has a unique maximum at point $\left(\frac{1}{J}, \frac{1}{J}, \dots, \frac{1}{J}\right)$;
2. ι has a unique minimum at points $(1,0,0,\dots,0)$, $(0,1,0,\dots,0)$, …, $(0,0,0,\dots,1)$;
3. ι is a symmetric function of $p_1, \dots, p_J$

Each function satisfying these conditions is called an impurity function. Given ι, define the *impurity measure* $i(t)$ for a node t as:

$$i(t) = \iota\left\{p(1|\,t), p(2|\,t), \dots, p(J|\,t)\right\} \tag{18.35}$$

Denote an arbitrary data split by s, then for a given node t which we will call a *parent node* two *child nodes* arise: t_L and t_R representing observations meeting and not meeting the split criterion s. A fraction p_L of data from t falls to the left child node and $p_R = 1 - p_L$ is the share of data in t_R.

A *quality measure* of how well split s works is:

$$\Delta i(s,t) = i(t) - p_L i(t_L) - p_R i(t_R) \tag{18.36}$$

The higher the value of $\Delta i(s,t)$ the better split we have since data impurity is reduced. In order to find an optimal split s it is natural to maximize $\Delta i(s,t)$. Note that in (18.36) for different splits s, the value of $i(t)$ remains constant, hence it is equivalent to find

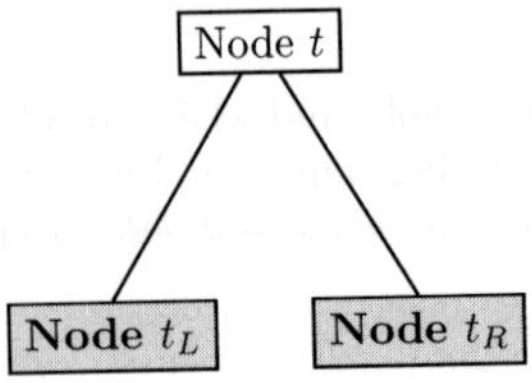

Figure 18.25. Parent and child nodes hierarchy

$$s^* = \underset{s}{\operatorname{argmax}}\, \Delta i\,(s,t) = \underset{s}{\operatorname{argmax}}\, \{-p_L i\,(t_L) - p_R i\,(t_R)\}$$

$$= \underset{s}{\operatorname{argmin}}\, \{p_L i\,(t_L) + p_R i\,(t_R)\} \tag{18.37}$$

where t_L and t_R are implicit functions of s. This splitting procedure is repeated until one arrives at a minimal bucket size. Classes are then assigned to terminal nodes using the following rule:

$$\text{If } p(j|\,t) = \max_i p(i|\,t), \;\; \text{then } j^*(t) = j \tag{18.38}$$

If the maximum is not unique, then $j^*(t)$ is assigned randomly to those classes for which $p(i|\,t)$ takes its maximum value. The crucial question is of course to define an impurity function $i\,(t)$. A natural definition of impurity is via a *variance* measure: Assign 1 to all observations at node t belonging to class j and 0 to others. A sample variance estimate for node t observations is $p(j|\,t)\,\{1 - p(j|\,t)\}$.

Summing over all J classes we obtain the *Gini index*:

$$i\,(t) = \sum_{j=1}^{J} p(j|\,t)\,\{1 - p(j|\,t)\} = 1 - \sum_{j=1}^{J} p^2(j|\,t) \tag{18.39}$$

The Gini index is an impurity function $\iota(p_1, \dots, p_J)$, $p_j = p(j|\,t)$. It is not hard see that the Gini index is a convex function. Since $p_L + p_R = 1$, we get:

$$\begin{aligned} i(t_L)p_L + i(t_R)p_R = &\; \iota\,\{p(1|\,t_L), \dots, p(J|\,t_L)\}\,p_L + \iota\,\{p(1|\,t_R), \dots, p(J|\,t_R)\}\,p_R \\ \leq &\; \iota\,\{p_L p(1|\,t_L) + p_R p(1|\,t_R), \dots, p_L p(J|\,t_L) + p_R p(J|\,t_R)\} \end{aligned}$$

where inequality becomes an equality in case $p(j|\,t_L) = p(j|\,t_R)$, $j = 1, \dots, J$.

Recall that

$$\frac{p(j, t_L)}{p(t)} = \frac{p(t_L)}{p(t)} \cdot \frac{p(j, t_L)}{p(t_L)} = p_L p(j|\,t_L)$$

and since

$$p(j|\,t) = \frac{p(j, t_L) + p(j, t_R)}{p(t)} = p_L p(j|\,t_L) + p_R p(j|\,t_R)$$

we can conclude that

$$i(t_L)p_L + i(t_R)p_R \le i(t) \tag{18.40}$$

Hence each variant of data split leads to $\Delta i(s,t) > 0$ unless $p(j|\,t_R) = p(j|\,t_L) = p(j|\,t)$ i.e. when no split decreases class heterogeneity.

Impurity measures can be defined in a number of different ways, for practical applications the so called *twoing rule* can also be considered. Instead of maximizing impurity change at a particular node, the twoing rule tries to balance as if the learning sample had only two classes. The reason for such an algorithm is that such a decision rule is able to distinguish observations between general factors on top levels of the tree and take into account specific data characteristics at lower levels.

If $S = \{1, \ldots, J\}$ is the set of learning sample classes, divide it into two subsets

$$S_1 = \{j_1, \ldots, j_n\}\,, \text{ and } S_2 = S \backslash S_1$$

All observations belonging to S_1 get dummy class 1, and the rest dummy class 2. The next step is to calculate $\Delta i(s,t)$ for different s *as if there were only two (dummy) classes.* Since actually $\Delta i(s,t)$ depends on S_1, the value $\Delta i(s,t,S_1)$ is maximized. Now apply a *two-step procedure*: first, find $s^*(S_1)$ maximizing $\Delta i(s,t,S_1)$ and second, find a *superclass* S_1^* maximizing $\Delta i\,\{s^*(S_1), t, S_1\}$. In other words the idea of twoing is to find a combination of superclasses at each node that maximizes the impurity increment for two classes.

This method provides one big advantage: it finds so called *strategic nodes* i.e. nodes filtering observations in the way that they are different to the maximum feasible extent. Although applying the twoing rule may seem to be desirable especially for data with a big number of classes, another challenge arises: computational speed. Let's assume that the learning sample has J classes, then a set S can be split into S_1 and S_2 by 2^{J-1} ways. For 11 classes data this will create more than 1000 combinations. Fortunately the following result helps to reduce drastically the amount of computations.

It can be proven (Breiman et al., 1984) that in a classification task with two classes and impurity measure $p(1|\,t)p(2|\,t)$ for an arbitrary split s a superclass $S_1(s)$ is determined by:

$$S_1(s) = \{j : p(j|\,t_L) \ge p(j|\,t_R)\}\,,$$

$$\max_{S_1} \Delta i(s,t,S_1) = \frac{p_L p_R}{4} \left\{ \sum_{j=1}^{J} |p(j|\,t_L) - p(j|\,t_R)| \right\}^2 \tag{18.41}$$

Hence the twoing rule can be applied in practice as well as Gini index, although the first criterion works a bit slower.

Gini Index and Twoing Rule in Practice

In this section we look at practical issues of using these two rules. Consider a learning dataset from Salford Systems with 400 observations characterizing automobiles: their make, type, color, technical parameters, age etc. The aim is to build a decision tree splitting different cars by their characteristics based on feasible relevant parameters. The classification tree constructed using the Gini index is given in Figure 18.26.

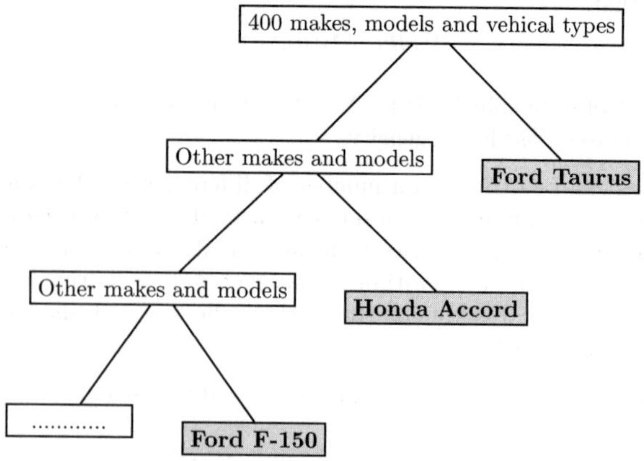

Figure 18.26. Classification tree constructed by Gini index

A particular feature here is that at each node observations belonging to one make are filtered out i.e. observations with most striking characteristics are separated. As a result a decision tree is able to pick out automobile makes quite easily.

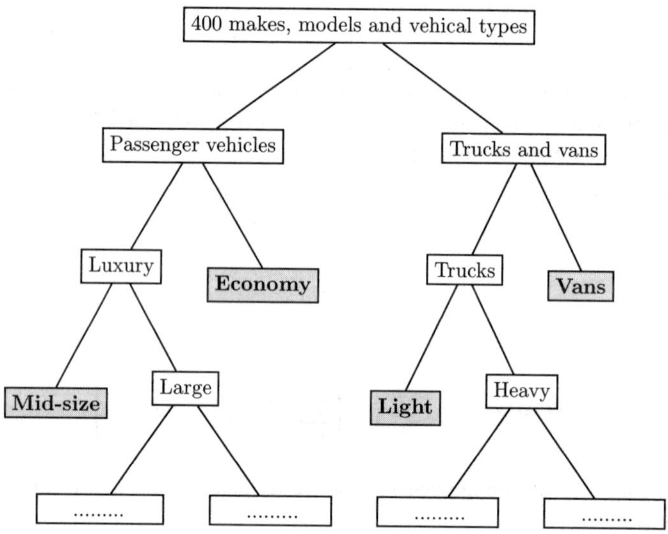

Figure 18.27. Classification tree constructed by twoing index

The twoing rule based tree Figure 18.27 for the same data is different. Instead of specifying particular car makes at each node, application of the twoing rule results in strategic nodes i.e. questions which distinguish between different car classes to the maximum extent. This feature can be vital when high-dimensional datasets with a big number of classes are processed.

Optimal Size of a Decision Tree

Up to now we were interested in determining the best split s^* at a particular node. The next and perhaps more important question is how to determine the optimal tree size i.e. when to *stop splitting.* If each terminal node has only class homogenous dataset, then every point of the learning sample can be flawlessly classified using this *maximum tree.* But can be such an approach fruitful?

The maximum tree is a case of overspecification. Some criterion is required to stop data splitting. Since tree building is dependent on $\Delta i(s, t)$, a criterion is to stop data splitting if

$$\Delta i(s, t) < \bar{\beta} \tag{18.42}$$

where $\bar{\beta}$ is some threshold value.

The value of $\bar{\beta}$ is to be chosen in a subjective way and this is unfortunately a drawback. Empirical simulations show that the impurity increment is frequently non-monotone, that is why even for small $\bar{\beta}$ the tree may be underparametrized. Setting even smaller values for $\bar{\beta}$ will probably remedy the situation but at the cost of tree overparametrization.

Another way to determine the adequate shape of a decision tree is to demand a minimum number of observations $\bar{N}$ (bucked size) at each terminal node. A disadvantage is that if at terminal node t the number of observations is higher

$$N(t) > \bar{N} \tag{18.43}$$

then this node is also being split as data are still not supposed to be clustered well enough.

Cross-validation for Tree Pruning

Cross-validation is a procedure which uses the bigger data part as a *training set* and the rest as a *test set.* Then the process is looped so that different parts of the data become learning and training set, so that at the end each datapoint was employed both as a member of test and learning sets. The aim of this procedure is to extract maximum information from the learning sample especially in the situations of data scarceness.

The procedure is implemented in the following way. First, the learning sample is *randomly* divided into V parts. Using the training set from the union of $(V - 1)$ subsets a decision tree is constructed while the test set is used to verify the tree quality. This procedure is looped over all possible subsets.

Unfortunately for small values of V cross-validation estimates can be *unstable* since each iteration a cluster of data is selected *randomly* and the number of iterations itself is relatively small, thus the overall estimation result is somewhat random. Nowadays cross-validation with $V = 10$ is an industry standard and for many applications a good balance between computational complexity and statistical precision.

Cost-complexity Function and Cross-validation

Another method taken into account is *tree complexity* i.e. the *number of terminal nodes.* The maximum tree will get a penalty for its big size, on the other hand it will be able to make perfect in-sample predictions. Small trees will, of course, get lower penalty for their size but their prediction abilities are limited. Optimization procedure based on such a trade-off criterion could determine a good decision tree.

Define *internal misclassification error* of an arbitrary observation at node t as $e(t) = 1 - \max_j p(j|\,t)$, define also $E(t) = e(t)p(t)$. Then *internal misclassification tree error* is $E(T) = \sum_{t\in\tilde{T}} E(t)$ where $\tilde{T}$ is a set of terminal nodes. The estimates are called *internal* because they are based solely on the learning sample. It may seem that $E(T)$ as a tree quality measure is sufficient but unfortunately it is not so. Consider the case of the maximum tree, here $E(T_{MAX}) = 0$ i.e. the tree is of best configuration.

For any subtree $T\,(\leq T_{MAX})$ define the number of terminal nodes $\left|\tilde{T}\right|$ as a measure of its complexity. The following cost-complexity function can be used:

$$E_\alpha(T) = E(T) + \alpha\left|\tilde{T}\right| \tag{18.44}$$

where $\alpha \geq 0$ is a complexity parameter and $\alpha\left|\tilde{T}\right|$ is a cost component. The more complex the tree (high number of terminal nodes) the lower is $E(T)$ but at the same time the higher is the penalty $\alpha\left|\tilde{T}\right|$ and vice versa.

The number of subtrees of T_{MAX} is finite. Hence pruning of T_{MAX} leads to creation of a subtree sequence $T_1, T_2, T_3, \ldots$ with a decreasing number of terminal nodes.

An important question is if a subtree $T \leq T_{MAX}$ for a given α minimizing $E_\alpha(T)$ always exists and whether it is unique?

In (Breiman et al., 1984) it is shown that for $\forall\alpha \geq 0$ there exists an optimal tree $T(\alpha)$ in the sense that

1. $E_\alpha\{T(\alpha)\} = \min\limits_{T\leq T_{MAX}} E_\alpha(T) = \min\limits_{T\leq T_{MAX}} \left\{E(T) + \alpha\left|\tilde{T}\right|\right\}$
2. if $E_\alpha(T) = E_\alpha\{T(\alpha)\}$ then $T(\alpha) \leq T$.

This result is a proof of existance, but also a proof of uniqueness: consider another subtree T' so that T and T' both minimize E_α and are not nested, then $T(\alpha)$ does not exist in accordance with second condition.

The idea of introducing cost-complexity function at this stage is to check only a subset of different subtrees of T_{MAX}: optimal subtrees for different values of α. The starting point is to define the first optimal subtree in the sequence so that $E(T_1) = E(T_{MAX})$ and the size of T_1 is minimum among other subtrees with the same cost level. To get T_1 out of T_{MAX} for each terminal node of T_{MAX} it is necessary to verify the condition $E(t) = E(t_L) + E(t_R)$ and if it is fulfilled – node t is pruned. The process is looped until no extra pruning is available – the resulting tree $T(0)$ becomes T_1.

Define a node t as an *ancestor* of t' and t' as *descendant* of t if there is a connected path down the tree leading from t to t'. Consider Figure 18.28 where nodes t_4, t_5, t_8, t_9, t_{10} and t_{11} are descendants of t_2 while nodes t_6 and t_7 are not descendants of t_2 although they are positioned lower since it is not possible to connect them with a path from t_2 to these nodes without engaging t_1. Nodes t_4, t_2 and t_1 are ancestors of t_9 and t_3 is not ancestor of t_9.

Define the *branch* T_t of the tree T as a subtree based on node t and all its descendants. An example is given in Figure 18.29. Pruning a branch T_t from a tree T means deleting all descendant nodes of t. Denote the transformed tree as $T - T_t$. Pruning the branch T_{t_2} results in the tree described in Figure 18.30.

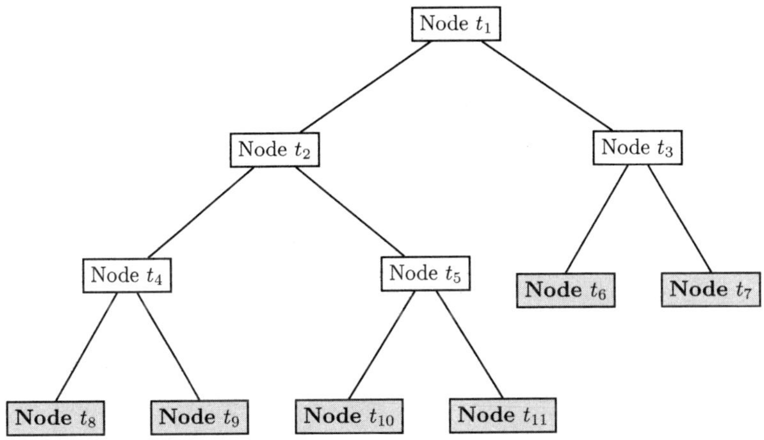

Figure 18.28. Decision tree hierarchy

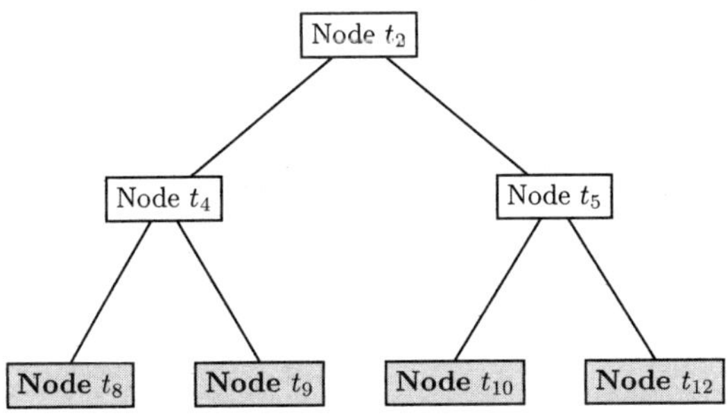

Figure 18.29. The branch T_{t_2} of the original tree T

For any branch T_t define the *internal misclassification estimate* as:

$$E(T_t) = \sum_{t' \in \tilde{T}_t} E(t') \tag{18.45}$$

where $\tilde{T}_t$ is the set of terminal nodes of T_t. Hence for an arbitrary node t of T_t:

$$E(t) > E(T_t) \tag{18.46}$$

Consider now the *cost-complexity misclassification estimate* for branches or single nodes. Define for a single node $\{t\}$:

$$E(\{t\}) = E(t) + \alpha \tag{18.47}$$

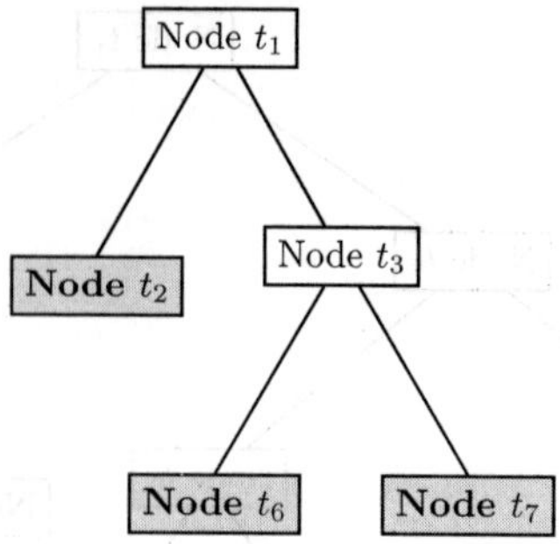

Figure 18.30. $T - T_{t_2}$ the pruned tree T

and for a branch:

$$E_\alpha(T_t) = E(T_t) + \alpha \left|\tilde{T}_t\right| \tag{18.48}$$

When $E_\alpha(T_t) < E_\alpha(\{t\})$ the branch T_t is preferred to a single node $\{t\}$ according to cost-complexity. For some α both (18.47) (18.48) will become equal. This critical value of α can be determined from:

$$E_\alpha(T_t) < E_\alpha(\{t\}) \tag{18.49}$$

which is equivalent to

$$\alpha < \frac{E(t) - E(T_t)}{\left|\tilde{T}_t\right| - 1} \tag{18.50}$$

where $\alpha > 0$ since $E(t) > E(T_t)$

To obtain the next member of the subtrees sequence i.e. T_2 out of T_1 a special node called *weak link* is determined. For this purpose a function $g_1(t)$, $t \in T_1$ is defined as

$$g_1(t) = \begin{cases} \frac{E(t)-E(T_t)}{\left|\tilde{T}_t\right|-1}, & t \notin \tilde{T}_1 \\ +\infty, & t \in \tilde{T}_1 \end{cases} \tag{18.51}$$

Node $\bar{t}_1$ is a weak link in T_1 if

$$g_1(\bar{t}_1) = \min_{t \in T_1} g_1(t) \tag{18.52}$$

and a new value for α_2 is defined as

$$\alpha_2 = g_1(\bar{t}_1) \tag{18.53}$$

A new tree $T_2 \prec T_1$ in the sequence is obviously defined by pruning the branch $T_{\bar{t}_1}$ i.e.

$$T_2 = T_1 - T_{\bar{t}_1} \tag{18.54}$$

The process is looped until root node $\{t_0\}$ – the final member of sequence – is reached. When there are multiple weak links detected, for instance $g_k(\bar{t}_k) = g_k(\bar{t}'_k)$, then both branches are pruned i.e. $T_{k+1} = T_k - T_{\bar{t}_k} - T_{\bar{t}'_k}$.

In this way it is possible to get the sequence of optimal subtrees $T_{MAX} \succ T_1 \succ T_2 \succ T_3 \succ \ldots \succ \{t_0\}$ for which it is possible to prove that the sequence $\{\alpha_k\}$ is increasing i.e. $\alpha_k < \alpha_{k+1}$, $k \geq 1$ and $\alpha_1 = 0$. For $k \geq 1$: $\alpha_k \leq \alpha < \alpha_{k+1}$ and $T(\alpha) = T(\alpha_k) = T_k$.

Practically this tells us how to implement the search algorithm. First, the maximum tree T_{MAX} is taken, then T_1 is found and a weak link $\bar{t}_1$ is detected and branch $T_{\bar{t}_1}$ is pruned off, α_2 is calculated and the process is continued.

When the algorithm is applied to T_1, the number of pruned nodes is usually quite significant. For instance, consider the following typical empirical evidence:

Tree	T_1	T_2	T_3	T_4	T_5	T_6	T_7	T_8	T_9	T_{10}	T_{11}	T_{12}	T_{13}
$\lvert\tilde{T}_k\rvert$	71	63	58	40	34	19	10	9	7	6	5	2	1

Table 18.7. Typical pruning speed

When the trees become smaller, the difference in the number of terminal nodes also gets smaller

Finally, it is worth mentioning that the sequence of optimally pruned subtrees is a subset of trees which might be constructed using direct method of internal misclassification estimator minimization given a fixed number of terminal nodes. Consider an example of tree $T(\alpha)$ with 7 terminal nodes, then there is no other subtree T with 7 terminal nodes having lower $E(T)$. Otherwise

$$E_\alpha(T) = E(T) + 7\alpha < E_\alpha\{T(\alpha)\} = \min_{T \leq T_{MAX}} E_\alpha(T)$$

which is impossible by definition.

Applying the method of V-fold cross-validation to the sequence $T_{MAX} \succ T_1 \succ T_2 \succ T_3 \succ \ldots \succ \{t_0\}$, an *optimal tree* is determined.

On the other hand it is frequently pointed out that choice of tree with minimum value of $E^{CV}(T)$ is not always adequate since $E^{CV}(T)$ is not too robust i.e. there is a whole range of values $E^{CV}(T)$ satisfying $E^{CV}(T) < E^{CV}_{MIN}(T) + \varepsilon$ for small $\varepsilon > 0$. Moreover, when $V < N$ a simpe change of random generator seed will definitely result in changed values of $\left|\tilde{T}_k\right|$ minimizing $\hat{E}(T_K)$. Hence a so called *one standard error* empirical rule is applied which states that if T_{k_0} is the tree minimizing $E^{CV}(T_{k_0})$ from the sequence $T_{MAX} \succ T_1 \succ T_2 \succ T_3 \succ \ldots \succ \{t_0\}$, then a value k_1 and a correspondent tree T_{k_1} are selected so that

$$\underset{k_1}{\operatorname{argmax}}\, \hat{E}(T_{k_1}) \leq \hat{E}(T_{k_0}) + \sigma\left\{\hat{E}(T_{k_0})\right\} \tag{18.55}$$

where $\sigma(\cdot)$ denotes sample estimate of standard error and $\hat{E}(\cdot)$ – the relevant sample estimators.

The dotted line on Figure 18.31 shows the area where the values of $\hat{E}(T_k)$ only slightly differ from $\min_{|\tilde{T}_k|} \hat{E}(T_k)$. The left edge which is roughly equivalent to 16 terminal nodes shows the application of one standard error rule. The use of one standard error rule allows not

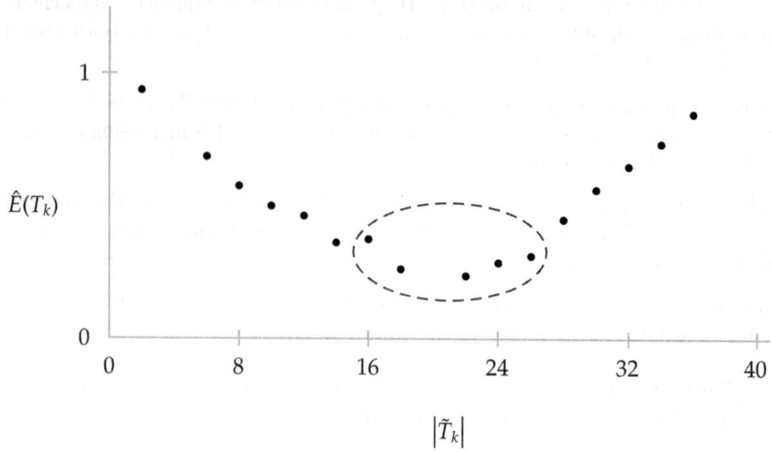

Figure 18.31. The example of relationship between $\hat{E}(T_k)$ and number of terminal nodes

only to achieve more robust results but also to get trees of lower complexity given the error comparable with $\min_{|\tilde{T}_k|} \hat{E}(T_k)$.

Regression Trees

Up to now we concentrate on classification trees. Although *regression trees* share a similar logical framework, there are some differences which need to be addressed. The important difference between classification and regression trees is the type of dependent variable Y. When Y is discrete, a decision tree is called a classification tree, a regression tree is a decision tree with a *continuous* dependent variable.

Gini index and twoing rule discussed in previous sections assume that the number of classes is finite and hence introduce some measures based mainly on $p(j|t)$ for arbitrary class j and node t. But since in case of continuous dependent variable there are no more classes, this approach cannot be used anymore unless groups of continuous values are effectively substituted with artificial classes. Since there are no classes anymore – how can be the maximum regression tree determined? Analogously with discrete case, absolute homogeneity can be then described only after some adequate impurity measure for regression trees is introduced.

Recall the idea of *Gini index*, then it becomes quite natural to use the *variance* as impurity indicator. Since for each node data variance can be easily computed, then splitting criterion for an arbitrary node t can be written as

$$s^* = \operatorname*{argmax}_s \left[p_L \text{var}\left\{t_L(s)\right\} + p_R \text{var}\left\{t_R(s)\right\}\right] \tag{18.56}$$

where t_L and t_R are emerging child nodes which are, of course, directly dependent on the choice of s^*.

Hence the maximum regression tree can be easily defined as a structure where each node has only the same predicted values. It is important to point out that since continuous data have much higher chances to take different values comparing with discrete ones, the size of maximum regression tree is usually very big.

When the maximum regression tree is properly defined, it is then of no problem to get an optimally-size tree. Like with classification trees, maximum regression tree is usually supposed to be upwardly pruned with the help of cost-complexity function and cross-validation. That is why the majority of results presented above is applied to regression trees as well.

Bankruptcy Analysis

This section provides a practical study on bankruptcy data involving decision trees. A dataset with 84 observations representing different companies is constituted by three variables:

- net income to total assets ratio
- total liabilities to total assets ratio
- company status (-1 if bankrupt and 1 if not)

The data is from (Securities and Exchange Commission, 2004)

The goal is to predict and describe the company status given the two primary financial ratios. Since no additional information like the functional form of possible relationship is available, the use of a *classification tree* is an active alternative.

The tree given in Figure 18.32 was constructed using the Gini index and a $\bar{N} = 30$ constraint i.e. the number of points in each of the terminal nodes can not be more than 30. Numbers in parentheses displayed on terminal nodes are observation quantities belonging to Class 1 and Class -1.

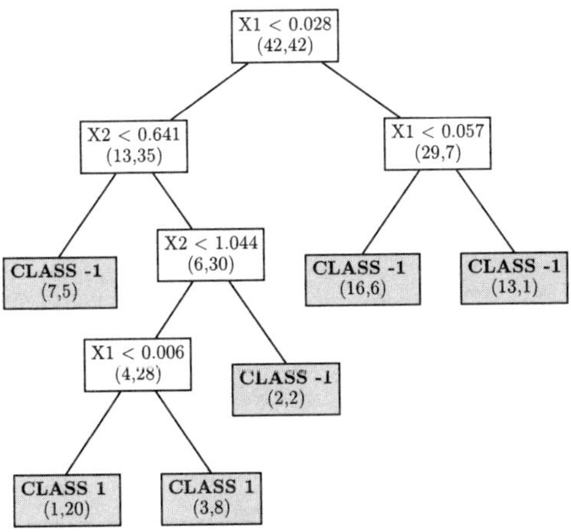

MVACARTBan1.xpl

Figure 18.32. Decision tree for bankruptcy dataset: Gini index, $\bar{N} = 30$

If we loose the constraint to $\bar{N} = 10$, the decision rule changes, see Figure 18.33. How exactly did the situation change? Consider the Class 1 terminal nodes of the tree on Figure 18.32. The first one contains 21 observations and thus was split for $\bar{N} = 10$. When it was split two new nodes of *different classes* emerged and for both of them the impurity measure has decreased.

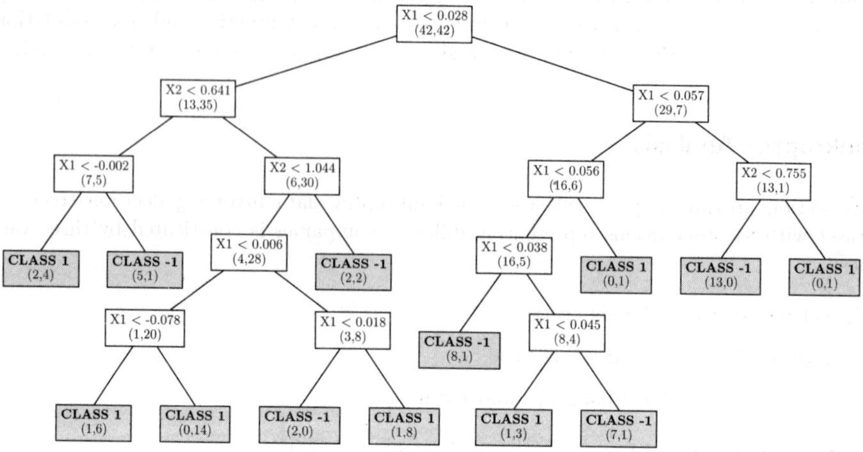

MVACARTBan2.xpl

Figure 18.33. Decision tree for bankruptcy dataset: Gini index, $\bar{N} = 10$

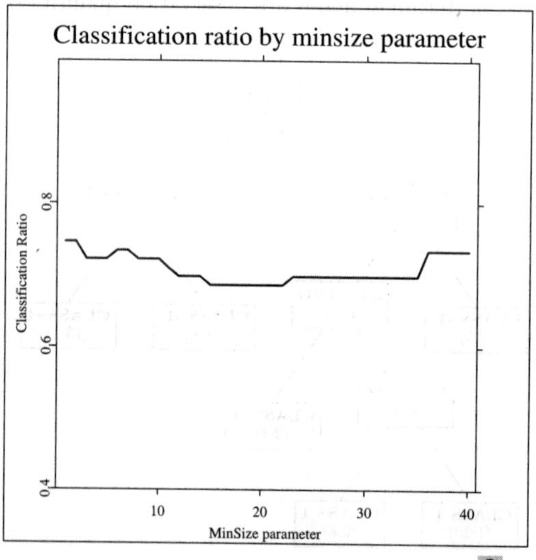

MVAbancrupcydis.xpl

Figure 18.34. Successful classification ratio dynamic over the number of terminal nodes: cross-validation

We may conclude that $\bar{N} \approx 10$ is a good choice and analyzing the tree produced we can state that for this particular example the net income to total assets (X_1) ratio appears to be an important class indicator.

For this example with relatively small sample size we construct two maximum trees – using the Gini and twoing rules. Looking at both decision trees we see that the choice of impurity measure is not so important as the right choice of tree size.

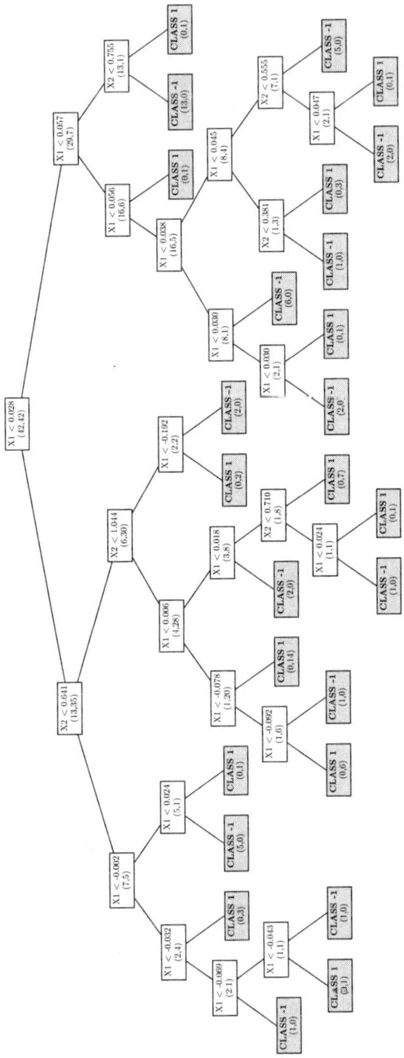

MVACARTGiniTree1.xpl

Figure 18.35. Maximum tree constructed employing Gini index

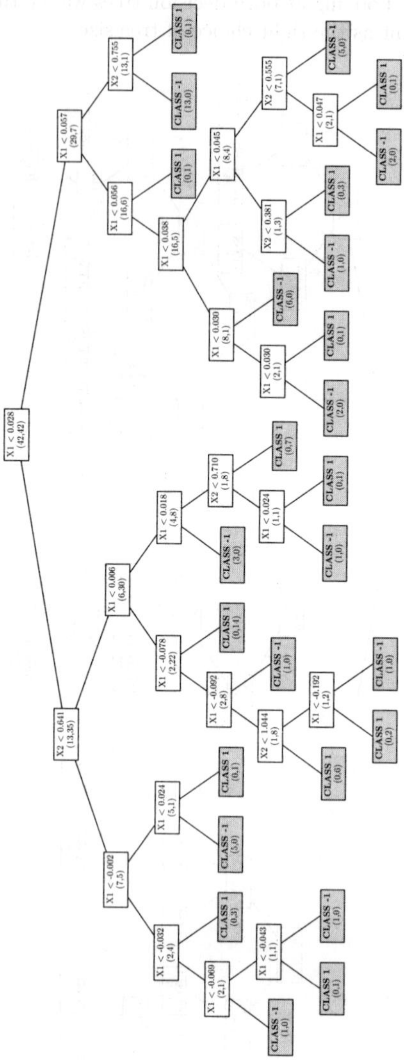

MVACARTTwoingTree1.xpl

Figure 18.36. Maximum tree constructed employing twoing rule

Summary

- ↪ CART is a tree based method splitting the data sequentially into a binary tree.
- ↪ CART determined the nodes by minimizing an impurity measure at each mode.
- ↪ CART is non-parametric:
  When no data structure hypotheses are available, non-parametric analysis becomes the single effective data mining tool. CART is a flexible nonparametric data mining tool.
- ↪ CART does not require variables to be selected in advance:
  From a learning sample CART will automatically select the most significant ones.
- ↪ CART is very efficient in computational terms:
  Although all possible data splits are analyzed, the CART architecture is flexible enough to do all of them quickly.
- ↪ CART is robust to the effect of outliers:
  Due to data-splitting nature of decision rules creation it is possible to distinguish between datasets with different characteristics and hence to neutralize outliers in separate nodes.
- ↪ CART can use any combination of continuous and categorical data:
  Researchers are no longer limited to a particular class of data and will be able to capture more real-life examples.

18.6 Boston Housing

Coming back to the Boston housing data set, we compare the results of exploratory projection pursuit on the original data $\mathcal{X}$ and the transformed data $\widehat{\mathcal{X}}$ motivated in Section 1.8. So we exclude X_4 (indicator of Charles River) from the present analysis.

The aim of this analysis is to see from a different angle whether our proposed transformations yield more normal distributions and whether it will yield data with less outliers. Both effects will be visible in our projection pursuit analysis.

We first apply the Jones and Sibson index to the non-transformed data with 50 randomly chosen 13-dimensional directions. Figure 18.37 displays the results in the following form. In the lower part, we see the values of the Jones and Sibson index. It should be constant for 13-dimensional normal data. We observe that this is clearly not the case. In the upper part of Figure 18.37 we show the standard normal density as a green curve and two densities corresponding to two extreme index values. The red, slim curve corresponds to the maximal value of the index among the 50 projections. The blue curve, which is close to the normal, corresponds to the minimal value of the Jones and Sibson index. The corresponding values of the indices have the same color in the lower part of Figure 18.37. Below the densities, a jitter plot shows the distribution of the projected points $\alpha^{\top}x_i$ $(i = 1, \ldots, 506)$. We conclude from the outlying projection in the red distribution that several points are in conflict with the normality assumption.

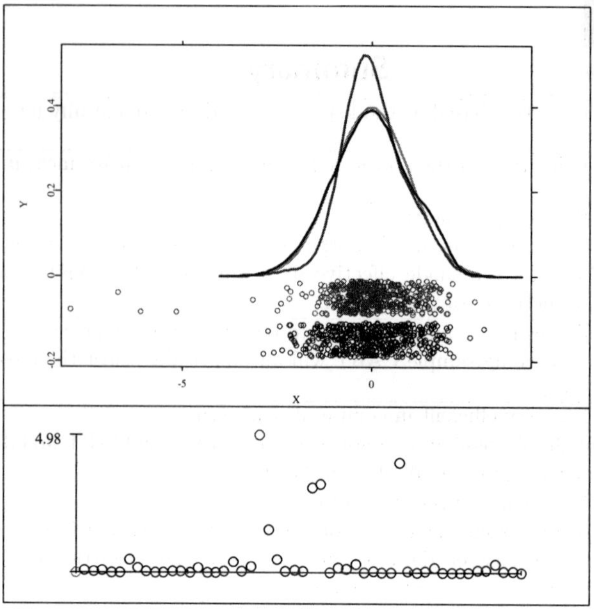

Figure 18.37. Projection Pursuit with the Sibson-Jones index with 13 original variables. MVAppsib

Figure 18.38 presents an analysis with the same design for the transformed data. We observe in the lower part of the figure values that are much lower for the Jones and Sibson index (by a factor of 10) with lower variability which suggests that the transformed data is closer to the normal. ("Closeness" is interpreted here in the sense of the Jones and Sibson index.) This is confirmed by looking to the upper part of Figure 18.38 which has a significantly less outlying structure than in Figure 18.37.

18.7 Exercises

EXERCISE 18.1 *Calculate the Simplicial Depth for the Swiss bank notes data set and compare the results to the univariate medians. Calculate the Simplicial Depth again for the genuine and counterfeit bank notes separately.*

EXERCISE 18.2 *Construct a configuration of points in* $\mathbb{R}^2$ *such that* $x_{med,j}$ *from (18.2) is not in the "center" of the scatterplot.*

EXERCISE 18.3 *Apply the SIR technique to the U.S. companies data with* $Y =$ *market value and* $X =$ *all other variables. Which directions do you find?*

EXERCISE 18.4 *Simulate a data set with* $X \sim N_4(0, I_4), Y = (X_1 + 3X_2)^2 + (X_3 - X_4)^4 + \varepsilon$ *and* $\varepsilon \sim N(0, (0.1)^2)$. *Use SIR and SIR II to find the EDR directions.*

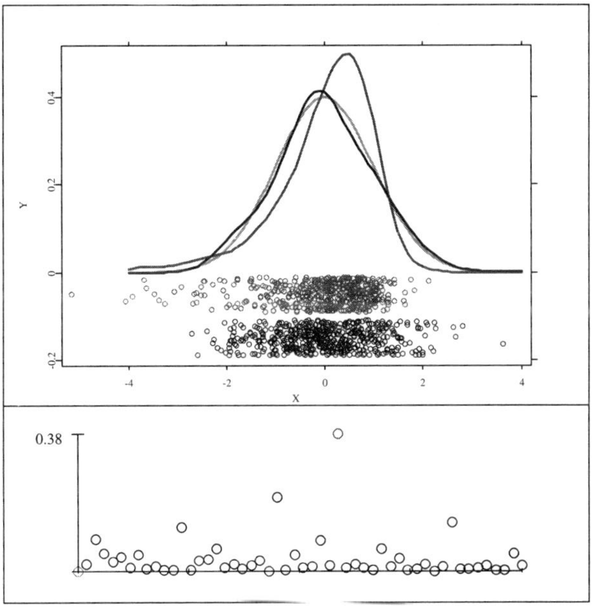

Figure 18.38. Projection Pursuit with the Sibson-Jones index with 13 transformed variables. MVAppsib

EXERCISE 18.5 *Apply the Projection Pursuit technique on the Swiss bank notes data set and compare the results to the PC analysis and the Fisher discriminant rule.*

EXERCISE 18.6 *Apply the SIR and SIR II technique on the car data set in Table B.3 with* $Y = price$.

EXERCISE 18.7 *Generate four regions on the two-dimensional unit square by sequentially cutting parallel to the coordinate axes. Generate 100 two-dimensional Uniform random variables and label them according to their presence in the above regions. Apply the CART algorithm to find the regions bound and to classify the observations.*

EXERCISE 18.8 *Modify Exercise 18.7 by defining the regions as lying above and below the main diagonal of the unit square. Make a CART analysis and comment on the complexity of the tree.*

EXERCISE 18.9 *Apply the SVM with different radial basis parameter* r *and different capacity parameter* c *in order to separate two circular datasets. This example is often called the Orange Peel exercise and involves two Normal distributions* $N(\mu, \Sigma_i)$, $i = 1, 2$, *with covariance matrices* $\Sigma_1 = 2\mathcal{I}_2$ *and* $\Sigma_2 = 0.5\mathcal{I}_2$.

EXERCISE 18.10 *The noisy spiral data set consists of two intertwining spirals that need to be separated by a nonlinear classification method. Apply the SVM with different radial basis parameter* r *and capacity parameter* c *in order to separate the two spiral datasets.*

EXERCISE 18.11 *Apply the SVM to separate the bankrupt from the surviving (profitable) companies using the profitability and leverage ratios given in the Bankruptcy data set in Table B.21.*

Part IV

Appendix

A Symbols and Notations

Basics

X, Y	random variables or vectors	
$X_1, X_2, \dots, X_p$	random variables	
$X = (X_1, \dots, X_p)^\top$	random vector	
$X \sim \cdot$	X has distribution $\cdot$	
$\mathcal{A}, \mathcal{B}$	matrices	41
Γ, Δ	matrices	46
$\mathcal{X}, \mathcal{Y}$	data matrices	62
Σ	covariance matrix	62
1_n	vector of ones $(\underbrace{1, \dots, 1}_{n\text{-times}})^\top$	42
0_n	vector of zeros $(\underbrace{0, \dots, 0}_{n\text{-times}})^\top$	42
$\boldsymbol{I}(.)$	indicator function, i.e. for a set M is $\boldsymbol{I} = 1$ on M, $\boldsymbol{I} = 0$ otherwise	
i	$\sqrt{-1}$	
$\Rightarrow$	implication	
$\Leftrightarrow$	equivalence	
$\approx$	approximately equal	
$\otimes$	Kronecker product	
iff	if and only if, equivalence	

Mathematical Abbreviations

$\text{tr}(\mathcal{A})$	trace of matrix $\mathcal{A}$
$\text{hull}(x_1, \dots, x_k)$	convex hull of points $\{x_1, \dots, x_k\}$
$\text{diag}(\mathcal{A})$	diagonal of matrix $\mathcal{A}$
$\text{rank}(\mathcal{A})$	rank of matrix $\mathcal{A}$
$\det(\mathcal{A})$	determinant of matrix $\mathcal{A}$

Samples

x, y	observations of X and Y	
$x_1, \dots, x_n = \{x_i\}_{i=1}^n$	sample of n observations of X	
$\mathcal{X} = \{x_{ij}\}_{i=1,\dots,n; j=1,\dots,p}$	$(n \times p)$ data matrix of observations of $X_1, \dots, X_p$ or of $X = (X_1, \dots, X_p)^T$	62
$x_{(1)}, \dots, x_{(n)}$	the order statistic of $x_1, \dots, x_n$	5
$\mathcal{H}$	centering matrix, $\mathcal{H} = \mathcal{I}_n - n^{-1} 1_n 1_n^\top$	71

Densities and Distribution Functions

$f(x)$	density of X	
$f(x, y)$	joint density of X and Y	
$f_X(x), f_Y(y)$	marginal densities of X and Y	
$f_{X_1}(x_1), \ldots, f_{X_p}(x_2)$	marginal densities of $X_1, \ldots, X_p$	
$\hat{f}_h(x)$	histogram or kernel estimator of $f(x)$	11
$F(x)$	distribution function of X	
$F(x, y)$	joint distribution function of X and Y	
$F_X(x), F_Y(y)$	marginal distribution functions of X and Y	
$F_{X_1}(x_1), \ldots, f_{X_p}(x_p)$	marginal distribution functions of $X_1, \ldots, X_p$	
$\varphi(x)$	density of the standard normal distribution	
$\Phi(x)$	standard normal distribution function	
$\varphi_X(t)$	characteristic function of X	
m_k	k-th moment of X	
κ_j	cumulants or semi-invariants of X	

Moments

EX, EY	mean values of random variables or vectors X and Y	61
$\sigma_{XY} = Cov(X, Y)$	covariance between random variables X and Y	61
$\sigma_{XX} = Var(X)$	variance of random variable X	61
$\rho_{XY} = \dfrac{Cov(X, Y)}{\sqrt{Var(X)\, Var(Y)}}$	correlation between random variables X and Y	65
$\Sigma_{XY} = Cov(X, Y)$	covariance between random vectors X and Y, i.e., $Cov(X, Y) = E(X - EX)(Y - EY)^\top$	
$\Sigma_{XX} = Var(X)$	covariance matrix of the random vector X	

Empirical Moments

$\overline{x} = \dfrac{1}{n}\sum_{i=1}^{n} x_i$	average of X sampled by $\{x_i\}_{i=1,\ldots,n}$	6
$s_{XY} = \dfrac{1}{n}\sum_{i=1}^{n}(x_i - \overline{x})(y_i - \overline{y})$	empirical covariance of random variables X and Y sampled by $\{x_i\}_{i=1,\ldots,n}$ and $\{y_i\}_{i=1,\ldots,n}$	62
$s_{XX} = \dfrac{1}{n}\sum_{i=1}^{n}(x_i - \overline{x})^2$	empirical variance of random variable X sampled by $\{x_i\}_{i=1,\ldots,n}$	62
$r_{XY} = \dfrac{s_{XY}}{\sqrt{s_{XX} s_{YY}}}$	empirical correlation of X and Y	65
$\mathcal{S} = \{s_{X_i X_j}\} = x^\top \mathcal{H} x$	empirical covariance matrix of $X_1, \ldots, X_p$ or of the random vector $X = (X_1, \ldots, X_p)^\top$	62, 71
$\mathcal{R} = \{r_{X_i X_j}\} = \mathcal{D}^{-1/2}\mathcal{S}\mathcal{D}^{-1/2}$	empirical correlation matrix of $X_1, \ldots, X_p$ or of the random vector $X = (X_1, \ldots, X_p)^\top$	66, 71

Distributions

$\varphi(x)$	density of the standard normal distribution	
$\Phi(x)$	distribution function of the standard normal distribution	
$N(0,1)$	standard normal or Gaussian distribution	
$N(\mu,\sigma^2)$	normal distribution with mean μ and variance σ^2	
$N_p(\mu,\Sigma)$	p-dimensional normal distribution with mean μ and covariance matrix Σ	
$\xrightarrow{\mathcal{L}}$	convergence in distribution	113
CLT	Central Limit Theorem	113
χ^2_p	χ^2 distribution with p degrees of freedom	
$\chi^2_{1-\alpha;p}$	$1-\alpha$ quantile of the χ^2 distribution with p degrees of freedom	
t_n	t-distribution with n degrees of freedom	
$t_{1-\alpha/2;n}$	$1-\alpha/2$ quantile of the t-distribution with n d.f.	
$F_{n,m}$	F-distribution with n and m degrees of freedom	
$F_{1-\alpha;n,m}$	$1-\alpha$ quantile of the F-distribution with n and m degrees of freedom	
$T^2(p,n)$	Hotelling T^2-distribution with p and n degrees of freedom	

B Data

All data sets are available on the Springer webpage or at the authors' home pages. More detailed information on the data sets may be found there.

B.1 Boston Housing Data

The Boston housing data set was collected by Harrison and Rubinfeld (1978). It comprise 506 observations for each census district of the Boston metropolitan area. The data set was analyzed in Belsley, Kuh and Welsch (1980).

X_1: per capita crime rate,
X_2: proportion of residential land zoned for large lots,
X_3: proportion of nonretail business acres,
X_4: Charles River (1 if tract bounds river, 0 otherwise),
X_5: nitric oxides concentration,
X_6: average number of rooms per dwelling,
X_7: proportion of owner-occupied units built prior to 1940,
X_8: weighted distances to five Boston employment centers,
X_9: index of accessibility to radial highways,
X_{10}: full-value property tax rate per \$10,000,
X_{11}: pupil/teacher ratio ,
X_{12}: $1000(B-0.63)^2\, \boldsymbol{I}(B < 0.63)$ where B is the proportion of African American,
X_{13}: % lower status of the population,
X_{14}: median value of owner-occupied homes in \$1000.

B.2 Swiss Bank Notes

Six variables measured on 100 genuine and 100 counterfeit old Swiss 1000-franc bank notes. The data stem from Flury and Riedwyl (1988). The columns correspond to the following 6 variables.

X_1: Length of the bank note,
X_2: Height of the bank note, measured on the left,
X_3: Height of the bank note, measured on the right,
X_4: Distance of inner frame to the lower border,
X_5: Distance of inner frame to the upper border,
X_6: Length of the diagonal.

Observations 1–100 are the genuine bank notes and the other 100 observations are the counterfeit bank notes.

Length	Height (left)	Height (right)	Inner Frame (lower)	Inner Frame (upper)	Diagonal
214.8	131.0	131.1	9.0	9.7	141.0
214.6	129.7	129.7	8.1	9.5	141.7
214.8	129.7	129.7	8.7	9.6	142.2
214.8	129.7	129.6	7.5	10.4	142.0
215.0	129.6	129.7	10.4	7.7	141.8
215.7	130.8	130.5	9.0	10.1	141.4
215.5	129.5	129.7	7.9	9.6	141.6
214.5	129.6	129.2	7.2	10.7	141.7
214.9	129.4	129.7	8.2	11.0	141.9
215.2	130.4	130.3	9.2	10.0	140.7
215.3	130.4	130.3	7.9	11.7	141.8
215.1	129.5	129.6	7.7	10.5	142.2
215.2	130.8	129.6	7.9	10.8	141.4
214.7	129.7	129.7	7.7	10.9	141.7
215.1	129.9	129.7	7.7	10.8	141.8
214.5	129.8	129.8	9.3	8.5	141.6
214.6	129.9	130.1	8.2	9.8	141.7
215.0	129.9	129.7	9.0	9.0	141.9
215.2	129.6	129.6	7.4	11.5	141.5
214.7	130.2	129.9	8.6	10.0	141.9
215.0	129.9	129.3	8.4	10.0	141.4
215.6	130.5	130.0	8.1	10.3	141.6
215.3	130.6	130.0	8.4	10.8	141.5
215.7	130.2	130.0	8.7	10.0	141.6
215.1	129.7	129.9	7.4	10.8	141.1
215.3	130.4	130.4	8.0	11.0	142.3
215.5	130.2	130.1	8.9	9.8	142.4
215.1	130.3	130.3	9.8	9.5	141.9
215.1	130.0	130.0	7.4	10.5	141.8
214.8	129.7	129.3	8.3	9.0	142.0
215.2	130.1	129.8	7.9	10.7	141.8
214.8	129.7	129.7	8.6	9.1	142.3
215.0	130.0	129.6	7.7	10.5	140.7
215.6	130.4	130.1	8.4	10.3	141.0
215.9	130.4	130.0	8.9	10.6	141.4
214.6	130.2	130.2	9.4	9.7	141.8
215.5	130.3	130.0	8.4	9.7	141.8
215.3	129.9	129.4	7.9	10.0	142.0
215.3	130.3	130.1	8.5	9.3	142.1
213.9	130.3	129.0	8.1	9.7	141.3
214.4	129.8	129.2	8.9	9.4	142.3
214.8	130.1	129.6	8.8	9.9	140.9
214.9	129.6	129.4	9.3	9.0	141.7
214.9	130.4	129.7	9.0	9.8	140.9
214.8	129.4	129.1	8.2	10.2	141.0
214.3	129.5	129.4	8.3	10.2	141.8
214.8	129.9	129.7	8.3	10.2	141.5
214.8	129.9	129.7	7.3	10.9	142.0
214.6	129.7	129.8	7.9	10.3	141.1
214.5	129.0	129.6	7.8	9.8	142.0
214.6	129.8	129.4	7.2	10.0	141.3
215.3	130.6	130.0	9.5	9.7	141.1
214.5	130.1	130.0	7.8	10.9	140.9
215.4	130.2	130.2	7.6	10.9	141.6
214.5	129.4	129.5	7.9	10.0	141.4
215.2	129.7	129.4	9.2	9.4	142.0
215.7	130.0	129.4	9.2	10.4	141.2
215.0	129.6	129.4	8.8	9.0	141.1
215.1	130.1	129.9	7.9	11.0	141.3
215.1	130.0	129.8	8.2	10.3	141.4
215.1	129.6	129.3	8.3	9.9	141.6
215.3	129.7	129.4	7.5	10.5	141.5
215.4	129.8	129.4	8.0	10.6	141.5
214.5	130.0	129.5	8.0	10.8	141.4
215.0	130.0	129.8	8.6	10.6	141.5
215.2	130.6	130.0	8.8	10.6	140.8
214.6	129.5	129.2	7.7	10.3	141.3
214.8	129.7	129.3	9.1	9.5	141.5
215.1	129.6	129.8	8.6	9.8	141.8
214.9	130.2	130.2	8.0	11.2	139.6
213.8	129.8	129.5	8.4	11.1	140.9
215.2	129.9	129.5	8.2	10.3	141.4
215.0	129.6	130.2	8.7	10.0	141.2

214.4	129.9	129.6	7.5	10.5	141.8
215.2	129.9	129.7	7.2	10.6	142.1
214.1	129.6	129.3	7.6	10.7	141.7
214.9	129.9	130.1	8.8	10.0	141.2
214.6	129.8	129.4	7.4	10.6	141.0
215.2	130.5	129.8	7.9	10.9	140.9
214.6	129.9	129.4	7.9	10.0	141.8
215.1	129.7	129.7	8.6	10.3	140.6
214.9	129.8	129.6	7.5	10.3	141.0
215.2	129.7	129.1	9.0	9.7	141.9
215.2	130.1	129.9	7.9	10.8	141.3
215.4	130.7	130.2	9.0	11.1	141.2
215.1	129.9	129.6	8.9	10.2	141.5
215.2	129.9	129.7	8.7	9.5	141.6
215.0	129.6	129.2	8.4	10.2	142.1
214.9	130.3	129.9	7.4	11.2	141.5
215.0	129.9	129.7	8.0	10.5	142.0
214.7	129.7	129.3	8.6	9.6	141.6
215.4	130.0	129.9	8.5	9.7	141.4
214.9	129.4	129.5	8.2	9.9	141.5
214.5	129.5	129.3	7.4	10.7	141.5
214.7	129.6	129.5	8.3	10.0	142.0
215.6	129.9	129.9	9.0	9.5	141.7
215.0	130.4	130.3	9.1	10.2	141.1
214.4	129.7	129.5	8.0	10.3	141.2
215.1	130.0	129.8	9.1	10.2	141.5
214.7	130.0	129.4	7.8	10.0	141.2
214.4	130.1	130.3	9.7	11.7	139.8
214.9	130.5	130.2	11.0	11.5	139.5
214.9	130.3	130.1	8.7	11.7	140.2
215.0	130.4	130.6	9.9	10.9	140.3
214.7	130.2	130.3	11.8	10.9	139.7
215.0	130.2	130.2	10.6	10.7	139.9
215.3	130.3	130.1	9.3	12.1	140.2
214.8	130.1	130.4	9.8	11.5	139.9
215.0	130.2	129.9	10.0	11.9	139.4
215.2	130.6	130.8	10.4	11.2	140.3
215.2	130.4	130.3	8.0	11.5	139.2
215.1	130.5	130.3	10.6	11.5	140.1
215.4	130.7	131.1	9.7	11.8	140.6
214.9	130.4	129.9	11.4	11.0	139.9
215.1	130.3	130.0	10.6	10.8	139.7
215.5	130.4	130.0	8.2	11.2	139.2
214.7	130.6	130.1	11.8	10.5	139.8
214.7	130.4	130.1	12.1	10.4	139.9
214.8	130.5	130.2	11.0	11.0	140.0
214.4	130.2	129.9	10.1	12.0	139.2
214.8	130.3	130.4	10.1	12.1	139.6
215.1	130.6	130.3	12.3	10.2	139.6
215.3	130.8	131.1	11.6	10.6	140.2
215.1	130.7	130.4	10.5	11.2	139.7
214.7	130.5	130.5	9.9	10.3	140.1
214.9	130.0	130.3	10.2	11.4	139.6
215.0	130.4	130.4	9.4	11.6	140.2
215.5	130.7	130.3	10.2	11.8	140.0
215.1	130.2	130.2	10.1	11.3	140.3
214.5	130.2	130.6	9.8	12.1	139.9
214.3	130.2	130.0	10.7	10.5	139.8
214.5	130.2	129.8	12.3	11.2	139.2
214.9	130.5	130.2	10.6	11.5	139.9
214.6	130.2	130.4	10.5	11.8	139.7
214.2	130.0	130.2	11.0	11.2	139.5
214.8	130.1	130.1	11.9	11.1	139.5
214.6	129.8	130.2	10.7	11.1	139.4
214.9	130.7	130.3	9.3	11.2	138.3
214.6	130.4	130.4	11.3	10.8	139.8
214.5	130.5	130.2	11.8	10.2	139.6
214.8	130.2	130.3	10.0	11.9	139.3
214.7	130.0	129.4	10.2	11.0	139.2
214.6	130.2	130.4	11.2	10.7	139.9
215.0	130.5	130.4	10.6	11.1	139.9
214.5	129.8	129.8	11.4	10.0	139.3
214.9	130.6	130.4	11.9	10.5	139.8
215.0	130.5	130.4	11.4	10.7	139.9
215.3	130.6	130.3	9.3	11.3	138.1
214.7	130.2	130.1	10.7	11.0	139.4
214.9	129.9	130.0	9.9	12.3	139.4
214.9	130.3	129.9	11.9	10.6	139.8

214.6	129.9	129.7	11.9	10.1	139.0
214.6	129.7	129.3	10.4	11.0	139.3
214.5	130.1	130.1	12.1	10.3	139.4
214.5	130.3	130.0	11.0	11.5	139.5
215.1	130.0	130.3	11.6	10.5	139.7
214.2	129.7	129.6	10.3	11.4	139.5
214.4	130.1	130.0	11.3	10.7	139.2
214.8	130.4	130.6	12.5	10.0	139.3
214.6	130.6	130.1	8.1	12.1	137.9
215.6	130.1	129.7	7.4	12.2	138.4
214.9	130.5	130.1	9.9	10.2	138.1
214.6	130.1	130.0	11.5	10.6	139.5
214.7	130.1	130.2	11.6	10.9	139.1
214.3	130.3	130.0	11.4	10.5	139.8
215.1	130.3	130.6	10.3	12.0	139.7
216.3	130.7	130.4	10.0	10.1	138.8
215.6	130.4	130.1	9.6	11.2	138.6
214.8	129.9	129.8	9.6	12.0	139.6
214.9	130.0	129.9	11.4	10.9	139.7
213.9	130.7	130.5	8.7	11.5	137.8
214.2	130.6	130.4	12.0	10.2	139.6
214.8	130.5	130.3	11.8	10.5	139.4
214.8	129.6	130.0	10.4	11.6	139.2
214.8	130.1	130.0	11.4	10.5	139.6
214.9	130.4	130.2	11.9	10.7	139.0
214.3	130.1	130.1	11.6	10.5	139.7
214.5	130.4	130.0	9.9	12.0	139.6
214.8	130.5	130.3	10.2	12.1	139.1
214.5	130.2	130.4	8.2	11.8	137.8
215.0	130.4	130.1	11.4	10.7	139.1
214.8	130.6	130.6	8.0	11.4	138.7
215.0	130.5	130.1	11.0	11.4	139.3
214.6	130.5	130.4	10.1	11.4	139.3
214.7	130.2	130.1	10.7	11.1	139.5
214.7	130.4	130.0	11.5	10.7	139.4
214.5	130.4	130.0	8.0	12.2	138.5
214.8	130.0	129.7	11.4	10.6	139.2
214.8	129.9	130.2	9.6	11.9	139.4
214.6	130.3	130.2	12.7	9.1	139.2
215.1	130.2	129.8	10.2	12.0	139.4
215.4	130.5	130.6	8.8	11.0	138.6
214.7	130.3	130.2	10.8	11.1	139.2
215.0	130.5	130.3	9.6	11.0	138.5
214.9	130.3	130.5	11.6	10.6	139.8
215.0	130.4	130.3	9.9	12.1	139.6
215.1	130.3	129.9	10.3	11.5	139.7
214.8	130.3	130.4	10.6	11.1	140.0
214.7	130.7	130.8	11.2	11.2	139.4
214.3	129.9	129.9	10.2	11.5	139.6

B.3 Car Data

The car data set (Chambers, Cleveland, Kleiner and Tukey, 1983) consists of 13 variables measured for 74 car types. The abbreviations in Table B.3 are as follows:

X_1: P Price,
X_2: M Mileage (in miles per gallone),
X_3: R78 Repair record 1978 (rated on a 5–point scale; 5 best, 1 worst),
X_4: R77 Repair record 1977 (scale as before),
X_5: H Headroom (in inches),
X_6: R Rear seat clearance (distance from front seat back to rear seat, in inches),
X_7: Tr Trunk space (in cubic feet),

X_8: W Weight (in pound),
X_9: L Length (in inches),
X_{10}: T Turning diameter (clearance required to make a U-turn, in feet),
X_{11}: D Displacement (in cubic inches),
X_{12}: G Gear ratio for high gear,
X_{13}: C Company headquarter (1 for U.S., 2 for Japan, 3 for Europe).

Model	P	M	R78	R77	H	R	Tr	W	L	T	D	G	C
AMC–Concord	4099	22	3	2	2.5	27.5	11	2930	186	40	121	3.58	1
AMC–Pacer	4749	17	3	1	3.0	25.5	11	3350	173	40	258	2.53	1
AMC–Spirit	3799	22	–	–	3.0	18.5	12	2640	168	35	121	3.08	1
Audi–5000	9690	17	5	2	3.0	27.0	15	2830	189	37	131	3.20	1
Audi–Fox	6295	23	3	3	2.5	28.0	11	2070	174	36	97	3.70	3
BMW–320i	9735	25	4	4	2.5	26.0	12	2650	177	34	121	3.64	3
Buick–Century	4816	20	3	3	4.5	29.0	16	3250	196	40	196	2.93	1
Buick–Electra	7827	15	4	4	4.0	31.5	20	4080	222	43	350	2.41	1
Buick–Le–Sabre	5788	18	3	4	4.0	30.5	21	3670	218	43	231	2.73	1
Buick–Opel	4453	26	–	–	3.0	24.0	10	2230	170	34	304	2.87	1
Buick–Regal	5189	20	3	3	2.0	28.5	16	3280	200	42	196	2.93	1
Buick–Riviera	10372	16	3	4	3.5	30.0	17	3880	207	43	231	2.93	1
Buick–Skylark	4082	19	3	3	3.5	27.0	13	3400	200	42	231	3.08	1
Cad.–Deville	11385	14	3	3	4.0	31.5	20	4330	221	44	425	2.28	1
Cad.–Eldorado	14500	14	2	2	3.5	30.0	16	3900	204	43	350	2.19	1
Cad.–Seville	15906	21	3	3	3.0	30.0	13	4290	204	45	350	2.24	1
Chev.–Chevette	3299	29	3	3	2.5	26.0	9	2110	163	34	231	2.93	1
Chev.–Impala	5705	16	4	4	4.0	29.5	20	3690	212	43	250	2.56	1
Chev.–Malibu	4504	22	3	3	3.5	28.5	17	3180	193	41	200	2.73	1
Chev.–Monte–Carlo	5104	22	2	3	2.0	28.5	16	3220	200	41	200	2.73	1
Chev.–Monza	3667	24	2	2	2.0	25.0	7	2750	179	40	151	2.73	1
Chev.–Nova	3955	19	3	3	3.5	27.0	13	3430	197	43	250	2.56	1
Datsun–200-SX	6229	23	4	3	1.5	21.0	6	2370	170	35	119	3.89	2
Datsun–210	4589	35	5	5	2.0	23.5	8	2020	165	32	85	3.70	2
Datsun–510	5079	24	4	4	2.5	22.0	8	2280	170	34	119	3.54	2
Datsun–810	8129	21	4	4	2.5	27.0	8	2750	184	38	146	3.55	2
Dodge–Colt	3984	30	5	4	2.0	24.0	8	2120	163	35	98	3.54	2
Dodge–Diplomat	5010	18	2	2	4.0	29.0	17	3600	206	46	318	2.47	1
Dodge–Magnum–XE	5886	16	2	2	3.5	26.0	16	3870	216	48	318	2.71	1
Dodge–St.–Regis	6342	17	2	2	4.5	28.0	21	3740	220	46	225	2.94	1
Fiat–Strada	4296	21	3	1	2.5	26.5	16	2130	161	36	105	3.37	3
Ford–Fiesta	4389	28	4	–	1.5	26.0	9	1800	147	33	98	3.15	1
Ford–Mustang	4187	21	3	3	2.0	23.0	10	2650	179	42	140	3.08	1
Honda–Accord	5799	25	5	5	3.0	25.5	10	2240	172	36	107	3.05	2
Honda–Civic	4499	28	4	4	2.5	23.5	5	1760	149	34	91	3.30	2
Linc.–Continental	11497	12	3	4	3.5	30.5	22	4840	233	51	400	2.47	1
Linc.–Cont–Mark–V	13594	12	3	4	2.5	28.5	18	4720	230	48	400	2.47	1
Linc.–Versailles	13466	14	3	3	3.5	27.0	15	3830	201	41	302	2.47	1
Mazda–GLC	3995	30	4	4	3.5	25.5	11	1980	154	33	86	3.73	1
Merc.–Bobcat	3829	22	4	3	3.0	25.5	9	2580	169	39	140	2.73	1
Merc.–Cougar	5379	14	4	3	3.5	29.5	16	4060	221	48	302	2.75	1
Merc.–Cougar–XR-7	6303	14	4	4	3.0	25.0	16	4130	217	45	302	2.75	1
Merc.–Marquis	6165	15	3	2	3.5	30.5	23	3720	212	44	302	2.26	1
Merc.–Monarch	4516	18	3	–	3.0	27.0	15	3370	198	41	250	2.43	1
Merc.–Zephyr	3291	20	3	3	3.5	29.0	17	2830	195	43	140	3.08	1
Olds.–98	8814	21	4	4	4.0	31.5	20	4060	220	43	350	2.41	1
Olds.–Cutlass	4733	19	3	3	4.5	28.0	16	3300	198	42	231	2.93	1
Olds.–Cutl–Supr	5172	19	3	4	2.0	28.0	16	3310	198	42	231	2.93	1
Olds.–Delta–88	5890	18	4	4	4.0	29.0	20	3690	218	42	231	2.73	1
Olds.–Omega	4181	19	3	3	4.5	27.0	14	3370	200	43	231	3.08	1
Olds.–Starfire	4195	24	1	1	2.0	25.5	10	2720	180	40	151	2.73	1
Olds.–Tornado	10371	16	3	3	3.5	30.0	17	4030	206	43	350	2.41	1
Peugeot–604–SL	12990	14	–	–	3.5	30.5	14	3420	192	38	163	3.58	3
Plym.–Arrow	4647	28	3	3	2.0	21.5	11	2360	170	37	156	3.05	1
Plym.–Champ	4425	34	5	4	2.5	23.0	11	1800	157	37	86	2.97	1
Plym.–Horizon	4482	25	3	–	4.0	25.0	17	2200	165	36	105	3.37	1
Plym.–Sapporo	6486	26	–	–	1.5	22.0	8	2520	182	38	119	3.54	1
Plym.–Volare	4060	18	2	2	5.0	31.0	16	3330	201	44	225	3.23	1
Pont.–Catalina	5798	18	4	4	4.0	29.0	20	3700	214	42	231	2.73	1
Pont.–Firebird	4934	18	1	2	1.5	23.5	7	3470	198	42	231	3.08	1
Pont.–Grand–Prix	5222	19	3	3	2.0	28.5	16	3210	201	45	231	2.93	1
Pont.–Le–Mans	4723	19	3	3	3.5	28.0	17	3200	199	40	231	2.93	1
Pont.–Phoenix	4424	19	–	–	3.5	27.0	13	3420	203	43	231	3.08	1
Pont.–Sunbird	4172	24	2	2	2.0	25.0	7	2690	179	41	151	2.73	1

Renault–Le–Car	3895	26	3	3	3.0	23.0	10	1830	142	34	79	3.72	3
Subaru	3798	35	5	4	2.5	25.5	11	2050	164	36	97	3.81	2
Toyota–Cecila	5899	18	5	5	2.5	22.0	14	2410	174	36	134	3.06	2
Toyota–Corolla	3748	31	5	5	3.0	24.5	9	2200	165	35	97	3.21	2
Toyota–Corona	5719	18	5	5	2.0	23.0	11	2670	175	36	134	3.05	2
VW–Rabbit	4697	25	4	3	3.0	25.5	15	1930	155	35	89	3.78	3
VW–Rabbit–Diesel	5397	41	5	4	3.0	25.5	15	2040	155	35	90	3.78	3
VW–Scirocco	6850	25	4	3	2.0	23.5	16	1990	156	36	97	3.78	3
VW–Dasher	7140	23	4	3	2.5	37.5	12	2160	172	36	97	3.74	3
Volvo–260	11995	17	5	3	2.5	29.5	14	3170	193	37	163	2.98	3

B.4 Classic Blue Pullovers Data

This is a data set consisting of 10 measurements of 4 variables. The story: A textile shop manager is studying the sales of "classic blue" pullovers over 10 periods. He uses three different marketing methods and hopes to understand his sales as a fit of these variables using statistics. The variables measured are

X_1: Numbers of sold pullovers,
X_2: Price (in EUR),
X_3: Advertisement costs in local newspapers (in EUR),
X_4: Presence of a sales assistant (in hours per period).

	Sales	Price	Advert.	Ass. Hours
1	230	125	200	109
2	181	99	55	107
3	165	97	105	98
4	150	115	85	71
5	97	120	0	82
6	192	100	150	103
7	181	80	85	111
8	189	90	120	93
9	172	95	110	86
10	170	125	130	78

B.5 U.S. Companies Data

The data set consists of measurements for 79 U.S. companies. The abbreviations in Table B.5 are as follows:

X_1: A Assets (USD),
X_2: S Sales (USD),
X_3: MV Market Value (USD),
X_4: P Profits (USD),
X_5: CF Cash Flow (USD),
X_6: E Employees.

Company	A	S	MV	P	CF	E	Sector
Bell Atlantic	19788	9084	10636	1092.9	2576.8	79.4	Communication
Continental Telecom	5074	2557	1892	239.9	578.3	21.9	Communication
American Electric Power	13621	4848	4572	485.0	898.9	23.4	Energy
Brooklyn Union Gas	1117	1038	478	59.7	91.7	3.8	Energy
Centra lIllinois Publ. Serv.	1633	701	679	74.3	135.9	2.8	Energy
Cleveland Electric Illum.	5651	1254	2002	310.7	407.9	6.2	Energy
Columbia Gas System	5835	4053	1601	-93.8	173.8	10.8	Energy
Florida Progress	3494	1653	1442	160.9	320.3	6.4	Energy
Idaho Power	1654	451	779	84.8	130.4	1.6	Energy
Kansas Power & Light	1679	1354	687	93.8	154.6	4.6	Energy
Mesa Petroleum	1257	355	181	167.5	304.0	0.6	Energy
Montana Power	1743	597	717	121.6	172.4	3.5	Energy
Peoples Energy	1440	1617	639	81.7	126.4	3.5	Energy
Phillips Petroleum	14045	15636	2754	418.0	1462.0	27.3	Energy
Publ. Serv. Coof New Mexico	3010	749	1120	146.3	209.2	3.4	Energy
San Diego Gas & Electric	3086	1739	1507	202.7	335.2	4.9	Energy
Valero Energy	1995	2662	341	34.7	100.7	2.3	Energy
American Savings Bank FSB	3614	367	90	14.1	24.6	1.1	Finance
Bank South	2788	271	304	23.5	28.9	2.1	Finance
H & R Block	327	542	959	54.1	72.5	2.8	Finance
California First Bank	5401	550	376	25.6	37.5	4.1	Finance
Cigna	44736	16197	4653	-732.5	-651.9	48.5	Finance
Dreyfus	401	176	1084	55.6	57.0	0.7	Finance
First American	4789	453	367	40.2	51.4	3.0	Finance
First Empire State	2548	264	181	22.2	26.2	2.1	Finance
First Tennessee National	5249	527	346	37.8	56.2	4.1	Finance
Marine Corp	3720	356	211	26.6	34.8	2.4	Finance
Mellon Bank	33406	3222	1413	201.7	246.7	15.8	Finance
National City	12505	1302	702	108.4	131.4	9.0	Finance
Norstar Bancorp	8998	882	988	93.0	119.0	7.4	Finance
Norwest	21419	2516	930	107.6	164.7	15.6	Finance
Southeast Banking	11052	1097	606	64.9	97.6	7.0	Finance
Sovran Financial	9672	1037	829	92.6	118.2	8.2	Finance
United Financial Group	4989	518	53	-3.1	-0.3	0.8	Finance
Apple Computer	1022	1754	1370	72.0	119.5	4.8	HiTech
Digital Equipment	6914	7029	7957	400.6	754.7	87.3	HiTech
Eg & G	430	1155	1045	55.7	70.8	22.5	HiTech
General Electric	26432	28285	33172	2336.0	3562.0	304.0	HiTech
Hewlett-Packard	5769	6571	9462	482.0	792.0	83.0	HiTech
IBM	52634	50056	95697	6555.0	9874.0	400.2	HiTech
NCR	3940	4317	3940	315.2	566.3	62.0	HiTech
Telex	478	672	866	67.1	101.6	5.4	HiTech
Armstrong World Industries	1093	1679	1070	100.9	164.5	20.8	Manufacturing
CBI Industries	1128	1516	430	-47.0	26.7	13.2	Manufacturing
Fruehauf	1804	2564	483	70.5	164.9	26.6	Manufacturing
Halliburton	4662	4781	2988	28.7	371.5	66.2	Manufacturing
LTV	6307	8199	598	-771.5	-524.3	57.5	Manufacturing
Owens-Corning Fiberglas	2366	3305	1117	131.2	256.5	25.2	Manufacturing
PPG Industries	4084	4346	3023	302.7	521.7	37.5	Manufacturing
Textron	10348	5721	1915	223.6	322.5	49.5	Manufacturing
Turner	752	2149	101	11.1	15.2	2.6	Manufacturing
United Technologies	10528	14992	5377	312.7	710.7	184.8	Manufacturing
Commun. Psychiatric Centers	278	205	853	44.8	50.5	3.8	Medical
Hospital Corp of America	6259	4152	3090	283.7	524.5	62.0	Medical
AH Robins	707	706	275	61.4	77.8	6.1	Medical
Shared Medical Systems	252	312	883	41.7	60.6	3.3	Medical
Air Products	2687	1870	1890	145.7	352.2	18.2	Other
Allied Signal	13271	9115	8190	-279.0	83.0	143.8	Other
Bally Manufactoring	1529	1295	444	25.6	137.0	19.4	Other
Crown Cork & Seal	866	1487	944	71.7	115.4	12.6	Other
Ex-Cell-0	799	1140	633	57.6	89.2	15.4	Other
Liz Claiborne	223	557	1040	60.6	63.7	1.9	Other
Warner Communications	2286	2235	2306	195.3	219.0	8.0	Other
Dayton-Hudson	4418	8793	4459	283.6	456.5	128.0	Retail
Dillard Department Stores	862	160l	1093	66.9	106.8	16.0	Retail
Giant Food	623	2247	797	57.0	93.8	18.6	Retail
Great A & P Tea	1608	6615	829	56.1	134.0	65.0	Retail
Kroger	4178	17124	2091	180.8	390.4	164.6	Retail
May Department Stores	3442	5080	2673	235.4	361.5	77.3	Retail
Stop & Shop Cos	1112	3689	542	30.3	96.9	43.5	Retail
Supermarkets General	1104	5123	910	63.7	133.3	48.5	Retail
Wickes Cos	2957	2806	457	40.6	93.5	50.0	Retail
FW Woolworth	2535	5958	1921	177.0	288.0	118.1	Retail
AMR	6425	6131	2448	345.8	682.5	49.5	Transportation
IU International	999	1878	393	-173.5	-108.1	23.3	Transportation
PanAm	2448	3484	1036	48.8	257.1	25.4	Transportation
Republic Airlines	1286	1734	361	69.2	145.7	14.3	Transportation
TWA	2769	3725	663	-208.4	12.4	29.1	Transportation
Western AirLines	952	1307	309	35.4	92.8	10.3	Transportation

B.6 French Food Data

The data set consists of the average expenditures on food for several different types of families in France (manual workers = MA, employees = EM, managers = CA) with different numbers of children (2,3,4 or 5 children). The data is taken from Lebart, Morineau and Fénelon (1982).

		bread	vegetables	fruits	meat	poultry	milk	wine
1	MA2	332	428	354	1437	526	247	427
2	EM2	293	559	388	1527	567	239	258
3	CA2	372	767	562	1948	927	235	433
4	MA3	406	563	341	1507	544	324	407
5	EM3	386	608	396	1501	558	319	363
6	CA3	438	843	689	2345	1148	243	341
7	MA4	534	660	367	1620	638	414	407
8	EM4	460	699	484	1856	762	400	416
9	CA4	385	789	621	2366	1149	304	282
10	MA5	655	776	423	1848	759	495	486
11	EM5	584	995	548	2056	893	518	319
12	CA5	515	1097	887	2630	1167	561	284
	$\overline{x}$	446,7	737,8	505,0	1886,7	803,2	358,2	368,6
	$s_{X_iX_i}$	102,6	172,2	158,1	378,9	238,9	112,1	68,7

B.7 Car Marks

The data are averaged marks for 24 car types from a sample of 40 persons. The marks range from 1 (very good) to 6 (very bad) like German school marks. The variables are:

X_1: A Economy,
X_2: B Service,
X_3: C Non-depreciation of value,
X_4: D Price, Mark 1 for very cheap cars
X_5: E Design,
X_6: F Sporty car,
X_7: G Safety,
X_8: H Easy handling.

Type	Model	Economy	Service	Value	Price	Design	Sport	Safety	Easy
Audi	100	3.9	2.8	2.2	4.2	3.0	3.1	2.4	2.8
BMW	5 series	4.8	1.6	1.9	5.0	2.0	2.5	1.6	2.8
Citroen	AX	3.0	3.8	3.8	2.7	4.0	4.4	4.0	2.6
Ferrari		5.3	2.9	2.2	5.9	1.7	1.1	3.3	4.3
Fiat	Uno	2.1	3.9	4.0	2.6	4.5	4.4	4.4	2.2
Ford	Fiesta	2.3	3.1	3.4	2.6	3.2	3.3	3.6	2.8
Hyundai		2.5	3.4	3.2	2.2	3.3	3.3	3.3	2.4
Jaguar		4.6	2.4	1.6	5.5	1.3	1.6	2.8	3.6
Lada	Samara	3.2	3.9	4.3	2.0	4.3	4.5	4.7	2.9
Mazda	323	2.6	3.3	3.7	2.8	3.7	3.0	3.7	3.1
Mercedes	200	4.1	1.7	1.8	4.6	2.4	3.2	1.4	2.4
Mitsubishi	Galant	3.2	2.9	3.2	3.5	3.1	3.1	2.9	2.6
Nissan	Sunny	2.6	3.3	3.9	2.1	3.5	3.9	3.8	2.4
Opel	Corsa	2.2	2.4	3.0	2.6	3.2	4.0	2.9	2.4
Opel	Vectra	3.1	2.6	2.3	3.6	2.8	2.9	2.4	2.4
Peugeot	306	2.9	3.5	3.6	2.8	3.2	3.8	3.2	2.6
Renault	19	2.7	3.3	3.4	3.0	3.1	3.4	3.0	2.7

Rover		3.9	2.8	2.6	4.0	2.6	3.0	3.2	3.0
Toyota	Corolla	2.5	2.9	3.4	3.0	3.2	3.1	3.2	2.8
Trabant	601	3.6	4.7	5.5	1.5	4.1	5.8	5.9	3.1
VW	Golf	3.8	2.3	1.9	4.2	3.1	3.6	1.6	2.4
VW	Passat	3.1	2.2	2.1	3.2	3.5	3.5	2.8	1.8
Wartburg	1.3	3.7	4.7	5.5	1.7	4.8	5.2	5.5	4.0

B.8 French Baccalauréat Frequencies

The data consist of observations of 202100 baccalauréats from France in 1976 and give the frequencies for different sets of modalities classified into regions. For a reference see Bourouche and Saporta (1980). The variables (modalities) are:

X_1: A Philosophy-Letters,
X_2: B Economics and Social Sciences,
X_3: C Mathematics and Physics,
X_4: D Mathematics and Natural Sciences,
X_5: E Mathematics and Techniques,
X_6: F Industrial Techniques,
X_7: G Economic Techniques,
X_8: H Computer Techniques.

Abbrev.	Region	A	B	C	D	E	F	G	H	total
ILDF	Ile-de-France	9724	5650	8679	9432	839	3353	5355	83	43115
CHAM	Champagne-Ardennes	924	464	567	984	132	423	736	12	4242
PICA	Picardie	1081	490	830	1222	118	410	743	13	4907
HNOR	Haute-Normandie	1135	587	686	904	83	629	813	13	4850
CENT	Centre	1482	667	1020	1535	173	629	989	26	6521
BNOR	Basse-Normandie	1033	509	553	1063	100	433	742	13	4446
BOUR	Bourgogne	1272	527	861	1116	219	769	1232	13	6009
NOPC	Nord - Pas-de-Calais	2549	1141	2164	2752	587	1660	1951	41	12845
LORR	Lorraine	1828	681	1364	1741	302	1289	1683	15	8903
ALSA	Alsace	1076	443	880	1121	145	917	1091	15	5688
FRAC	Franche-Comté	827	333	481	892	137	451	618	18	3757
PAYL	Pays de la Loire	2213	809	1439	2623	269	990	1783	14	10140
BRET	Bretagne	2158	1271	1633	2352	350	950	1509	22	10245
PCHA	Poitou-Charentes	1358	503	639	1377	164	495	959	10	5505
AQUI	Aquitaine	2757	873	1466	2296	215	789	1459	17	9872
MIDI	Midi-Pyrénées	2493	1120	1494	2329	254	855	1565	28	10138
LIMO	Limousin	551	297	386	663	67	334	378	12	2688
RHOA	Rhônes-Alpes	3951	2127	3218	4743	545	2072	3018	36	19710
AUVE	Auvergne	1066	579	724	1239	126	476	649	12	4871
LARO	Languedoc-Roussillon	1844	816	1154	1839	156	469	993	16	7287
PROV	Provence-Alpes-Côte d'Azur	3944	1645	2415	3616	343	1236	2404	22	15625
CORS	Corse	327	31	85	178	9	27	79	0	736
	total	45593	21563	32738	46017	5333	19656	30749	451	202100

B.9 Journaux Data

This is a data set that was created from a survey completed in the 1980's in Belgium questioning people's reading habits. They were asked where they live (10 regions comprised of 7 provinces and 3 regions around Brussels) and what kind of newspaper they read on a regular basis. The 15 possible answers belong to 3 classes: Flemish newspapers (first letter v), French newspapers (first letter f) and both languages (first letter b).

X_1: WaBr Walloon Brabant
X_2: Brar Brussels area
X_3: Antw Antwerp
X_4: FlBr Flemish Brabant
X_5: OcFl Occidental Flanders
X_6: OrFl Oriental Flanders
X_7: Hain Hainaut
X_8: Lièg Liège
X_9: Limb Limburg
X_{10}: Luxe Luxembourg

	WaBr	Brar	Antw	FlBr	OcFl	OrFl	Hain	Lièg	Limb	Luxe
v_a	1.8	7.8	9.1	3.0	4.3	3.9	0.1	0.3	3.3	0.0
v_b	0.1	3.4	17.8	1.0	0.7	4.1	0.0	0.0	0.2	0.0
v_c	0.1	9.4	4.6	7.7	4.4	5.8	1.6	0.1	1.4	0.0
v_d	0.5	15.6	6.1	12.0	10.5	10.2	0.7	0.3	5.4	0.0
v_e	0.1	5.2	3.3	4.8	1.6	1.4	0.1	0.0	3.5	0.0
f_f	5.6	13.7	3.1	2.4	0.5	1.7	1.9	2.3	0.2	0.2
f_g	4.1	16.5	1.9	1.0	1.0	0.9	2.4	3.2	0.1	0.3
f_h	8.3	29.5	1.8	7.3	0.8	0.4	5.1	3.2	0.2	0.3
f_i	0.9	7.8	0.2	2.6	0.1	0.1	5.6	3.8	0.1	0.8
b_j	6.1	18.2	10.8	4.1	4.5	5.3	2.0	2.6	3.4	0.2
b_k	8.3	35.4	6.2	11.0	5.0	6.1	5.5	3.3	1.5	0.3
b_l	4.4	9.9	6.7	3.4	1.1	3.9	2.1	1.5	2.1	0.0
v_m	0.3	11.6	14.2	4.7	5.1	7.9	0.3	0.5	3.0	0.0
f_n	5.1	21.0	1.3	3.4	0.2	0.2	2.3	4.4	0.0	0.4
f_o	2.2	9.8	0.1	0.3	0.0	0.7	2.3	3.0	0.3	1.0

B.10 U.S. Crime Data

This is a data set consisting of 50 measurements of 7 variables. It states for one year (1985) the reported number of crimes in the 50 states of the U.S. classified according to 7 categories (X_3–X_9).

X_1: land area (land)
X_2: population 1985 (popu 1985)
X_3: murder (murd)
X_4: rape
X_5: robbery (robb)
X_6: assault (assa)
X_7: burglary (burg)
X_8: larcery (larc)
X_9: autothieft (auto)
X_{10}: US states region number (reg)
X_{11}: US states division number (div)

division numbers		*region numbers*	
New England	1	Northeast	1
Mid Atlantic	2	Midwest	2
E N Central	3	South	3
W N Central	4	West	4
S Atlantic	5		
E S Central	6		
W S Central	7		
Mountain	8		
Pacific	9		

abb. of state	land area	popu 1985	murd	rape	robb	assa	burg	larc	auto	reg	div
ME	33265	1164	1.5	7.0	12.6	62	562	1055	146	1	1
NH	9279	998	2.0	6.0	12.1	36	566	929	172	1	1
VT	9614	535	1.3	10.3	7.6	55	731	969	124	1	1
MA	8284	5822	3.5	12.0	99.5	88	1134	1531	878	1	1
RI	1212	968	3.2	3.6	78.3	120	1019	2186	859	1	1
CT	5018	3174	3.5	9.1	70.4	87	1084	1751	484	1	1
NY	49108	17783	7.9	15.5	443.3	209	1414	2025	682	1	2
NJ	7787	7562	5.7	12.9	169.4	90	1041	1689	557	1	2
PA	45308	11853	5.3	11.3	106.0	90	594	11	340	1	2
OH	41330	10744	6.6	16.0	145.9	116	854	1944	493	2	3
IN	36185	5499	4.8	17.9	107.5	95	860	1791	429	2	3
IL	56345	11535	9.6	20.4	251.1	187	765	2028	518	2	3
MI	58527	9088	9.4	27.1	346.6	193	1571	2897	464	2	3
WI	56153	4775	2.0	6.7	33.1	44	539	1860	218	2	3
MN	84402	4193	2.0	9.7	89.1	51	802	1902	346	2	4
IA	56275	2884	1.9	6.2	28.6	48	507	1743	175	2	4
MO	69697	5029	10.7	27.4	2.8	167	1187	2074	538	2	4
ND	70703	685	0.5	6.2	6.5	21	286	1295	91	2	4
SD	77116	708	3.8	11.1	17.1	60	471	1396	94	2	4
NE	77355	1606	3.0	9.3	57.3	115	505	1572	292	2	4
KS	82277	2450	4.8	14.5	75.1	108	882	2302	257	2	4
DE	2044	622	7.7	18.6	105.5	196	1056	2320	559	3	5
MD	10460	4392	9.2	23.9	338.6	253	1051	2417	548	3	5
VA	40767	5706	8.4	15.4	92.0	143	806	1980	297	3	5
WV	24231	1936	6.2	6.7	27.3	84	389	774	92	3	5
NC	52669	6255	11.8	12.9	53.0	293	766	1338	169	3	5
SC	31113	3347	14.6	18.1	60.1	193	1025	1509	256	3	5
GA	58910	5976	15.3	10.1	95.8	177	9.0	1869	309	3	5
FL	58664	11366	12.7	22.2	186.1	277	1562	2861	397	3	5
KY	40409	3726	11.1	13.7	72.8	123	704	1212	346	3	6
TN	42144	4762	8.8	15.5	82.0	169	807	1025	289	3	6
AL	51705	4021	11.7	18.5	50.3	215	763	1125	223	3	6
MS	47689	2613	11.5	8.9	19.0	140	351	694	78	3	6
AR	53187	2359	10.1	17.1	45.6	150	885	1211	109	3	7
LA	47751	4481	11.7	23.1	140.8	238	890	1628	385	3	7
OK	69956	3301	5.9	15.6	54.9	127	841	1661	280	3	7
TX	266807	16370	11.6	21.0	134.1	195	1151	2183	394	3	7
MT	147046	826	3.2	10.5	22.3	75	594	1956	222	4	8
ID	83564	15	4.6	12.3	20.5	86	674	2214	144	4	8
WY	97809	509	5.7	12.3	22.0	73	646	2049	165	4	8
CO	104091	3231	6.2	36.0	129.1	185	1381	2992	588	4	8
NM	121593	1450	9.4	21.7	66.1	196	1142	2408	392	4	8
AZ	1140	3187	9.5	27.0	120.2	214	1493	3550	501	4	8
UT	84899	1645	3.4	10.9	53.1	70	915	2833	316	4	8
NV	110561	936	8.8	19.6	188.4	182	1661	3044	661	4	8
WA	68138	4409	3.5	18.0	93.5	106	1441	2853	362	4	9
OR	97073	2687	4.6	18.0	102.5	132	1273	2825	333	4	9
CA	158706	26365	6.9	35.1	206.9	226	1753	3422	689	4	9
AK	5914	521	12.2	26.1	71.8	168	790	2183	551	4	9
HI	6471	1054	3.6	11.8	63.3	43	1456	3106	581	4	9

B.11 Plasma Data

In Olkin and Veath (1980), the evolution of citrate concentration in the plasma is observed at 3 different times of day, X_1 (8 am), X_2 (11 am) and X_3 (3 pm), for two groups of patients. Each group follows a different diet.

X_1: 8 am
X_2: 11 am
X_3: 3 pm

Group	(8 am)	(11 am)	(3 pm)
	125	137	121
	144	173	147
I	105	119	125
	151	149	128
	137	139	109
	93	121	107
	116	135	106
II	109	83	100
	89	95	83
	116	128	100

B.12 WAIS Data

Morrison (1990) compares the results of 4 subtests of the Wechsler Adult Intelligence Scale (WAIS) for 2 categories of people: in group 1 are $n_1 = 37$ people who do not present a senile factor, group 2 are those ($n_2 = 12$) presenting a senile factor.

WAIS subtests:
X_1: information
X_2: similarities
X_3: arithmetic
X_4: picture completion

Group II				
subject	information	similarities	arithmetic	picture completion
1	9	5	10	8
2	10	0	6	2
3	8	9	11	1
4	13	7	14	9
5	4	0	4	0
6	4	0	6	0
7	11	9	9	8
8	5	3	3	6
9	9	7	8	6
10	7	2	6	4
11	12	10	14	3
12	13	12	11	10
Mean	8.75	5.33	8.50	4.75

Group I				
subject	information	similarities	arithmetic	picture completion
1	7	5	9	8
2	8	8	5	6
3	16	18	11	9
4	8	3	7	9
5	6	3	13	9
6	11	8	10	10
7	12	7	9	8
8	8	11	9	3
9	14	12	11	4
10	13	13	13	6
11	13	9	9	9

12	13	10	15	7
13	14	11	12	8
14	15	11	11	10
15	13	10	15	9
16	10	5	8	6
17	10	3	7	7
18	17	13	13	7
19	10	6	10	7
20	10	10	15	8
21	14	7	11	5
22	16	11	12	11
23	10	7	14	6
24	10	10	9	6
25	10	7	10	10
26	7	6	5	9
27	15	12	10	6
28	17	15	15	8
29	16	13	16	9
30	13	10	17	8
31	13	10	17	10
32	19	12	16	10
33	19	15	17	11
34	13	10	7	8
35	15	11	12	8
36	16	9	11	11
37	14	13	14	9
Mean	12.57	9.57	11.49	7.97

B.13 ANOVA Data

The yields of wheat have been measured in 30 parcels which have been randomly attributed to 3 lots prepared by one of 3 different fertilizers A, B, and C.

X_1: fertilizer A
X_2: fertilizer B
X_3: fertilizer C

Fertilizer Yield	A	B	C
1	4	6	2
2	3	7	1
3	2	7	1
4	5	5	1
5	4	5	3
6	4	5	4
7	3	8	3
8	3	9	3
9	3	9	2
10	1	6	2

B.14 Timebudget Data

In Volle (1985), we can find data on 28 individuals identified according to sex, country where they live, professional activity and matrimonial status, which indicates the amount of time each person spent on ten categories of activities over 100 days (100·24h = 2400 hours total in each row) in the year 1976.

X_1: prof : professional activity
X_2: tran : transportation linked to professional activity
X_3: hous : household occupation
X_4: kids : occupation linked to children
X_5: shop : shopping
X_6: pers : time spent for personal care
X_7: eat : eating
X_8: slee : sleeping
X_9: tele : watching television
X_{10}: leis : other leisures

maus: active men in the U.S.
waus: active women in the U.S.
wnus: nonactive women in the U.S.
mmus: married men in U.S.
wmus: married women in U.S.
msus: single men in U.S.
wsus: single women in U.S.
mawe: active men from Western countries
wawe: active women from Western countries
wnwe: nonactive women from Western countries
mmwe: married men from Western countries
wmwe: married women from Western countries
mswe: single men from Western countries
wswe: single women from Western countries
mayo: active men from yugoslavia
wayo: active women from yugoslavia
wnyo: nonactive women from yugoslavia
mmyo: married men from yugoslavia
wmyo: married women from yugoslavia
msyo: single men from yugoslavia
wsyo: single women from yugoslavia
maes: active men from eastern countries
waes: active women from eastern countries
wnes: nonactive women from eastern countries
mmes: married men from eastern countries
wmes: married women from eastern countries
mses: single men from eastern countries
wses: single women from eastern countries

	prof	tran	hous	kids	shop	pers	eat	slee	tele	leis
maus	610	140	60	10	120	95	115	760	175	315
waus	475	90	250	30	140	120	100	775	115	305
wnus	10	0	495	110	170	110	130	785	160	430
mmus	615	140	65	10	115	90	115	765	180	305
wmus	179	29	421	87	161	112	119	776	143	373
msus	585	115	50	0	150	105	100	760	150	385
wsus	482	94	196	18	141	130	96	775	132	336
mawe	653	100	95	7	57	85	150	808	115	330
wawe	511	70	307	30	80	95	142	816	87	262

wnwe	20	7	568	87	112	90	180	843	125	368
mmwe	656	97	97	10	52	85	152	808	122	321
wmwe	168	22	528	69	102	83	174	824	119	311
mswe	643	105	72	0	62	77	140	813	100	388
wswe	429	34	262	14	92	97	147	849	84	392
mayo	650	140	120	15	85	90	105	760	70	365
wayo	560	105	375	45	90	90	95	745	60	235
wnyo	10	10	710	55	145	85	130	815	60	380
mmyo	650	145	112	15	85	90	105	760	80	358
wmyo	260	52	576	59	116	85	117	775	65	295
msyo	615	125	95	0	115	90	85	760	40	475
wsyo	433	89	318	23	112	96	102	774	45	408
maea	650	142	122	22	76	94	100	764	96	334
waea	578	106	338	42	106	94	92	752	64	228
wnea	24	8	594	72	158	92	128	840	86	398
mmea	652	133	134	22	68	94	102	763	122	310
wmea	436	79	433	60	119	90	107	772	73	231
msea	627	148	68	0	88	92	86	770	58	463
wsea	434	86	297	21	129	102	94	799	58	380

B.15 Geopol Data

This data set contains a comparison of 41 countries according to 10 different political and economic parameters.

X_1: popu population
X_2: giph Gross Internal Product per habitant
X_3: ripo rate of increase of the population
X_4: rupo rate of urban population
X_5: rlpo rate of illiteracy in the population
X_6: rspo rate of students in the population
X_7: eltp expected lifetime of people
X_8: rnnr rate of nutritional needs realized
X_9: nunh number of newspapers and magazines per 1000 habitants
X_{10}: nuth number of television per 1000 habitants

AFS South Africa
ALG Algeria
BRD Germany
GBR Great Britain
ARS Saudi Arabia
ARG Argentine
AUS Australia
AUT Austria
BEL Belgium
CAM Cameroon
CAN Canada
CHL Chile
CHN China
CUB Cuba
DAN Denmark
EGY Egypt
ESP Spain
FRA France
GAB Gabun
GRE Greece
HOK Hong Kong
HON Hungary
IND India
IDO Indonesia
ISR Israel
ITA Italia
JAP Japan
KEN Kenia
MAR Marocco
MEX Mexico
NOR Norway
PER Peru
POL Poland
POR Portugal
SUE Sweden
SUI Switzerland
THA Tailand
URS USSR
USA USA
VEN Venezuela
YOU Yugoslavia

	popu	giph	ripo	rupo	rlpo	rspo	eltp	rnnr	nunh	nuth
AFS	37	2492	2	58.9	44	1.08	60	120	48	98
ALG	24.6	1960	3	44.7	50.4	0.73	64	112	21	71
BRD	62	19610	0.4	86.4	2	2.72	72	145	585	759
GBR	57.02	14575	0.04	92.5	2.2	1.9	75	128	421	435
ARS	14.4	5980	2.7	77.3	48.9	0.91	63	125	34	269
ARG	32.4	2130	1.6	86.2	6.1	2.96	71	136	82	217
AUS	16.81	16830	1.4	85.5	5	2.5	76	125	252	484
AUT	7.61	16693	0	57.7	1.5	2.52	74	130	362	487
BEL	9.93	15243	0.2	96.9	3	2.56	74	150	219	320
CAM	11	1120	2.7	49.4	58.8	0.17	53	88	6	12
CAN	26.25	20780	0.9	76.4	1	6.89	77	129	321	586
CHL	12.95	1794	1.6	85.6	8.9	1.73	71	106	67	183
CHN	1119	426	1.1	21.4	34.5	0.16	69	111	36	24
CUB	10.5	1050	0.8	74.9	3.8	2.38	75	135	129	203
DAN	5.13	20570	0.4	86.4	1.5	2.38	75	131	359	526
EGY	52.52	665	2.5	48.8	61.8	1.67	59	132	39	84
ESP	39.24	9650	0.4	78.4	4.2	2.55	77	137	75	380
FRA	56.1	16905	0.4	74.1	2	2.63	76	130	193	399
GAB	1.1	3000	4	45.7	60	0.36	52	107	14	23
GRE	10	5370	0.3	62.60	9.5	1.89	76	147	102	175
HOK	5.75	10900	0	100	22.7	1.34	77	121	521	247
HON	10.6	2330	-0.1	60.3	1.1	0.93	70	135	273	404
IND	810	317	1.9	28	59.2	0.55	57	100	28	7
IDO	179	454	2	28.8	32.7	0.55	60	116	21	41
ISR	4.47	9800	1.4	91.6	8.2	2.62	75	118	253	276
ITA	57.55	15025	0.1	68.6	3.5	2.25	75	139	105	419
JAP	123.2	22825	0.6	77	3	2.1	78	122	566	589
KEN	23.88	400	3.8	23.6	69	0.11	58	92	13	6
MAR	24.51	800	2.2	48.5	78.6	0.86	61	118	12	55
MEX	84.3	2096	2.5	72.6	17	1.55	68	120	124	124
NOR	4.2	22060	0.3	74.4	2	2.74	77	124	551	350
PER	21.75	1899	2.1	70.2	18.1	2.04	61	93	31	85
POL	38	1740	0.9	63.2	1.2	1.3	71	134	184	263
POR	10.5	4304	0.6	33.3	20.6	1.99	74	128	70	160
SUE	8.47	22455	0.1	84	1.5	2.21	77	113	526	395
SUI	6.7	26025	0.5	59.6	1	1.87	77	128	504	408
THA	55.45	1130	1.9	22.6	12	1.59	65	105	46	104
URS	289	6020	0.8	67.5	2	1.76	69	133	474	319
USA	247.5	20765	1	74	0.5	5.01	75	138	259	812
VEN	19.2	3220	2.5	90	15.3	2.6	69	102	164	147
YOU	23.67	2599	0.7	50.2	10.4	1.44	72	139	100	179

B.16 U.S. Health Data

This is a data set consisting of 50 measurements of 13 variables. It states for one year (1985) the reported number of deaths in the 50 states of the U.S. classified according to 7 categories.

X_1: land area (land)
X_2: population 1985 (popu)
X_3: accident (acc)
X_4: cardiovascular (card)
X_5: cancer (canc)
X_6: pulmonar (pul)
X_7: pneumonia flu (pnue)
X_8: diabetis (diab)
X_9: liver (liv)
X_{10}: Doctors (doc)
X_{11}: Hospitals (hosp)
X_{12}: U.S. states region number (r)
X_{13}: U.S. states division number (d)

division numbers		*region numbers*	
New England	1	Northeast	1
Mid Atlantic	2	Midwest	2
E N Central	3	South	3
W N Central	4	West	4
S Atlantic	5		
E S Central	6		
W S Central	7		
Mountain	8		
Pacific	9		

state	land	popu 1985	acc	card	canc	pul	pneu	diab	liv	doc	hosp	r	d
ME	33265	1164	37.7	466.2	213.8	33.6	21.1	15.6	14.5	1773	47	1	1
NH	9279	998	35.9	395.9	182.2	29.6	20.1	17.6	10.4	1612	34	1	1
VT	9614	535	41.3	433.1	188.1	33.1	24.0	15.6	13.1	1154	19	1	1
MA	8284	5822	31.1	460.6	219	24.9	29.7	16.0	13.0	16442	177	1	1
RI	1212	968	28.6	474.1	231.5	27.4	17.7	26.2	13.4	2020	21	1	1
CT	5018	3174	35.3	423.8	205.1	23.2	22.4	15.4	11.7	8076	65	1	1
NY	49108	17783	31.5	499.5	209.9	23.9	26.0	17.1	17.7	49304	338	1	2
NJ	7787	7562	32.2	464.7	216.3	23.3	19.9	17.3	14.2	15120	131	1	2
PA	45308	11853	34.9	508.7	223.6	27.0	20.1	20.4	12.0	23695	307	1	2
OH	41330	10744	33.2	443.1	198.8	27.4	18.0	18.9	10.2	18518	236	2	3
IN	36185	5499	37.7	435.7	184.6	27.2	18.6	17.2	8.4	7339	133	2	3
IL	56345	11535	32.9	449.6	193.2	22.9	21.3	15.3	12.5	22173	279	2	3
MI	58527	9088	34.3	420.9	182.3	24.2	18.7	14.8	13.7	15212	231	2	3
WI	56153	4775	33.8	444.3	189.4	22.5	21.2	15.7	8.7	7899	163	2	3
MN	84402	4193	35.7	398.3	174	23.4	25.6	13.5	8.1	8098	181	2	4
IA	56275	2884	38.6	490.1	199.1	31.2	28.3	16.6	7.9	3842	140	2	4
MO	69697	5029	42.2	475.9	211.1	29.8	25.7	15.3	9.6	8422	169	2	4
ND	70703	685	48.2	401	173.7	18.2	25.9	14.9	7.4	936	58	2	4
SD	77116	708	53.0	495.2	182.1	30.7	32.4	12.8	7.2	833	68	2	4
NE	77355	1606	40.8	479.6	187.4	31.6	28.3	13.5	7.8	2394	110	2	4
KS	82277	2450	42.9	455.9	183.9	32.3	24.9	16.9	7.8	3801	165	2	4
DE	2044	622	38.8	404.5	202.8	25.3	16.0	25.0	10.5	1046	14	3	5
MD	10460	4392	35.2	366.7	195	23.4	15.8	16.1	9.6	11961	85	3	5
VA	40767	5706	37.4	365.3	174.4	22.4	20.3	11.4	9.2	9749	135	3	5
MV	24231	1936	46.7	502.7	199.6	35.2	20.1	18.4	10.0	2813	75	3	5
NC	52669	6255	45.4	392.6	169.2	22.6	19.8	13.1	10.2	9355	159	3	5
SC	31113	3347	47.8	374.4	156.9	19.6	19.2	14.8	9.0	4355	89	3	5
GA	58910	5976	48.2	371.4	157.9	22.6	20.5	13.2	10.4	8256	191	3	5
FL	58664	11366	46.0	501.8	244	34.0	18.3	16.1	17.2	18836	254	3	5
KY	40409	3726	48.8	442.5	194.7	29.8	22.9	15.9	9.1	5189	120	3	6
TN	42144	4762	45.0	427.2	185.6	27.0	20.8	12.0	8.3	7572	162	3	6
AL	51705	4021	48.9	411.5	185.8	25.5	16.8	16.1	9.1	5157	146	3	6
MS	47689	2613	59.3	422.3	173.9	21.7	19.5	14.0	7.1	2883	118	3	6

AR	53187	2359	51.0	482	202.1	29.0	22.7	15.0	8.7	2952	97	3	7
LA	47751	4481	52.3	390.9	168.1	18.6	15.8	17.8	8.3	7061	158	3	7
OK	69956	3301	62.5	441.4	182.4	27.6	24.5	15.3	9.6	4128	143	3	7
TX	266807	16370	48.9	327.9	146.5	20.7	17.4	12.1	8.7	23481	562	3	7
MT	147046	826	59.0	372.2	170.7	33.4	25.1	14.4	11.1	1058	67	4	8
ID	83564	15.0	51.5	324.8	140.4	29.9	22.3	12.4	9.2	1079	52	4	8
WY	97809	509	67.6	264.2	112.2	27.7	18.5	9.2	9.2	606	31	4	8
CO	104091	3231	44.7	280.2	125.1	29.9	22.8	9.6	9.5	5899	98	4	8
NM	121593	1450	62.3	235.6	137.2	28.7	17.8	17.5	13.1	2127	56	4	8
AZ	1140	3187	48.3	331.5	165.6	36.3	21.2	12.6	13.1	5137	79	4	8
UT	84899	1645	39.3	242	93.7	17.6	14.5	11.1	7.3	2563	44	4	8
NV	110561	936	57.3	299.5	162.3	32.3	13.7	11.1	15.4	1272	26	4	8
WA	68138	4409	41.4	358.1	171	31.1	21.2	13.0	10.9	7768	122	4	9
OR	97073	2687	41.6	387.8	179.4	33.8	23.1	11.2	10.4	4904	83	4	9
CA	158706	26365	40.3	357.8	173	26.9	22.2	10.7	16.7	57225	581	4	9
AK	5914	521	85.8	114.6	76.1	8.3	12.4	3.4	11.0	545	26	4	9
HI	6471	1054	32.5	216.9	125.8	16.0	16.8	12.7	6.2	1953	26	4	9

B.17 Vocabulary Data

This example of the evolution of the vocabulary of children can be found in Bock (1975). Data are drawn from test results on file in the Records Office of the Laboratory School of the University of Chicago. They consist of scores, obtained from a cohort of pupils from the eighth through eleventh grade levels, on alternative forms of the vocabulary section of the Coorperative Reading Test. It provides the following scaled scores shown for the sample of 64 subjects (the origin and units are fixed arbitrarily).

	Grade				
Subjects	8	9	10	11	Mean
1	1.75	2.60	3.76	3.68	2.95
2	0.90	2.47	2.44	3.43	2.31
3	0.80	0.93	0.40	2.27	1.10
4	2.42	4.15	4.56	4.21	3.83
4	−1.31	−1.31	−0.66	−2.22	−1.38
6	−1.56	1.67	0.18	2.33	0.66
7	1.09	1.50	0.52	2.33	1.36
8	−1.92	1.03	0.50	3.04	0.66
9	−1.61	0.29	0.73	3.24	0.66
10	2.47	3.64	2.87	5.38	3.59
11	−0.95	0.41	0.21	1.82	0.37
12	1.66	2.74	2.40	2.17	2.24
13	2.07	4.92	4.46	4.71	4.04
14	3.30	6.10	7.19	7.46	6.02
15	2.75	2.53	4.28	5.93	3.87
16	2.25	3.38	5.79	4.40	3.96
17	2.08	1.74	4.12	3.62	2.89
18	0.14	0.01	1.48	2.78	1.10
19	0.13	3.19	0.60	3.14	1.77
20	2.19	2.65	3.27	2.73	2.71
21	−0.64	−1.31	−0.37	4.09	0.44
22	2.02	3.45	5.32	6.01	4.20
23	2.05	1.80	3.91	2.49	2.56
24	1.48	0.47	3.63	3.88	2.37
25	1.97	2.54	3.26	5.62	3.35
26	1.35	4.63	3.54	5.24	3.69
27	−0.56	−0.36	1.14	1.34	0.39
28	0.26	0.08	1.17	2.15	0.92
29	1.22	1.41	4.66	2.62	2.47

30	−1.43	0.80	−0.03	1.04	0.09
31	−1.17	1.66	2.11	1.42	1.00
32	1.68	1.71	4.07	3.30	2.69
33	−0.47	0.93	1.30	0.76	0.63
34	2.18	6.42	4.64	4.82	4.51
35	4.21	7.08	6.00	5.65	5.73
36	8.26	9.55	10.24	10.58	9.66
37	1.24	4.90	2.42	2.54	2.78
38	5.94	6.56	9.36	7.72	7.40
39	0.87	3.36	2.58	1.73	2.14
40	−0.09	2.29	3.08	3.35	2.15
41	3.24	4.78	3.52	4.84	4.10
42	1.03	2.10	3.88	2.81	2.45
43	3.58	4.67	3.83	5.19	4.32
44	1.41	1.75	3.70	3.77	2.66
45	−0.65	−0.11	2.40	3.53	1.29
46	1.52	3.04	2.74	2.63	2.48
47	0.57	2.71	1.90	2.41	1.90
48	2.18	2.96	4.78	3.34	3.32
49	1.10	2.65	1.72	2.96	2.11
50	0.15	2.69	2.69	3.50	2.26
51	−1.27	1.26	0.71	2.68	0.85
52	2.81	5.19	6.33	5.93	5.06
53	2.62	3.54	4.86	5.80	4.21
54	0.11	2.25	1.56	3.92	1.96
55	0.61	1.14	1.35	0.53	0.91
56	−2.19	−0.42	1.54	1.16	0.02
57	1.55	2.42	1.11	2.18	1.82
58	0.04	0.50	2.60	2.61	1.42
59	3.10	2.00	3.92	3.91	3.24
60	−0.29	2.62	1.60	1.86	1.45
61	2.28	3.39	4.91	3.89	3.62
62	2.57	5.78	5.12	4.98	4.61
63	−2.19	0.71	1.56	2.31	0.60
64	−0.04	2.44	1.79	2.64	1.71
Mean	1.14	2.54	2.99	3.47	2.53

B.18 Athletic Records Data

This data set provides data on Men's athletic records for 55 countries in 1984 Olympic Games.

Country	100m (s)	200m (s)	400m (s)	800m (s)	1500m (min)	5000m (min)	10000m (min)	Marathon (min)
Argentina	10.39	20.81	46.84	1.81	3.70	14.04	29.36	137.71
Australia	10.31	20.06	44.84	1.74	3.57	13.28	27.66	128.30
Austria	10.44	20.81	46.82	1.79	3.60	13.26	27.72	135.90
Belgium	10.34	20.68	45.04	1.73	3.60	13.22	27.45	129.95
Bermuda	10.28	20.58	45.91	1.80	3.75	14.68	30.55	146.61
Brazil	10.22	20.43	45.21	1.73	3.66	13.62	28.62	133.13
Burma	10.64	21.52	48.30	1.80	3.85	14.45	30.28	139.95
Canada	10.17	20.22	45.68	1.76	3.63	13.55	28.09	130.15

Chile	10.34	20.80	46.20	1.79	3.71	13.61	29.30	134.03
China	10.51	21.04	47.30	1.81	3.73	13.90	29.13	133.53
Columbia	10.43	21.05	46.10	1.82	3.74	13.49	27.88	131.35
Cook Is	12.18	23.20	52.94	2.02	4.24	16.70	35.38	164.70
Costa Rica	10.94	21.90	48.66	1.87	3.84	14.03	28.81	136.58
Czech	10.35	20.65	45.64	1.76	3.58	13.42	28.19	134.32
Denmark	10.56	20.52	45.89	1.78	3.61	13.50	28.11	130.78
Dom Rep	10.14	20.65	46.80	1.82	3.82	14.91	31.45	154.12
Finland	10.43	20.69	45.49	1.74	3.61	13.27	27.52	130.87
France	10.11	20.38	45.28	1.73	3.57	13.34	27.97	132.30
GDR	10.12	20.33	44.87	1.73	3.56	13.17	27.42	129.92
FRG	10.16	20.37	44.50	1.73	3.53	13.21	27.61	132.23
GB	10.11	20.21	44.93	1.70	3.51	13.01	27.51	129.13
Greece	10.22	20.71	46.56	1.78	3.64	14.59	28.45	134.60
Guatemala	10.98	21.82	48.40	1.89	3.80	14.16	30.11	139.33
Hungary	10.26	20.62	46.02	1.77	3.62	13.49	28.44	132.58
India	10.60	21.42	45.73	1.76	3.73	13.77	28.81	131.98
Indonesia	10.59	21.49	47.80	1.84	3.92	14.73	30.79	148.83
Ireland	10.61	20.96	46.30	1.79	3.56	13.32	27.81	132.35
Israel	10.71	21.00	47.80	1.77	3.72	13.66	28.93	137.55
Italy	10.01	19.72	45.26	1.73	3.60	13.23	27.52	131.08
Japan	10.34	20.81	45.86	1.79	3.64	13.41	27.72	128.63
Kenya	10.46	20.66	44.92	1.73	3.55	13.10	27.80	129.75
Korea	10.34	20.89	46.90	1.79	3.77	13.96	29.23	136.25
P Korea	10.91	21.94	47.30	1.85	3.77	14.13	29.67	130.87
Luxemburg	10.35	20.77	47.40	1.82	3.67	13.64	29.08	141.27
Malaysia	10.40	20.92	46.30	1.82	3.80	14.64	31.01	154.10
Mauritius	11.19	33.45	47.70	1.88	3.83	15.06	31.77	152.23
Mexico	10.42	21.30	46.10	1.80	3.65	13.46	27.95	129.20
Netherlands	10.52	29.95	45.10	1.74	3.62	13.36	27.61	129.02
NZ	10.51	20.88	46.10	1.74	3.54	13.21	27.70	128.98
Norway	10.55	21.16	46.71	1.76	3.62	13.34	27.69	131.48
Png	10.96	21.78	47.90	1.90	4.01	14.72	31.36	148.22
Philippines	10.78	21.64	46.24	1.81	3.83	14.74	30.64	145.27
Poland	10.16	20.24	45.36	1.76	3.60	13.29	27.89	131.58
Portugal	10.53	21.17	46.70	1.79	3.62	13.13	27.38	128.65
Rumania	10.41	20.98	45.87	1.76	3.64	13.25	27.67	132.50
Singapore	10.38	21.28	47.40	1.88	3.89	15.11	31.32	157.77
Spain	10.42	20.77	45.98	1.76	3.55	13.31	27.73	131.57
Sweden	10.25	20.61	45.63	1.77	3.61	13.29	27.94	130.63
Switzerland	10.37	20.45	45.78	1.78	3.55	13.22	27.91	131.20
Taipei	10.59	21.29	46.80	1.79	3.77	14.07	30.07	139.27
Thailand	10.39	21.09	47.91	1.83	3.84	15.23	32.56	149.90
Turkey	10.71	21.43	47.60	1.79	3.67	13.56	28.58	131.50
USA	9.93	19.75	43.86	1.73	3.53	13.20	27.43	128.22
USSR	10.07	20.00	44.60	1.75	3.59	13.20	27.53	130.55

W Samoa	10.82	21.86	49.00	2.02	4.24	16.28	34.71	161.83

B.19 Unemployment Data

This data set provides unemployment rates in all federal states of Germany in November 2005.

No.	Federal state	Unemployment rate
1	Schleswig-Holstein	10.7
2	Hamburg	11.3
3	Mecklenburg-Vorpommern	17.9
4	Niedersachsen	11.0
5	Bremen	15.5
6	Nordrhein-Westfalen	11.6
7	Hessen	9.4
8	Rheinland-Pfalz	8.1
9	Saarland	9.9
10	Baden-Württemberg	6.7
11	Bayern	7.0
12	Berlin	17.8
13	Brandenburg	16.5
14	Sachsen-Anhalt	17.4
15	Thüringen	15.2
16	Sachsen	16.6

B.20 Annual Population Data

The data shows yearly average population rates for Former territory of the Federal Republic of Germany incl. Berlin-West (given in 1000 inhabitants).

Year	Inhabitants	Unemployed
1960	55958	271
1961	56589	181
1962	57247	155
1963	57865	186
1964	58587	169
1965	59297	147
1966	59793	161
1967	59948	459
1968	60463	323
1969	61195	179
1970	61001	149
1971	61503	185
1972	61809	246
1973	62101	273
1974	61991	582
1975	61645	1074
1976	61442	1060
1977	61353	1030
1978	61322	993
1979	61439	876
1980	61658	889

1981	61713	1272
1982	61546	1833
1983	61307	2258
1984	61049	2266
1985	61020	2304
1986	61140	2228
1987	61238	2229
1988	61715	2242
1989	62679	2038
1990	63726	1883
1991	64485	1689
1992	65289	1808
1993	65740	2270
1994	66007	2556
1995	66342	2565
1996	66583	2796
1997	66688	3021

B.21 Bankruptcy Data

The data are the profitability, leverage, and bankruptcy indicators for 84 companies.

The data set contains information on 42 of the largest companies that filed for protection against creditors under Chapter 11 of the U.S. Bankruptcy Code in 2001–2002 after the stock market crash of 2000. The bankrupt companies were matched with 42 surviving companies with the closest capitalizations and the same US industry classification codes available through the Division of Corporate Finance of the Securities and Exchange Commission (CF SEC, 2004).

The information for each company was collected from the annual reports for 1998–1999 (CF SEC, 2004), i.e., three years prior to the defaults of the bankrupt companies. The following data set contains profitability and leverage ratios calculated, respectively, as the ratio of net income (NI) and total assets (TA) and the ratio of total liabilities (TL) and total assets (TA).

Profitability (NI/TA)	Leverage (TL/TA)	Bankruptcy
0.022806	0.7816	1
−0.063584	0.7325	1
−0.227860	1.2361	1
0.021364	0.7350	1
0.042058	0.6339	1
0.021662	0.8614	1
0.023952	0.6527	1
0.000005	0.7385	1
0.020702	0.8954	1
−0.006640	0.7009	1
0.021634	0.7338	1
−0.023206	0.8226	1
−0.263667	0.9085	1
−0.047161	0.9275	1
−0.098931	0.7617	1
0.140857	0.7802	1
−0.018031	0.6187	1
0.047647	0.5294	1
−0.267393	1.0289	1
−0.105816	0.6542	1
−0.013929	0.7181	1
0.001215	0.8653	1
−0.012698	1.0000	1
−0.021990	0.8811	1

0.039147	0.5987	1
−0.000300	0.9213	1
−0.071546	1.0254	1
0.004256	0.7058	1
−0.003599	0.4086	1
−0.029814	0.6017	1
0.030197	0.7866	1
−0.016912	0.9428	1
0.026682	0.5173	1
−0.052413	0.4983	1
−0.408583	0.6821	1
−0.015960	0.6915	1
0.022807	0.8348	1
0.055888	0.6986	1
0.025634	1.0152	1
−0.599016	1.4633	1
0.044064	0.4001	1
−0.121531	0.8697	1
−0.000172	0.6131	−1
−0.156216	1.0584	−1
0.012473	0.6254	−1
0.081731	0.2701	−1
0.080826	0.5593	−1
0.033538	0.7468	−1
0.036645	0.5338	−1
0.052686	0.7101	−1
0.122404	0.2700	−1
0.068682	0.7182	1
0.030576	0.6175	−1
0.094346	0.7293	−1
0.091535	0.4425	−1
0.058916	0.6997	−1
0.186226	0.7254	−1
0.072777	0.5797	−1
0.101209	0.4526	−1
0.015374	0.8504	−1
0.047247	0.5918	−1
−0.085583	0.5945	−1
0.033137	0.5160	−1
0.016055	1.1353	−1
0.008357	0.9068	−1
0.034960	0.7169	−1
0.046514	0.3473	−1
−0.084510	0.8422	−1
0.029492	0.8319	−1
0.045271	0.5813	−1
0.041463	0.3619	−1
0.030059	0.9479	−1
0.023445	0.6856	−1
0.046705	0.4164	−1
0.127897	0.3694	−1
0.050956	0.6073	−1
0.020425	0.5295	−1
0.035311	0.6796	−1
0.066434	0.5303	−1
0.066550	0.7194	−1
0.055333	0.8063	−1
0.015738	0.6294	−1
−0.034455	0.3446	−1
0.004824	0.5705	−1

Bibliography

Andrews, D. (1972). Plots of high-dimensional data, *Biometrics* **28**: 125–136.

Backhaus, K., Erichson, B., Plinke, W. and Weiber, R. (1996). *Multivariate Analysemethoden*, Springer, Berlin.

Bartlett, M. S. (1939). A note on tests of significance in multivariate analysis, *Proceedings of the Cambridge Philosophical Society* **35**: 180–185.

Bartlett, M. S. (1954). A note on multiplying factors for various chi-squared approximations, *JRSSB* **16**: 296–298.

Belsley, D. A., Kuh, E. and Welsch, R. E. (1980). *Regression Diagnostics*, Wiley.

Berndt, E. (1990). *The Practice of Econometics: Classic and Contemporary*, Addison-Wesley, Massacusetts.

Blæsild, P. and Jensen, J. (1981). Multivariate distributions of hyperbolic type, *Statistical Distributions in Scientific Work – Proceedings of the NATO Advanced Study Institute held at the Università degli studi di Trieste* **Vol.4**: 45–66.

Bock, R. D. (1975). *Multivariate Statistical Methods In Behavioral Research*, Mc Graw-Hill, New York.

Bouroche, J.-M.and Saporta, G. (1980). *L'analyse des données*, Presses Universitaires de France, Paris.

Breiman, L. (1973). *Statistics: With a view towards application*, Houghton Mifflin Company, Boston.

Breiman, L., Friedman, J. H., Olshen, R. and Stone, C. J. (1984). *Classification and Regression Trees*, Wadsworth, Belmont.

CF SEC (2004). Division of Corporate Finance of the Securities and Exchange Commission: Standard industrial classification (SIC) code list, *http://www.sec.gov/info/edgar/siccodes.htm* .

Chambers, J. M., Cleveland, W. S., Kleiner, B. and Tukey, P. A. (1983). *Graphical Methods for Data Analysis*, Duxbury Press, Boston.

Chen, Y., Härdle, W. and Jeong, S.-O. (2005). Nonparametric risk management with generalized hyperbolic distributions, *Working Paper, Humboldt-Universiät zu Berlin* .

Chernoff, H. (1973). Using faces to represent points in k-dimensional space graphically, *Journal of the American Statistical Association* **68**: 361–368.

Cook, R. D. and Weisberg, S. (1991). Comment on "sliced inverse regression for dimension reduction", *Journal of the American Statistical Association* **86**(414): 328–332.

Dillon, W. R. and Goldstein, M. (1984). *Multivariate Analysis*, John Wiley & Sons, New York.

Duan, N. and Li, K.-C. (1991). Slicing regression: A link-free regression method, *Annals of Statistics* **19**(2): 505–530.

Embrechts, P., McNeil, A. and Straumann, D. (1999). Correlation and dependence in risk management: Properties and pitfalls. Preprint ETH Zürich.

EUNITE (2001). Electricity load forecast competition of the EUropean Network on Intelligent TEchnologies for Smart Adaptive Systems, *http://neuron.tuke.sk/competition/* .

Fahrmeir, L. and Hamerle, A. (1984). *Multivariate Statistische Verfahren*, De Gruyter, Berlin.

Fang, K. T., Kotz, S. and Ng, K. W. (1990). *Symmetric Multivariate and Related Distributions*, Chapman and Hall, London.

Feller, W. (1966). *An Introduction to Probability Theory and Its Application*, Vol. 2, Wiley & Sons, New York.

Fengler, M. R., Härdle, W. and Villa, C. (2003). The dynamics of implied volatilities: A common principal components approach, *Review of Derivative Research* **6**: 179–202.

Flury, B. (1988). *Common Principle Components Analysis and Related Multivariate Models*, John Wiley and Sons, New York.

Flury, B. and Gautschi, W. (1986). An algorithm for simultaneous orthogonal transformation of several positive definite symmetric matrices to nearly diagonal form, *SIAM Journal on Scientific and Statistical Computing* **7**: 169–184.

Flury, B. and Riedwyl, H. (1988). *Multivariate Statistics, A practical Approach*, Cambridge University Press.

Franke, J., Härdle, W. and Hafner, C. (2004). *Introduction to Statistics of Financial Markets*, Springer, Heidelberg.

Friedman, J. H. and Stuetzle, W. (1981). Projection pursuit classification. unpublished manuscript.

Friedman, J. H. and Tukey, J. W. (1974). A projection pursuit algorithm for exploratory data analysis, *IEEE Transactions on Computers* **C 23**: 881–890.

Gale, D., Kuhn, H. W. and Tucker, A. W. (1951). *Linear Programming and the Theory of Games, in Activity Analysis of Production and Allocation, T. C. Koopmans (ed.)*, John Wiley & Sons, New York, NY.

Gibbins, R. (1985). *Canonical Analysis. A Review with Application in Ecology*, Springer-Verlag, Berlin.

Giri, N. C. (1996). *Multivariate Statistical Analysis*, Marcel Dekker, New York.

Gosset, W. S. (1908). The probable error of a mean, *Biometrika* **Vol.6**: 1–25.

Hall, P. (1992). *The Bootstrap and Edgeworth Expansion*, Statistical Series, Springer, New York.

Hall, P. and Li, K.-C. (1993). On almost linearity of low dimensional projections from high dimensional data, *Annals of Statistics* **21**(2): 867–889.

Härdle, W. (1991). *Smoothing Techniques, With Implementations in S*, Springer, New York.

Härdle, W., Kleinow, T. and Stahl, G. (2002). *Applied Quantitative Finance*, Springer, Heidelberg.

Härdle, W., Müller, M., Sperlich, S. and Werwatz, A. (2003). *Non- and Semiparametric Models*, Springer, Heidelberg.

Härdle, W. and Scott, D. (1992). Smoothing by weighted averaging of rounded points, *Computational Statistics* **7**: 97–128.

Harrison, D. and Rubinfeld, D. L. (1978). Hedonic prices and the demand for clean air, *J. Environ. Economics & Management* **5**: 81–102.

Hoaglin, W., Mosteller, F. and Tukey, J. (1983). *Understanding Robust and Exploratory Data Analysis*, Whiley, New York.

Hodges, J. L. and Lehman, E. L. (1956). The efficiency of some non–parametric competitors of the t-test, *Annals of Mathematical Statistics* **27**: 324–335.

Hotelling, H. (1935). The most predictable criterion, *Journal of Educational Psychology* **26**: 139–142.

Hotelling, H. (1953). New light on the correlation coefficient and its transform, *Journal of the Royal Statistical Society, Series B* **15**: 193–232.

Huber, P. (1985). Projection pursuit, *Annals of Statistics* **13**(2): 435–475.

Johnson, R. A. and Wichern, D. W. (1998). *Applied Multivariate Analysis, 4th ed.*, Prentice Hall, Englewood Cliffs, New Jersey.

Jones, M. C. and Sibson, R. (1987). What is projection pursuit? (with discussion), *Journal of the Royal Statistical Society, Series A* **150**(1): 1–36.

Kaiser, H. F. (1985). The varimax criterion for analytic rotation in factor analysis, *Psychometrika* **23**: 187–200.

Kendall, K. and Stuart, S. (1977). *Distribution Theory*, Vol. 1 of *The advanced Theory of Statistics*, Griffin, London.

Klinke, S. and Polzehl (1995). Implementation of kernel based indices in XGobi, *Discussion paper 47*, SFB 373, Humboldt-University of Berlin.

Kruskal, J. B. (1965). Analysis of factorial experiments by estimating a monotone transformation of data., *Journal of the Royal Statistical Society, Series B* **27**: 251–263.

Kruskal, J. B. (1969). Toward a practical method which helps uncover the structure of a set of observations by finding the line tranformation which optimizes a new "index of condensation", *in* R. C. Milton and J. A. Nelder (eds), *Statistical Computation*, Academic Press, New York, pp. 427–440.

Kruskal, J. B. (1972). Linear transformation of multivariate data to reveal clustering, *in* R. N. Shepard, A. K. Romney and S. B. Nerlove (eds), *Multidimensional scaling: Theory and Applications in the Behavioural Sciences*, Vol. 1, Seminar Press, London, pp. 179–191.

Lachenbruch, P. A. and Mickey, M. R. (1968). Estimation of error rates in discriminant analysis, *Technometrics* **10**: 1–11.

Laplace, P.-S. (1774). Mémoire sur la probabilité des causes par les événements, *Savants étranges* **Vol.6**: 621–656.

Li, K.-C. (1991). Sliced inverse regression for dimension reduction (with discussion), *Journal of the American Statistical Association* **86**(414): 316–342.

Li, K.-C. (1992). On principal Hessian directions for data visualization and dimension reduction: another application of Stein's lemma, *Journal of the American Statistical Association* **87**: 1025–1039.

Lorenz, M. O. (1905). Methods of measuring the concentration of wealth, *Journal of the American Statistical Association* **9**: 209–219.

Mardia, K. V., Kent, J. T. and Bibby, J. M. (1979). *Multivariate Analysis*, Academic Press, Duluth, London.

Mercer, J. (1909). Functions of positive and negative type and their connection with the theory of integral equations, *Philosophical Transactions of the Royal Society of London* **209**: 415–446.

Morrison, D. F. (1990). *Multivariate statistical methods*, McGraw–Hill, New-York.

Muirhead, R. J. (1982). *Aspects of Multivariate Statistics*, John Wiley and Sons, New York.

Nelsen, R. B. (1999). *An Introduction to Copulas*, Springer, New York.

Olkin, I. and Veath, M. (1980). Maximum likelihood estimation in a two-way analysis with correlated errors in one classification, *Biometrika* **68**: 653–660.

Parzen, E. (1962). On estimating of a probability density and mode, *Annals of Mathematical Statistics* **35**: 1065–1076.

Rosenblatt, M. (1956). Remarks on some nonparametric estimates of a density function, *Annals of Mathematical Statistics* **27**: 832–837.

Schott, J. R. (1994). Determining the dimensionality in sliced inverse regression, *Journal of the American Statistical Association* **89**(425): 141–148.

Scott, D. (1985). Averaged shifted histograms: Effective nonparametric density estimation in several dimensions, *Annals of Statistics* **13**: 1024–1040.

Securities and Exchange Commission (2004). Securities and Exchange Commission: Archive of historical documents, *http://www.sec.gov/cgi-bin/srch-edgar* .

Silverman, B. W. (1986). *Density Estimation for Statistics and Data Analysis*, Vol. 26 of *Monographs on Statistics and Applied Probability*, Chapman and Hall, London.

Sklar, A. (1959). Fonctions de répartition à n dimensions et leurs marges, *Publ. Inst. Statist. Univ. Paris 8* pp. 229–231.

Tufte, E. (1983). *The Visual Display of Quantitative Information*, Graphics Press.

Vapnik (1995). *The Nature of Statistical Learning Theory*, Springer, New York, NY.

Volle, V. (1985). *Analyse des Données*, Economica, Paris.

Whittle, P. (1958). On the smoothing of probability density functions, *Journal of the Royal Statistical Society, Series B* **55**: 549–557.

Index